Richard
Kuelemeyer

IND. 47905

Richard Koelemeyer
330 W. Sylvia W. Lay

MULTIVARIATE CALCULUS

By the same author:

Calculus with an Introduction to Vectors, Wiley, 1972.

MULTIVARIATE CALCULUS
with Linear Algebra

Philip C. Curtis, Jr.
University of California, Los Angeles

John Wiley and Sons, Inc., New York · London · Sydney · Toronto

Copyright © 1972, by John Wiley & Sons, Inc.

All rights reserved. Published simultaneously in Canada.

No part of this book may be reproduced by any means, nor transmitted, nor translated into a machine language without the written permission of the publisher.

Library of Congress Catalogue Card Number: 77-175791

ISBN 0-471-18993-6

Printed in the United States of America

10 9 8 7 6 5 4 3 2 1

PREFACE

My objective is, first, to develop the geometric and algebraic tools of linear algebra and then to use them in the subsequent discussion of the calculus of functions of several variables. This approach has the dual advantage of providing many geometrical applications for the ideas of linear algebra and of giving deeper insight into the linear aspects of analysis, particularly the approximation of complicated functions by affine or linear functions. The book is designed for second-year college students who have completed a one-year, single-variable, calculus course.

The first chapter introduces vector techniques, and these ideas are used to describe the linear aspects of Euclidean geometry. Chapters 2 and 3 develop the theory of linear transformations and matrices. The problem of finding eigenvalues for matrices motivates the study of determinants in Chapter 4. The application of these ideas begins with a discussion of certain linear differential equations in Chapter 4. The geometric aspects of the theory are emphasized in Chapter 5 with a discussion of orthogonal transformations and the diagonalization of quadratic forms.

Applications to the calculus begin in Chapter 6. The emphasis here is on the ability to provide linear approximations to general functions. This idea is further developed in the discussion of the chain rule. Taylor's theorem and extreme value problems are studied from a geometric point of view in Chapter 7, using the ideas developed earlier in the study of eigenvectors.

Multiple integrals are introduced and discussed in Chapter 8. The notion of an integral is further expanded with the discussion of line integrals in Chapter 9 and surface integrals in Chapter 10. The discussion of the theorems of Green, Stokes, and Gauss and the related ideas from vector field theory conclude the book.

For schools that use this book on a semester basis, the first five chapters form a one-semester course in linear algebra. Chapters 6 to 10 then provide the material for a semester course in several-variable calculus. For schools that are on a quarter plan, Chapters 1 to 4, 5 to 7, and 8 to 10 provide a natural division of the material in three parts. If students have had an introduction to vector algebra and geometry in their first-year calculus course, then Chapter 1 will constitute a review of this material.

I have tried to strike a balance between the theoretical development and applications. Proofs of harder results are often developed in starred exercises or are occasionally omitted. Reference is often made to my volume, *Calculus with an Introduction to Vectors* (Wiley, 1972) for results from single-variable calculus. Other standard texts may be used equally well. I have included copious exercise sections wherever possible. Answers to all problems of a computational nature are included at the end of the book.

I thank many people for their help in completing this project. Fred Corey at Wiley and Jack Hoey, formerly of Wiley, were a source of constant support. I particularly thank Elaine Stafford for many hours of expert and painstaking typing. And finally, I especially thank my students at UCLA for their comments and criticisms of the material in preliminary form. Their efforts in revealing obscure parts of the exposition as well as errors in the text and the answer sections are greatly appreciated.

Philip C. Curtis, Jr.

Los Angeles, California, October, 1971

CONTENTS

GLOSSARY OF SYMBOLS AND FORMULAS   x

CHAPTER ONE   VECTORS AND VECTOR ALGEBRA

1.1	Definition of Vectors	1
1.2	Geometric Representation of Vectors	4
1.3	Geometric Vector Addition and Scalar Multiplication	10
1.4	The Inner Product of Two Vectors	15
1.5	The Schwarz Inequality and the Triangle Inequality	18
1.6	Lines in $V_n(R)$	21
1.7	Planes in $V_n(R)$	28
1.8	The Vector Product	37
1.9	Linear Independence of Vectors	44

CHAPTER TWO   VECTOR SPACES

2.1	Examples and Motivation	50
2.2	Vector Spaces and Subspaces	51
2.3	Linear Independence	55
2.4	Tests for Independence	58
2.5	Bases and Dimension	60
2.6	Linear Equations	65
2.7*	Linear Equations Continued	73

CHAPTER THREE   MATRICES AND LINEAR TRANSFORMATIONS

3.1	Matrices	76
3.2	Matrix Multiplication	81
3.3	Linear Transformations	85
3.4	The Matrix of a Linear Transformation	92
3.5	The Space $L(V)$	101
3.6	Invertible Matrices	108
3.7	Change of Basis Matrices	115
3.8	Similarity of Matrices; Eigenvalues and Eigenvectors	119

CHAPTER FOUR   DETERMINANTS

4.1	Definition of a Determinant Function	124
4.2	Existence of a Determinant Function	128
4.3	Properties of Determinant Functions; Uniqueness	132
4.4	The Transpose of a Matrix; Permutation Matrices	140

4.5	Further Properties of the Determinant Function	145
4.6	Eigenvalues and Applications to Differential Equations	149
4.7*	Further Applications to Differential Equations	155

CHAPTER FIVE SYMMETRIC MATRICES AND QUADRATIC FORMS

5.1	Orthogonal Transformations	160
5.2	Orthogonal Transformations in $V_3(R)$	165
5.3	Symmetric Matrices	170
5.4	Quadratic Forms	174
5.5	Quadric Surfaces	180

CHAPTER SIX THE CALCULUS OF VECTOR FUNCTIONS

6.1	Vector Functions	189
6.2	Limits and Continuity	194
6.3	Partial Derivatives	202
6.4	Differentiable Functions	207
6.5	Tangent Lines and Planes, Tangent Spaces	213
6.6	Directional Derivatives, the Mean Value Theorem	219
6.7	The Gradient	227
6.8	The Chain Rule	229
6.9	More on the Chain Rule; Coordinate Transformations	236
6.10	Inverse and Implicit Functions	246
6.11*	Implicit Functions — Continued	253

CHAPTER SEVEN TAYLOR POLYNOMIALS AND EXTREMAL PROBLEMS

7.1	Approximation by Polynomials	262
7.2	Taylor's Theorem	266
7.3	The Geometric Interpretation of the Second Differential	270
7.4	Extreme Values	277
7.5	Constrained Extremal Problems	283

CHAPTER EIGHT MULTIPLE INTEGRALS

8.1	Introduction	291
8.2	Step Functions and the Integral	293
8.3	The Integral of a Bounded Function	298
8.4	Applications	305
8.5	General Multiple Integrals	315
8.6	Change of Variable in Multiple Integrals	319

CHAPTER NINE LINE INTEGRALS

9.1	Line Integrals of Scalar Functions	330
9.2	Applications	335

9.3	Vector Fields and Line Integrals	337
9.4	Properties of Line Integrals	344
9.5	Fundamental Theorems of Calculus for Line Integrals	348
9.6	Green's Theorem	357

CHAPTER TEN   VECTOR FIELDS IN SPACE

10.1	Divergence and Curl of a Vector Field	367
10.2	Surfaces	374
10.3	Surface Area	379
10.4	Surface Integrals	384
10.5	Stokes' Theorem	393
10.6	Gauss' Theorem	399
10.7*	Applications to Fluid Mechanics	405

ANSWERS TO EXERCISES 407

INDEX 423

Glossary of Symbols and Formulas

$\mathbf{A} + \mathbf{B}$ (vector sum)	1
$a\mathbf{A}$ (scalar multiplication)	2
(A,B)	15
$\|A\|$	3
$\|A\|$	21
$A \times B$	37
$\|(A,B)\| \leq \|A\|\|B\|$	16
$\|A+B\| \leq \|A\| + \|B\|$	19
$\|A \times B\| = \|A\|\|B\| \sin \theta$	40
$\|A \times B\|^2 = \|A\|^2\|B\|^2 - (A,B)^2$	39
(a_{ij})	76
A_T	92
A^{-1}	84
A^*	145
A^t	140
$A_f = \begin{pmatrix} \dfrac{\partial^2 f}{\partial x^2} & \dfrac{\partial^2 f}{\partial x\,\partial y} \\ \dfrac{\partial^2 f}{\partial x\,\partial y} & \dfrac{\partial^2 f}{\partial y^2} \end{pmatrix}$	273
$(A\|I)$	112
$AA^* = D(A)I$	146
$A = \iint_R \left\|\dfrac{\partial \varphi}{\partial u} \times \dfrac{\partial \varphi}{\partial v}\right\| du\, dv$	380
$\begin{vmatrix} a_1 a_2 \\ b_1 b_2 \end{vmatrix}$	38
$\begin{vmatrix} a_1 a_2 a_3 \\ b_1 b_2 b_3 \\ c_1 c_2 c_3 \end{vmatrix}$	38
$\begin{vmatrix} a_{11} & a_{12} \\ a_{21} & a_{22} \end{vmatrix} = D\begin{pmatrix} a_{11} & a_{12} \\ a_{21} & a_{22} \end{pmatrix} = a_{11}a_{22} - a_{21}a_{12}$	124
$(b_{jl})(a_{lk}) = \left(\sum_{l=1}^{m} b_{jl}a_{lk}\right)$	81
$\cos \theta = \dfrac{(A,B)}{\|A\|\|B\|}$	17
$\operatorname{curl} F = D\begin{pmatrix} I_1 & I_2 & I_3 \\ \dfrac{\partial}{\partial x_1} & \dfrac{\partial}{\partial x_2} & \dfrac{\partial}{\partial x_3} \\ F_1 & F_2 & F_3 \end{pmatrix}$	368
$D(A) = D(A^t)$	141
$D(A) = \sum_{i=1}^{n} (-1)^{i+1} a_{i1} D(A_{i1})$	136

$D(A) = \sum_{i=1}^{n} (-1)^{i+j} a_{ij} D(A_{ij}), j = 1, 2, \ldots, n$ 142

$D(A) = \sum_{j=1}^{n} (-1)^{i+j} a_{ij} D(A_{ij}), i = 1, 2, \ldots, n$ 143

$D(A_{ij})$ 130

$D(A - xI)$ 139

$D_i(a)$ 109

$D(f'(X)) = \dfrac{\partial(f_1, \ldots, f_n)}{\partial(x_1, \ldots, x_n)}(X)$ 247

$\dfrac{\partial f}{\partial Y}(x) = \lim_{h \to 0} \dfrac{f(X + hY) - f(X)}{h}$ 219

$\dfrac{\partial f}{\partial(aY)}(X) = a \dfrac{\partial f}{\partial Y}(X)$ 220

$D_{x_k}(f)$ 206

$\dfrac{\partial f}{\partial Y}(X_0) = f'(X_0) Y$ 222

$\nabla f(X) = \left(\dfrac{\partial f}{\partial x_1}(X), \ldots, \dfrac{\partial f}{\partial x_n}(X) \right)$ 227

$\dfrac{\partial f}{\partial Y}(X) = (\nabla f(X), Y)$ 227

$\dfrac{\partial^2 f}{\partial x_j \, \partial x_k}$ 205

$\left(\dfrac{\partial h_i}{\partial x_j} \right)(X) = \left(\dfrac{\partial f_i}{\partial y_k} \right)(g(X)) \left(\dfrac{\partial g_k}{\partial x_j} \right)(X)$ if $h = f \cdot g$ 229

$d_{X_0} f$ 208

$d_X h = d_{g(X)} f \cdot d_X g$ if $h = f \cdot g$ 232

$d_{f(X)} f^{-1} \cdot d_X f = I$ 246

$\dfrac{\partial f}{\partial x_1} \times \dfrac{\partial f}{\partial x_2}$ 216

$\dfrac{\partial \varphi}{\partial u} \times \dfrac{\partial \varphi}{\partial v}$ 375

$(\text{div } F)(X) = \dfrac{\partial F_1}{\partial x_1} + \dfrac{\partial F_2}{\partial x_2} + \dfrac{\partial F_3}{\partial x_3}$ 367

$\dfrac{\partial \rho}{\partial t} + \text{div}(\rho F) = 0$ 406

ϵ 1

E_{ij} 109

$F_{ij}(a)$ 110

f_{x_k} 206

$f'(X_0) = \left(\dfrac{\partial f_i}{\partial x_j} \right)(X_0)$ 209

$f'(X_0)(X - X_0) = \dfrac{\partial f}{\partial x_1} dx_1 + \cdots + \dfrac{\partial f}{\partial x_n} dx_n$ 211

$f(B) - f(A) = f'(X_0)(B - A)$ 224

$g \cdot f$ 192

$I_{j,k}$ 77

$\displaystyle\int_R f$ 298

$$\iint_R f \qquad 298$$

$$\int_T f(x,y)\,dx\,dy = \int_S f(\Phi(u,v)) \left|\frac{\partial(\varphi_1,\varphi_2)}{\partial(u,v)}\right| du\,dv \qquad 321$$

$$\int \cdots \int_S f(x_1, \ldots, x_n)\,dx_1 \cdots dx_n \qquad 315$$

$$\int_\alpha f = \int_a^b f(\alpha(t))|\alpha'(t)|\,dt \qquad 332$$

$$\int_\alpha F = \int_a^b \{F[\alpha(t)], \alpha'(t)\}\,dt \qquad 339$$

$$\int_{-\alpha} F = -\int_\alpha F \qquad 347$$

$$\iint_S \left(\frac{\partial F_2}{\partial x} - \frac{\partial F_1}{\partial y}\right) dx\,dy = \oint_\alpha F \qquad 359$$

$$\iint_\varphi F = \iint_E \left(F(\varphi(u,v)), \frac{\partial \varphi}{\partial u} \times \frac{\partial \varphi}{\partial v}\right) du\,dv \qquad 387$$

$$\iint_\varphi \operatorname{curl} F = \oint_{\partial \varphi} F \qquad 393$$

$$\iiint_U \operatorname{div} F = \iint_{\partial U} F \qquad 402$$

$L(V,W)$ 87

$$l = \int_a^b |\alpha'(t)|\,dt \qquad 331$$

$\lim_{X \to X_0} f(X) = L$ 196

$\lim_{X \to X_0} f(X) = f(X_0)$ 196

$$\lim_{h \to 0} \frac{f(X + hI_k) - f(X)}{h} = \frac{\partial f}{\partial x_k}(X) = \frac{\partial f}{\partial I_k}(X) \qquad 202$$

M_R 299

$1:1$ 89

$$p(x) \sum_{m=0}^n \frac{f^{(m)}(x_0)}{m!}(x - x_0)^m \qquad 263$$

$$p_{X_0}(X) = \sum_{m=0}^n \frac{(\partial_{X_0}^m f)(X - X_0)}{m!} \qquad 268$$

$R_{m,n}$ 76

$\operatorname{sgn}(P)$ 143

T^{-1} 103

$$\left(x\frac{\partial}{\partial x} + y\frac{\partial}{\partial y}\right)_{X_0} f \qquad 264$$

$$(x_1 \ldots x_n)^{1/n} \leq \frac{x_1 + \cdots + x_n}{n} \qquad 283$$

$X = A + tB$ 22

$X = C + sA + tB$ 28

$V_n(R)$ 1

$W_1 + W_2$ (sum of two subspaces)   63

$W_1 \oplus W_2$ 64

MULTIVARIATE CALCULUS

CHAPTER ONE

VECTORS AND VECTOR ALGEBRA

1.1 Definition of Vectors

One might say that the real numbers were invented to obtain quantitative measurements of physical phenomena. We have met many instances of this: measurements of distance, speed, area, pressure, and volume, to mention only a few. However, it is also apparent that in many situations one number alone is not adequate to describe the phenomenon under consideration. Consider a space vehicle in flight. To specify the location of the rocket at a given time three numbers are required. Three more are necessary to specify the velocity. If we wish to specify the acceleration, we need three more. Thus to describe the position, velocity, and acceleration at time t an array of nine numbers is needed.

To measure complicated physical phenomena in quantitative terms we are forced to use arrays of more than one number. Such an array we call a vector. To be precise we define a *vector* to be an n-tuple of real numbers, that is, an ordered array of n real numbers $(a_1, a_2, \ldots, a_n)$. The ith number entering in this array is called the *ith component* of the vector. When we say that a vector is an *ordered* array of real numbers, we mean that if $A = (a_1, \ldots, a_n)$ and $B = (b_1, \ldots, b_n)$ are two vectors, then $A = B$ if and only if $a_1 = b_1$, $a_2 = b_2, \ldots$, and $a_n = b_n$. Thus the vector $A = (a_2, \ldots, a_n)$ is not to be confused with the set whose members are the numbers $a_1, \ldots, a_n$ which we write $\{a_1, \ldots, a_n\}$. The set of all such vectors will be written $V_n(R)$. If X is a vector from $V_n(R)$ then we write $X \in V_n(R)$. The symbol $\in$ stands for 'is a member of' or 'belongs to'.

In our discussion of vectors we shall always denote vectors by capital letters A, B, C, etc., and real numbers by small letters a, b, c, etc. (To further distinguish between vectors and real numbers the latter are often called *scalars*.) We shall formulate the definitions and theorems of vector algebra in terms of vectors from $V_n(R)$. However, for examples and calculations we will assume for the most part that $n = 2$ or $n = 3$. We are of course limited to these cases when we give spatial interpretation to the notion of vectors.

Certain aspects of the algebra of real numbers extend readily to vectors. Without attempting to give any geometrical or physical motivation for the idea, it is reasonably apparent that addition between vectors in $V_n(R)$ can be readily defined. Indeed if $A = (a_1, \ldots, a_n)$ and $B = (b_1, \ldots, b_n)$, we define the vector sum by the formula

$$A + B = (a_1 + b_1, a_2 + b_2, \ldots, a_n + b_n).$$

Thus the sum of two vectors is formed by adding the respective components. It is easily checked that vector addition has all the properties possessed by addition of real numbers. It is commutative and associative. There is a zero vector, $0 = (0, 0, \ldots, 0)$, and each vector has an inverse with respect to addition.

Indeed if $A = (a_1, \ldots, a_n)$, then $-A = (-a_1, \ldots, -a_n)$. This implies that subtraction is always possible since

$$A - B = (a_1 - b_1, \ldots, a_n - b_n).$$

Since addition extends so readily to vectors, it is somewhat of a shock to learn that in general it is impossible to define a multiplication for vectors which possesses all the properties enjoyed by multiplication of real numbers. Indeed this is only possible when $n = 2$. (See Exercise 10, p. 3). However, it is possible to define a "multiplication" of a vector by a real number. This operation yields a vector as a result. The multiplication, called *scalar multiplication* is defined by the following formula. If a is a real number, or scalar, and $A = (a_1, \ldots, a_n) \in V_n(R)$, then we define

$$aA = (aa_1, aa_2, \ldots, aa_n).$$

We emphasize again that scalar multiplication is not a multiplication between vectors but is a multiplication of a vector by a real number or scalar which yields a vector as the product. If a vector B can be written $B = aA$, then B is called a *scalar multiple* of A. Thus $(2, -4, 6)$ is a scalar multiple of $(1, -2, 3)$ since

$$(2, -4, 6) = 2(1, -2, 3).$$

It may be easily checked that scalar multiplication satisfies the following distributive laws

(1) $\qquad\qquad a(A + B) = aA + aB$

(2) $\qquad\qquad (a + b)A = aA + bA,$

and the associative law

(3) $\qquad\qquad (ab)A = a(bA).$

If k vectors $A_1, \ldots, A_k$ are given and there exist k real numbers (scalars) $a_1, \ldots, a_k$ so that the vector B can be written

$$B = a_1 A_1 + \cdots + a_k A_k,$$

then B is called a *linear combination* of the vectors $A_1, \ldots, A_k$.

Example 1. Express $(-1, 2)$ as a linear combination of the vectors $(1, 2)$ and $(2, 3)$. We seek scalars x_1, x_2 so that

$$(-1, 2) = x_1(1, 2) + x_2(2, 3).$$

But

$$x_1(1, 2) + x_2(2, 3) = (x_1 + 2x_2, 2x_1 + 3x_2).$$

So

$$(-1, 2) = (x_1 + 2x_2, 2x_1 + 3x_2)$$

if and only if the components of the two vectors are equal. Consequently

$$-1 = x_1 + 2x_2$$
$$2 = 2x_1 + 3x_2.$$

This pair of linear equations can be immediately solved by elimination yielding $x_1 = 7$ and $x_2 = -4$. Thus

$$(-1, 2) = 7(1, 2) - 4(2, 3).$$

Finally we define the *length* or *norm* $|A|$ of a vector A by the formula
$$|A| = (a_1^2 + \cdots + a_n^2)^{1/2}$$
where
$$A = (a_1, \ldots, a_n).$$
It may be immediately checked that for a scalar a
$$|aA| = |a||A|$$
where $|a|$ is of course the absolute value of the scalar a.

EXERCISES

1. If $A = (-1, 4)$, $B = (2, 3)$, $C = (2, -1)$, compute $A + B$, $B - 2C$, $A - B + 2C$, $2(A + 3B)$, $|3A + 4B|$.
2. If $A = (3, 1, -2)$, $B = (2, 0, -1)$, $C = (1, 1, -2)$, compute $A - 2B$, $3(A + 3B)$, $2(B - A + C)$, $|2B - 4A + C|$.
3. Verify that vector addition possesses all the properties of addition of real numbers. That is, show that vector addition, is associative, and commutative. Show that there is an identity for vector addition and that every vector has an additive inverse.
4. Show that scalar multiplication satisfies the distributive laws (1) and (2) as well as the associative law (3). Also verify the formula
$$|aA| = |a||A|.$$
5. By analogy with vector addition one is tempted to define the "product" of two vectors $A = (a_1, \ldots, a_n)$, $B = (b_1, \ldots, b_n)$, by the formula
$$A \# B = (a_1 b_1, a_2 b_2, \ldots, a_n b_n).$$
Verify that this "product" is commutative, associative, and distributive with respect to vector addition. Show however that not all of the properties of multiplication of real numbers are possessed by this product. (To show that the commutative, associative, and distributive laws hold one must check that $A \# B = B \# A$, $A \# (B \# C) = (A \# B) \# C$ and $A \# (B + C) = A \# B + A \# C$ for vectors A, B, C. Show also that the vector $I = (1, 1, \ldots, 1)$ is an identity with respect to this product, i.e. $I \# X = X$ for each vector X. Show, however, that if $X \neq 0$ it does not necessarily follow that X has an inverse, i.e. a vector X^{-1} satisfying $X \# X^{-1} = I$.
6. Express $(2, -4)$ as a linear combination of $(1, 1)$ and $(2, 3)$.
7. Express $(1, -1, 2)$ as a linear combination of $(1, 0, -1)$, $(2, 2, 1)$, $(2, -1, 3)$.
8. Can $(-1, 1)$ be expressed as a linear combination of $(1, 4)$ and $(-2, -8)$?
9. Can $(4, 1, 1)$ be expressed as a linear combination of $(1, -2, 1)$, $(2, 1, 1)$ and $(3, -1, 2)$?
10. A multiplication may be defined in $V_2(R)$ having all of the properties of multiplication of real numbers (that is, $V_2(R)$ provided with this multiplication is a field) in the following way. If $A = (a_1, a_2)$ and $B = (b_1, b_2)$ then we define

(4) $$A \cdot B = (a_1 b_1 - a_2 b_2, a_2 b_1 + a_1 b_2).$$

Verify that the set $V_2(R)$ is a field when provided with this multiplication. This field is called the field of *complex numbers*. (Show that $A \cdot B$ is commutative, associative and distributive with respect to vector addition. These verifications are virtually automatic. Show also that if $A \neq 0$, then A has an inverse with respect to the multiplication defined by (4). To prove this show first that $I = (1, 0)$ is an identity element with respect to the multiplication. Then to determine a formula for the inverse A^{-1} solve the equation $A \cdot X = I$.)

4 VECTORS AND VECTOR ALGEBRA

1.2 Geometric Representation of Vectors

The geometric representation of vectors from $V_3(R)$ in space, and of vectors from $V_2(R)$ in the plane, is extremely useful both for giving spatial or geometric meaning to these algebraic ideas as well as for providing geometric models for applications. We construct the spatial representation for vectors in $V_3(R)$ as follows. First construct a coordinate system in space consisting of three mutually perpendicular lines l_1, l_2, l_3 which intersect at a point which we take to be the origin. Next choose positive directions along each of these lines. Which direction we take to be positive and in which order we label the lines is of course arbitrary. However, in this book we shall number the axes and choose directions as in Figure 1.

The coordinates of P will be the numbers corresponding to the points of intersection of the coordinate axes and lines through P each perpendicular to the axes l_1, l_2, l_3. We shall label these axes the first, second, and third coordinate axes, rather than the x, y, z axes. Hence the coordinates of a point P will be written (x_1, x_2, x_3) rather than x, y, z. The planes determined by pairs of the coordinate axes will be called the *coordinate planes*. We label the plane determined by the axes l_1, l_2 as the 1, 2 plane or the x_1, x_2 plane. Similarly the 2, 3 and 1, 3 planes are determined by the axes l_2, l_3 and l_1, l_3, respectively. Thus to each point in space corresponds a unique triple of real numbers (x_1, x_2, x_3). Conversely, we assume that each triple of real numbers (x_1, x_2, x_3) defines a unique point in space with these numbers x_1, x_2, x_3 as first, second, and third coordinates.

One might expect that the appropriate geometric representation of a vector $A = (a_1, a_2, a_3)$ would consist just of the point in space having coordinates (a_1, a_2, a_3). This unfortunately is not general enough to give an adequate geometric interpretation to the operations we perform on vectors such as vector addition and scalar multiplication.

Instead we represent the vector A by a family of directed line segments. One line segment of this family consists of the segment drawn from the origin to the point with coordinates (a_1, a_2, a_3). The remainder consist of all directed line segments having the same length and the same direction as the first one. This is illustrated in Figure 2. To indicate the direction of a directed line segment from a point P to a point Q we draw an arrow beginning at P and ending at Q. If the directed line segment PQ is a representation for the vector

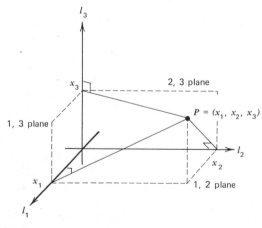

Fig. 1

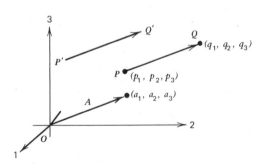

Fig. 2 Geometric representations for the vector A consist of all directed line segments having the same length and the same direction as the line segment from the origin to the point with coordinates (a_1, a_2, a_3).

A, then the point P is called the *initial point* of the representation and Q is the *terminal point*. The points Q and P are also called the head and tail of the representation for A.

Now the length $|PQ|$ of the segment PQ is given by

$$\sqrt{(q_1-p_1)^2+(q_2-p_2)^2+(q_3-p_3)^2}$$

and the length $|A|$ of the vector A is given by $\sqrt{a_1^2+a_2^2+a_3^2}$. Hence if PQ is a representation for the vector A we must have

(1) $\quad \sqrt{a_1^2+a_2^2+a_3^2} = \sqrt{(q_1-p_1)^2+(q_2-p_2)^2+(q_3-p_3)^2}.$

Next we must give mathematical meaning to the statement that the directed line segment PQ has the same direction as the directed line segment from the origin to the point with coordinates (a_1, a_2, a_3). We do this by introducing the notions of *direction angles* and *direction cosines* of a directed line segment.

Let us first consider a line segment from the origin to a point P with coordinates (p_1, p_2, p_3). The *direction angles* of the line segment OP are the angles $\alpha_1, \alpha_2, \alpha_3$ between 0 and π that this directed line segment makes with the positive first, second, and third coordinate axes. Since the length of this segment is $\sqrt{p_1^2+p_2^2+p_3^2}$, the cosines of each of these angles are given by the following formulas. (See Figure 3.)

$$\cos \alpha_1 = \frac{p_1}{\sqrt{p_1^2+p_2^2+p_3^2}}$$

$$\cos \alpha_2 = \frac{p_2}{\sqrt{p_1^2+p_2^2+p_3^2}}$$

$$\cos \alpha_3 = \frac{p_3}{\sqrt{p_1^2+p_2^2+p_3^2}}.$$

The numbers $\cos \alpha_1, \cos \alpha_2, \cos \alpha_3$ are called the *direction cosines* of the directed line segment OP. Now a line segment OP' will have the same direction as OP if and only if OP' lies along OP. But this means that the direction angles of the two line segments are equal. Equivalently their direction cosines are equal. (See Figure 4.)

To define the direction angles of an arbitrary directed line segment PQ we translate our coordinate system so that the origin is at the point P. The new

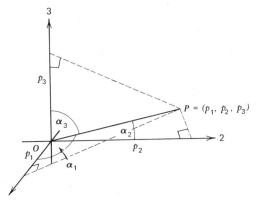

Fig. 3

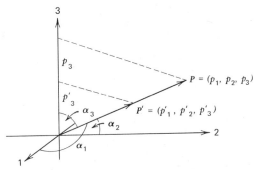

Fig. 4 Directed line segments emanating from the origin have the same direction if their direction cosines are equal.

6 VECTORS AND VECTOR ALGEBRA

coordinate axes l_1', l_2', l_3' emanating from P have the same direction as the original first, second, and third coordinate axes. The direction angles of the segment PQ are the angles between 0 and π that this segment makes with the translated positive coordinate axes. Since the coordinates of the points P and Q are (p_1, p_2, p_3) and (q_1, q_2, q_3), respectively, the cosines of these angles are given by

$$\cos \alpha_1 = \frac{q_1 - p_1}{|PQ|}$$

$$\cos \alpha_2 = \frac{q_2 - p_2}{|PQ|}$$

$$\cos \alpha_3 = \frac{q_3 - p_3}{|PQ|}$$

where $|PQ| = \sqrt{(q_1 - p_1)^2 + (q_2 - p_2)^2 + (q_3 - p_3)^2}$. (See Figure 5.)

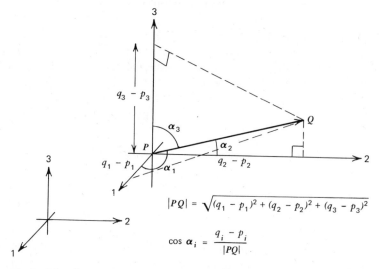

Fig. 5 Direction cosines of the directed line segment PQ.

Now two directed line segments PQ and $P_1 Q_1$ have the same direction if whenever one line segment, say PQ, is translated parallel to itself so that the point P goes into the point P_1, then the translate of PQ will lie along $P_1 Q_1$. (See Figure 6.)

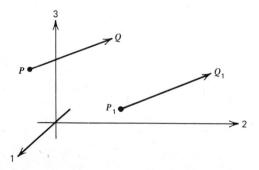

Fig. 6 Line segments having the same direction.

In terms of direction cosines this just means that PQ and P_1Q_1 will have the same direction if and only if the two segments have the same direction cosines.

Example 1. If $P = (2, 1, 3)$, $Q = (1, -1, 0)$, $P_1 = (1, 3, 4)$ and $Q_1 = (-1, -1, -2)$, show that the directed line segments PQ and P_1Q_1 have the same direction. Let α_i be the direction angles for PQ and β_i be the direction angles for P_1Q_1. Clearly $|PQ| = \sqrt{14}$ and $|P_1Q_1| = \sqrt{56}$. Moreover,

$$\cos \alpha_1 = \frac{-1}{\sqrt{14}}, \quad \cos \alpha_2 = \frac{-2}{\sqrt{14}}, \quad \cos \alpha_3 = \frac{-3}{\sqrt{14}}$$

$$\cos \beta_1 = \frac{-2}{\sqrt{56}} = \frac{-1}{\sqrt{14}}, \quad \cos \beta_2 = \frac{-4}{\sqrt{56}} = \frac{-2}{\sqrt{14}}, \quad \text{and } \cos \beta_3 = \frac{-6}{\sqrt{56}} = \frac{-3}{\sqrt{14}}.$$

Since the direction cosines are equal the directed line segments have the same direction.

We define the *direction cosines* of a *vector* A to be just the direction cosines of the geometric representations for A.

For each value of i these cosines may be written

$$(2) \qquad \cos \alpha_i = \frac{a_i}{|A|} = \frac{q_i - p_i}{|PQ|}$$

if $A = (a_1, a_2, a_3)$ and if the directed line segment PQ is a representation for A. Notice that the direction cosines satisfy the following important identity

$$\cos^2 \alpha_1 + \cos^2 \alpha_2 + \cos^2 \alpha_3 = \frac{a_1^2 + a_2^2 + a_3^2}{|A|^2} = 1.$$

Also if we observe that $|A| = |PQ|$, then (2) implies that for each value of i, $a_i = q_i - p_i$, or equivalently $A = Q - P$. Thus if the directed line segment PQ is a geometric representation for the vector A, then the vectors A, P, Q satisfy $A = Q - P$. (See Figure 7.)

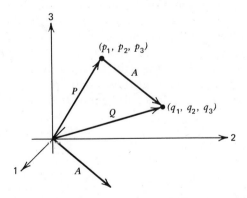

Fig. 7 The directed line segment PQ is a representation for the vector A if and only if $A = Q - P$.

The converse of this statement is also true. Namely, if $A = Q - P$ then the segment PQ is a representation for the vector A. We leave the verification of this as an exercise (Exercise 8).

In practice we only compute with the direction cosines of a vector A. Usually it is impractical and unnecessary to determine the angles α_i.

It is convenient to define the *direction* of a vector A to be the vector $(\cos \alpha_1, \cos \alpha_2, \cos \alpha_3)$ where $\cos \alpha_i$ are the direction cosines of A. Since

$$(\cos \alpha_1, \cos \alpha_2, \cos \alpha_3) = \left(\frac{a_1}{|A|}, \frac{a_2}{|A|}, \frac{a_3}{|A|}\right)$$

$$= \frac{1}{|A|}(a_1, a_2, a_3) = \frac{1}{|A|}A,$$

we see that the direction of a nonzero vector A is just the vector $(1/|A|)A$ which we write $A/|A|$. Thus two vectors A and B have the same direction if

$$\frac{A}{|A|} = \frac{B}{|B|}.$$

A direction, such as $A/|A|$, clearly has length equal to one since

$$\left|\frac{1}{|A|}A\right| = \frac{|A|}{|A|} = 1.$$

Such a vector is commonly called a *unit* vector to emphasize this fact. Thus "B is a unit vector in the direction of A" and "B is the direction of A" are synonomous statements. They both mean that

$$B = \frac{A}{|A|}.$$

The expression $A/|A|$ only makes sense for nonzero vectors, hence we adopt the position that the *zero vector has no direction*.

In space the directions of the first, second, and third coordinate axes are given by the vectors $I_1 = (1, 0, 0)$, $I_2 = (0, 1, 0)$, and $I_3 = (0, 0, 1)$, respectively. The vectors I_1, I_2, I_3 are called the unit coordinate vectors, and we will always reserve I_k for this purpose. In $V_2(R)$ the unit coordinate vectors are $I_1 = (1, 0)$ and $I_2 = (0, 1)$. The direction of a vector $A = (a_1, a_2)$ is the vector $A/|A|$, and the two direction cosines of A are $a_1/\sqrt{a_1^2 + a_2^2}$, and $a_2/\sqrt{a_1^2 + a_2^2}$. These are the cosines of the angles made by the vector A with the first and second coordinate axes, respectively. (See Figure 8.)

It is clear that the algebraic aspects of the above discussion are valid in general for vectors belonging to $V_n(R)$. If $A = (a_1, \ldots, a_n)$, then $A/|A|$ is the direction of A and $a_k/|A|$ is the kth direction cosine for A. Since $-1 \leq a_k/|A| \leq 1$, there is a unique angle α_k, satisfying $0 \leq \alpha_k \leq \pi$, and $\cos \alpha_k = a_k/|A|$. This angle α_k we define to be the kth direction angle of the

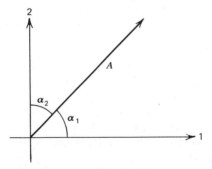

Fig. 8 Direction angles of a vector in the plane.

GEOMETRIC REPRESENTATION OF VECTORS

vector A. The kth coordinate vector I_k is the vector all of whose coordinates are zero except the kth which is 1. We note in passing that every vector A in $V_n(R)$ can be written as a linear combination of the unit coordinate vectors $I_1, \ldots, I_n$. This is clear since $A = (a_1, \ldots, a_n) = a_1(1, 0, \ldots, 0) + \cdots + a_n(0, 0, \ldots, 0, 1) = a_1 I_1 + \cdots + a_n I_n$.

EXERCISES

1. Determine the direction cosines of the following directed line segments PQ if the points P and Q have coordinates as follows.
   (a) $P = (2, 1, -1)$ $Q = (3, 2, 4)$
   (b) $P = (1, -1, 0)$ $Q = (4, 1, -1)$
   (c) $P = (4, 0, -1)$ $Q = (-1, 2, 1)$.

2. Determine the direction of the vector A if
   (a) $A = (3, 2, -1)$
   (b) $A = (4, -1, 2)$
   (c) $A = (0, -1, 2)$.

3. Let $A = (2, -1, 0)$, $B = (1, 1, 5)$. Determine the direction of the vectors $A, B, A + 2B, A - B$.

4. Let $A = (2, -1)$, $B = (3, 4)$. Represent A and B geometrically as directed line segments emanating from the origin. Determine $A + B$, $A - B$, $3A + 2B$ and represent these vectors as directed line segments emanating from the origin. Determine the direction of these three vectors.

5. Let A be the vector $(2, 3, 1)$. Determine the coordinates of a point Q such that the directed line segment PQ is a geometric representation for A if P has coordinates
   (a) $(2, -1, 0)$
   (b) $(3, 0, -2)$
   (c) $(4, 1, 3)$.

6. Which of the directed line segments PQ are geometric representations for the vector $A = (2, 3, -2)$ if the coordinates of P and Q are given by
   (a) $P = (2, 0, -1)$ $Q = (4, 1, 3)$
   (b) $P = (-2, 1, 0)$ $Q = (1, -1, -2)$.
   (c) $P = (4, -2, 1)$ $Q = (6, 1, -1)$.

7. Let A be the vector $(4, 2, -1)$. Determine the coordinates of a point P such that the directed line segment PQ is a geometric representation for the vector A if Q has coordinates
   (a) $(2, -1, 3)$
   (b) $(-1, -1, 2)$
   (c) $(-3, 2, 1)$.

8. Let $A = (a_1, a_2, a_3)$, $P = (p_1, p_2, p_3)$, $Q = (q_1, q_2, q_3)$ be vectors from $V_3(R)$. If $A = Q - P$ show that the directed line segment PQ is a geometric representation for the vector A.

9. If $A = (a_1, \ldots, a_n)$ is a nonzero vector in $V_n(R)$ with direction cosines $\cos \alpha_1, \ldots, \cos \alpha_n$, show that
$$\cos^2 \alpha_1 + \cdots + \cos^2 \alpha_n = 1.$$

10. Let A and B be nonzero vectors in $V_3(R)$. If $A = tB$ where t is a positive scalar, show that A and B have the same direction.

11. If A and B are nonzero vectors in $V_3(R)$ having the same direction show that $A = tB$ for some positive scalar t.

12. Let A and B be nonzero vectors in $V_n(R)$. Show that they have the same direction if and only if one is a positive scalar multiple of the other.

13*. It was shown in 1877 by the German mathematician Frobenius that $V_n(R)$ can be made into a field only in the case $n = 2$. For $n = 4$ however, a multiplication can be defined in $V_4(R)$ having all the properties of multiplication in a field except that it is noncommutative. This multiplication, invented by the Irish mathematician W. R. Hamilton, is defined as follows. Let

$$X = (x_1, x_2, x_3, x_4) = x_1 I_1 + x_2 I_2 + x_3 I_3 + x_4 I_4$$

where

$$I_1 = (1, 0, 0, 0), \ldots, I_4 = (0, 0, 0, 1),$$

are the unit coordinate vectors in $V_4(R)$. We denote our multiplication by $X \cdot Y$ and postulate that for all vectors $X, Y, Z \in V_4(R)$ and real numbers a

(2) $$X \cdot (Y + Z) = X \cdot Y + X \cdot Z$$

(3) $$a(X \cdot Y) = (aX) \cdot Y = X \cdot (aY).$$

Since each vector $X \in V_4(R)$ may be written $X = x_1 I_1 + \cdots + x_4 I_4$ we may infer from (2) and (3) that to define $X \cdot Y$ we need only define how the unit vectors $I_1, \ldots, I_4$ multiply together. Define

(4) $$I_1 \cdot I_k = I_k \cdot I_1 = I_k \quad \text{for} \quad k = 1, 2, 3, 4.$$

(5) $$I_2^2 = I_3^2 = I_4^2 = -I_1.$$

(6) $$\begin{cases} I_2 \cdot I_3 = -I_3 \cdot I_2 = I_4 \\ I_3 \cdot I_4 = -I_4 \cdot I_3 = I_2 \\ I_4 \cdot I_2 = -I_2 \cdot I_4 = I_3 \end{cases}$$

This multiplication is clearly noncommutative. Show that it is associative and show that if we define for $X \neq 0$,

$$X^{-1} = \frac{1}{|X|^2} (x_1, -x_2, -x_3, -x_4)$$

then

(7) $$X \cdot X^{-1} = X^{-1} \cdot X = I_1.$$

The set $V_4(R)$ provided with this multiplication is called the set of *quarternions*. For $n = 8$ a multiplication may be defined such that division is possible, i.e. so that (7) holds. However, in this case the multiplication is nonassociative as well as noncommutative. It was finally shown around 1960 by Raoul Bott, John W. Milnor, and others that only in these cases, $n = 2, 4, 8$ could a multiplication be defined in $V_n(R)$ for which division by nonzero vectors is always possible. This fact is regarded by many mathematicians as one of the truly outstanding accomplishments of 20th-century mathematics.

1.3 Geometric Vector Addition and Scalar Multiplication

We turn next to the geometric interpretation of scalar multiplication and vector addition for vectors in $V_3(R)$. If $B = tA$ for a scalar t, then $|B| = |tA| = |t||A|$, or the length of B is the product of this length of A and the absolute value of t. It can be immediately verified that if $t > 0$, then B has the same direction as A, and if $t < 0$, then the direction of B is the negative of that of A. On the other hand the vectors A and B are parallel if A and B have the same direction or if A and B have opposite directions. Hence we see that nonzero vectors A and B are parallel if and only if one is a nonzero scalar multiple of the other. (See Figure 9.)

The geometric interpretation of vector addition may be determined equally easily. Indeed if A and B are represented as in Figure 10, then, $A + B$ is represented by the directed line segment from the origin to the fourth vertex of the parallelogram determined by the vectors A and B. This follows because

GEOMETRIC VECTOR ADDITION AND SCALAR MULTIPLICATION 11

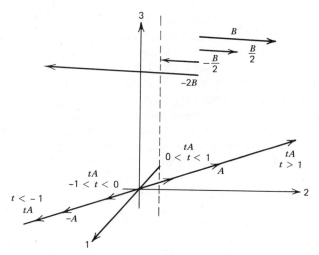

Fig. 9 Geometric representation of scalar multiplication.

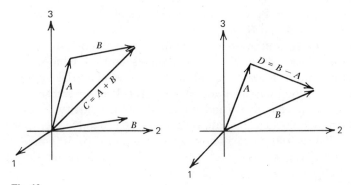

Fig. 10

if (c_1, c_2, c_3) is the fourth vertex of this parallelogram, then the line segment from (a_1, a_2, a_3) to (c_1, c_2, c_3) must be a representation for B since the line segment has the same direction as B as well as the same length. Therefore

$$B = C - A$$

or

$$C = A + B.$$

When the vector B is represented as the directed line segment from the point (a_1, a_2, a_3) to the point (c_1, c_2, c_3), we have noted that it is convenient to refer to the point (a_1, a_2, a_3) as the initial point or tail of the vector B and (c_1, c_2, c_3) as the terminal point or head of the vector. These points are not uniquely defined of course, but depend on the representation of B. Using this terminology a very convenient recipe for performing vector addition geometrically can be given. Translate B parallel to itself attaching the tail of B to the head of A. The directed line segment from the tail of A to the head of B is then a representation for the vector $A + B$.

In the plane we have the following very useful characterization of nonparallel vectors.

THEOREM 1.1. *Let A and B be nonzero vectors in $V_2(R)$. Then each vector $X \in V_2(R)$ is a linear combination of the vectors A and B if and only if these vectors are not parallel.*

Proof. We leave the proof that if each vector X can be written $X = sA + tB$, then A and B must be nonparallel as an exercise. We shall give a geometric proof of the converse statement.

Let A and B be nonparallel vectors represented as directed line segments emanating from the origin. Then the collection of scalar multiples $\{sA\}$ and $\{tB\}$ determine lines through the origin which we label l_1, l_2. (See Figure 11.) Draw lines l'_1, l'_2 through the point with coordinates (x_1, x_2) parallel to l_1 and l_2. The line l'_2 intersects l_1 at P and l'_1 intersects l_2 at Q. But OP is parallel to A. Hence $OP = sA$ for some scalar s. Similarly $OQ = tB$ for some scalar t. Since X is the diagonal of the parallelogram determined by OP and OQ, we infer that $X = sA + tB$.

To determine the scalars s and t we solve the equations

$$x_1 = sa_1 + tb_1$$

and

$$x_2 = sa_2 + tb_2$$

where $X = (x_1, x_2)$, $A = (a_1, a_2)$ and $B = (b_1, b_2)$. When each vector $X \in V_2(R)$ can be written as a linear combination of the vectors A and B, we say that A and B *span* $V_2(R)$.

Much of the importance of vectors in applications stems from the fact that vector addition describes the way that quantities, which are represented by vectors, interact with one another. For example, the velocity of a particle moving in the plane is a vector in $V_2(R)$. It can be observed that if an airplane has a velocity V and the wind has a velocity V' then the resulting velocity of the airplane with respect to the ground is the vector sum $V + V'$. Thus if V is 400 miles an hour northeast and V' is 100 miles an hour southeast, then the resulting velocity $V + V'$ of the airplane can be constructed geometrically as in Figure 12.

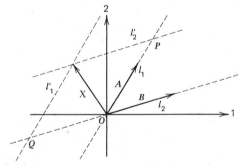
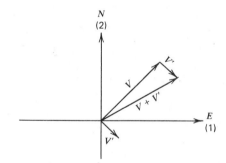

Fig. 11 **Fig. 12**

Thus $V = 400(1/\sqrt{2}, 1/\sqrt{2})$ and $V' = 100(1/\sqrt{2}, -(1/\sqrt{2}))$. Hence

$$V + V' = \left(\frac{500}{\sqrt{2}}, \frac{300}{\sqrt{2}}\right) = 50\sqrt{2}(5, 3).$$

The direction of $V + V'$ is $1/\sqrt{34}(5, 3)$.

Another example is provided by forces acting on a particle in space. If the various forces acting on the particle are denoted by the vectors $F_1, \ldots, F_n$, then Newton's second law of motion states that the motion of the particle is described by the vector equation

$$mA = F_1 + \cdots + F_n$$

GEOMETRIC VECTOR ADDITION AND SCALAR MULTIPLICATION

where m is the mass of the particle, and A is the acceleration. In this equation the mass m is a scalar and the acceleration A is a vector. Thus if the particle is at rest, the vector sum of the forces is zero. That is,

$$F_1 + \cdots + F_n = 0.$$

Example 1. If $F_1 = (5, 6)$, and $F_2 = (2, -3)$ are vectors describing two forces acting on a particle P in the plane, what is the third force F_3 acting on P if the particle is at rest? By Newton's law

$$F_1 + F_2 + F_3 = 0.$$

Therefore

$$F_3 = -F_1 - F_2 = -(7, 3).$$

This is represented geometrically in Figure 13.

When we use vectors to describe physical phenomena it is convenient to make a choice between the many geometrical representations of the same vector. For example, suppose we are using vectors to describe a particle moving in the plane or in space. Suppose at time t the coordinates of the particle are given by (x_1, x_2, x_3). Then the directed line segment from the origin to the point with coordinates (x_1, x_2, x_3) would be taken to represent the position vector $X = (x_1, x_2, x_3)$. Suppose for the same value of t the velocity of the particle is given by $V = (v_1, v_2, v_3)$ and the acceleration by $A = (a_1, a_2, a_3)$. Then it is natural to take the geometric representation of these vectors to be directed line segments emanating from the point with coordinates (x_1, x_2, x_3) since this is the position at which the velocity and acceleration are taking place. If for all values of t the position of the particle is given by a curve, then the velocity and acceleration vectors would be represented by vectors attached or emanating from the appropriate points on the curve. Thus if at time t, $X(t) = (x_1(t), x_2(t), x_3(t))$ is the position vector and $V(t) = (v_1(t), v_2(t), v_3(t))$ is the velocity and $A(t)$ the acceleration, these vectors would be drawn as in Figure 14.

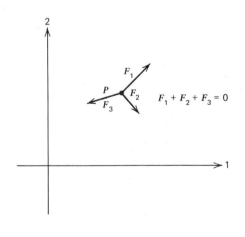

Fig. 13

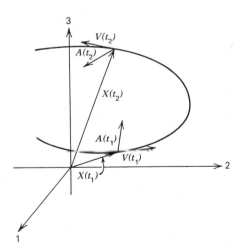

Fig. 14 Representation of motion of a particle by vectors.

Similarly if F_1, F_2, F_3 are forces acting on a particle whose position has coordinates (x_1, x_2, x_3), the geometric representations of the vectors F_1, F_2, F_3 would be chosen as line segments emanating from this point.

EXERCISES

1. Let $A = (2, 1)$, $B = (3, -1)$. Represent A and B as directed line segments emanating from the origin. Represent geometrically the vectors $2A$, $-3A$, $A + B$, $A - B$, $A - 2B$.

2. Let $A = (3, 1)$, $B = (2, 1)$, $C = (0.3)$. Represent geometrically the vectors $A - B + C$, $A + B - C$, and $2A - \frac{1}{2}B + 2C$.

3. Let $A = (1, 2, 3)$, $B = (1, -1, 1)$, $C = (2, 1, -3)$. Determine the vectors $A + B$, $2A - B$, $A + B - C$, and their respective directions. Represent these vectors as directed line segments.

4. Let $A = (2, 1)$, $B = (3, 4)$. For a real number x let $C = A + xB$. Compute C for three different values of x, and draw geometric representations as line segments with common initial point. What is the locus of the terminal points of all such vectors C?

5. Solve Exercise 4 if $A = (-1, 1)$, $B = (2, -3)$.

6. Let $A = (0, 1)$, $B = (-1, 2)$. What is the locus of all terminal points of the set of vectors $xA + (1 - x)B$ if $0 \leq x \leq 1$?

7. Let $A = (1, 2, 3)$, $B = (3, 1, 2)$. What is the locus of the set of terminal points of the vectors $xA + B$? What is the locus for the set of vectors $x(B - A) + A$ if $0 \leq x \leq 1$? Draw the loci involved.

8. Let $A = (1, 3)$, $B = (3, -1)$. Let $C = (c_1, c_2)$ be an arbitrary vector in the plane. Determine real numbers x, y so that $C = xA + yB$. For $C = 0 = (0, 0)$ what are the values of x and y? Represent this geometrically if $C = (2, 4)$.

9. Answer Exercise 8 if $A = (-1, 2)$, $B = (2, 3)$.

10. Let A and B be vectors in $V_2(R)$. If either A or B is zero, or if A and B are parallel show that not every $X \in V_2(R)$ is a linear combination of A and B.

11. Let $A = (1, 2, 3)$, $B = (1, 0, 1)$, $C = (-1, 1, 0)$. What is the locus of terminal points of vectors of the form $A + xB + yC$, x, y real numbers? Can an arbitrary vector $D = (d_1, d_2, d_3)$ be written in this form? For what values of d can the vector $D = (1, -1, d)$ be so represented?

12. Solve Exercise 11 if $A = (1, 0, -1)$, $B = (-1, 1, 1)$, $C = (0, 0, -1)$.

13. For A, B, C in Exercise 11 can we find for an arbitrary vector $D = (d_1, d_2, d_3)$ real numbers x, y, z so that

$$D = xA + yB + zC?$$

If $D = (-1, 0, -1)$ what are the values of x, y, z? What are they for $D = (0, 0, 0)$?

14. Answer Exercise 13 for the vectors A, B, C in Exercise 12.

15. An airplane is flying 500 miles per hour northwest, and there is an easterly cross wind blowing of 50 miles per hour. What is the resulting velocity of the airplane with respect to the ground? What is the direction? (Choose a coordinate system so that the first and second coordinate axes correspond to the directions east and north, respectively.)

16. Let π be a regular polygon with center at the origin. Let $A_1, \ldots, A_n$ be the vectors from the origin to each of the vertices. Show (geometrically) that $A_1 + \cdots + A_n = 0$. (This is obvious if the number n of vertices is even. If n is odd, let $B = A_1 + \cdots + A_n$ and now rotate B through a suitable angle.)

17. A 100-pound weight is suspended on a five foot flexible cord joining two pegs two feet apart. Determine the resulting forces at each peg if the coordinate system is chosen as in the following figure. Determine the forces if the right-hand peg is moved up two feet. Assume the weight is free to move along the cord. (See Figure 15.)

THE INNER PRODUCT OF TWO VECTORS   15

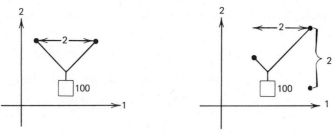

Fig. 15

1.4 The Inner Product of Two Vectors

One additional operation can be defined on vectors in $V_n(R)$ which has very useful geometric interpretations. This is the *inner product* of the vectors A and B.

Definition. *If $A = (a_1, \ldots, a_n)$ and $B = (b_1, \ldots, b_n)$, then the inner product of A and B, written (A, B) is defined by the formula*

$$(A, B) = a_1 b_1 + \cdots + a_n b_n.$$

For example if $A = (2, 1, 3)$ and $B = (-1, 1, -2)$, then $(A, B) = -2 + 1 - 6 = -7$. This inner product is an operation on vectors which *does not* yield a vector as a result but instead yields a scalar. It is often referred to as the *scalar product*, or *dot product*. If the latter terminology is used, then (A, B) is written $A \cdot B$. We shall adhere to the parenthesis terminology, however.

The following algebraic properties of the inner product are readily established.

(1) $(A, B) = (B, A)$ (commutative law)

(2) $(A, B + C) = (A, B) + (A, C)$ (distributive law)

(3)   If c is a scalar, then $c(A, B) = (cA, B) = (A, cB)$.

The verification of these three properties is immediate and is left as an exercise. The student should note that the associative law does not make sense for the inner product since (A, B) is a scalar, not a vector. We note finally the following useful connection with the length of A:

(4) $(A, A) = a_1^2 + \cdots + a_n^2 = |A|^2.$

Therefore $(A, A) \geq 0$. Furthermore $(A, A) = 0$ if and only if $A = 0$.

For vectors in the plane or in space the inner product of vectors A and B may be given an immediate geometric interpretation. Representing the vector $A - B$ as the directed line segment from the point with coordinates (b_1, b_2, b_3) to the point with coordinates (a_1, a_2, a_3) as in Figure 16, the law

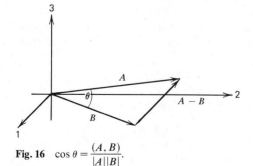

Fig. 16 $\cos \theta = \dfrac{(A, B)}{|A||B|}.$

of cosines gives the following formula for the angle θ between the vectors A and B.

(5) $$|A|^2 + |B|^2 - |A-B|^2 = 2|A||B| \cos \theta.$$

But applying property (4) of the inner product we see that

(6) $$|A|^2 + |B|^2 - |A-B|^2 = (A,A) + (B,B) - (A-B, A-B).$$

Using (1)–(3) we infer that

$$\begin{aligned}(A-B, A-B) &= (A-B, A) - (A-B, B) \quad \text{(properties (2) and (3))} \\ &= (A, A-B) - (B, A-B) \quad \text{(property (1))} \\ &= (A,A) - (A,B) - (B,A) + (B,B) \quad \text{(by (2) and (3))} \\ &= (A,A) - 2(A,B) + (B,B) \quad \text{(by (1))}.\end{aligned}$$

Therefore
$$(A,A) + (B,B) - (A-B, A-B) = 2(A,B).$$

Combining this with (5) and (6) we see that

$$(A,B) = |A||B| \cos \theta$$

or

(7) $$\cos \theta = \frac{(A,B)}{|A||B|}.$$

For example if $A = (1,2,3)$, $B = (-1,-1,2)$, then the angle θ between these vectors satisfies

$$\cos \theta = \frac{(A,B)}{|A||B|} = \frac{3}{\sqrt{14}\sqrt{6}} = \frac{3}{2\sqrt{21}}.$$

Consequently,

$$\theta = \arccos \frac{3}{2\sqrt{21}}.$$

For vectors in $V_2(R)$ or $V_3(R)$ formula (7) shows that

$$-1 \leq (A,B)/|A||B| \leq 1.$$

A natural question is to ask if, this formula holds in general for vectors in $V_n(R)$. It does and the inequality is called the *Schwarz inequality*. It may be written in the following two equivalent forms

$$|(A,B)| \leq |A||B|$$

or

$$(A,B)^2 \leq (A,A)(B,B).$$

Once this formula has been established for vectors in $V_n(R)$ it gives us the means for *defining* the angle between two such vectors A and B. Indeed we define θ to be that number θ, satisfying $0 \leq \theta \leq \pi$, such that

(8) $$\cos \theta = \frac{(A,B)}{|A||B|}.$$

This is legitimate once we have shown that

$$-1 \leq \frac{(A,B)}{|A||B|} \leq 1.$$

After this has been done, it can be immediately established that the law of cosines (5) holds for vectors in $V_n(R)$. The argument is exactly the same as the one that was used to prove (7) for vectors in space assuming the law of cosines.

This situation is very typical. A geometric fact which is proved for vectors in the plane or in space is taken to be the definition of a corresponding notion for vectors in $V_n(R)$. For example, if A and B are nonzero vectors in $V_3(R)$, then they are perpendicular if the angle θ between them is $\pi/2$. Applying (7) we have
$$0 = \cos \theta = (A, B)/|A||B|$$
or
$$(A, B) = 0.$$

We now use this fact as the definition of perpendicularity for vectors in $V_n(R)$.

Definition. *If A and B are vectors in $V_n(R)$, then A and B are* orthogonal *if*
$$(A, B) = 0.$$

If in addition A and B are different from zero, then A and B are called perpendicular.

Another instance of this is the following. We saw in Section 8.3 that nonzero vectors in space were parallel if and only if one was a nonzero scalar multiple of the other. For vectors in $V_n(R)$ we take this fact as the definition.

Definition. *If A and B are nonzero vectors in $V_n(R)$, they are* parallel *if $A = tB$ for some nonzero scalar t.*

Note that under these definitions the notions of parallelism and perpendicularity do not apply to the zero vector. This is consistent with our position that the zero vector has no direction.

EXERCISES

1. Let $A = (1, 2, 1)$, $B = (2, -1, 0)$, $C = (0, 1, -1)$. Compute: (A, A), (A, B), $(C, A - B)$.
2. Using the vectors of Exercise 1 for which value of t is C perpendicular to $A + tB$? Represent the vectors involved as directed line segments.
3. If $A = (-1, 0, 2)$, $B = (1, -1, 1)$, $C = (-1, 2, 3)$, find a scalar t so that A is orthogonal to $B + tC$. Represent the vectors involved as directed line segments.
4. If $A = (4, 2, -4)$, $B = (2, -1, 0)$, $C = (0, 1, -1)$, find a scalar t so that A is parallel to $B + tC$. Draw the representation of the vectors involved. (*Hint:* vectors X and Y are parallel if $X = sY$ for some nonzero scalar s.
5. Solve Exercise 4 if $A = (4, -1, 2)$, $B = (2, 1, 1)$, $C = (2, -1, 1)$.
6. If $A \neq 0$ and $(A, B) = (A, C)$ for vectors B and C, does it follow that $B = C$? If not give examples which satisfy the equation.
7. Let $A = (1, -1, 0)$, $B = (2, 0, -1)$. Find a nonzero vector C orthogonal to both A and B. (Write $C = (c_1, c_2, c_3)$ and apply orthogonality relationship.) Are all such vectors C parallel to each other?
8. Solve Exercise 7 if $A = (1, 1, 1)$, $B = (2, 1, -1)$.
9. If $A = (-1, 1, 1)$, $B = (0, 0, 1)$ find all vectors C of unit length perpendicular to both A and B.

10. Solve Exercise 9 if $A = (1, 0, 1)$, $B = (1, 1, 1)$.
11. If $A = (2, -1)$, $B = (1, 1)$, determine scalars x and y so that $xA + yB$ is perpendicular to A. Determine the scalars so that in addition $xA + yB$ has unit length. Represent the vectors involved as directed line segments.
12. Let $A = (1, -1, 0)$, $B = (-1, 0, 1)$. Solve Exercise 11 for these vectors.
13. Verify properties (1)–(3) for the inner product of vectors in $V_n(R)$.
14. Show that vectors A and B in $V_n(R)$ are orthogonal if and only if

$$|A + B|^2 = |A|^2 + |B|^2$$

(Theorem of Pythagoras).

15. Let A and B be vectors in $V_n(R)$. Using properties (1)–(4) of the inner product prove that
    (a) $|A + B|^2 - |A - B|^2 = 4(A, B)$.
    (b) $|A + B|^2 + |A - B|^2 = 2\{|A|^2 + |B|^2\}$.
16. Let A and B be vectors in $V_3(R)$ and consider the parallelogram formed by A and B. Interpret 15(b) above as a geometric statement about this parallelogram.
17. Using 15(a) give a condition on the diagonals of a parallelogram which is necessary and sufficient for the parallelogram to be a rectangle.
18. A pirate captain sailing along the coast of Florida has recently plundered a Spanish galleon. He has a sizable amount of treasure which must be hidden to avoid detection. The captain chooses a stretch of coast with two large rocks about a quarter of a mile apart which will identify the spot. Inland between the rocks is a solitary oak. The captain gives the following directions for burying the treasure. Two men stand at the tree and march to the respective rocks pacing off the distances. They both then make right angle turns inland and proceed to pace off a distance equal to what they have just walked. They then face each other and at a point half way between them the treasure is buried. Years later the pirate captain returns for his treasure. He finds the spot, guided by the two rocks, but is horrified to see that the oak has disappeared. Can he find his treasure without digging up the entire coast line? (It might help if the captain had studied vector algebra. Of course, it had not yet been invented, but you can ignore that difficulty.)

1.5 The Schwarz Inequality and the Triangle Inequality

THEOREM 1.2. *(Schwarz Inequality). Let A and B be vectors in $V_n(R)$. Then*

(1) $$(A, B)^2 \leq (A, A)(B, B).$$

Equality holds in the above expression if and only if one vector is a scalar multiple of the other.

Proof. If either A or B is the zero vector, then the theorem clearly holds. So suppose $B \neq 0$, and consider for a real number x the function

$$p(x) = (A + xB, A + xB) = |A + xB|^2.$$

For all values of x, $p(x) \geq 0$, and if we expand the above inner product we get

(2) $$p(x) = (B, B)x^2 + 2(A, B)x + (A, A).$$

Therefore $p(x)$ is a polynomial in x of degree 2. If we complete the square on the right-hand side of (2), we obtain

$$p(x) = (B,B)\left[x^2 + \frac{2(A,B)x}{(B,B)} + \frac{(A,B)^2}{(B,B)^2}\right] - \frac{(A,B)^2}{(B,B)} + (A,A)$$

(3)

$$= (B,B)\left[x + \frac{(A,B)}{(B,B)}\right]^2 + \frac{(A,A)(B,B) - (A,B)^2}{(B,B)}.$$

Set $x_0 = -\dfrac{(A,B)}{(B,B)}$. Then

(4)
$$0 \leq p(x_0) = 0 + \frac{(A,A)(B,B) - (A,B)^2}{(B,B)}.$$

Since $(B,B) > 0$, this implies

$$(A,A)(B,B) - (A,B)^2 \geq 0$$

or

$$(A,B)^2 \leq (A,A)(B,B).$$

Thus we have established (1).

To finish the proof, suppose $(A,B)^2 = (A,A)(B,B)$. Then from (4)

$$p(x_0) = 0$$

where $x_0 = -(A,B)/(B,B)$. But $p(x_0) = |A + x_0 B|^2$. Hence $A + x_0 B = 0$ or A is a scalar multiple of B. It remains only to show that if one vector is a scalar multiple of the other, then $(A,B)^2 = (A,A)(B,B)$. This can be immediately verified.

COROLLARY. *Let A and B be nonzero vectors in $V_n(R)$. These vectors are parallel if and only if*

$$|(A,B)| = |A||B|.$$

As an immediate consequence of the Schwarz inequality we have the following.

THEOREM 1.3 (Triangle Inequality). *Let A, B be vectors in $V_n(R)$. Then*

$$|A + B| \leq |A| + |B|.$$

Furthermore $|A + B| = |A| + |B|$ if and only if one vector is a nonnegative scalar multiple of the other.

Proof. Before beginning the proof let us observe that if we consider vectors in the plane or in space then the triangle inequality just expresses the familiar fact that the sum of the lengths of two sides of a triangle exceeds the length of the third side. (See Figure 17.)

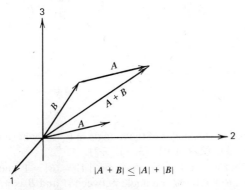

Fig. 17 The triangle inequality.

To establish the triangle inequality for vectors in $V_n(R)$ we perform the following computation.

$$\begin{aligned}
|A+B|^2 &= (A+B, A+B) \\
&= (A,A) + 2(A,B) + (B,B) \\
&\leq (A,A) + 2|(A,B)| + (B,B) \\
&\leq (A,A) + 2|A||B| + (B,B) \text{ (Schwartz Inequality)} \\
&= |A|^2 + 2|A||B| + |B|^2 = (|A| + |B|)^2.
\end{aligned}$$

Taking nonnegative square roots we have

$$|A+B| \leq |A| + |B|.$$

If we examine the above computation, we see that $|A+B| = |A| + |B|$ if and only if $(A,B) = |A||B|$. But by Theorem 1.2 this can hold if and only if one vector is a scalar multiple of the other. But if $B = tA$ and $(A,B) = |A||B|$, then $0 \leq (A,B) = t(A,A)$. Therefore if $A \neq 0$, it must follow that $t \geq 0$.

EXERCISES

1. Let $A = (2, 3, -1)$, $B = (-4, -6, 2)$, $C = (1, 1, -1)$, $D = (1, 2, 8)$. Which pairs of these vectors are orthogonal and which are parallel?
2. Determine the cosines of the angles of the triangle whose vertices are the points with coordinates $(2, 1, 3)$, $(3, 1, 2)$, $(2, -1, 1)$.
3. For what values of x is $(3, x, 1)$ perpendicular to $(1, 2, -1)$? Is it possible to choose x so that these vectors are parallel?
4. Answer Exercise 3 if the vectors are $(x, 2, -3)$ and $(-2, 4, -6)$.
5. Deduce from the triangle inequality that if A and B in $V_n(R)$, then

$$||A| - |B|| \leq |A \pm B| \leq |A| + |B|.$$

   (Write $A = B + (A - B)$, etc., and apply Theorem 1.3.)
6. Prove that if A and B are parallel vectors in $V_n(R)$, then

$$|(A, B)| = |A||B|.$$

7. Show by vector methods that a triangle inscribed in a semicircle is a right triangle. (See Figure 18.)

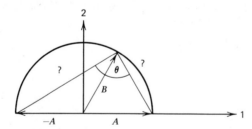

Fig. 18

8. Determine the cosines of the angles of the triangle in $V_4(R)$ whose vertices have coordinates $(1, 2, -1, 3)$, $(2, 1, 4, 0)$, $(3, -1, 2, 2)$.
9. Let A, B, C be vectors in $V_n(R)$. Assume $|A| = |B| = 1$ and $|C| = 4$. If $|A - B + C| = |A + 2B + C|$ and the angle between A and B is $\pi/4$, determine the angle between B and C.

10. Let A, B, C be nonzero vectors and assume the angle between A and C equals that between B and C. For what value of t is C perpendicular to
$$D = |B|A + tB?$$

11. Give an alternate proof of the Schwarz inequality by evaluating the polynomial $p(x) = |A + xB|^2$ at the point $x = -((A,A)/(A,B))$. (Of course, we must assume $(A, B) \neq 0$. What happens if $(A, B) = 0$?)

12. Let A and B be nonzero vectors in $V_n(R)$. For what value of x does the quadratic polynomial $p(x) = |A + xB|^2$ take on its minimum value? What is the value of this minimum?

13. We have defined length or norm of a vector $A = (a_1, \ldots, a_n)$ by the following formula
$$|A| = (a_1^2 + \cdots + a_n^2)^{1/2}.$$

More properly $|A|$ should be called the Euclidean length or norm of the vector A. The function $|A|$ is a real valued function defined on the vectors in $V_n(R)$, and we have shown that the following four properties hold for $|A|$
   (i) $|A + B| \leq |A| + |B|$
   (ii) $|aA| = |a||A|$ for scalars a
   (iii) $|A| > 0$ if $A \neq 0$
   (iv) $|A| = 0$ if $A = 0$.

Now any real valued function defined on the vectors in $V_n(R)$ which satisfies (i)–(iv) could be called a length or norm for the vector A. Such norms, if different from $|A|$, we shall call non-Euclidean. Verify that
$$\|A\| = |a_1| + \cdots + |a_n|$$
and
$$\|\|A\|\| = \max(|a_1|, \ldots, |a_n|)$$

are two examples of non-Euclidean norms for the vectors of order n. (That is show that properties (i), (iv) above hold for $\|A\|$ and $\|\|A\|\|$.) If we define $|A|_1 = |a_1 + \cdots + a_n|$, is this a non-Euclidean norm for A?

14. If A is a vector in $V_2(R)$, then $\{A : |A| \leq 1\}$ is represented in the plane by the unit disc of radius 1. (See Figure 19.)

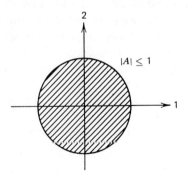

Fig. 19

Represent in the plane the sets
$$\{A \in V_2(R) : \|A\| \leq 1\} \quad \text{and} \quad \{A \in V_2(R) : \|\|A\|\| \leq 1\},$$
where $\|A\|$, $\|\|A\|\|$ are the non-Euclidean norms of Exercise 13.

1.6 Lines in $V_n(R)$.

Let A and B be two vectors in $V_n(R)$ and assume $B \neq 0$. By the *line l through A parallel to B* we shall mean the set of all vectors X which can be written

(1) $$X = A + tB$$

for some scalar t. Equation (1) is called the vector equation of the line l.

To describe geometrically a line in $V_3(R)$, we represent all of the vectors X satisfying (1) as directed line segments emanating from the origin. (See Figure 20.) Then a point (x_1, x_2, x_3) in space lies on the line l with equation (1) if and only if this point is a terminal point of a directed line segment representing one of the vectors X. With this choice of the representations for X we see that a point with coordinates (p_1, p_2, p_3) lies on the line l if and only if the vector $P = (p_1, p_2, p_3)$ satisfies the equation of the line. If we represent the vectors A and B in space, then the geometric representation of the line l is illustrated in Figure 20.

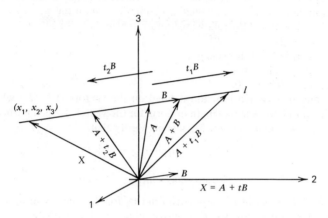

Fig. 20

If the vector P satisfies the equation of the line, it is convenient to say that P lies on the line. One should notice that this does *not* say that all points lying on the directed line segment from the origin to the point with coordinates (p_1, p_2, p_3) lie on the line. It is clear that this happens if and only if the line passes through the origin, that is, if $A = 0$. In general only the terminal point of the directed line segment from the origin which represents P lies on the line.

In addition to the vector equation (1), the line l is also characterized by scalar equations obtained from (1) by equating the respective components. Thus in $V_3(R)$ equation (1) is equivalent to three scalar equations. For if

$$X = (x_1, x_2, x_3), \quad A = (a_1, a_2, a_3), \quad B = (b_1, b_2, b_3)$$

then $X = A + tB$ implies

(2) $$\begin{cases} x_1 = a_1 + tb_1 \\ x_2 = a_2 + tb_2 \\ x_3 = a_3 + tb_3. \end{cases}$$

These equations are often called *the parametric equations* of the line l.

Solving for the parameter t we obtain the three equations

$$t = \frac{x_i - a_i}{b_i} \qquad i = 1, 2, 3.$$

Eliminating t between these equations we have the following symmetric form of the equation for a line

$$\frac{x_1 - a_1}{b_1} = \frac{x_2 - a_2}{b_2} = \frac{x_3 - a_3}{b_3}.$$

If one of the components of B is zero, say $b_1 = 0$, this pair of scalar equations becomes

$$x_1 = a_1; \qquad \frac{x_2 - a_2}{b_2} = \frac{x_3 - a_3}{b_3}.$$

The vector and scalar equations of the line l may be determined in a variety of ways by specifying geometric conditions.

Example 1. Determine the vector and scalar equation of the line passing through $A = (1, 2, -1)$ and $B = (2, 1, 1)$. Now the vector equation will be $X = A + tC$ for some vector C. But B lies on the line, hence for some scalar t_0

$$B = A + t_0 C$$

or

$$C = \frac{B - A}{t_0}.$$

Thus we may take C to be any vector parallel to $B - A = (1, -1, 2)$. In particular if we take $C = B - A$, the vector equation becomes

$$X = (x_1, x_2, x_3) = A + t(B - A)$$
$$= (1, 2, -1) + t(1, -1, 2).$$

The resulting scalar equations are

$$x_1 = 1 + t$$
$$x_2 = 2 - t$$
$$x_3 = -1 + 2t.$$

Eliminating t we have the symmetric equation of the line

$$x_1 - 1 = \frac{x_2 - 2}{-1} = \frac{x_3 + 1}{2}.$$

For vectors $A = (a_1, a_2)$ and $B = (b_1, b_2)$ in the plane the scalar equation of the line l through A parallel to B takes the form

$$\frac{x_2 - a_2}{b_2} = \frac{x_1 - a_1}{b_1}.$$

Writing this as

$$\frac{x_2 - a_2}{x_1 - a_1} = \frac{b_2}{b_1}$$

we see that the slope of the line is b_2/b_1.

Example 2. Determine the vector equation of the line passing through $(1, 2)$ with slope 3. This line has scalar equation

$$\frac{x_2 - 2}{x_1 - 1} = 3$$

or

$$\frac{x_2 - 2}{3} = \frac{x_1 - 1}{1}.$$

Therefore the line passes through $(1, 2)$ and is parallel to $(1, 3)$. Hence the vector equation is

$$X = (1, 2) + t(1, 3).$$

If the lines l_1 and l_2 have vector equations $X_1 = A_1 + tB_1$ and $X_2 = A_2 + tB_2$, respectively, we say that these lines are perpendicular or parallel according to whether the vectors B_1 and B_2 are perpendicular or parallel.

Lines in the plane may be described in terms of vectors perpendicular to a fixed vector N. If $X = A + tB$ is the equation of a line l in $V_2(R)$ and $B = (b_1, b_2)$, then let us set $N = (-b_2, b_1)$. It is clear that B and N are perpendicular since $(B, N) = -b_1 b_2 + b_1 b_2 = 0$. Moreover, if X lies on the line l, then $X - A$ is perpendicular to N. This is clear since

$$(X - A, N) = (tB, N) = t(B, N) = 0.$$

Suppose conversely that X is a vector satisfying $(X - A, N) = 0$. We assert that the vector X lies on the line l. To see this, observe that since B and N are not parallel, we know by Theorem 1.1 that $X - A = tB + sN$ for some scalars s and t. If we take the inner product of this equation with the vector N, we obtain

$$0 = (X - A, N) = (tB + sN, N)$$
$$= t(B, N) + s(N, N) = s(N, N)$$

since $(B, N) = 0$. However, $N \neq 0$. Hence $(N, N) \neq 0$, and therefore $s = 0$. Thus it follows that $X = A + tB$.

As a result we conclude that in $V_2(R)$ the line through A parallel to B may be characterized as the set of all vectors X, such that $(X - A, N) = 0$ where $N = (-b_2, b_1)$ and $B = (b_1, b_2)$. Hence we may speak of l as the line through A perpendicular to N. The vector equation of this line is then

(3) $$(X - A, N) = 0.$$

The vector N, or any nonzero scalar multiple of N, is called a *normal vector* to the line l. (See Figure 21.)

If $X = (x_1, x_2)$ and $N = (n_1, n_2)$, then we may write (3) as

(4) $$n_1 x_1 + n_2 x_2 = c, \text{ where } c = (N, A).$$

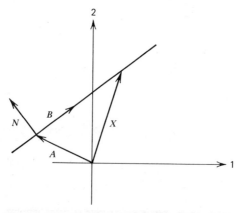

Fig. 21 The line through A perpendicular to N has the equation $(X - A, N) = 0$.

Thus if the equation of a line in the plane is written in the form (4), then (n_1, n_2) is a normal vector to the line and the constant c equals the inner product of N with any vector lying on the line. For example if the nonvertical straight line has equation $y = mx + b$ where m is the slope and b the y intercept, then $(-m, 1)$ is a normal vector to this line. The constant b is the inner product of the vector $(-m, 1)$ with any vector X lying on the line.

Example 3. A line l in the plane passes through the points $(2, 1)$ and $(3, -2)$. Determine a normal vector to the line and the equation of the line in the form (4). The line clearly has equation
$$\frac{x_2 - 1}{x_1 - 2} = \frac{-3}{1}$$
or
$$3x_1 + x_2 = 7.$$
Therefore $(3, 1)$ is a normal vector to the line. The vector equation (1) to this line is
$$X = (2, 1) + t(-1, 3).$$
(See Figure 22.)

We turn next to the determination of the distance d from the origin to the line l through A parallel to the vector B. This distance d is defined to be the minimum length of all the vectors X lying on the line l. This is of course just the minimum length of all vectors of the form $A + tB$. But by the Pythagorean theorem $|A + t_0 B|$ will be a minimum if $A + t_0 B$ is perpendicular to B. (See Figure 23.) To see this, note that if $(A + t_0 B, B) = 0$, then for any value of t
$$|A + tB|^2 = |A + t_0 B + (t - t_0)B|^2$$
$$= |A + t_0 B|^2 + |(t - t_0)B|^2$$
$$\geq |A + t_0 B|^2.$$

The second equality holds since $A + t_0 B$ and $(t - t_0)B$ are perpendicular. To compute the value of d note that if $(A + t_0 B, B) = 0$, then
$$t_0 = -\frac{(A, B)}{(B, B)}.$$

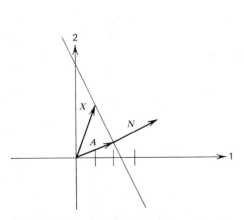

Fig. 22 $3x_1 + x_2 = 7$ is the equation of the line through $(2, 1)$ perpendicular to $(3, 1)$.

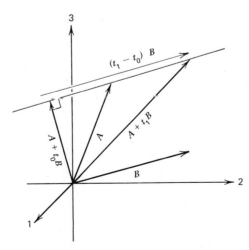

Fig. 23 $|A + t_0 B|$ will be a minimum if $A + t_0 B$ is perpendicular to B.

26 VECTORS AND VECTOR ALGEBRA

Hence
$$d^2 = |A + t_0 B|^2 = (A, A) + 2t_0(A, B) + t_0^2(B, B)$$
$$= (A, A) - \frac{2(A, B)^2}{(B, B)} + \frac{(A, B)^2}{(B, B)}$$
$$= \frac{(A, A)(B, B) - (A, B)^2}{(B, B)}$$

and consequently

(5) $$d = \frac{\sqrt{|A|^2 |B|^2 - (A, B)^2}}{|B|}.$$

The numerator in (5) has a simple geometric description. If θ is the angle between the vectors A and B, then we know from section 8.4 that

$$(A, B) = |A||B| \cos \theta.$$

Therefore

$$\sqrt{|A|^2 |B|^2 - (A, B)^2} = \sqrt{|A|^2 |B|^2 (1 - \cos^2 \theta)} = |A||B| \sin \theta.$$

Hence

(6) $$d = |A| \sin \theta.$$

Thus the distance from the origin to the line through A parallel to B is just the product of the length of A with the sine of the angle between A and B. (See Figure 24.)

This discussion is valid for lines in $V_n(R)$ as well, since we have used only the fundamental properties of the inner product to derive (5) and (6). For lines in the plane the distance d from the origin to a line l may be expressed in terms of a normal vector N to the line. Indeed this distance d will be the length of the vector $t_0 N$ where $t_0 N$ is chosen so that the terminal point of the line segment representing $t_0 N$ lies on the line l. (See Figure 25.) When this holds, $X = t_0 N$ must satisfy (3). Hence $(t_0 N - A, N) = 0$ where A is some vector lying on the line l. Solving for t_0 we obtain

$$t_0 = \frac{(A, N)}{(N, N)}. \text{ Therefore } d = |t_0 N| = \frac{|(A, N)|}{|N|}.$$

If the equation of l has the form $n_1 x_1 + n_2 x_2 = c$, then $(A, N) = c$, and

$$d = \frac{|c|}{|N|} = \frac{|c|}{\sqrt{n_1^2 + n_2^2}}.$$

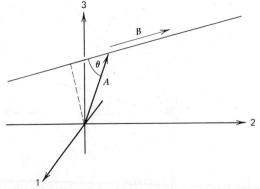

Fig. 24 The distance from origin to line l is $|A| \sin \theta$.

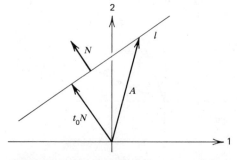

Fig. 25 The distance from line l to origin is $|t_0 N|$ where $t_0 N$ lies on l.

EXERCISES

1. Determine equations for the following lines in the plane in both the form $X = A + tB$ and $(X - A, N) = 0$.
   (a) Line through $A = (1, -1)$ and parallel to $B = (2, 3)$.

   (b) Line through $A = (2, -3)$ and through $B = (-1, 2)$.
   (c) Line through $A = (4, 1)$ and perpendicular to $B = (1, -2)$.
   (d) Line through $A = (1, 3)$ and parallel to the line through $B = (-1, 4)$ and $C = (2, -1)$.

2. Determine the vector equations of the medians of the triangle formed by $(1, 2)$, $(-1, 4)$, $(2, -1)$.

3. Let $A = (1, 3)$, $B = (2, 1)$, $C = (1, 4)$. Determine the vector equation of the line through A bisecting the line segment between B and C.

4. Determine normal vectors to the sides of the triangle with vertices $(3, 0)$, $(2, -1)$, and $(-3, 4)$.

5. Determine the distance from the origin to the line $y = 2x + 5$.

6. In space determine vector and scalar equations of the lines defined by the following conditions.
   (a) Through $A = (1, -1, 1)$ and parallel to $B = (2, 0, -3)$.
   (b) Through $A = (2, 1, -2)$ and through $B = (1, 2, 3)$.
   (c) Through $A = (1, -1, 1)$ and parallel to the line through $B = (1, 1, 1)$ and $C = (1, 4, 3)$.

7. Show that the line through $A = (3, 5, -1)$ and $B = (4, 3, -1)$ is perpendicular to the vector $C = (2, 1, 1)$.

8. Show that the points $A = (0, 1, 3)$, $B = (-1, 0, 2)$, $C = (-1, 1, 4)$ are the vertices of a right triangle.

9. Show that the vectors $A = (5, 3, -2)$, $B = (4, 1, -1)$, $C = (2, -3, 1)$ are collinear.

10. Determine the vector equations of the lines defined by

    (a) $\dfrac{x-5}{2} = \dfrac{y+3}{-4} = \dfrac{z-1}{6}$

    (b) $3x + 2y - 2z + 3 = 0$
        $2x + y - 2z + 2 = 0$.

11. What is the distance from the origin to the line through $(2, 1, -1)$ and $(1, -1, 3)$?

12. Show that the distance from the point with coordinates (c_1, c_2, c_3) to the line through A parallel to B is given by

$$d = \frac{\sqrt{|A - C|^2 |B|^2 - (A - C, B)^2}}{|B|}$$

where C is the vector (c_1, c_2, c_3). Give a geometric interpretation of d analogous to formula (6). (This distance will be the minimum length of the vectors $A + tB - C$. Using the same argument as in the text, show that this vector will have minimum length if $A + tB - C$ is perpendicular to B.)

13. If l is a line in the plane with equation $c_1 x_1 + c_2 x_2 = d$ show that the distance from a point (d_1, d_2) to l is given by

$$\frac{|c_1 d_1 + c_2 d_2 - d|}{\sqrt{c_1^2 + c_2^2}}.$$

(Make a sketch and convince yourself that this minimum distance will be the length of the vector $t_0 C$ if $t_0 C + D$ lies on the line l. D is the vector (d_1, d_2).)

14. Compute the distance from $(1, 2)$ to $2x + y + 4 = 0$.
15. What are the lengths of the altitudes of the triangle formed by $(1, -1, 1)$, $(2, 3, 0)$, and $(1, 0, -1)$?
16. Determine the distance from $(1, -1, 2, 1)$ to the line through $(2, 0, 0, 1)$ and $(1, -1, 1, -1)$.
17. Let l_1 be the line through $(1, -1, 1)$ parallel to $(0, 1, 1)$ and l_2 be the line through $(2, 1, 1)$ parallel to $(1, -1, 0)$. What is the distance between these lines? (If X_1 lies on l_1 and X_2 on l_2, what is the geometric condition for $|X_1 - X_2|$ to be a minimum?)

1.7 Planes in $V_n(R)$.

We showed in Theorem 1.1 that every vector X in $V_2(R)$ is a linear combination of the vectors A and B if and only if A and B are *not* parallel. When every vector X in $V_2(R)$ can be written

$$X = sA + tB$$

we say that A and B *span* $V_2(R)$.

If now A and B are nonzero vectors in $V_n(R)$ which are not parallel, then in an analogous fashion we define the *plane spanned by A and B* to be the set of vectors X which can be written

$$X = sA + tB.$$

where s and t are scalars. More generally if C is a third vector in $V_n(R)$, then the set of vectors

$$(1) \qquad X = C + sA + tB$$

will be called the plane *through C spanned by A and B*. Equation (1) is called the vector equation of this plane.

If the plane π is in $V_3(R)$, then we represent the vectors X satisfying (1) as line segments emanating from the origin. Then the representation of π as a set of points will consist of all those points which are terminal points of the line segments representing the vectors X. Representing the vectors X in this way we see that a point with coordinates (p_1, p_2, p_3) lies on the plane if and only if the vector $P = (p_1, p_2, p_3)$ satisfies the equation of the plane.

Just as in the case of a line l we say that a vector X lies on the plane π if X satisfies the equation of the plane. Notice that this does *not* say that all points on the directed line segment representing X lie on the plane. This will only be true if the plane π passes through the origin. The geometric representation of the plane π in space is given in Figure 26.

Planes in $V_3(R)$ may be characterized in much the same way as lines in $V_2(R)$. If π is the plane through C spanned by A and B, we assert first that there is a nonzero vector N perpendicular to both A and B. Second, this plane π may be characterized as precisely those vectors X such that $X - C$ is perpendicular to N. In terms of the inner product this just means that

$$(2) \qquad (X - C, N) = 0.$$

As a result we may describe π as the plane through C perpendicular to N.

In order to construct this normal vector N we must first derive two results on simultaneous linear equations. If $x_1, \ldots, x_n$ are unknowns and $a_1, \ldots, a_n, b$ are scalars, then

$$a_1 x_1 + \cdots + a_n x_n = b$$

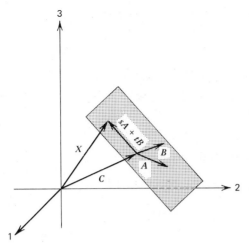

Fig. 26 Plane π with vector equation $X = C + sA + tB$.

is called a *linear equation* in the unknowns $x_1, \ldots, x_n$. If $b = 0$, then the equation is called a *homogeneous* linear equation. Now one or more homogeneous equations

$$a_1 x_1 + \cdots + a_n x_n = 0$$

always have one solution, namely, the solution $x_1 = x_2 = \cdots = x_n = 0$. We call this the *trivial* solution to the equations. The problem that we need to solve is to determine conditions that guarantee that solutions exist to these homogeneous equations which are not identically equal to zero. Such a solution we call nontrivial. We shall not try to give the most general result at this time but shall instead limit our consideration to what will be required in our discussion of planes.

First, one equation in two unknowns

(3) $$a_1 x_1 + a_2 x_2 = 0$$

certainly has a nontrivial solution. This is clear since if $a_1 = a_2 = 0$, then any pair of numbers x_1, x_2 is a solution. If one of the coefficients, say a_1, is different from zero, we may divide by it and obtain

(4) $$x_1 = -\frac{a_2}{a_1} x_2.$$

Thus if x_2 is any nonzero real number and x_1 is then computed by (4), the resulting pair (x_1, x_2) is a nontrivial solution to (3). Notice that this solution to (3) is not unique. Since in (4) we may take x_2 to be any nonzero real number, there must be infinitely many nontrivial solutions to (3).

We may now extend this result by the familiar process of elimination of unknowns.

THEOREM 1.4 *Two homogeneous linear equations in three unknowns*

(5)
$$a_1 x_1 + a_2 x_2 + a_3 x_3 = 0$$
$$b_1 x_1 + b_2 x_2 + b_3 x_3 = 0$$

always have a nontrivial solution x_1, x_2, x_3.

Before proving this result, let us consider an example. To find a solution to

(6)
$$2x_1 - x_2 + x_3 = 0$$
$$x_1 - 4x_2 + 3x_3 = 0$$

different from the solution $x_1 = x_2 = x_3 = 0$, solve the first equation for x_3 obtaining

(7) $$x_3 = -2x_1 + x_2.$$

Substitute this in the second obtaining

$$x_1 - 4x_2 + 3(-2x_1 + x_2) = 0$$

or

$$-5x_1 - x_2 = 0$$

then

(8) $$x_2 = -5x_1.$$

Now for any nonzero value of x_1, x_2 is determined by (8) and x_3 by (7). For example if $x_1 = 1$, then $x_2 = -5$ and $x_3 = -7$. There are, of course, infinitely many nontrivial solutions to (6). Each corresponds to a nonzero value of x_1 in (8).

To prove Theorem 1.4 observe first that if all of the coefficients a_1, a_2, a_3, b_1, b_2, b_3 are zero, then any triple x_1, x_2, x_3 is a solution of (5). Therefore assume one of the coefficients, for example, a_1, is different from zero. Then dividing by a_1 we have

(9) $$x_1 = -\frac{a_2}{a_1}x_2 - \frac{a_3}{a_1}x_3.$$

If we now substitute this in the second equation of (5), the unknown x_1 is eliminated, and we have one equation

(10) $$c_1 x_2 + c_2 x_3 = 0$$

in the two unknowns x_2, x_3.

We know by what we proved on page 29 that there is a nontrivial solution to (10). Hence if we then determine x_1 by (9), we have a nontrivial solution to (5). In fact there must exist infinitely many nontrivial solutions to (5).

It is readily apparent that we may extend this result to three equations in four unknowns. Indeed using induction the result may be extended to n equations in $n+1$ unknowns. We state this result as a theorem and leave the proof as an exercise.

THEOREM 1.5 *The n homogeneous linear equations in the $n+1$ unknowns $x_1, \ldots, x_{n+1}$ defined by*

$$a_{1,1}x_1 + \cdots + a_{1,n+1}x_{n+1} = 0$$
$$a_{2,1}x_1 + \cdots + a_{2,n+1}x_{n+1} = 0$$
$$\vdots$$
$$a_{n,1}x_1 + \cdots + a_{n,n+1}x_{n+1} = 0$$

always have a nontrivial solution, that is, a solution different from the solution $x_1 = x_2 = \cdots = x_{n+1} = 0$.

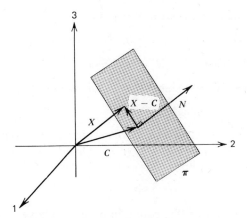

Fig. 27 Plane through C perpendicular to N.

We now apply these results to describe the plane through C spanned by the nonparallel vectors A and B. (See Figure 27.) First we assert there is a nonzero vector $N = (n_1, n_2, n_3)$ perpendicular to both $A = (a_1, a_2, a_3)$ and $B = (b_1, b_2, b_3)$. This just means that

$$(N, A) = n_1 a_1 + n_2 a_2 + n_3 a_3 = 0$$

and

$$(N, B) = n_1 b_1 + n_2 b_2 + n_3 b_3 = 0.$$

But by Theorem 1.4 this pair of homogeneous linear equations must have a nontrivial solution $(n_1, n_2, n_3) = N$.

Next observe that $(N, A) = 0 = (N, B)$ implies $(N, sA + tB) = 0$ for any scalar s and t. Hence if X is a vector in the plane through C spanned by A and B, then

$$X - C = sA + tB$$

and consequently

(11) $$(X - C, N) = 0.$$

We now must show that every vector X in $V_3(R)$ satisfying (11) lies in the plane through C spanned by A and B. This is tantamount to showing that every vector Y perpendicular to N must be a linear combination of A and B. To establish this we must first show that N, A, B together span all of $V_3(R)$. That is we must show that any vector Y in $V_3(R)$ may be written

(12) $$Y = rN + sA + tB.$$

Now by Theorem 1.5 there exist scalars u, v, w, x not all zero such that

(13) $$uY + vN + wA + xB = 0$$

since this vector equation represents three homogeneous equations, one for each component, in the four unknowns, u, v, w, x. By Theorem 1.5 these equations must have a nontrivial solution. If in equation (13) $u \neq 0$, we may divide by it, solve for Y, and obtain (12). Let us show that it cannot be the case that $u = 0$ in equation (13). We argue by contradiction. If $u = 0$ then from (13) we have

(14) $$vN + wA + xB = 0.$$

Taking the inner product of (14) with N we obtain

$$0 = (0, N) = (vN + wA + xB, N)$$
$$= v(N, N) + w(A, N) + x(B, N) = v(N, N),$$

since $(A, N) = (B, N) = 0$. But since $N \neq 0$, this implies that $v = 0$, or that

$$wA + xB = 0.$$

This is clearly impossible since A and B are assumed to be nonparallel. We therefore have proved that $u \neq 0$ and have established (12).

To prove that $(Y, N) = 0$ implies $Y = sA + tB$ write

$$Y = rN + sA + tB.$$

If we take the inner product of this equation by N, we have

$$0 = (Y, N) = (rN + sA + tB, N)$$
$$= r(N, N) + s(A, N) + t(B, N) = r(N, N).$$

Since $N \neq 0$, this implies $r = 0$. Hence $Y = sA + tB$ and we are done.

We have now shown that any vector $X - C$ satisfying $(X - C, N) = 0$ may be written

$$X - C = sA + tB$$

for appropriate scalars s and t. Thus the plane through C perpendicular to N is precisely the plane through C spanned by A and B. We note also that equation (12) shows that all vectors M perpendicular to the plane π are parallel to N. This is, of course, geometrically clear, and an arithmetic proof may be easily given. We leave this verification for the exercises (Exercise 22).

The vector N, or any nonzero scalar multiple of N, is called a *normal vector* for the plane π. Two planes π_1 and π_2 in $V_3(R)$ are said to be parallel or perpendicular according to whether their normal vectors N_1 and N_2 are parallel or perpendicular.

The vector equation (2) for the plane π through C perpendicular to N may be translated into a scalar equation by expanding the inner product. Thus

$$0 = (X - C, N) = (x_1 - c_1)n_1 + (x_2 - c_2)n_2 + (x_3 - c_3)n_3$$

or

$$x_1 n_1 + x_2 n_2 + x_3 n_3 = c_1 n_1 + c_2 n_2 + c_3 n_3 = c.$$

Conversely, the scalar equation

(15) $$x_1 n_1 + x_2 n_2 + x_3 n_3 = c$$

defines the plane through the points $(c/n_1, 0, 0)$, $(0, c/n_2, 0)$, $(0, 0, c/n_3)$, perpendicular to (n_1, n_2, n_3). Indeed any vector lies in the plane π if and only if its components satisfy (15). For example,

$$3x_1 - 4x_2 + x_3 = 5$$

is a scalar equation of the plane perpendicular (normal) to the vector $(3, -4, 1)$ and passing through $(5/3, 0, 0)$.

The numbers c/n_1, c/n_2, c/n_3 are the intercepts of the plane, that is the coordinates of the points of intersection of the plane with the first, second, and third coordinate axis, respectively. (See Figure 28.)

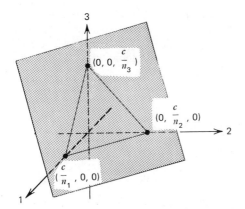

Fig. 28 Intercepts of a plane.

Example 1. Determine a normal vector for the plane π through $C = (2, 1, 1)$ spanned by $A = (1, -1, 1)$ and $B = (2, 0, 1)$. Determine a scalar equation for this plane. $N = (n_1, n_2, n_3)$ will be a normal vector for π if $(N, A) = (N, B) = 0$. This implies
$$n_1 - n_2 + n_3 = 0$$
and
$$2n_1 + n_3 = 0.$$
Setting $n_1 = 1$ we have $n_3 = -2$ and $n_2 = -1$. Therefore $N = (1, -1, -2)$ is one such normal vector. All others are, of course, parallel to N. Now $C = (2, 1, 1)$ is assumed to lie in the plane. So translating the vector equation $(X - C, N) = 0$ into scalar form we have
$$(x_1 - 2)1 + (x_2 - 1)(-1) + (x_3 - 1)(-2) = 0$$
or
$$x_1 - x_2 - 2x_3 + 1 = 0.$$
This is a scalar equation for the plane π as is any scalar multiple of this equation.

Example 2. Determine two vectors A and B which span the plane determined by the scalar equation
$$2x_1 - x_2 - x_3 = 5.$$
Now $(2, -1, -1)$ is a normal vector for this plane. Hence any two nonparallel (necessarily nonzero) vectors A and B which are perpendicular to $(2, -1, 1)$ will span the plane π. Two such vectors may be easily computed by inspecting the equation
$$2x_1 - x_2 - x_3 = 0$$
which must be satisfied by both of these vectors. One choice is $A = (1, 2, 0)$, and $B = (1, 0, 2)$. They both are perpendicular to N and are clearly not parallel.

Example 3. Determine a scalar equation for the plane
$$\pi = \{(1, 3, 2) + s(2, 1, 1) + t(0, 1, -1)\}.$$
This plane has vector equation
$$X = (1, 3, 2) + s(2, 1, 1) + t(0, 1, -1).$$
Since $X = (x_1, x_2, x_3)$, this vector equation is equivalent to three scalar equations

(16)
$$x_1 = 1 + 2s$$
$$x_2 = 3 + s + t$$
$$x_3 = 2 + s - t.$$

These scalar equations are often called *parametric* equations of the plane. The scalar equation for the plane may be determined as in Example 1 by computing a vector normal to the vectors $(2, 1, 1)$ and $(0, 1, -1)$. This equation also may be determined by eliminating the parameters s and t in the equations (16). Indeed solving the first equation for s we obtain

$$s = \frac{x_1 - 1}{2}.$$

Solving for t in the second equation we obtain

$$t = x_2 - 3 - s = x_2 - 3 - \frac{x_1 - 1}{2}.$$

Substituting these values of s and t in the third equation we obtain

$$x_3 = 2 + \frac{x_1 - 1}{2} - \left(x_2 - 3 - \frac{x_1 - 1}{2}\right),$$

or

$$x_1 - x_2 - x_3 + 4 = 0$$

which is the desired scalar equation of the plane.

For $n > 3$ the set of vectors $X - C$ in $V_n(R)$ satisfying

(10) $$(X - C, N) = 0$$

cannot be spanned by two nonparallel vectors A and B. Hence for $n > 3$ equation (10) does not determine a plane. It can be shown however that there are $n - 1$ vectors $A_1, \ldots, A_{n-1}$ which span this set of vectors. For $n > 3$ the set of vectors X satisfying (10) is often called the *hyperplane* through C perpendicular to N.

Next we determine a formula for the distance from a point in space to the plane through C perpendicular to the vector N. In $V_3(R)$ the distance from a point with coordinates (d_1, d_2, d_3) to the plane π described by (10) is clearly the minimum of the quantity

$$|X - D|$$

where X ranges over all the vectors of the plane and $D = (d_1, d_2, d_3)$. This minimum will be obtained when $X - D$ is normal to the plane or in other words, when $X - D = tN$ for some scalar t. But since X lies in the plane, $(X - C, N) = 0$. Combining these two facts we have

$$(D + tN - C, N) = (X - C, N) = 0.$$

Solving for t we obtain

$$t = \frac{(C - D, N)}{|N|^2}.$$

Thus the distance from the point with coordinates (d_1, d_2, d_3) to π is

$$|tN| = |t||N| = \frac{|(C - D, N)|}{|N|}.$$

See Figure 29.

Example 4. Determine the distance from $(1, -1, 2)$ to the plane with scalar equation $2x_1 + x_2 - x_3 = 5$. This plane has normal vector $N = (2, 1, -1)$ and $C = (0, 5, 0)$ lies in the plane. Therefore letting $D = (1, -1, 2)$ we have $C - D = (-1, 6, -2)$. The required distance is

$$\frac{|(C - D, N)|}{|N|} = \frac{6}{\sqrt{6}} = \sqrt{6}.$$

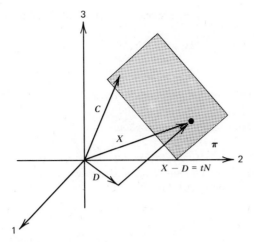

Fig. 29 If $X = D + tN$ lies in π, and N is normal to the plane π, then $|tN|$ is the distance from the point with coordinates (d_1, d_2, d_3) to the plane π.

For $n > 3$ exactly the same procedure determines the distance from a point with coordinates $(d_1, \ldots, d_n)$ to the hyperplane in $V_n(R)$ through C perpendicular to N.

EXERCISES

1. If the plane π has scalar equation
$$2x_1 - x_2 + 3x_3 = 5,$$
determine
   (a) The intercepts of the plane.
   (b) A normal vector to the plane of unit length (length = 1).
   (c) The vector equation of the plane.
   (d) Two vectors which span the plane.

2. Let π be the plane through $(1, 1, 3)$ perpendicular to $(1, -1, 1)$. Determine
   (a) A scalar equation for this plane.
   (b) Two vectors which span the plane.
   (c) The intercepts of the plane.
   (d) The distance from the origin to the plane.

3. Determine the Cartesian equation of the plane passing through the terminal points of the vectors $A = (-1, 2, 4)$; $B = (2, 1, 0)$, $C = (3, -1, 2)$. Determine a unit normal vector to this plane. Determine two vectors spanning the plane as well as the distance from $(4, 1, 1)$ to the plane.

4. Let $\pi = \{C + sA + tB: s \text{ and } t \text{ real numbers}\}$, where $C = (-1, 1, 2), A = (2, 1, -2)$, $B = (1, 1, 3)$. Which of the following vectors lie in the plane π?
   (a) $(-1, 2, 0)$
   (b) $(3, 4, 6)$
   (c) $(3, -2, 7)$
   (d) $(0, 1, -3)$.

5. Determine a scalar equation for the plane $\pi = \{C + sA + tB: s \text{ and } t \text{ real numbers}\}$ if $C = (2, 3, 7)$, $A = (1, 0, -1)$, $B = (1, -1, 2)$.

6. Determine scalar equations for the two planes parallel to the plane through $C = (1, -1, 2)$ spanned by $A = (1, 1, 5)$ and $B = (2, 0, 1)$ and at a distance 5 from this plane.

7. Let π_1, π_2 be two planes and let A, B be two vectors belonging to the intersection of these planes. Prove that the line through A and B is contained in $\pi_1 \cap \pi_2$.

8. Determine the distance from the point $(-1, 2, 1)$ to the line of intersection of the planes defined by
$$2x_1 - x_2 + x_3 = 2$$
and
$$x_1 + x_2 - 3x_3 = 1.$$

9. Determine the equation of the plane parallel to $2x_1 + x_2 - x_3 = 5$ if the point with coordinates $(3, 1, 1)$ is equidistant from the two planes.

10. Find the distance between the parallel planes
    (a) $2x_1 - 3x_2 + x_3 = 5 \qquad 4x_1 - 6x_2 + 2x_3 = 1$
    (b) $x_1 - x_2 + 3x_3 = 2 \qquad -x_1 + x_2 - 3x_3 = 6.$

11. Prove that if a plane π in $V_3(R)$ has scalar equation $ax_1 + bx_2 + cx_3 = d$, then the distance between the origin and this plane is
$$\frac{|d|}{\sqrt{a^2 + b^2 + c^2}}$$

12. Prove that if a plane π in $V_3(R)$ has scalar equation $ax_1 + bx_2 + cx_3 = d$, then the distance from the point (x_1, x_2, x_3) to the plane π is given by
$$\frac{|d - ax_1 - bx_2 - cx_3|}{\sqrt{a^2 + b^2 + c^2}}.$$

13. Determine a scalar equation for the plane containing the point $(1, -2, 1)$ as well as the line through $(1, 4, 2)$ parallel to the vector $(3, 1, 7)$.

14. In $V_4(R)$ if π is the hyperplane through $C = (c_1, c_2, c_3, c_4)$ perpendicular to $N = (n_1, n_2, n_3, n_4)$, determine a scalar equation for this hyperplane. (Argue by analogy with the case $n = 3$.) If $C = (1, -1, 2, 0)$ and $N = (1, 0, -1, -1)$ what is the distance from the origin to this hyperplane?

15. Determine a nontrivial solution to the equations
$$x_1 - 2x_2 + x_3 = 0$$
$$2x_1 + x_2 - 3x_3 = 0.$$

16. Determine a nontrivial solution to the equations
$$2x_1 + x_2 - x_3 + 3x_4 = 0$$
$$x_1 - x_2 + x_3 - 2x_4 = 0$$
$$2x_1 - x_2 - x_3 + x_4 = 0.$$

17. Using Theorem 1.4 prove that three homogeneous linear equations in four unknowns always possess a nontrivial solution.

18. Using induction and Theorem 1.4 prove that n homogeneous linear equations in $n + 1$ unknowns always have a nontrivial solution.

19. Deduce from Exercise 18 that k homogeneous linear equations in n unknowns always has a nontrivial solution provided that $n > k$.

20. Use the fact that two homogeneous equations in three unknowns always possess a nontrivial solution to prove that two nonparallel vectors A and B will span $V_2(R)$.

21. Using the fact that one homogeneous equation in two unknowns always has a nontrivial solution, show that if A and B span $V_2(R)$, then A and B are not parallel.

22. Let N be a normal vector to the plane through C spanned by A and B. If M is a vector perpendicular to A and B show that M is parallel to N. (We must show $M = rN$ for some scalar r. By the discussion on pp. 31–32 the vectors A, B, N span $V_3(R)$. Hence $M = rN + sA + tB$. By taking the inner product of this equation with the vectors A, B, and N show that $s = t = 0$ and $r = [(M, N)/|N|^2]$.)

1.8 The Vector Product

Even though it is impossible to define a multiplication between vectors in $V_3(R)$ for which division is always possible, a multiplication which has many useful and interesting properties can be defined. This multiplication is called the *vector product or cross product* and it is written $A \times B$. We first require that the cross product of parallel vectors always vanishes. Hence in particular for the unit coordinate vectors we define

(1) $$I_1 \times I_1 = I_2 \times I_2 = I_3 \times I_3 = 0.$$

Second, if I and J are orthogonal unit vectors, then $I \times J$ is to be a unit vector orthogonal to both I and J. There are two choices for $I \times J$ satisfying this requirement. We take $I \times J$ to be the vector K such that if the triple I, J, and K are rotated so that I lies along I_1, and J lies along I_2, then $K = I \times J$ will lie along I_3. (See Figure 30.)

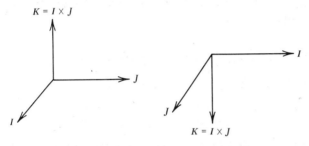

Fig. 30 Orientation of the vectors $I, J, I \times J$.

This geometric condition immediately gives us a multiplication table for the unit coordinate vectors. Accordingly we define

(2) $$\begin{aligned} I_1 \times I_2 &= -I_2 \times I_1 = I_3 \\ I_2 \times I_3 &= -I_3 \times I_2 = I_1 \\ I_3 \times I_1 &= -I_1 \times I_3 = I_2. \end{aligned}$$

We extend the cross product to all vectors $A, B \in V_3(R)$ by requiring in addition that the product $\times$ satisfy the distributive law

(3) $$A \times (B+C) = A \times B + A \times C,$$

and that for each real scalar a,

(4) $$a(A \times B) = (aA) \times B = A \times (aB).$$

To obtain a formula for $A \times B$ we need only multiply out the expression

$$A \times B = (a_1 I_1 + a_2 I_2 + a_3 I_3) \times (b_1 I_1 + b_2 I_2 + b_3 I_3)$$

using (1)–(4). This yields

(5) $$\begin{aligned} A \times B &= (a_1 I_1 + a_2 I_2 + a_3 I_3) \times (b_1 I_1 + b_2 I_2 + b_3 I_3) \\ &= (a_2 b_3 - a_3 b_2) I_1 + (a_3 b_1 - a_1 b_3) I_2 + (a_1 b_2 - a_2 b_1) I_3. \end{aligned}$$

We note a few properties of this cross product. First, $A \times B$ always defines a vector in $V_3(R)$. Second, (2) shows that the cross product is noncommutative. It is in nonassociative as well since by (1) and (2).

38 VECTORS AND VECTOR ALGEBRA

$$I_1 \times (I_1 \times I_2) = I_1 \times I_3 = -I_2.$$

On the other hand, $(I_1 \times I_1) \times I_2 = 0$. Furthermore there is no identity since it is easy to check (Exercise 6) that if $A \times A = A$, then $A = 0$.

Formula (5) is most easily remembered and used if determinant notation is introduced. Recall from high school algebra that a 2×2 determinant is defined by the formula

$$\begin{vmatrix} a_1 & a_2 \\ b_1 & b_2 \end{vmatrix} = a_1 b_2 - a_2 b_1.$$

A 3×3 determinant is defined by the formula

(6) $\quad \begin{vmatrix} a_1 & a_2 & a_3 \\ b_1 & b_2 & b_3 \\ c_1 & c_2 & c_3 \end{vmatrix} = a_1 b_2 c_3 - a_1 b_3 c_2 + a_2 b_3 c_1 - a_2 b_1 c_3 + a_3 b_1 c_2 - a_3 b_2 c_1.$

We may simplify the right-hand side of this expression by using 2×2 determinants as follows. By (6)

$$\begin{vmatrix} a_1 & a_2 & a_3 \\ b_1 & b_2 & b_3 \\ c_1 & c_2 & c_3 \end{vmatrix} = a_1(b_2 c_3 - b_3 c_2) - a_2(b_1 c_3 - b_3 c_1) + a_3(b_1 c_2 - b_2 c_1)$$

(7)
$$= a_1 \begin{vmatrix} b_2 & b_3 \\ c_2 & c_3 \end{vmatrix} - a_2 \begin{vmatrix} b_1 & b_3 \\ c_1 & c_3 \end{vmatrix} + a_3 \begin{vmatrix} b_1 & b_2 \\ c_1 & c_2 \end{vmatrix}.$$

Formula (7) is called the *Lagrange expansion* of a 3×3 determinant. It may perhaps best be remembered in the following way. Each term a_1, a_2, a_3 in the first row is multiplied by the 2×2 determinant obtained by striking out the first row and the column containing the appropriate coefficient a_1, a_2, a_3. The resulting terms

$$a_1 \begin{vmatrix} b_2 & b_3 \\ c_2 & c_3 \end{vmatrix}, \quad a_2 \begin{vmatrix} b_1 & b_3 \\ c_1 & c_3 \end{vmatrix}, \quad a_3 \begin{vmatrix} b_1 & b_2 \\ c_1 & c_2 \end{vmatrix},$$

are then multiplied by $1, -1, 1$, respectively, and added together.

Now observe that formula (5) just states that

$$A \times B = \begin{vmatrix} a_2 & a_3 \\ b_2 & b_3 \end{vmatrix} I_1 - \begin{vmatrix} a_1 & a_3 \\ b_1 & b_3 \end{vmatrix} I_2 + \begin{vmatrix} a_1 & a_2 \\ b_1 & b_2 \end{vmatrix} I_3.$$

Hence it is quite reasonable to adopt the terminology

(8) $\quad A \times B = \begin{vmatrix} I_1 & I_2 & I_3 \\ a_1 & a_2 & a_3 \\ b_1 & b_2 & b_3 \end{vmatrix},$

where in accord with (7) we define

$$\begin{vmatrix} I_1 & I_2 & I_3 \\ a_1 & a_2 & a_3 \\ b_1 & b_2 & b_3 \end{vmatrix} = \begin{vmatrix} a_2 & a_3 \\ b_2 & b_3 \end{vmatrix} I_1 - \begin{vmatrix} a_1 & a_3 \\ b_1 & b_3 \end{vmatrix} I_2 + \begin{vmatrix} a_1 & a_2 \\ b_1 & b_2 \end{vmatrix} I_3.$$

Thus if $A = (2, -1, 1)$, and $B = (-1, 1, 2)$, then

$$A \times B = \begin{vmatrix} I_1 & I_2 & I_3 \\ 2 & -1 & 1 \\ -1 & 1 & 2 \end{vmatrix} = -3I_1 - 5I_2 + I_3 = (-3, -5, 1).$$

It follows immediately from the definition of a determinant that if two rows of a determinant are interchanged then the sign is reversed. Using (8) we immediately conclude that the vector product is anti commutative. That is,
$$A \times B = -B \times A.$$

The determinant notation gives a very concise formula for the inner product of $A \times B$ with a third vector $C = (c_1, c_2, c_3)$. If
$$A \times B = \begin{vmatrix} a_2 & a_3 \\ b_2 & b_3 \end{vmatrix} I_1 - \begin{vmatrix} a_1 & a_3 \\ b_1 & b_3 \end{vmatrix} I_2 + \begin{vmatrix} a_1 & a_2 \\ b_1 & b_2 \end{vmatrix} I_3$$
and
$$C = c_1 I_1 + c_2 I_2 + c_3 I_3,$$
then
$$(C, A \times B) = \begin{vmatrix} a_2 & a_3 \\ b_2 & b_3 \end{vmatrix} c_1 - \begin{vmatrix} a_1 & a_3 \\ b_1 & b_3 \end{vmatrix} c_2 + \begin{vmatrix} a_1 & a_2 \\ b_1 & b_2 \end{vmatrix} c_3 = \begin{vmatrix} c_1 & c_2 & c_3 \\ a_1 & a_2 & a_3 \\ b_1 & b_2 & b_3 \end{vmatrix}.$$

Thus if $A = (0, 1, -1)$, $B = (1, 1, 1)$ and $C = (2, -1, -1)$,
$$(C, A \times B) = \begin{vmatrix} 2 & -1 & -1 \\ 0 & 1 & -1 \\ 1 & 1 & 1 \end{vmatrix} = 2 \begin{vmatrix} 1 & -1 \\ 1 & 1 \end{vmatrix} + \begin{vmatrix} 0 & -1 \\ 1 & 1 \end{vmatrix} - \begin{vmatrix} 0 & 1 \\ 1 & 1 \end{vmatrix} = 4 + 1 + 1 = 6.$$

The inner product $(C, A \times B) = (A \times B, C)$ is called the *scalar triple product* of the vectors A, B and C. It follows immediately from the definition that a determinant has value zero if two rows are the same. Using this fact we deduce that

(9) $\qquad\qquad (A, A \times B) = (B, A \times B) = 0.$

Hence if $A \times B \neq 0$, the vector $A \times B$ is perpendicular to both A and B.

Next we investigate what it means for $A \times B$ to be equal to the zero vector. We note first that if A and B are parallel, then $A \times B = 0$. This is clear, since if A and B are parallel, then $B = aA$ for some scalar a and
$$A \times B = A \times (aA) = a(A \times A) = a \begin{vmatrix} I_1 & I_2 & I_3 \\ a_1 & a_2 & a_3 \\ a_1 & a_2 & a_3 \end{vmatrix} = 0.$$

This discussion shows that if $A, B \in V_3(R)$ and one vector is a scalar multiple of the other, then $A \times B = 0$. The converse to this statement is also valid. The proof requires the following identity due to Lagrange.

THEOREM 1.6 *If $A, B \in V_3(R)$, then*
$$|A \times B|^2 = |A|^2 |B|^2 - (A, B)^2.$$

Proof. The identity results from the following calculation. Letting $A = (a_1, a_2, a_3)$, $B = (b_1, b_2, b_3)$ we have
$$|A|^2 |B|^2 - (A, B)^2 = (a_1^2 + a_2^2 + a_3^2)(b_1^2 + b_2^2 + b_3^2) - (a_1 b_1 + a_2 b_2 + a_3 b_3)^2.$$

Observe that each of the squares in $(a_1 b_1 + a_2 b_2 + a_3 b_3)^2$ cancels with an appropriate term of $(a_1^2 + a_2^2 + a_3^2)(b_1^2 + b_2^2 + b_3^2)$. Hence we have

$$\begin{aligned}
|A|^2|B|^2 - (A,B)^2 &= a_1^2(b_2^2+b_3^2) + a_2^2(b_1^2+b_3^2) + a_3^2(b_1^2+b_2^2) \\
&\quad - 2(a_1a_2b_1b_2 + a_1a_3b_1b_3 + a_2a_3b_2b_3) \\
&= (a_1b_2 - a_2b_1)^2 + (a_1b_3 - a_3b_1)^2 + (a_2b_3 - a_3b_2)^2 \\
&= \begin{vmatrix} a_1 & a_2 \\ b_1 & b_2 \end{vmatrix}^2 + \begin{vmatrix} a_1 & a_3 \\ b_1 & b_3 \end{vmatrix}^2 + \begin{vmatrix} a_2 & a_3 \\ b_2 & b_3 \end{vmatrix}^2 = |A \times B|^2.
\end{aligned}$$

COROLLARY. *$A \times B = 0$, if and only if one vector is a scalar multiple of the other.*

Proof. By Theorem 1.6, $A \times B = 0$ if and only if $|(A, B)| = |A||B|$. But by the Schwarz inequality, Theorem 1.2, this is true if and only if one vector is a scalar multiple of the other.

If A and B are vectors and one is a scalar multiple of the other, then it is easy to see that this is equivalent to the assertion that there exist scalars a and b not both equal to zero such that $aA + bB = 0$. (The student should convince himself that these two assertions are equivalent). When either of these assertions hold for vectors A and B, these vectors are said to be *linearly dependent*. Thus the corollary to Theorem 1.6 says that $A \times B = 0$ if and only if A and B are linearly dependent. We will have more to say about this notion of linear dependence in the next section.

Next we turn our attention to some geometric interpretations of Lagrange's identity. First recall that if $B \neq 0$ and l is the line through A parallel to B, then the distance d from the origin to l is given by the formula

$$d = \frac{\sqrt{|A|^2|B|^2 - (A,B)^2}}{|B|}$$

(p. 26). In view of Theorem 1.6 this may be written

$$d = \frac{|A \times B|}{|B|}.$$

Secondly if A and B are nonzero vectors and θ is the angle between them, then

$$\begin{aligned}
|A \times B|^2 &= |A|^2|B|^2 - (A,B)^2 \\
&= |A|^2|B|^2 - |A|^2|B|^2 \cos^2 \theta \\
&= |A|^2|B|^2 \sin^2 \theta.
\end{aligned}$$

Therefore

$$|A \times B| = |A||B| \sin \theta,$$

and it is easily verified that this is the area of the parallelogram formed by A and B. (See Figure 31.)

Thus if A and B are not parallel, the vector $A \times B$ is perpendicular to the plane π spanned by A and B and has length $|A \times B| = |A||B| \sin \theta$. To give a complete geometric description of $A \times B$ it only remains to determine the direction of $A \times B$. This direction is of course reversed if we interchange the order of A and B.

Now the direction of $A \times B$ is determined by the choice of the initial coordinate axes. Fixing one axis, for example I_3 to be vertical, the axes I_1, I_2 may be specified in essentially two different ways. By different, we mean that

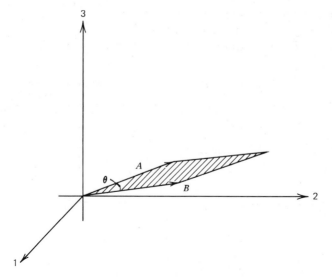

Fig. 31 Parallelogram formed by A and B has area $= |A \times B|$.

one choice cannot be transformed into the other by a combination of rotations and translations. These two sets of axes are called right-hand and left-hand, respectively. (See Figure 32.)

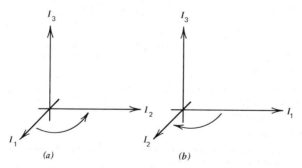

Fig. 32 (a) Right hand coordinate system. (b) Left hand coordinate system.

If we have chosen a right-hand coordinate system, as has been the convention in this book, then the direction of $A \times B$ is given by the right-hand rule. That is to say, if the direction of rotation from A to B is given by the clasped fingers of the *right* hand, then the direction of $A \times B$ is in the direction of the thumb of the right hand. (See Figure 33.) If we had chosen a left hand coordinate system, then the direction of $A \times B$ would be just the reverse. The student may object that this description of the orientation of a coordinate system in three space is highly nonmathematical. It is, but it is not worthwhile at this point to erect the necessary machinery to put these ideas on a firm mathematical foundation.

We summarize the properties of the vector product in the following theorem.

THEOREM 1.7 *Let $A, B, C \in V_3(R)$ and $a \in R$. If $A \times B$ denotes the vector product of A and B, then*

(i) $A \times (B+C) = A \times B + A \times C$

(ii) $a(A \times B) = (aA) \times B = A \times (aB)$
(iii) $(A \times B) = -(B \times A)$
(iv) $A \times B = 0$, if and only if A and B are linearly dependent.

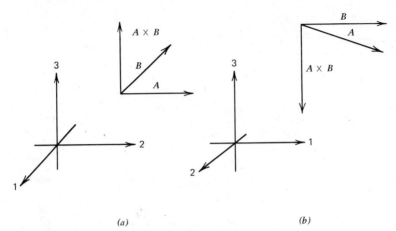

Fig. 33 (a) Direction of $A \times B$ in right-hand coordinate system. (b) Direction of $A \times B$ in left-hand coordinate system.

EXERCISES

1. If $A = (2, -1, 3)$, $B = (1, -1, -1)$, $C = (2, 2, -1)$ compute
    (a) $A \times (B \times C)$
    (b) $(A \times B) \times C$
    (c) $(A \times A) \times B$
    (d) $A \times (A \times B)$
    (e) $(A - B) \times (B - C)$
    (f) $(A, B \times C)$
    (g) $(B, C \times A)$
    (h) $(A - B, A \times B)$
    (i) $(A \times B, A \times C)$.

2. Using the vector product determine a normal vector to the planes spanned by
    (a) $(2, -1, 3)$ and $(4, 6, -2)$
    (b) $(-1, -1, 2)$ and $(2, -1, 1)$
    (c) $(1, 0, 1)$ and $(1, 1, 1)$.

3. Using the vector product determine a normal vector for the plane passing through
    (a) $(1, -1, 1)$, $(2, 2, -1)$, $(1, 0, 1)$
    (b) $(2, -1, 3)$, $(1, -1, -1)$, $(1, 2, -1)$
    (c) $(2, -1, 0)$, $(1, 3, 4)$, $(0, -1, 5)$.

4. Let P, Q, R be three points in space. Determine a formula for the area of the triangle determined by these points which involves the vector product of appropriate vectors.

5. Using the vector product determine the area of the triangle determined by the following triples of points. (Use Exercise 4.)
    (a) $(1, 2, -3)$, $(1, 1, 1)$, $(2, 0, -1)$
    (b) $(2, 0, 1)$, $(-1, 5, 2)$, $(1, 1, -1)$
    (c) $(3, 2, 1)$, $(-1, 1, 0)$ $(0, -1, 1)$.

6. Show that if $A \times A = A$, then $A = 0$. Also show that $0 \times A = 0$ for each $A \in V_3(R)$. Using these two facts deduce that there is no identity element for the vector product.

7. Using Lagrange's identity deduce a necessary and sufficient geometric condition on nonzero vectors A and B so that
$$|A \times B| = |A||B|.$$
Prove the condition.

8. Vectors A, B, C in $V_3(R)$ determine a parallelepiped in space as in Figure 34.

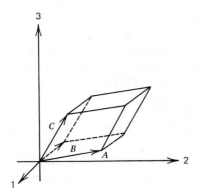

Fig. 34

The volume of this parallelepiped is clearly the product of the area of the base and the altitude. Assuming the base is determined by the vectors A and B, show that the altitude equals
$$\frac{|(A \times B, C)|}{|A \times B|}.$$
Hence the volume of the parallelepiped is given by $|(A \times B, C)|$. (What is the cosine of the angle between C and a normal to the plane spanned by A and B?)

9. Let $A = (1, 3, 4)$, $B = (2, -1, 1)$, $C = (1, 1, -2)$. For what value of t is the volume of the parallelepiped determined by A, B, tC equal to 1?

10. Show directly from the definition of $A \times B$ that
$$A \times B = -B \times A.$$

11. Derive the following properties of the scalar triple product.
    (a) $(A \times B, C) = (A, B \times C)$
    (b) $(A, B \times C) = -(B, A \times C)$
    (c) $(A \times B, C) = -(C, B \times A)$
    (d) $(A \times B, C) = (B \times C, A)$.

12. If $A, B, C, D \in V_3(R)$ prove that
    (i) $A \times (B \times C) = (A, C)B - (A, B)C$
    (ii) $A \times (B \times C) + B \times (C \times A) + C \times (A \times B) = 0$
    (iii) $(A \times B, C \times D) = (A, C)(B, D) - (A, D)(B, C)$.
    [Derive (i) directly from the definition; (ii) and (iii) now follow from (i).]

13. Let l be the line through the points with coordinates (a_1, a_2, a_3) and (b_1, b_2, b_3). Show that the distance from the point (c_1, c_2, c_3) to l is given by
$$\frac{|(C - A) \times (A - B)|}{|A - B|}$$
where A, B, and C are the vectors (a_1, a_2, a_3), (b_1, b_2, b_3), and (c_1, c_2, c_3).

14. Let A and B be nonparallel vectors in $V_3(R)$. Show that if real numbers a, b, c satisfy
$$aA + bB + cA \times B = 0$$
then $a = b = c = 0$. (Compute the inner product of this equation with the vector $A \times B$.)

15. Let $A, B, C \in V_3(R)$ and suppose $(A, B \times C) \neq 0$. Show that if real numbers a, b, c exists satisfying $aA + bB + cC = 0$, then $a = b = c = 0$. (Compute the vector product of the equation $aA + bB + cC = 0$ with one vector then the inner product of the resulting equation with another vector.)

1.9 Linear Independence of Vectors

Theorem 1.1 states that two vectors A and B span $V_2(R)$ if and only if A and B are not parallel. Let us consider the analagous question in $V_3(R)$. The problem is this. What are necessary and sufficient conditions on three vectors A, B, C in $V_3(R)$ so that each vector $D \in V_3(R)$ can be written

$$D = x_1 A + x_2 B + x_3 C$$

for appropriate scalars x_1, x_2, x_3. What we need is an appropriate generalization of the notion of parallelism.

If A and $B \in V_n(R)$, then A and B are parallel if the vectors are nonzero and if one is a scalar multiple of the other. The latter condition means that there exist scalars a, b not both zero so that

$$aA + bB = 0.$$

When this condition holds we have said that A and B are *linearly dependent*. The only difference between parallelism and linear dependence for two vectors is that parallelism only applies to nonzero vectors, whereas, linear dependence applies to the zero vector as well. It is this notation of linear dependence which we shall generalize.

Definition. *Vectors $A_1, \ldots, A_k$ in $V_n(R)$ are said to be* linearly dependent *if there exist scalars $a_1, \ldots, a_k$ not all equal to zero such that*

$$a_1 A_1 + \cdots + a_k A_k = 0.$$

If no such scalars different from zero exist with this property, then the vectors $A_1, \ldots, A_k$ are called linearly independent.

Thus to verify that vectors $A_1, \ldots, A_k$ are linearly independent one must prove that if scalars $a_1, \ldots, a_k$ exist so that

$$a_1 A_1 + \cdots + a_k A_k = 0,$$

then it must follow that

$$a_1 = a_2 = \cdots = a_k = 0.$$

Example 1. The unit coordinate vectors $I_1 = (1, 0, \ldots, 0)$, $I_2 = (0, 1, 0, \ldots, 0)$, $\ldots, I_n = (0, 0, \ldots, 0, 1)$ in $V_n(R)$ are linearly independent since if $a_1 I_1 + \cdots + a_n I_n = 0$, this means that $(a_1, \ldots, a_n)$ is the zero vector. Hence $a_1 = a_2 = \cdots = a_n = 0$.

Example 2. If one of the vectors $A_1, \ldots, A_k$ is the zero vector then these vectors are linearly dependent. For if $A_j = 0$, then

$$0 \cdot A_1 + 0 \cdot A_2 + \cdots + 1 \cdot A_j + \cdots + 0 \cdot A_k = 0,$$

and not all the scalars in the above equation are equal to zero.

Example 3. The vectors $A = (1, -1, 1)$, $B = (-1, 4, 2)$, $C = (1, 2, 3)$ are linearly dependent since

$$2A + B - C = 0.$$

Example 4. Two vectors A and B in $V_n(R)$ are linearly independent if and only if they are not parallel. (Why?)

Example 5. If $A_1, \ldots, A_k$ are nonzero vectors in $V_n(R)$ which are pairwise orthogonal, i.e. $(A_i, A_j) = 0$ if $i \neq j$, then $A_1, \ldots, A_k$ are linearly independent. This follows easily since if

(1) $$a_1 A_1 + \cdots + a_k A_k = 0,$$

LINEAR INDEPENDENCE OF VECTORS

then taking the inner product of the equation with A_1, we get

$$0 = a_1(A_1, A_1) + a_2(A_1, A_2) + \cdots + a_k(A_1, A_k) = a_1(A_1, A_1).$$

Since $A_1 \neq 0$, this implies $a_1 = 0$. Taking the inner product of (1) successively with the vectors $A_2, \ldots, A_k$ we deduce that

$$a_2(A_2, A_2) = 0 = a_3(A_3, A_3) = \cdots = a_k(A_k, A_k).$$

Since for $j = 2, \ldots, k$

$$(A_j, A_j) = |A_j|^2 \neq 0$$

we have

$$a_2 = \cdots = a_k = 0.$$

Example 6. In $V_3(R)$ if A and B are linearly independent, then $A, B, A \times B$ are linearly independent (Exercise 14, Section 1.8).

Whenever a vector B can be written as a linear combination of vectors $A_1, \ldots, A_k$, we say that B is in the *span* of the vectors $A_1, \ldots, A_k$. It follows immediately from the definition of linear independence that vectors $A_1, \ldots, A_k$ are linearly independent if no vector A_i in this set is in the span of any of the other vectors of the set. We leave the verification of this fact as an exercise.

THEOREM 1.8 *Three vectors A, B, C will span $V_3(R)$ if and only if they are linearly independent.*

Furthermore the scalar triple product is precisely the tool which determines if the vectors are linearly independent.

THEOREM 1.9 *Vectors A, B, C in $V_3(R)$ are linearly independent if and only if*

$$(A, B \times C) \neq 0.$$

To prove 1.8 we need to know that any set of four vectors in $V_3(R)$ is linearly dependent. This follows readily from Theorem 1.5 but since the proof of 1.5 was left as an exercise we state and give the proof of the result we need.

LEMMA 1.10 *Any set of four vectors A, B, C, D in $V_3(R)$ is linearly dependent.*

Proof. We seek scalars x_1, x_2, x_3, x_4 not all zero satisfying

$$x_1 A + x_2 B + x_3 C + x_4 D = 0.$$

Writing $A = (a_1, a_2, a_3)$, etc., the above vector equation gives rise to three scalar equations

(2)
$$\begin{aligned} x_1 a_1 + x_2 b_1 + x_3 c_1 + x_4 d_1 &= 0 \\ x_1 a_2 + x_2 b_2 + x_3 c_2 + x_4 d_2 &= 0 \\ x_1 a_3 + x_2 b_3 + x_3 c_3 + x_4 d_3 &= 0. \end{aligned}$$

If all the vectors A, B, C, D are zero, the set is certainly linearly dependent. So assume one vector, A say, is nonzero. Therefore one of the real numbers a_1, a_2, a_3 is different from zero. Assume $a_3 \neq 0$. Then

(3)
$$x_1 = -\frac{1}{a_3}(x_2 b_3 + x_3 c_3 + x_4 d_3).$$

Substitute this value of x_1 in the first two equations of (2). This yields two equations

(4) $$x_2 b'_1 + x_3 c'_1 + x_4 d'_1 = 0 \\ x_2 b'_2 + x_3 c'_2 + x_4 d'_2 = 0$$

with three unknowns x_2, x_3, x_4 and certain new constants b'_i, c'_i, d'_i, $i = 1, 2$. However, the statement that two homogeneous linear equations in three unknowns always possesses a nontrivial solution is just Theorem 1.4. Therefore (4) must have a nonzero solution x_2, x_3, x_4. Defining x_1 by (3) we have a nonzero solution to (2), and hence we have shown that the vectors A, B, C, D are linearly dependent.

We are now ready to prove 1.8. First we show that if A, B, C are linearly independent, then A, B, C span $V_3(R)$. If $D \in V_3(R)$, we know that A, B, C, D are linearly dependent. Hence there are scalars, x_1, x_2, x_3, x_4 not all zero such that

(5) $$x_1 A + x_2 B + x_3 C + x_4 D = 0.$$

Now since A, B, C are linearly independent, it follows that in equation (5) $x_4 \neq 0$. (Why?) Thus dividing by x_4 we have

$$-\frac{x_1}{x_4} A - \frac{x_2}{x_4} B - \frac{x_3}{x_4} C = D.$$

Consequently A, B, C span $V_3(R)$.

To finish the proof of 1.8 we must show that if A, B, C are linearly dependent, then these vectors do not span $V_3(R)$. So suppose

$$x_1 A + x_2 B + x_3 C = 0, \text{ and}$$

at least one of the numbers x_1, x_2, x_3 is not zero. Assume $x_3 \neq 0$. Then

(6) $$C = -\frac{x_1}{x_3} A - \frac{x_2}{x_3} B.$$

Now we saw in Section 1.7 that if A and B were any two vectors in $V_3(R)$ there exists a nonzero vector N orthogonal to A and B. Indeed if A and B are linearly independent, we may take $N = A \times B$. But $(N, A) = (N, B) = 0$ implies by (6) that $(N, C) = 0$. This vector N is not in the span of A, B, C for if $N = xA + yB + zC$ then

$$(N, N) = x(A, N) + y(B, N) + z(C, N) = 0.$$

This is impossible since $N \neq 0$. Thus we have proved that A, B, C do not span $V_3(R)$.

We now turn to the proof of 1.9. First let us prove that if $(A, B \times C) \neq 0$, then A, B, C are linearly independent. Suppose therefore

(7) $$aA + bB + cC = 0.$$

We must show $a = b = c = 0$.
Taking the cross product of (7) with C we have since $C \times C = 0$

(8) $$aA \times C + bB \times C = 0.$$

Taking the inner product of (8) with A we obtain

$$a(A, A \times C) + b(A, B \times C) = 0.$$

Since $(A, A \times C) = 0$ and $(A, B \times C) \neq 0$, this implies that $b = 0$. In a similar fashion we deduce that $a = c = 0$ (Exercise 11).

To complete the proof of 1.9 we must show that if $(A \times B, C) = 0$, then A, B, C are linearly dependent. If $A \times B = 0$, then we proved in Section 1.8 that A and B are linearly dependent. Hence A, B, C certainly are linearly dependent. If $A \times B \neq 0$, then $A \times B$ is a normal vector to the plane through the origin spanned by A and B. Moreover, we showed in Section 1.7 that any vector C perpendicular to this normal vector must lie in that plane. This, however, is just the statement that A, B, C are linearly dependent, and the proof of Theorem 1.9 is complete.

If A, B, C are linearly independent in $V_3(R)$ and D is a fourth vector, then using the scalar triple product we may determine scalars x_1, x_2, x_3 so that

(9) $$x_1 A + x_2 B + x_3 C = D.$$

If we write this vector equation as three scalar equations,

(10) $$\begin{aligned} a_1 x_1 + b_1 x_2 + c_1 x_3 &= d_1 \\ a_2 x_1 + b_2 x_2 + c_2 x_3 &= d_2 \\ a_3 x_1 + b_3 x_2 + c_3 x_3 &= d_3 \end{aligned}$$

then this amounts to exhibiting the solution to this triple of linear equations. Indeed taking the vector product of (9) by B and then taking the inner product of the resulting equation by C we obtain first

$$x_1 A \times B + x_3 C \times B = D \times B$$

since $B \times B = 0$. Hence

$$x_1 (A \times B, C) = (D \times B, C)$$

or

$$x_1 = \frac{(D \times B, C)}{(A \times B, C)}.$$

In a similar fashion we may obtain

$$x_2 = \frac{(A \times D, C)}{(A \times B, C)}$$

and

$$x_3 = \frac{(A \times B, D)}{(A \times B, C)}.$$

Using the determinant notation for the scalar triple product we have

$$x_1 = \frac{\begin{vmatrix} d_1 & d_2 & d_3 \\ b_1 & b_2 & b_3 \\ c_1 & c_2 & c_3 \end{vmatrix}}{\begin{vmatrix} a_1 & a_2 & a_3 \\ b_1 & b_2 & b_3 \\ c_1 & c_2 & c_3 \end{vmatrix}} \qquad x_2 = \frac{\begin{vmatrix} a_1 & a_2 & a_3 \\ d_1 & d_2 & d_3 \\ c_1 & c_2 & c_3 \end{vmatrix}}{\begin{vmatrix} a_1 & a_2 & a_3 \\ b_1 & b_2 & b_3 \\ c_1 & c_2 & c_3 \end{vmatrix}} \qquad x_3 = \frac{\begin{vmatrix} a_1 & a_2 & a_3 \\ b_1 & b_2 & b_3 \\ d_1 & d_2 & d_3 \end{vmatrix}}{\begin{vmatrix} a_1 & a_2 & a \\ b_1 & b_2 & b_3 \\ c_1 & c_2 & c_3 \end{vmatrix}}.$$

The determination of the scalars x_1, x_2, x_3 by the above formulas is called *Cramer's rule* for solving the linear equations (9). Note that each of the numerators in the above fractions are obtained by replacing the first, second,

and third row, respectively, in
$$(A \times B, C) = \begin{vmatrix} a_1 & a_2 & a_3 \\ b_1 & b_2 & b_3 \\ c_1 & c_2 & c_3 \end{vmatrix}$$
by the components of the vector (d_1, d_2, d_3).

Example 7. Show that the vectors $A = (1, -1, 1)$, $B = (0, 4, 3)$ and $C = (-1, 1, 0)$ are linearly independent. If $D = (1, 1, 1)$ determine x_1, x_2, x_3 so that $x_1 A + x_2 B + x_3 C = D$. To show that the vectors are linearly independent compute

$$(A \times B, C) = \begin{vmatrix} 1 & -1 & 1 \\ 0 & 4 & 3 \\ -1 & 1 & 0 \end{vmatrix} = 4 \neq 0.$$

By Cramer's rule

$$x_1 = \frac{\begin{vmatrix} 1 & 1 & 1 \\ 0 & 4 & 3 \\ -1 & 1 & 0 \end{vmatrix}}{4} = -\frac{2}{4} = -\frac{1}{2}$$

$$x_2 = \frac{\begin{vmatrix} 1 & -1 & 1 \\ 1 & 1 & 1 \\ -1 & 1 & 0 \end{vmatrix}}{4} = \frac{2}{4} = \frac{1}{2}$$

$$x_3 = \frac{\begin{vmatrix} 1 & -1 & 1 \\ 0 & 4 & 3 \\ 1 & 1 & 1 \end{vmatrix}}{4} = -\frac{6}{4} = -\frac{3}{2}.$$

Therefore $D = -\dfrac{A}{2} + \dfrac{B}{2} - \dfrac{3C}{2}$.

EXERCISES

1. Determine which of the following sets of vectors are linearly independent
   (a) $A = (1, 1, 0)$, $B = (0, 1, 1)$, $C = (1, 0, 1)$
   (b) $A = (1, 1, -1)$, $B = (0, 1, 3)$, $C = (2, 0, -5)$
   (c) $A = (1, -1, 3)$, $B = (2, 2, 1)$, $C = (1, -2, 0)$
   (d) $A = (1, 1, 0)$, $B = (0, 2, 3)$, $C = (1, -1, -3)$.

2. For what values of t, if any, are the following sets of vectors linearly dependent.
   (a) $A = (-1, 1, t)$, $B = (2, -1, 3)$, $C = (1, 1, -1)$
   (b) $A = (1, 2t, t^2)$, $B = (-1, 2, 3)$, $C = (0, -1, 1)$.
   (c) $A = (1, -t, t)$, $B = (t, -1, 1)$, $C = (1, 1, -1)$.

3. For the following choices of A, B, C, D determine scalars x_1, x_2, x_3 by Cramer's rule so that $x_1 A + x_2 B + x_3 C = D$ or explain why this rule is not applicable.
   (a) $A = (-1, 1, 4)$, $B = (2, 0, -3)$, $C = (1, 1, 1)$, $D = (-1, 1, 0)$
   (b) $A = (1, 4, 3)$, $B = (-1, 1, 2)$, $C = (3, 7, 4)$, $D = (1, 0, -1)$
   (c) $A = (-1, 0, 1)$, $B = (2, 3, 5)$, $C = (2, -1, 0)$, $D = (1, 1, 1)$
   (d) $A = (1, 2, 3)$, $B = (-1, 4, 2)$, $C = (-1, 2, 1)$, $D = (0, 0, 1)$.

4. For the following choices of A, B, C determine scalars x_1, x_2, x_3 by Cramer's rule so that
$$x_1 A + x_2 B + x_3 A \times B = C$$
or explain why this rule is not applicable.
   (a) $A = (1, 4, 3)$, $B = (-1, 1, 2)$, $C = (1, 0, -1)$
   (b) $A = (-1, 1, 1)$, $B = (2, -2, -2)$, $C = (-1, 0, 1)$
   (c) $A = (4, 3, 2)$, $B = (-1, 1, 1)$, $C = (0, 0, 1)$
   (d) $A = (2, -1, 1)$, $B = (6, -3, 3)$, $C = (4, -2, 2)$.

5. If A, B, C are nonzero vectors which are orthogonal, prove that they span $V_3(R)$.
6. If the zero vector in $V_n(R)$ can be written in more than one way as a linear combination of vectors $A_1, \ldots, A_k$, prove that the vectors $A_1, \ldots, A_k$ are linearly dependent.
7. If one vector B in $V_n(R)$ can be written in more than one way as a linear combination of vectors $A_1, \ldots, A_k$, prove that $A_1, \ldots, A_k$ are linearly dependent.
8. Let $A_1, \ldots, A_k$ be vectors in $V_n(R)$. If no vector in this set is in the span of the remaining vectors, show that the vectors $A_1, \ldots, A_k$ are linearly independent.
9. Let $A, B, C \in V_3(R)$. If each vector D in the span of A, B, C can be written in precisely one way as a linear combination of A, B, C prove that A, B, C span $V_3(R)$.
10. Prove that any set of k vectors in $V_3(R)$ is linearly dependent if $k > 3$.
11. Finish the proof of Theorem 1.8 showing that if $(A, B \times C) \neq 0$, and $aA + bB + cC = 0$, then $a = b = c = 0$.
12. Derive the formulas for x_2 and x_3 in Cramer's rule if
$$x_1 A + x_2 B + x_3 C = D.$$
13. Show that
$$\begin{vmatrix} a_1 & a_2 & a_3 \\ b_1 & b_2 & b_3 \\ c_1 & c_2 & c_3 \end{vmatrix} = \begin{vmatrix} a_1 & b_1 & c_1 \\ a_2 & b_2 & c_2 \\ a_3 & b_3 & c_3 \end{vmatrix}$$

hence Cramer's rule may be written

$$x_1 = \frac{(D \times B, C)}{(A \times B, C)} = \frac{\begin{vmatrix} d_1 & b_1 & c_1 \\ d_2 & b_2 & c_2 \\ d_3 & b_3 & c_3 \end{vmatrix}}{\begin{vmatrix} a_1 & b_1 & c_1 \\ a_2 & b_2 & c_2 \\ a_3 & b_3 & c_3 \end{vmatrix}}.$$

What are the analagous formulas for x_2, x_3?

The following exercises sketch the analogue of Theorem 1.8 for $V_n(R)$. Namely, that $A_1, \ldots, A_n$ span $V_n(R)$ if and only if they are linearly independent. The proof follows the case $n = 3$.

14. Prove Theorem 1.5. That is, prove that for each positive integer n the n equations in $n + 1$ unknowns
$$a_{1,1}x_1 + \cdots + a_{1,n+1}x_{n+1} = 0$$
$$a_{2,1}x_1 + \cdots + a_{2,n+1}x_{n+1} = 0$$
$$\cdot \qquad \qquad \cdot$$
$$\cdot \qquad \qquad \cdot$$
$$\cdot \qquad \qquad \cdot$$
$$a_{n,1}x_1 + \cdots + a_{n,n+1}x_n = 0$$
always have a solution $x_1, \ldots, x_{n+1}$ not identically zero. (Use induction and Theorem 1.4.)
15. Prove that $n + 1$ vectors in $V_n(R)$ are linearly dependent.
16. Prove that if $A_1, \ldots, A_n$ are linearly independent vectors in $V_n(R)$, then these vectors span $V_n(R)$.
17. Prove that if $A_1, \ldots, A_n$ are linearly dependent, there is a nonzero vector $B \in V_n(R)$ orthogonal to $A_1, \ldots, A_n$.
18. Prove that if $B \neq 0$ and is orthogonal to $A_1, \ldots, A_n$ then B is not in the span of $A_1, \ldots, A_n$.
19. Prove that if $A_1, \ldots, A_n$ are linearly dependent, they do not span $V_n(R)$.

CHAPTER TWO

VECTOR SPACES

2.1 Examples and Motivation

In Chapter 1 we introduced the space $V_n(R)$ of n-tuples of real numbers, and showed that such n-tuples or vectors were necessary to describe many complicated phenomena in a mathematical way. However, our use of vectors to describe such phenomena has one drawback, and this is the dependence on a coordinate system. For example if we wish to represent the forces acting on a body or the velocity of a particle moving in space by vectors, then the vector or n-tuple chosen depends on the coordinate system used. On the other hand, the actual force or velocity is independent of this coordinate system. To get around this difficulty we shall generalize our notion of vectors and introduce what is called a vector space. First let us recall the properties of addition and scalar multiplication in $V_n(R)$.

If $X = (x_1, \ldots, x_n)$ and $Y = (y_1, \ldots, y_n)$ then we defined the sum of these two vectors by the formula

$$X + Y = (x_1 + y_1, \ldots, x_n + y_n).$$

The set $V_n(R)$ is *closed* under the operation $+$. That is, whenever X and Y belong to $V_n(R)$, then the sum $X + Y$ also belongs to $V_n(R)$. The addition operation $+$ is commutative and associative. Furthermore there is a zero vector and each vector X has an inverse with respect to addition.

Scalar multiplication of vectors by real numbers is defined by the formula

$$aX = (ax_1, \ldots, ax_n).$$

We have verified that $V_n(R)$ is *closed* under this operation of scalar multiplication. That is, if $X \in V_n(R)$ and $a \in R$, then $aX \in V_n(R)$. Furthermore scalar multiplication satisfies the following properties:
For each $a, b \in R$ and $X, Y \in V_n(R)$
  (i) $(a+b)X = aX + bX$
  (ii) $a(X+Y) = aX + aY$
  (iii) $a(bX) = (ab)X$
  (iv) $1 \cdot (X) = X$.

Let us consider some examples of sets V which are closed under a sum operation having the properties listed above and which are closed under a scalar multiplication which satisfies (i)–(iv)

Example 1. The first example we have met before. Choose a point P in space and consider the family V of all directed line segments emanating from P. If X and Y are two such line segments, we may define the sum $X + Y$ as the directed line segment from P to the fourth vertex of the parallelogram formed by X and Y. (See Figure 1.)

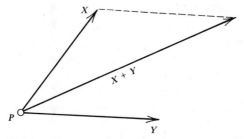

Fig. 1

To define scalar multiplication of a directed line segment X by a real number a we define aX to be that line segment having length equal to the length of X multiplied by $|a|$ and direction the same as X if $a > 0$ and opposite to that of X if $a < 0$. Of course aX must be defined if $a = 0$, so we postulate the existence of a line segment $0 \in V$ with length equal to zero. Then $0X = 0$. Now it can be checked, and this we have done in Chapter 1, that this vector sum has all the properties of ordinary addition. Furthermore the scalar multiplication defined above satisfies properties (i)-(iv).

Example 2. Let V be the set of real valued continuous functions defined on an interval (a, b). If we define the *sum* of two functions by the formula

(1) $$(f + g)(x) = f(x) + g(x)$$

and *scalar multiplication* by the formula

(2) $$(cf)(x) = cf(x),$$

then it is an elementary theorem of calculus that V is closed under these two operations. In fact this is just the familiar statement that if f and g are continuous on the interval (a, b) and c is a real number, then both $f + g$ and cf are continuous on (a, b). In addition it may be easily checked that this function addition defined by (1) has all the properties possessed by addition of real numbers. Furthermore, the scalar multiplication defined by (2) satisfies (i)-(iv).

Example 3. Our third example will be somewhat different. Let a_0, a_1, a_2 be given real numbers, and consider the second order differential equation

(3) $$a_2 y'' + a_1 y' + a_0 y = 0.$$

A function $y = f(x)$ is a solution to this equation on the interval I if for all $x \in I$

$$a_2 f''(x) + a_1 f'(x) + a_0 f(x) = 0.$$

Let V be the set of all solutions to (3). If $f, g \in V$, then we may define the sum $f + g$ by (1) and scalar multiplication af by (2). Let us verify that V is closed under these two operations. Both of these results are consequences of the familiar linearity property of the derivative. Since $(f + g)' = f' + g'$ and $(af)' = a(f)'$, we have that

$$a_2(f + g)''(x) + a_1(f + g)'(x) + a_0(f + g)(x)$$
$$= a_2 f''(x) + a_2 g''(x) + a_1 f'(x) + a_1 g'(x) + a_0 f(x) + a_0 g(x)$$
$$= a_2 f''(x) + a_1 f'(x) + a_0 f(x) + a_2 g''(x) + a_1 g'(x) + a_0 g(x)$$
$$= 0$$

since f and g are assumed to be members of V, the set of solutions to equation (1). It is equally trivial to verify that if $f \in V$, then $af \in V$.

2.2 Vector Spaces and Subspaces

With the examples of Section 2.1 in mind we now give the definition of a vector space V over the field R of real numbers.

Definition. Let V be a nonempty set which is closed under an operation $+$. That is, if $X, Y \in V$, then $X+Y \in V$. In addition assume that to each $X \in V$ and $a \in R$ there is a unique element $Y \in V$ which we denote by $Y = aX$. Then V is a vector space over the real number field if the following properties hold.

1. The operation $+$ is commutative and associative.
2. There is an identity element 0 with respect to the operation $+$, that is, $X+0 = X$ for each $X \in V$.
3. Each $X \in V$ has an inverse, with respect to the operation $+$, that is, there is an element $Y \in V$ such that $X+Y = 0$.
4. For each $a, b \in R$ and $X, Y \in V$
   (i) $(a+b)X = aX + bX$
   (ii) $a(X+Y) = aX + aY$
   (iii) $a(bX) = (ab)X$
   (iv) $1X = X$.

Statements 1, 2, and 3 comprise all the properties of ordinary addition of real numbers, and indeed the set of real numbers is a vector space over itself. It is reasonable, therefore, to call the operation $+$ vector addition and to call $X+Y$ the sum of the vectors X and Y. From now on we shall use the term vector to denote any member of a vector space. Although Statement 2 asserts the existence of an identity element 0 for addition, (a zero vector) it does not assert that there is only one. This, however, follows easily from Statements 1 and 2. For if 0 and $0'$ are zero vectors, we have by Statement 2 that
$$0' + 0 = 0'$$
and
$$0 + 0' = 0.$$
Since by Statement 1 addition is commutative, it follows that $0 = 0'$.

In a similar way it can be shown that for a given $X \in V$ the vector $Y \in V$ satisfying $X+Y = 0$ is unique. Therefore we may write
$$Y = -X.$$
However, it is not automatic that for each $X \in V$
$$-X = (-1)X.$$
This must be proved, and we leave it as an exercise.

Just as for n-tuples of real numbers the operation of associating to each $a \in R$ and $X \in V$ a vector $Y \in V$, written $Y = aX$, is called *scalar multiplication*. As in the case of $V_n(R)$, the elements of the field R are called scalars to distinguish them from the elements of the vector space V. Our practice will be to denote vectors by capital Latin letters and scalars by small Latin letters.

Property (iv) of 4 states that $X = 1X$ for each $X \in V$. Let us verify that $0X = \tilde{0}$ where for the moment we use $\tilde{0}$ to denote the zero vector and 0 to denote the zero scalar. To prove this observe that from 4 it follows that for each $X \in V$
$$X = 1X = (1+0)X = 1X + 0X = X + 0X.$$
Adding $-X$ to both sides of this equation, we see that
$$\tilde{0} = -X + X = -X + (X + 0X) = (-X + X) + 0X$$
$$= \tilde{0} + 0X = 0X,$$

or
$$\tilde{0} = 0X.$$

(The student should explain the reason for each of the above equalities.)

To use the same symbol 0 for both the zero vector and the real number zero is an abuse of terminology, but it should cause no confusion.

The statement "V is a vector space over the field R of real numbers" is conveniently abbreviated by just saying "V is a real vector space." The theory which we shall develop, however, holds equally well for vector spaces where the scalars are drawn from an arbitrary field F. (Consult *Calculus, with an Introduction to Vectors*, p. 5, for the definition of a field.) Such a vector space is called a *vector space over the field F*. However, in our development we shall always assume that the scalars are real numbers. Consequently when we speak of a vector space we shall always mean a real vector space.

We note in passing that scalar multiplication does not possess all the properties of multiplication of real numbers. In particular if X and Y are nonzero vectors in V, then it is not true that there is an $a \in R$ satisfying

$$Y = aX.$$

(Give an example!)

One problem that arises very frequently in the study of vector spaces is the following. If V is a vector space and S is a nonempty subset of V, when is S closed under the operations of addition and scalar multiplication present in V? When this is true we say that S is a *subspace* of V. To determine if S is a subspace, we must verify two things. First, for each pair $X, Y \in S$ the sum $X + Y$, which necessarily belongs to V, belongs also to S. Second, for each $a \in R$ and $X \in S$, the vector aX in V belongs also to S. Whenever S is a subspace of the vector space V, and $S \neq V$, then S is called a *proper* subspace of V. If $S = \{0\}$, then S is called the *zero subspace*.

Examples of subspaces are very common. In $V_2(R)$ the subspaces are the zero subspace, the whole space $V_2(R)$, and all lines through the origin. In $V_3(R)$ the subspaces are the lines and planes through the origin together with the zero space and the space $V_3(R)$. If V is the set of all real valued functions defined on an interval I and addition and scalar multiplication are defined by formulas (1) and (2) p. 51, then V is a vector space. The subset S of those continuous functions forms a subspace of V as does the set of differentiable functions. Another subspace of the space of real valued functions is the space of polynomials. This is the set of functions f such that for all values of x

$$f(x) = a_0 + a_1 x + \cdots + a_n x^n$$

where $a_0, a_1, \ldots, a_n$ are fixed real numbers.

If S is a subspace of a vector space V, then it is easy to show that S is a vector space in its own right under the operations of addition and scalar multiplication present in V. We leave this verification to the exercises (Exercise 17).

EXERCISES

1. Give four examples of vector spaces different from those discussed in the text.
2. Let V be the set of pairs (x_1, x_2) satisfying
$$2x_1 - x_2 = 0$$
$$4x_1 - 2x_2 = 0.$$

Verify that V is a subspace of $V_2(R)$. Is V a proper subspace? Is V the zero subspace?

3. Let V be the set of pairs (x_1, x_2) satisfying
$$3x_1 + 2x_2 = 1$$
$$x_1 - x_2 = 0.$$
Is V a subspace of $V_2(R)$?

4. Let V be the set of differentiable functions and S be the set of functions $y = f(x)$ satisfying
$$y' - 3y = 0.$$
Is S a subspace of V? Is S a proper subspace of V?

5. If p and q are continuous functions, when do the set of solutions to the differential equation
$$y' + p(x) \cdot y = q(x)$$
form a vector space?

6. Consider the following linear equations
$$3x_1 - 2x_2 + x_3 = 0$$
$$2x_1 + x_2 - x_3 = 0$$
$$5x_1 - 8x_2 + 5x_3 = 0.$$
Show that the set S of solutions to these equations forms a subspace of $V_3(R)$. Is S a proper subspace? Why? Is S just the zero subspace?

7. Let $A_1 = (a_{11}, a_{12}, a_{13})$, $A_2 = (a_{21}, a_{22}, a_{23})$ and $A_3 = (a_{31}, a_{32}, a_{33})$ be three vectors from $V_3(R)$. Give a necessary and sufficient condition on these vectors which will insure that the set of solutions of the equations
$$a_{11}x_1 + a_{12}x_2 + a_{13}x_3 = 0$$
$$a_{21}x_1 + a_{22}x_2 + a_{23}x_3 = 0$$
$$a_{31}x_1 + a_{32}x_2 + a_{33}x_3 = 0$$
consists only of the zero subspaces of $V_3(R)$? Justify your result. (The vector $X = (x_1, x_2, x_3)$ is a solution to the above equations if and only if X is orthogonal to the vectors A_1, A_2, A_3.)

8. Let V be the vector space of all directed line segments in the plane emanating from a fixed point. Classify all the subspaces of V.

9. Let V be the vector space of directed line segments emanating from a fixed point in space. Classify all the subspaces of V.

10. Let V be a vector space. Show that each $X \in V$ has a unique inverse with respect to addition.

11. Let V be a real vector space. If $X \in V$ and $a \in R$, verify
   (i) $(-1) \cdot X = -X$
   (ii) $(-a) \cdot (-X) = a \cdot X$.

12. Let V be a real vector space. Let $a \in R$ and $X \in V$.
   (i) Show that $a \cdot 0 = 0$ where 0 is the zero vector in V.
   (ii) Show that if $aX = 0$ and $X \neq 0$, then $a = 0$.

13. Let S and T be subspaces of the vector space V. Show that $S \cap T$ is also a subspace of V. Is $S \cup T$ a subspace of V? Justify your answer. (Recall that if E and F are sets then $E \cap F$ is the set of those elements belonging to both E and F. The set $E \cup F$ is the set of those elements belonging either to E or to F or to both.)

14. If S and T are subspaces of V, then $S + T$ is defined to be the set of all vectors $X + Y$ where $X \in S$ and $Y \in T$. Verify that $S + T$ is a subspace of V.

15. Let V be the space of continuous functions defined on an interval $[a, b]$. Which of the following subsets of V are subspaces?

   (i) $\{f \in V: \int_a^b f(x)\,dx = 0\}$

   (ii) $\{f \in V: \int_a^b f(x)\,dx \neq 0\}$

   (iii) $\{f \in V: f' = 0\}$

   (iv) $\{f \in V; f'(a) > 0\}$

   (v) $\{f \in V: f$ is differentiable at all points of $[a, b]$ except possibly for a finite subset$\}$.

   (Recall that if P is a given property and E is a set, then $\{x \in E: P\}$ stands for the set of those elements $x \in E$ which satisfy property P.)

16. Consider the following set of m linear equations in n unknowns $x_1, \ldots, x_n$

$$a_{11}x_1 + \cdots + a_{1n}x_n = 0$$
$$\vdots$$
$$a_{m1}x_1 + \cdots + a_{mn}x_n = 0.$$

   Show that the set of vectors $X = (x_1, \ldots, x_n)$ which are solutions to these equations forms a subspace of $V_n(R)$.

17. Let S be a subspace of the vector space V. Show that S is a vector space under the operations of addition and scalar multiplication present in V.

18. Let S be a nonempty subset of $V_n(R)$. Define $S^\perp = \{X \in V_n(R): (X, Y) = 0$ for each $Y \in S\}$. Show that $S^\perp$ is a subspace of $V_n(R)$. The subspace $S^\perp$ is called the *orthogonal complement* of the set S.

2.3 Linear Independence

Definition. *Let V be a real vector space, and let $X_1, \ldots, X_n$ be vectors from V. A vector $X \in V$ which can be written*

$$X = a_1 X_1 + \cdots + a_n X_n,$$

for appropriate scalars $a_1, \ldots, a_n \in R$ is called a linear combination *of the vectors $X_1, \ldots, X_n$. If S is a nonempty subset of V, then the totality of vectors each of which is a linear combination of vectors from S is called the* span *of S and is denoted by* $\mathrm{sp}(S)$.

Clearly $\mathrm{sp}(S)$ is the smallest subspace of V containing S. If it happens that $V = \mathrm{sp}(S)$, then we say that the set S *spans* V or *generates* V, and S is called a *set of generators* for V.

For example, if $V = V_n(R)$, then the set of vectors

$$\{I_1 = (1, 0, \ldots, 0), I_2 = (0, 1, 0, \ldots, 0), \ldots, I_n = (0, 0, \ldots, 0, 1)\}$$

spans V. This is clear since $(a_1, \ldots, a_n) = a_1 I_1 + \cdots + a_n I_n$. Thus $V_n(R) = \mathrm{sp}\{I_1, \ldots, I_n\}$. If V is the space of polynomials, then the powers of x, $\{1, x, x^2, \ldots\}$, span V since each polynomial $p(x) = a_0 + a_1 x + \cdots + a_n x^n$ is a linear combination of powers of x. As a third example, let V be the set of solutions to the differential equation

(1) $\qquad\qquad\qquad y'' + y = 0.$

To specify a set of generators for V we must appeal to a theorem proved in *Calculus, with an Introduction to Vectors*, Chapter 7 to the effect that each solution of (1) is a linear combination of the functions $\sin x$ and $\cos x$. Thus V is spanned by the set $S = \{\sin x, \cos x\}$.

It is convenient to have $\text{sp}(S)$ defined for all subsets of the vector space V. Therefore, if $\emptyset$ is the null set, we agree to define

$$\text{sp}(\emptyset) = \{0\}.$$

Thus the span of the null set is the zero subspace. Of course the zero subspace is spanned by the zero vector as well.

If a vector space V is spanned by a finite set S, then V is said to be *finitely generated*. The space $V_n(R)$ and the set of solutions to $y'' + y = 0$ are finitely generated whereas the set of polynomials is not. For spaces which are finitely generated an important problem is to determine sets of generators having the smallest number of elements. Fundamental to this problem is the notion of linear dependence. We have already defined linear dependence for sets of vectors in $V_n(R)$. The definition for general vector spaces is exactly the same.

Definition. *A finite set of vectors $\{X_1, \ldots, X_n\}$ from a vector space V is called* linearly dependent *if there exist scalars $a_1, \ldots, a_n$ not all zero such that*

(2) $$a_1 X_1 + \cdots + a_n X_n = 0.$$

If no such scalars different from zero can be found satisfying (2), *then the set $\{X_1, \ldots, X_n\}$ is called* linearly independent.

To verify that a set $\{X_1, \ldots, X_n\}$ is linearly independent one must verify the following conditional statement. If

$$a_1 X_1 + \cdots + a_n X_n = 0,$$

then

$$a_1 = a_2 = \cdots = a_n = 0.$$

Example 1. In the space $V_n(R)$ the unit coordinate vectors

$$I_1 = (1, 0, \ldots, 0), \ldots, I_n = (0, 0, \ldots, 0, 1)$$

are linearly independent.

Example 2. In $V_3(R)$ if $A = (1, -1, 2)$, $B = (0, 1, 3)$ and $C = (2, -3, 1)$, then the set $\{A, B, C\}$ is linearly dependent since $2A - B - C = 0$.

Example 3. In the space of all continuous functions the set of functions $\{1, x, \ldots, x^n\}$ is linearly independent but the set $\{1, \sin^2 x, \cos^2 x\}$ is linearly dependent.

The following facts follow readily from the definitions. The proofs are left as exercises.

1. If S is a linearly independent subset of a vector space V, then any subset of S is linearly independent. In particular the null set is linearly independent.
2. If S is a finite set of vectors and $0 \in S$, then S is linearly dependent.
3. If $S = \{X\}$, then S is linearly independent if and only if $X \neq 0$.
4. If $S = \{X, Y\}$ where X, Y are two vectors from V, then S is linearly dependent if and only if one vector of S is a scalar multiple of the other.

This last statement may be readily generalized to arbitrary finite sets of vectors. We state the result as a theorem.

THEOREM 2.1 *Let V be a vector space and let S be a finite subset of V having more than one member. Then S is linearly dependent if and only if some vector $X \in S$ is a linear combination of the other remaining vectors of S.*

Proof. Let $S = \{X_1, \ldots, X_n\}$, and suppose first that S is linearly dependent. Then
$$a_1 X_1 + \cdots + a_n X_n = 0,$$
for some scalars $a_1, \ldots, a_n$, and not all of the scalars are zero. If $a_k \neq 0$, then
$$-a_k X_k = a_1 X_1 + \cdots + a_{k-1} X_{k-1} + a_{k+1} X_{k+1} + \cdots + a_n X_n,$$
and dividing by a_k we obtain
$$X_k = \left(-\frac{a_1}{a_k}\right) X_1 + \cdots + \left(-\frac{a_{k-1}}{a_k}\right) X_{k-1} + \left(-\frac{a_{k+1}}{a_k}\right) X_{k+1} + \cdots + \left(-\frac{a_n}{a_k}\right) X_n.$$

Thus X_k is a linear combination of the vectors $X_1, \ldots, X_{k-1}, X_{k+1}, \ldots, X_n$. Conversely if X_k is a linear combination of $X_1, \ldots, X_{k-1}, X_{k+1}, \ldots, X_n$, then
$$X_k = a_1 X_1 + \cdots + a_{k-1} X_{k-1} + a_{k+1} X_{k+1} + \cdots + a_n X_n.$$

Hence
$$a_1 X_1 + \cdots + a_{k-1} X_{k-1} - X_k + a_{k+1} X_{k+1} + \cdots + a_n X_n = 0$$
and not all the coefficients $a_1, \ldots, a_{k-1}, -1, a_{k+1}, \ldots, a_n$ are zero. Hence $\{X_1, \ldots, X_n\}$ are linearly dependent. This completes the proof.

If $S = \{X_1, \ldots, X_n\}$ is a finite set of generators for V which is linearly dependent and X_n is a linear combination of $X_1, \ldots, X_{n-1}$ say, then
$$\text{sp}(\{X_1, \ldots, X_n\}) = \text{sp}(\{X_1, \ldots, X_{n-1}\})$$
and V is now generated by the $n-1$ vectors $X_1, \ldots, X_{n-1}$. Continuing this process inductively it follows that V is spanned by a linearly independent subset of S. This is the content of the next theorem.

THEOREM 2.2 *Let V be a vector space and assume that V is spanned by a finite set S. Then V is spanned by a linearly independent subset of S.*

Proof. We may first assume that V is not the zero subspace. For if $V = \{0\}$, then V is spanned by the null set which is a linearly independent subset of S. The proof now proceeds by induction on the number of generators for V. Indeed we must verify the following statement. "Each vector space V which has n generators $\{X_1, \ldots, X_n\}$; $n \geq 1$, is spanned by a linearly independent subset of these generators." Suppose $n - 1$, and assume $V = \text{sp}\{X_1\}$. Since $V \neq \{0\}$, the vector $X_1 \neq 0$ and the set $S = \{X_1\}$ is linearly independent. Hence the theorem is true for $n = 1$.

We now assume the statement for $n = k$ and attempt to deduce it for $n = k+1$. Let S be a set of $k+1$ generators for V. If S is linearly independent, there is nothing to prove. If not, then some vector $X \in S$ is a linear combination of the other k vectors of S. Let T be the set remaining when X is removed. Then $\text{sp}(T) = \text{sp}(S) = V$ and T has only k members. By our inductive assumption V is spanned by a linearly independent subset of T which is also a subset of S. Thus our statement is true for $n = k+1$. Applying the principle of induction it follows that our statement is true for all positive integers n, and the proof of the theorem is complete.

Now a finitely generated vector space V has many sets of linearly independent generators, that is if $V \neq \{0\}$. However each set of linearly independent generators must have the same number of elements. This number is called the *dimension* of the space V. This very important fact we shall verify in Section 2.5.

If a vector space V is not finitely generated, then it is still true that V is generated by a linearly independent set. However, the proof which depends on the "axiom of choice" of axiomatic set theory will not concern us. (The interested student is referred to the book, *Naive Set Theory*, by P. R. Halmos.)

2.4 Tests for Independence

It is often important to know if the vectors $X_1, \ldots, X_n$ from the vector space V are linearly independent or not. Many tests are available depending on the nature of the space V. For example, if $V = V_3(R)$, then it was proved in Chapter 1, that the vectors

$$A = (a_1, a_2, a_3), \quad B = (b_1, b_2, b_3), \quad C = (c_1, c_2, c_3)$$

are linearly independent if and only if the determinant

$$\begin{vmatrix} a_1 & a_2 & a_3 \\ b_1 & b_2 & b_3 \\ c_1 & c_2 & c_3 \end{vmatrix} \neq 0.$$

This result we shall generalize to $V_n(R)$ in due course, but for the moment we content ourselves with a more elementary test. First we need some terminology.

Definition. *The vectors $X_1, \ldots, X_m$ of $V_n(R)$ are said to be in* echelon *form if for each k the first nonzero component of $X_k = (x_{k1}, x_{k2}, \ldots, x_{kn})$, reading from left to right, precedes the first nonzero component of X_{k+1}.*

Thus if $X_1, \ldots, X_m$ are in echelon form and $x_{k,l}$ and $x_{k+1,m}$ are the first nonzero components of X_k and X_{k+1} respectively, then $l < m$. For example the vectors $(1, 0, 0, 2)$, $(0, 0, 1, 1)$, $(0, 0, 0, 3)$ are in echelon form, whereas, the vectors $(1, 0, 1)$, $(1, 1, 0)$, $(0, 1, 1)$ are not.

THEOREM 2.3 *A finite set of vectors from $V_n(R)$ which can be arranged to be in echelon form is linearly independent.*

Proof. Assume $X_1, \ldots, X_m$ are in echelon form and suppose $a_1 X_1 + \cdots + a_m K_m = 0$. We must show that $a_1 = a_2 = \cdots = a_m = 0$. Assume the first nonzero component of X_1 is in the kth position. Then $X_1 = (0, 0, \ldots, x_{1k}, x_{1,k+1}, \ldots, x_{1n})$ and $x_{1k} \neq 0$. Since the vectors $X_1, \ldots, X_m$ are in echelon form, the kth components of the vectors $X_2, \ldots, X_m$ are all zero. Hence

$$a_1 X_1 + \cdots + a_m X_m = 0$$

implies $a_1 x_{1k} = 0$ or $a_1 = 0$. Similarly if the first nonzero component of X_2 is in the lth position, then the lth components of the vectors $X_3, \ldots, X_m$ all vanish. Hence $a_2 X_2 + \cdots + a_m X_m = 0$ implies $a_2 x_{2l} = 0$ where x_{2l} is the lth component of X_2. Since $x_{2l} \neq 0$ we infer that $a_2 = 0$. Continuing inductively it follows that $a_3 = a_4 = \cdots = a_m = 0$.

If V is a subspace of $V_n(R)$ spanned by the vectors $X_1, \ldots, X_k$, then we may always replace the set of generators $\{X_1, \ldots, X_k\}$ by a linearly independent set of vectors $\{Y_1, \ldots, Y_l\}$ in echelon form. The reason for this is the following *replacement principle*. In any vector space if a vector X is a linear combination of vectors $X_1, \ldots, X_m$, i.e., $X = a_1 X_1 + \cdots + a_m X_m$, and if the kth coefficient $a_k \neq 0$, then we may solve for X_k and write

$$X_k = -\frac{a_1}{a_k} X_1 + \cdots + \frac{1}{a_k} X + \cdots + -\frac{a_m}{a_k} X_m.$$

As a consequence we may replace the vector X_k by X without changing the span of the vectors $X_1, \ldots, X_m$ that is

$$\text{sp}(X_1, \ldots, X_m) = \text{sp}(X_1, \ldots, X_{k-1}, X, X_{k+1}, \ldots, X_m).$$

The use of this principle to construct a set of generators in echelon form is best illustrated by an example.

Example 1. Construct a set of generators in echelon form for the subspace of $V_4(R)$ spanned by $X_1 = (2, 1, 0, -1)$, $X_2 = (1, -1, 1, 2)$ and $X_3 = (0, 2, 1, -3)$. By the replacement principle if $Y = a_1 X_1 + a_2 X_2$ and $a_2 \neq 0$ we may replace X_2 by Y and $\text{sp}(X_1, X_2, X_3) = \text{sp}(X_1, Y, X_3)$. We choose a_1 and a_2 so that Y has first component equal to zero. To do this we take $a_1 = 1$ and $a_2 = -2$. Then

$$Y = (2, 1, 0, -1) - 2(1, -1, 1, 2) = (0, 3, -2, -5).$$

Now let $Z = b_1 Y + b_2 X_3$ and choose b_1, b_2 so that Z has a zero in the second component. Clearly if $b_1 = 2$ and $b_2 = -3$, then

$$Z = 2(0, 3, -2, -5) - 3(0, 2, 1, -3) = (0, 0, -7, -1).$$

The vectors $X_1 = (2, 1, 0, -1)$, $Y = (0, 3, -2, -5)$ and $Z = (0, 0, -7, -1)$ are in echelon form and they have the same span as the vectors X_1, X_2, X_3.

EXERCISES

1. If S is a finite set of vectors, show that S is linearly dependent if $0 \in S$.
2. If $S = \{X_1, \ldots, X_n\}$ show that S is linearly dependent if for some $i \neq j$, $X_i = X_j$.
3. Prove that $S = \{X\}$ is linearly independent if and only if $X \neq 0$.
4. If S is a set of two vectors, show that S is linearly dependent if and only if one of the vectors is a scalar multiple of the other.
5. Prove that any subset of a linearly independent set of vectors is linearly independent. What does this say about the empty set?
6. Let $S = \{X_1, \ldots, X_n\}$ and V be the span of S. Show that S is linearly independent if and only if each vector X in V can be written in precisely one way as a linear combination of vectors from S.
7. Let V be a vector space which has at most n linearly independent vectors. If $S = \{X_1, \ldots, X_n\}$ is a linearly independent subset of V show that $V = \text{sp}(S)$.
8. Let V be the subspace of $V_3(R)$ spanned by $X_1 = (-1, 0, 1)$, $X_2 = (2, 1, 3)$, $X_3 = (0, 1, -2)$. Construct a set of generators for V in echelon form.
9. Construct a set of generators in echelon form for the space spanned by $(1, -1, 2)$, $(-1, 1, 1)$, and $(2, 0, -1)$.
10. Construct a set of generators in echelon form for the space V in $V_4(R)$ spanned by $(1, 0, -1, 2)$, $(3, -1, 0, 0)$, and $(2, -1, 2, 1)$.
11. Do the same for the space spanned by $(1, -2, 3, 1)$, $(2, -4, 1, -3)$, and $(-1, 2, 1, 1)$.
12. Do the same for the space V in $V_5(R)$ spanned by $(1, -1, 0, 0, 2)$, $(-2, 0, 0, 1, 3)$, $(0, -4, 2, 1, 3)$, $(2, -1, 1, 0, 0)$.

13. Consider the m linear equations in n unknowns defined by
$$a_{11}x_1 + \cdots + a_{1n}x_n = 0$$
$$a_{21}x_1 + \cdots + a_{2n}x_n = 0$$
$$\vdots$$
$$a_{m1}x_1 + \cdots + a_{mn}x_n = 0.$$

Clearly $x_1 = x_2 = \cdots = x_n = 0$ is one solution to this set of equations. Show that there is another solution if and only if the vectors $A_1 = (a_{11}, a_{21}, \ldots, a_{m1})$, $A_2 = (a_{12}, a_{22}, \ldots, a_{m2}), \ldots, A_n = (a_{1n}, a_{2n}, \ldots, a_{mn})$ are linearly dependent.

14. Show that the m linear equations
$$a_{11}x_1 + \cdots + a_{1n}x_n = b_1$$
$$a_{21}x_1 + \cdots + a_{2n}x_n = b_2$$
$$\vdots$$
$$a_{m1}x_1 + \cdots + a_{mn}x_n = b_m$$

will have a solution if and only if the vector $B = (b_1, \ldots, b_m)$ belongs to the span of $A_1 = (a_{11}, \ldots, a_{m1}), \ldots, A_n = (a_{1n}, \ldots, a_{mn})$.

15. Formulate a necessary and sufficient condition that the linear equations of Exercise 14 will have more than one solution. Prove your result.

16. Let $X_1, X_2, \ldots, X_m$ be nonzero vectors in $V_n(R)$ which are pairwise orthogonal. That is, $(X_i, X_j) = 0$ if $i \neq j$. Show that the set $\{X_1, \ldots, X_m\}$ is linearly independent.

17. Complete the inductive argument of Theorem 2.3.

18. Give a formal proof for the replacement principle discussed on p. 59.

2.5 Bases and Dimension

The following fundamental replacement theorem which generalizes the replacement principle of the previous section allows us to assert that all linearly independent sets of generators for a finitely generated vector space must have the same number of generators.

THEOREM 2.4 *Let $S = \{X_1, \ldots, X_n\}$ be a set of generators for the vector space V, and let $T = \{Y_1, \ldots, Y_m\}$ be a linearly independent subset of V. Then $n \geq m$. Furthermore we may replace m of the vectors of S by the vectors from T and still have a set of generators for V.*

Proof. The proof proceeds by induction on the number of elements in T which we have denoted by m. If $m = 1$, then always $n \geq 1$. Now if $T = \{Y_1\}$, then since S generates V

(1) $$Y_1 = a_1 X_1 + \cdots + a_n X_n.$$

Since T is linearly independent, $Y_1 \neq 0$, and not all the coefficients a_j in (1) are zero. Let a_k be the first nonzero coefficient. Replace the vector X_k by Y_1. Then by the replacement principle

(2) $$\operatorname{sp}\{Y_1, X_1, \ldots, X_{k-1}, X_{k+1}, \ldots, X_n\} = \operatorname{sp}\{X_1, \ldots, X_n\} = V.$$

Hence we have proved the theorem for $m = 1$.

Now let $m = 2$. Then by (2)
$$Y_2 = b_1 Y_1 + a_1 X_1 + \cdots + a_{k-1} X_{k-1} + a_{k+1} X_{k+1} + \cdots + a_n X_n.$$

Again not all the coefficients a_j can equal zero since the vectors $\{Y_1, Y_2\}$ are linearly independent. Hence if $a_l \neq 0$ we may replace X_l by Y_2, and the resulting set of vectors still spans V.

Continuing inductively we conclude that for each vector Y_j in T there is a corresponding vector X_j in S which may be replaced by Y_j with the resulting set of vectors still spanning V. Hence $n \geq m$ and the proof is complete.

It is an interesting exercise for the student to supply the missing part of the above induction. This is Exercise 17 at the end of the section.

COROLLARY. *All linearly independent sets of generators in a finitely generated vector space have the same number of elements.*

Proof. Let $\{X_1, \ldots, X_n\}$ and $\{Y_1, \ldots, Y_m\}$ be two linearly independent sets of generators. Then by Theorem 2.4 $m \leq n$. Reversing the roles of the two sets we conclude that $n \leq m$. Hence $n = m$.

The number of linearly independent generators for a finitely generated vector space V is called the *dimension* of the space V. In place of the phrase "finitely generated" vector space V, we often substitute "finite-dimensional" vector space. We abbreviate dimension of V by dim V.

A linearly independent set of generators for a vector space V is called a *basis* for the space V.

For example, in $V_n(R)$ the vectors $I_1 = (1, 0, \ldots, 0)$, $I_2 = (0, 1, 0, \ldots, 0), \ldots, I_n = (0, 0, \ldots, 0, 1)$ are linearly independent and generate the space. Therefore $V_n(R)$ has dimension n, and $\{I_1, \ldots, I_n\}$ is a basis. Indeed it follows from Theorem 2.3 that any set of n vectors from $V_n(R)$ which are in echelon form constitute a basis for $V_n(R)$. The basis $\{I_1, \ldots, I_n\}$ in $V_n(R)$ will be called the *canonical basis* for this space.

The following result now follows immediately from Theorem 2.2.

THEOREM 2.5 *Every finitely generated vector space has a basis.*

Next we prove that a vector space of dimension n is not spanned by a dependent set of n vectors.

THEOREM 2.6 *Assume the vector space V has dimension n. If V is generated by n vectors $\{X_1, \ldots, X_n\}$, then these vectors are linearly independent.*

Proof. By Theorem 2.2, V is spanned by a linear independent subset S of the vectors $\{X_1, \ldots, X_n\}$. Since the dimension of V is n, there must be n vectors in S. Therefore $S = \{X_1, \ldots, X_n\}$ which proves that the vectors $\{X_1, \ldots, X_n\}$ are linearly independent.

THEOREM 2.7 *Let $T = \{Y_1, \ldots, Y_k\}$ be a linearly independent set in the finite-dimensional vector space V. If T is not a basis for V, then T may be enlarged to form a basis for V.*

Proof. By Theorem 2.2 the space V is spanned by a linearly independent set $S = \{X_1, \ldots, X_m\}$. By Theorem 2.4, $m \geq k$. Furthermore k of the vectors X_i may be replaced by the vectors from T and the resulting set of m vectors still generate V. This set of generators must be linearly independent by Theorem 2.6. Hence it is a basis.

The basis constructed in Theorem 2.7 is called an *extension* of the linearly independent set T.

Example 1. If $X_1 = (2, -1, 0, 1)$ and $X_2 = (0, -2, 1, 1)$, extend $\{X_1, X_2\}$ to be a basis for $V_4(R)$. Since the set $\{X_1, X_2\}$ is in echelon form, it is independent. Hence to extend $\{X_1, X_2\}$ to be a basis for $V_4(R)$ we need only construct vectors Y_1, Y_2 such that $\{X_1, X_2, Y_1, Y_2\}$ is linearly independent. This will be true if, for example, $Y_1 = (0, 0, 1, 0)$ and $Y_2 = (0, 0, 0, 1)$ since the resulting set is in echelon form.

The characterization of a basis given by the next theorem is very useful in applications. The proof is easy and we leave it as an exercise.

THEOREM 2.8 *Let $S = \{X_1, \ldots, X_n\}$ be a finite subset of the vector space V. Then S is a basis for V if and only if each vector $X \in V$ can be expressed uniquely as a linear combination of the vectors in S.*

If W is a subspace spanned by vectors $X_1, \ldots, X_k$ and these vectors are linearly dependent, it is not always easy to construct a basis which is a subset of the given set of generators. Even in $V_n(R)$ it is tedious to determine the largest linearly independent subset of a given set of vectors $\{X_1, \ldots, X_k\}$. However, just to find a basis for a subspace in $V_n(R)$ is relatively straightforward. If $X_1, \ldots, X_k$ span the subspace W in question, we may use the replacement technique of the previous section to construct vectors $Y_1, \ldots, Y_j$ in W which are in echelon form and which span W. Such a set of vectors is necessarily a basis.

Thus if we wish to construct a basis in echelon form for the subspace of $V_n(R)$ spanned by the vectors

$$X_j = (a_{j1}, \ldots, a_{jn}) \qquad j = 1, \ldots, k,$$

we first group together all the vectors X, with nonzero first component. If $X_1, \ldots, X_l$ are these vectors, select one, say X_1, to be the first element of the basis. Then define

$$Y_2 = \frac{1}{a_{11}} X_1 - \frac{1}{a_{21}} X_2$$

$$\vdots$$

$$Y_l = \frac{1}{a_{11}} X_1 - \frac{1}{a_{l1}} X_l.$$

The vectors $Y_2, \ldots, Y_l, X_{l+1}, \ldots, X_k$, all have zero as first coordinate and $\text{sp}(X_1, \ldots, X_k) = \text{sp}(X_1, Y_2, \ldots, Y_l, X_{l+1}, \ldots, X_k)$. We now take those vectors in the set $\{Y_2, \ldots, Y_l, X_{l+1}, \ldots, X_k\}$ which have nonzero second component, and repeat the above construction. Continuing inductively we have the desired basis to V in echelon form. We summarize this result in the following theorem.

THEOREM 2.9 *Each nonzero subspace of $V_n(R)$ has a basis consisting of vectors which are in echelon form.*

Example 2. Construct a basis in echelon form for the subspace V of $V_4(R)$, spanned by the vectors $X_1 = (2, 1, 0, 1)$, $X_2 = (1, 0, -1, 1)$, $X_3 = (0, 0, 1, -1)$, $X_4 = (0, 2, 0, 2)$, and determine the dimension of this space. We take X_1 as the first element of the basis. Since the basis is to be in echelon form, all the remaining vectors must have 0 as their first component. If we let

$$Y_2 = \frac{1}{2} X_1 - X_2,$$

then

$$Y_2 = \left(1, \frac{1}{2}, 0, \frac{1}{2}\right) - (1, 0, -1, 1) = \left(0, \frac{1}{2}, 1, -\frac{1}{2}\right)$$

and the vectors X_1 and Y_2 are in echelon form. Next we replace X_4 by a vector Y_3 having 0 as both first and second component. This will be accomplished if we take

$$Y_3 = 2Y_2 - \frac{1}{2}X_4 = (0, 1, 2, -1) + (0, -1, 0, -1) = (0, 0, 2, -2).$$

Next observe that since

$$X_3 = \frac{1}{2}Y_3, \text{ it follows that } X_3 \in \text{sp}(X_1, Y_2, Y_3),$$

and hence

$$V = \text{sp}(X_1, X_2, X_3, X_4) = \text{sp}(X_1, Y_2, Y_3).$$

Since the vectors X_1, Y_2, Y_3 are in echelon form, they are independent and hence constitute a basis for V. Thus the dimension of the subspace V equals three.

To close this section we determine the dimension of the smallest subspace W of a vector space V containing two given subspaces W_1 and W_2. This space W is the space of all sums $X + Y$ where $X \in W_1$ and $Y \in W_2$. Hence we write $W = W_1 + W_2$. The next result shows how a basis for $W_1 + W_2$ may be expressed in terms of a basis for W_1 and a basis for W_2. Recall that if W_1 and W_2 are subspaces, then the intersection, or common part of these subspaces, $W_1 \cap W_2$, is also a subspace.

THEOREM 2.10 *Let W_1 and W_2 be subspaces of the finite-dimensional vector space V. Let S be a basis for $W_1 \cap W_2$. If S_1 is an extension of S which is a basis for W_1 and S_2 is an extension of S which is a basis for W_2, then $S_1 \cup S_2$ is a basis for $W_1 + W_2$. In particular if $W_1 \cap W_2 = 0$, and S_1 is a basis for W_1 and S_2 is a basis for W_2, then $S_1 \cup S_2$ is a basis for $W_1 + W_2$.*

Proof. We must verify that

(3) $$W_1 + W_2 = \text{sp}(S_1 \cup S_2)$$

and furthermore that $S_1 \cup S_2$ is linearly independent. Since $W_1 = \text{sp}(S_1)$ and $W_2 = \text{sp}(S_2)$, (3) follows immediately. To show that $S_1 \cup S_2$ is linearly independent, let

$$S = \{X_1, \ldots, X_n\}$$
$$S_1 = \{X_1, \ldots, X_n, X'_{n+1}, \ldots, X'_l\}$$
$$S_2 = \{X_1, \ldots, X_n, X''_{n+1}, \ldots, X''_m\}$$

and suppose that

(4) $$a_1 X_1 + \cdots + a_n X_n + a'_{n+1} X'_{n+1} + \cdots + a'_l X'_l + a''_{n+1} X''_{n+1} + \cdots + a''_m X''_m = 0.$$

Since S_1 is linearly independent, it suffices to show that all of the coefficients $a''_{n+1}, \ldots, a''_m$ are zero. To show this set

(5) $$X = a''_{n+1} X''_{n+1} + \cdots + a''_m X''_m.$$

Then we may infer from (4) that $X \in \text{sp}(S_1) = W_1$. But (5) implies that $X \in W_2$ since $X''_{n+1}, \ldots, X''_m \in W_2$. Therefore $X \in W_1 \cap W_2$. Since S is a basis for $W_1 \cap W_2$, $X = a''_{n+1} X''_{n+1} + \cdots + a''_m X''_m = b_1 X_1 + \cdots + b_n X_n$ for suitable coefficients $b_1, \ldots, b_n$. Thus $a''_{n+1} X''_{n+1} + \cdots + a''_m X''_m - b_1 X_1 - \cdots - b_n X_n = 0$. But $X_1, \ldots, X_n, X''_{n+1}, \ldots, X''_m$ are linearly independent. Therefore $a''_{n+1} = \cdots = a''_m = b_1 = \cdots = b_n = 0$, and we are done.

We may interpret the result of 2.10 in terms of the dimension of the spaces involved. If $\dim W_1 = l$, $\dim W_2 = m$ and $\dim (W_1 \cap W_2) = n$, then since the union of the bases S_1 and S_2 for W_1 and W_2 is a basis for $W_1 + W_2$, we see that
$$\dim (W_1 + W_2) = l + m - n.$$
We state this result as a corollary to Theorem 2.10.

COROLLARY. *If W_1 and W_2 are subspaces of the finite-dimensional vector space V, then*
$$\dim W_1 + \dim W_2 = \dim (W_1 \cap W_2) + \dim (W_1 + W_2).$$

In particular, if $W_1 + W_2 = V$, then $\dim V = \dim W_1 + \dim W_2$ if and only if $W_1 \cap W_2 = \{0\}$.

When $W_1 + W_2 = V$ and $W_1 \cap W_2 = \{0\}$ then V is said to be the *direct sum* of the spaces W_1 and W_2. In this case we write $V = W_1 \oplus W_2$.

EXERCISES

1. If $X_1 = (1, -1, 1)$, $X_2 = (0, 4, 1)$, extend $\{X_1, X_2\}$ to be a basis for $V_3(R)$.
2. If $X_1 = (4, 0, 1, -1)$, $X_2 = (0, 2, 3, 0)$, extend $\{X_1, X_2\}$ to be a basis of $V_4(R)$.
3. If $X_1 = (1, 4, -2)$, $X_2 = (-1, 0, 3)$, extend $\{X_1, X_2\}$ to be a basis for $V_3(R)$.
4. If $X_1 = (-1, 1, 1, -1)$, $X_2 = (1, 2, 0, 3)$, extend $\{X_1, X_2\}$ to be a basis for $V_4(R)$.
5. Construct a basis in echelon form for the subspaces of $V_n(R)$ spanned by the following sets of vectors $\{X_i\}$. Also determine the dimension of the subspace.
    (a) $X_1 = (1, -1)$, $X_2 = (-3, 3)$
    (b) $X_1 = (1, 2, -1)$, $X_2 = (2, -1, 4)$, $X_3 = (0, 5, -6)$
    (c) $X_1 = (-1, 0, 1)$, $X_2 = (2, 1, 3)$, $X_3 = (3, 2, 7)$
    (d) $X_1 = (1, 2, -1)$, $X_2 = (3, 1, 0)$, $X_3 = (0, 2, 4)$
    (e) $X_1 = (1, 2, 3)$, $X_2 = (1, 0, 1)$, $X_3 = (2, 4, 6)$
    (f) $X_1 = (1, -1, 1, -1)$, $X_2 = (2, 0, 1, 1)$, $X_3 = (1, -1, 0, 0)$, $X_4 = (1, 1, 1, 1)$
    (g) $X_1 = (1, 1, -1, -1)$, $X_2 = (1, -1, 1, -1)$, $X_3 = (1, -3, 3, -1)$,
    $X_4 = (3, -3, 3, -3)$.
6. Prove that the space of all polynomials of degree $\leq n$ has dimension $n + 1$.
7. Show that the polynomials $p_0(x) = 1 + x + \cdots + x^n$, $p_1(x) = x + x^2 + \cdots + x^n$, $\ldots, p_n(x) = x^n$ form a basis for the space of polynomials of degree $\leq n$.
8. A one-dimensional subspace of a vector space V is called a line; a two-dimensional subspace is called a plane. Must the intersection of two planes always be a line in a finite-dimensional vector space? Give an example.
9. Show that any set of k vectors in $V_n(R)$ must be linearly dependent if $k > n$.
10. Let V be a vector space such that the maximum number of linearly independent vectors is finite. If m denotes this maximum, prove that V is finitely generated and that $m = \dim V$.
11. Let V be a vector space such that every pair of distinct two-dimensional subspaces always intersect in a one-dimensional subspace. Prove that $\dim V = 3$.
12. If $\dim V = n$, then every $(n-1)$-dimensional subspace is called a *hyperplane*. Prove that a finite-dimensional vector space of dimension greater than two is never the direct sum of two hyperplanes.
13. Prove that $V = \{f : f^{(n)} = 0\}$ is an n-dimensional subspace of the space of all continuous functions defined on the real line.
14. Determine the dimensions of the spaces
    (a) $V = \{f : f' + kf = 0,\ k\ \text{constant}\}$.
    (b) $W = \{f : f'' + k^2 f = 0,\ k\ \text{constant}\}$.
    Justify your answer.

15. Let S be a finite set of generators for the vector space V such that each $X \in V$ can be written uniquely as a linear combination of vectors from S. Prove that S is a basis for V. (Theorem 2.8.)

16. Let S be a basis for the vector space V. Show that each vector X in V can be written *uniquely* as a linear combination of vectors from S. (Theorem 2.8.)

17. Complete the induction proof of Theorem 2.4. That is, assume that k of the vectors from S may be replaced by vectors $Y_1, \ldots, Y_k$ so that the resulting set still spans V. Then prove the analagous statement for $k+1$ vectors $Y_1, \ldots, Y_{k+1}$.

18. If W is a subspace of the finite-dimensional vector space V prove that dim W $\leq$ dim V and that dim $W =$ dim V if and only if $W = V$.

19. Let V and W be subspaces of a vector space U. Verify the following facts concerning the sum of the subspaces V and W.
    (1) $V + V = V$
    (2) $V + W = W$ if and only if $V \subset W$.
    (For definition of $V + W$ see p. 63).

20. If V is a finite-dimensional vector space, and W_1 and W_2 are subspaces such that

$$V = W_1 \oplus W_2$$

prove that for each $X \in V$ there exist unique vectors $X_1 \in W_1$, $X_2 \in W_2$ such that $X = X_1 + X_2$.

21. A basis $\{X_1, \ldots, X_n\}$ for $V_n(R)$ consisting of vectors X_i satisfying $(X_i, X_j) = 0$, $i \neq j$ is called an *orthogonal basis*. Show that if $X_1 \neq 0$ is given, then an orthogonal basis can be constructed for $V_n(R)$ with X_1 as the first vector. (What linear equations must be satisfied to define the second vector in the basis? Continue inductively.)

22. If $\{X_1, \ldots, X_n\}$ is an orthogonal basis for $V_n(R)$ and we wish to write a vector $X \in V_n(R)$ as a linear combination of the vectors $\{X_1, \ldots, X_n\}$, then the coefficients a_k in the equation $X = a_1 X_1 + \cdots + a_n X_n$ can be very easily determined. Indeed taking the inner product of this equation with the vectors $X_1, \ldots, X_n$ in turn and using the fact that $(X_i, X_j) = 0$ if $i \neq j$ we see that

$$a_k = \frac{(X, X_k)}{(X_k, X_k)} \qquad k = 1, \ldots, n.$$

Write the following vectors X as linear combinations of the given orthogonal basis vectors $\{X_1, \ldots, X_n\}$.
(a) $X = (2, -1)$, $X_1 = (1, 1)$, $X_2 = (1, -1)$
(b) $X = (3, 1)$, $X_1 = (2, 1)$, $X_2 = (1, -2)$
(c) $X = (2, 1, -1)$, $X_1 = (1, 0, 1)$, $X_2 = (1, -1, -1)$, $X_3 = (1, -2, -1)$
(d) $X = (4, 2, -1)$, $X_1 = (1, 1, 1)$, $X_2 = (1, 2, -3)$, $X_3 = (-5, 4, 1)$
(e) $X = (3, 2, -1, 1)$, $X_1 = (1, 0, 1, 1)$, $X_2 = (0, 1, 1, -1)$, $X_3 = (1, 1, -1, 0)$, $X_4 = (-1, 1, 0, 1)$.

2.6 Linear Equations

As an application of the results of the previous sections we consider the problem of determining all solutions to a set of linear equations. A vector $Y = (y_1, \ldots, y_n) \in V_n(R)$ is a solution to m linear equations

(1)
$$\begin{aligned} a_{11}x_1 + \cdots + a_{1n}x_n &= b_1 \\ \vdots \qquad\qquad \vdots \qquad &\;\; \vdots \\ a_{m1}x_1 + \cdots + a_{mn}x_n &= b_m \end{aligned}$$

if each of the questions is valid when $y_1, \ldots, y_n$ are substituted for the unknowns $x_1, \ldots, x_n$, respectively. The $m \times n$ array of real numbers

(2)
$$\begin{pmatrix} a_{11} & \cdots & a_{1n} \\ \vdots & & \vdots \\ a_{m1} & \cdots & a_{mn} \end{pmatrix}$$

is called the *coefficient matrix* of the system (1). Each row of this matrix is a vector in $V_n(R)$ and each column is a vector in $V_m(R)$. Let us denote the m rows of (2) by $R_i = (a_{i1}, \ldots, a_{in})$ and the n columns by

$$C_j = \begin{pmatrix} a_{1j} \\ \vdots \\ a_{mj} \end{pmatrix}.$$

The vectors R_i and C_j are called the *row vectors* and *column vectors* respectively of the $m \times n$ matrix

$$A = \begin{pmatrix} a_{11} & \cdots & a_{1n} \\ \vdots & & \vdots \\ a_{m1} & \cdots & a_{mn} \end{pmatrix}.$$

To further distinguish between the rows and columns of A we shall write the column vectors vertically instead of horizontally.

If we rewrite (1) in vector form, we may immediately give necessary and sufficient conditions that (1) has a solution. Indeed if $C_1, \ldots, C_n$ denote the column vectors of (2) and

$$B = \begin{pmatrix} b_1 \\ \vdots \\ b_m \end{pmatrix},$$

then the system of equations (1) just says

(3) $\qquad\qquad x_1 C_1 + \cdots + x_n C_n = B.$

Thus equations (1) have a solution if and only if the vector B is a linear combination of the vectors $C_1, \ldots, C_n$. This result may be phrased in many equivalent ways. For emphasis we state the result as a theorem.

THEOREM 2.11 *The m equations*

(1)
$$\begin{aligned} a_{11} x_1 + \cdots + a_{1n} x_n &= b_1 \\ &\vdots \\ a_{m1} x_1 + \cdots + a_{mn} x_n &= b_m \end{aligned}$$

in the n unknowns $x_1, \ldots, x_n$ have a solution if and only if the vector

$$B = \begin{pmatrix} b_1 \\ \vdots \\ b_m \end{pmatrix}$$

in $V_m(R)$ belongs to the span of the vectors

$$C_j = \begin{pmatrix} a_{1j} \\ \vdots \\ a_{mj} \end{pmatrix}, j = 1, \ldots, m.$$

Equivalently the equations have a solution if and only if

(4) $\qquad \dim \mathrm{sp}(C_1, \ldots, C_n) = \dim \mathrm{sp}(C_1, \ldots, C_n, B).$

The proof that (4) is necessary and sufficient for (1) to have a solution we leave as an exercise.

If A is an arbitrary $m \times n$ matrix and $C_1, \ldots, C_n$ denote the columns of A, then the dimension of the space spanned by the columns C_j is called the *rank* of A. Thus if in addition to the coefficient matrix A for the system (1) we form the $m \times n+1$ matrix

$$A' = \begin{pmatrix} a_{11} & \cdots & a_{1n} b_1 \\ \vdots & & \\ a_{m1} & \cdots & a_{mn} b_m \end{pmatrix},$$

then system (1) has a solution if and only if

$$\mathrm{rank}\, A = \mathrm{rank}\, A'.$$

The matrix A' formed by adjoining the column vector $B = \begin{pmatrix} b_1 \\ \vdots \\ b_m \end{pmatrix}$ to the matrix A is called the *augmented matrix* for the system (1).

Example 1. Show that the equations

$$\begin{aligned} 2x_1 + x_2 + x_3 &= 1 \\ x_1 \phantom{{} + x_2} + x_3 &= 0 \\ x_2 - x_3 &= 0 \end{aligned}$$

have no solution. In this example the coefficient matrix is $\begin{pmatrix} 2 & 1 & 1 \\ 1 & 0 & 1 \\ 0 & 1 & -1 \end{pmatrix}$ and the augmented matrix is $\begin{pmatrix} 2 & 1 & 1 & 1 \\ 1 & 0 & 1 & 0 \\ 0 & 1 & -1 & 0 \end{pmatrix}$. If $C_1 = \begin{pmatrix} 2 \\ 1 \\ 0 \end{pmatrix}, C_2 = \begin{pmatrix} 1 \\ 0 \\ 1 \end{pmatrix}, C_3 = \begin{pmatrix} 1 \\ 1 \\ -1 \end{pmatrix}$ and $B = \begin{pmatrix} 1 \\ 0 \\ 0 \end{pmatrix}$ it may be easily verified that $C_3 = C_1 - C_2$. Also the vectors C_1, C_2, B are linearly independent. Therefore

$$\dim \mathrm{sp}(C_1, C_2, C_3) = 2 < \dim \mathrm{sp}(C_1, C_2, B) = 3$$

and the equations have no solution.

We turn next to the problem of determining all solutions to equations (1). If $Y = (y_1, \ldots, y_n)$ and $Z = (z_1, \ldots, z_n)$ are both solutions of (1), then writing this fact in vector form we have

$$y_1 C_1 + \cdots + y_n C_n = B$$

and

$$z_1 C_1 + \cdots + z_n C_n = B.$$

Subtracting these equations we have $(y_1 - z_1)C_1 + \cdots + (y_n - z_n)C_n = 0$. Thus the difference $Y - Z$ of two solutions to (1) is a solution to the system of equations

(5)
$$a_{11}x_1 + \cdots + a_{1n}x_n = 0$$
$$\vdots \qquad \vdots \qquad \vdots$$
$$a_{m1}x_1 + \cdots + a_{mn}x_n = 0$$

These equations have the equivalent vector form

$$x_1 C_1 + \cdots + x_n C_n = 0$$

where as usual $C_1, \ldots, C_n$ denote the columns of the coefficient matrix.

If in equation (1), the vector $B = \begin{pmatrix} b_1 \\ \vdots \\ b_m \end{pmatrix}$ is the zero vector, then the equations (1) are called *homogeneous*. If $B \neq 0$, then the equations are called *inhomogeneous*. For a given system (1) of equations the homogeneous system (5) obtained by setting the right hand side of (1) equal to zero is called the *associated* homogeneous system of equations. We may phrase the solution to our problem of determining all solutions to (1) in terms of this homogeneous system of equations.

THEOREM 2.12 *If $Y = (y_1, \ldots, y_n)$ and $Z = (z_1, \ldots, z_n)$ are two solutions of the equations (1), the differences $Y - Z$ is a solution of the associated homogeneous system (5). Conversely if Y is a solution of (1) and X is a solution of (5), then $Z = Y + X$ is a solution of (1).*

Thus in reality the problem of determining all solutions of (1) is equivalent to the problem of determining all solutions to the associated homogeneous system (5) and one solution to (1). If we write (5) in the vector form

(6) $$x_1 C_1 + \cdots + x_n C_n = 0,$$

it is clear that if Y and Z are solutions of (6) then so is each linear combination

$$aY + bZ.$$

Hence the set of all solutions to (6) is a subspace of $V_n(R)$. Thus we have shown that the set of all solutions to (1) consists of one vector from $V_n(R)$ plus all vectors from an appropriate subspace of $V_n(R)$. Such a collection of vectors is called an *affine* subspace of $V_n(R)$. Thus an affine subspace is a subspace translated by one fixed vector. The dimension of an affine subspace is the dimension of the associated subspace. Lines and planes are examples of one- and two-dimensional affine subspaces of general vector spaces.

We shall now consider the problem of determining all solutions to the homogeneous system of equations. If we denote this subspace of solutions by W, then our problem is solved if we can determine a basis for W. To see how to construct this basis we must first make some observations on the rows of the augmented matrix

(7)
$$\begin{pmatrix} a_{11} & \cdots & a_{1n} & b_1 \\ \vdots & & \vdots & \\ a_{m1} & \cdots & a_{mn} & b_m \end{pmatrix}.$$

If $R_1, \ldots, R_m$ are the rows of the augmented matrix (7) and V is the subspace of $V_{n+1}(R)$ spanned by these rows, then we know from the discussion in Sections 2.4 and 2.5 that a basis can be constructed for V consisting of vectors which are in echelon form. Let $S_1, \ldots, S_k$ be the vectors of this basis and suppose

$$S_j = (a'_{j1}, \ldots, a'_{jn}, b'_j)$$

where now $a'_{jl} = 0$ if $j > l$. Now each vector S_j is a linear combination of the vectors $R_1, \ldots, R_m$. Therefore if $y_1, \ldots, y_n$ is a solution to equations (1), it follows that $Y = (y_1, \ldots, y_n)$ is a solution to the set of linear equations

(8)
$$\begin{aligned} a'_{11}x_1 + a'_{12}x_2 + \cdots \cdots \cdots + a'_{1n}x_n &= b'_1 \\ a'_{22}x_2 + \cdots \cdots \cdots + a'_{2n}x_n &= b'_2 \\ &\vdots \\ a'_{kk}x_k + \cdots + a'_{kn}x_n &= b'_k. \end{aligned}$$

This is clear since if $S = aR_1 + bR_2$ and

$$\begin{aligned} a_{11}y_1 + \cdots + a_{1n}y_n &= b_1 \\ a_{21}y_1 + \cdots + a_{2n}y_n &= b_2 \end{aligned}$$

then

$$(aa_{11} + ba_{21})y_1 + \cdots + (aa_{1n} + ba_{2n})y_n = ab_1 + bb_2.$$

Also each solution of equations (8) is a solution of (1) since each of the vectors $R_1, \ldots, R_m$ is a linear combination of the vectors $S_1, \ldots, S_k$. The augmented matrix for the system (8) is now

(9)
$$\begin{pmatrix} a'_{11} & \cdot & \cdot & \cdot & a'_{1n} & b'_1 \\ 0 & & & & \vdots & \vdots \\ \vdots & & & & \vdots & \vdots \\ 0 \cdots 0 & a'_{kk} & \cdots & a'_{kn} & b'_k \end{pmatrix}$$

To find all solutions to (8) we must find one solution to (8) and then all solutions to the associated homogeneous system of equations. Since the rows of the augmented matrix are in echelon form, (8) will have a solution if and only if the first nonzero entry in the last vector S_k precedes b'_k. When this is the case, we may determine one solution to (8) by inspection. If a'_{kl} is the first nonzero entry in

$$S_k = (0, 0, \ldots, a'_{kl}, a'_{k,l+1}, \ldots, a'_{kn}, b'_k)$$

where $l \geqslant k$ then set $x_{l+1} = \cdots = x_n = 0$ and $x_l = b'_k/a'_{kl}$. If $a'_{k-1,p}$ is the first nonzero entry in S_{k-1} then set $x_{p+1} = x_{p+2} = \cdots = x_{l-1} = 0$. Then

$$a'_{k-1,p}x_p + a'_{k-1,l}x_l = b'_{k-1}$$

Consequently

$$x_p = \frac{1}{a'_{k-1,p}}\left[b'_{k-1} - \frac{a'_{k-1,l}}{a'_{k,l}}b'_k\right].$$

Continuing inductively we construct one solution to (8) by inspection. We illustrate this with an example.

Example 2. Find one solution to the equations

(10)
$$\begin{aligned} x_1 - x_2 + 3x_3 + x_4 - x_5 &= 2 \\ 2x_3 - x_4 + x_5 &= 1 \\ x_5 &= -1. \end{aligned}$$

The augmented matrix to this system is

$$\begin{pmatrix} 1 & -1 & 3 & 1 & -1 & 2 \\ 0 & 0 & 2 & -1 & 1 & 1 \\ 0 & 0 & 0 & 0 & 1 & -1 \end{pmatrix}$$

and the rows are in echelon form. Hence setting $x_5 = -1$ and $x_4 = 0$ implies $x_3 = 1$. If now $x_2 = 0$, then $x_1 = -2$. Therefore one solution to the equation (10) is the vector $(-2, 0, 1, 0, -1)$.

Next we shall give a computational recipe for determining all solutions to the homogeneous system

(11)
$$\begin{aligned} a'_{11}x_1 + \cdots \cdots \cdots \cdots a'_{1n}x_n &= 0 \\ \vdots \quad\quad\quad \vdots \quad\quad\quad \vdots & \\ a'_{kk}x_k + \cdots + a'_{kn}x_n &= 0 \end{aligned}$$

when the rows of the coefficient matrix

(12)
$$\begin{pmatrix} a'_{11} & \cdot & \cdot & \cdot & a'_{1n} \\ 0 & & & & \vdots \\ \vdots & & & & \vdots \\ 0 \cdots 0 & a'_{kk} & \cdots & a'_{kn} \end{pmatrix}$$

are in echelon form. Such a matrix we call an *echelon* matrix. To do this let $D_1, \ldots, D_n$ denote the columns of the matrix (12). Note first that if $a'_{11}, a'_{2j_2}, \ldots, a'_{kj_k}$ are the first nonzero entries in each of the rows $S_1, \ldots, S_k$, then the columns $D_1, D_{j_2}, \ldots, D_{j_k}$ form a basis for $\mathrm{sp}(D_1, \ldots, D_n)$. Hence the rank of the echelon matrix (12) equals k. This is also the dimension of the space spanned by the rows $S_1, \ldots, S_k$. Hence for echelon matrices the rank of the matrix is the dimension of the space spanned by the rows as well as the dimension of the space spanned by the column. This fact is true for arbitrary matrices, and we shall give the proof in the next section.

Next observe that each column D_l of (12) is a linear combination of only those columns $D_1, D_{j_2}, \ldots, D_{j_l}$ where $j_l \leq l$. Indeed for each $l \neq 1, j_2, \ldots, j_k$ there exist scalars $c_{l_1}, \ldots, c_{lj_l}, c_l$ such that

$$c_{l_1}D_1 + c_{lj_2}D_{j_2} + \cdots + c_{lj_l}D_{j_l} + c_l D_l = 0.$$

Now it is a fact, which we shall prove in the next section, that the $n-k$ vectors

(13)
$$(c_{l_1}, 0, \ldots, c_{lj_2}, 0, \ldots, c_{lj_l}, 0, \ldots, c_l, 0 \ldots 0)$$

form a basis to the space of solutions to (11). Moreover, these basis vectors can be determined by inspection.

Example 3. Find all solutions to

(14)
$$\begin{aligned} x_1 - x_2 + 3x_3 + x_4 - x_5 &= 0 \\ 2x_3 - x_4 + x_5 &= 0 \\ x_5 &= 0. \end{aligned}$$

The coefficient matrix is the echelon matrix

$$\begin{pmatrix} 1 & -1 & 3 & 1 & -1 \\ 0 & 0 & 2 & -1 & 1 \\ 0 & 0 & 0 & 0 & 1 \end{pmatrix}.$$

The columns $D_1 = \begin{pmatrix} 1 \\ 0 \\ 0 \end{pmatrix}$, $D_3 = \begin{pmatrix} 3 \\ 2 \\ 0 \end{pmatrix}$ and $D_5 = \begin{pmatrix} -1 \\ 1 \\ 1 \end{pmatrix}$ form a basis for the space spanned by the columns $D_1, \ldots, D_5$. Also

$$D_1 + D_2 = 0$$

and

$$-5D_1 + 1D_3 + 2D_4 = 0.$$

Hence the vectors $(1, 1, 0, 0, 0)$ and $(-5, 0, 1, 2, 0)$ form a basis for the set of all solutions to (14). Each solution X to (10) must therefore have the form

$$X = (-2, 0, 1, 0, -1) + a(1, 1, 0, 0, 0) + b(-5, 0, 1, 2, 0).$$

Example 4. Determine all solutions to the equations

(15)
$$\begin{aligned} x_1 \quad - x_3 + x_4 &= 1 \\ -x_1 + x_2 \quad\quad + x_4 &= 0 \\ x_2 \quad - x_4 &= 1. \end{aligned}$$

The augmented matrix for this system is

(16)
$$\begin{pmatrix} 1 & 0 & -1 & 1 & 1 \\ -1 & 1 & 0 & 1 & 0 \\ 0 & 1 & 0 & -1 & 1 \end{pmatrix}.$$

Denoting the rows by R_1, R_2, R_3 we first construct an echelon matrix, the rows of which generate the space spanned by R_1, R_2, R_3.

Since $R_1 + R_2 = (0, 1, -1, 2, 1)$ and $R_3 - (0, 1, -1, 2, 1) = (0, 0, 1, -3, 0)$ the desired echelon matrix will be

(17)
$$\begin{pmatrix} 1 & 0 & -1 & 1 & 1 \\ 0 & 1 & -1 & 2 & 1 \\ 0 & 0 & 1 & -3 & 0 \end{pmatrix}.$$

Matrix (17) is the augmented matrix for the system

(18)
$$\begin{aligned} x_1 \quad - x_3 + x_4 &= 1 \\ x_2 - x_3 + 2x_4 &= 1 \\ x_3 - 3x_4 &= 0. \end{aligned}$$

Moreover a vector (x_1, x_2, x_3, x_4) is a solution to (18) if and only if it is a solution to (15). We may determine one solution to (18) by inspection. Setting $x_3 = x_4 = 0$ we have $x_2 = 1$ and $x_1 = 1$. Therefore $(1, 1, 0, 0)$ is one solution to (18). To find a basis for the solution space of the associated homogeneous equations

(19)
$$\begin{aligned} x_1 \quad - x_3 + x_4 &= 0 \\ x_2 - x_3 + 2x_4 &= 0 \\ x_3 - 3x_4 &= 0. \end{aligned}$$

we examine the coefficient matrix

(20)
$$\begin{pmatrix} 1 & 0 & -1 & 1 \\ 0 & 1 & -1 & 2 \\ 0 & 0 & 1 & -3 \end{pmatrix}.$$

We note that the columns $D_1 = \begin{pmatrix} 1 \\ 0 \\ 0 \end{pmatrix}$, $D_2 = \begin{pmatrix} 0 \\ 1 \\ 0 \end{pmatrix}$, and $D_3 = \begin{pmatrix} -1 \\ -1 \\ 1 \end{pmatrix}$ are a basis for the space spanned by the columns of (20). Moreover if $D_4 = \begin{pmatrix} 1 \\ 2 \\ -3 \end{pmatrix}$ then

$$2D_1 + D_2 + 3D_3 + D_4 = 0.$$

Hence the solution space to (19) is one-dimensional and the vector $(2, 1, 3, 1)$ is a basis. All solutions to (15) are therefore given by the vectors $X = (1, 1, 0, 0) + a(2, 1, 3, 1)$. Thus the set of solutions to (15) is a line in $V_4(R)$.

EXERCISES

Determine if the following systems of linear equations have a solution. If so, find one solution.

1. $2x_1 + x_2 + 3x_3 = 0$
   $x_2 + x_3 = 1$
   $2x_3 = 1$

2. $x_1 - x_2 - x_3 = 1$
   $2x_2 + 4x_3 = 0$
   $x_2 + 2x_3 = 1$

3. $x_1 + x_2 + x_3 = 3$
   $-x_1 \qquad - 3x_3 = 7$
   $x_1 + x_2 + x_3 = 1$

Determine bases for the solution spaces of the following systems of homogeneous equations.

4. $2x_1 + x_2 + 3x_3 = 0$
   $x_2 + x_3 + x_4 = 0$

5. $x_1 - x_2 + x_3 - x_4 + x_5 = 0$
   $2x_2 - x_3 + x_4 - 2x_5 = 0$

6. $x_1 + x_2 - x_3 - x_4 + x_5 - x_6 = 0$
   $x_2 + x_3 - x_4 - x_5 + x_6 = 0$

7. $x_1 - x_2 + x_3 + x_4 - x_5 = 0$
   $x_2 - x_3 - x_4 + x_5 = 0$
   $x_3 - x_4 + x_5 = 0$

Determine all solutions to the following systems of linear equations.

8. $x_1 - x_2 + x_3 - x_4 = 1$
   $x_2 - x_3 - x_4 = -1$

9. $x_1 - x_2 + x_3 - x_4 = 0$
   $x_1 + x_2 - x_3 - x_4 = 1$

10. $x_1 - x_2 - 2x_3 + x_4 - x_5 = 1$
    $x_1 \qquad -x_4 + x_5 = 0$
    $x_3 + x_4 + x_5 = -1$

Determine all solutions to the following systems of linear equations. The rows of the augmented matrix are in echelon form.

11. $x_1 - x_2 + x_3 - x_4 = 1$
    $x_3 + x_4 = -1$

12. $2x_1 - x_2 + x_3 + x_4 - x_5 = 1$
    $x_3 - x_4 - x_5 = -1$

13. $x_1 + x_2 - x_3 - x_4 + x_5 = 2$
    $x_2 + x_3 + x_4 - x_5 = -1$
    $x_4 - x_5 = 1$

Construct a basis in echelon form for the space spanned by the rows of the augmented matrices for the following systems of equations. Determine all solutions if there are any.

14. $x_1 + x_2 - x_3 + x_4 = 1$
    $x_1 - x_2 + x_3 - x_4 = 0$
    $2x_1 - x_2 - x_3 + x_4 = -1$

15. $x_1 - x_2 + x_3 = 0$
    $x_1 + x_2 + 3x_3 = 0$
    $2x_1 - x_2 + 5x_3 = 1$

16. $x_1 - x_2 - x_3 + x_4 + x_5 = 1$
    $x_1 + x_2 + x_3 - x_4 - x_5 = -1$
    $x_1 + x_2 - x_3 - x_4 + x_5 = 1$

17. If $C_1, \ldots, C_n, B$ are the columns of the augmented matrix for the system of equations (1), prove that the system has a solution if and only if

    $$\dim \text{sp}(C_1, \ldots, C_n) = \dim \text{sp}(C_1, \ldots, C_n, B).$$

18. Show that n linear equations in n unknowns have exactly one solution if and only if the rank of the coefficient matrix $= n$.

19. Prove that m homogeneous equations in n unknowns always have a solution different from the zero solution if $n > m$.

2.7*. Linear Equations—*Continued*

Our next result will justify the construction on p. 70 of a basis for the space W of solutions to a homogeneous system of equations. First, recall that the *rank* of a matrix is the dimension of the space spanned by the columns.

THEOREM 2.13 *Let W be the space of all solutions to the homogeneous equations*

(1) $$a_{11}x_1 + \cdots + a_{1n}x_n = 0$$
$$\vdots \qquad \qquad \vdots \qquad \qquad \vdots$$
$$a_{m1}x_1 + \cdots + a_{m,n}x_n = 0.$$

Let $C_1, \ldots, C_n$ denote the columns of the coefficient matrix

(2) $$\begin{pmatrix} a_{11} & \cdots & a_{1n} \\ \vdots & & \vdots \\ a_{m1} & \cdots & a_{mn} \end{pmatrix}$$

and assume that this matrix has rank r. Then the dimension of W equals $n - r$. Furthermore if the first r columns $C_1, \ldots, C_r$ form a basis for $\text{sp}(C_1, \ldots, C_n)$, then there exist scalars $y_{1k}, \ldots, y_{rk}, z_k$ where $z_k \neq 0$ satisfying

$$y_{1k}C_1 + \cdots + y_{rk}C_r + z_k C_k = 0.$$

Moreover the vectors

(3) $\quad Y_k = (y_{1k}, \ldots, y_{rk}, 0, \ldots, z_k, 0, \ldots, 0) \qquad k = r+1, \ldots, n$

form a basis for the solution space W.

Proof. Note first that if $\dim \text{sp}(C_1, \ldots, C_n) = r$, then it follows from Theorem 2.2 and the corollary to Theorem 2.4 that r of the columns C_i are linearly independent and span $\text{sp}(C_1, \ldots, C_n)$. Therefore we may rearrange the columns $C_1, \ldots, C_n$ in such a way that the first r columns form a basis for $\text{sp}(C_1, \ldots, C_n)$. Of course, if this is done then the unknowns $\{x_1, \ldots, x_n\}$ must be renumbered accordingly.

To prove the theorem it suffices to show that the vectors $Y_k, k = r+1, \ldots, n$ defined by (3) form a basis for W. Since there are $n - r$ of these vectors, this proves that $\dim W = n - r$. To establish that $\{Y_{r+1}, \ldots, Y_n\}$ is a basis for W observe first that if the first r columns $C_1, \ldots, C_r$ of the matrix (2) form a basis for the space spanned by $C_1, \ldots, C_n$, then for each $k = r+1, \ldots, n$ the vector C_k is a linear combination of the vectors $C_1, \ldots, C_r$. Hence there exist scalars $y_{1k}, \ldots, y_{rk} z_k$ where $z_k \neq 0$ such that

(3) $$y_{1k}C_1 + \cdots + y_{rk}C_r + z_kC_k = 0.$$

From this it follows that the vectors

$$\begin{aligned}
Y_{r+1} &= (y_{1,r+1}, \ldots, y_{r,r+1}, z_{r+1}, 0, \ldots, \quad 0) \\
Y_{r+2} &= (y_{1,r+2}, \ldots, y_{r,r+2}, 0, z_{r+2}, 0, \ldots, 0) \\
&\vdots \\
Y_n &= (y_{1n}, \ldots, \quad y_{rn}, 0, \ldots, 0, \ldots, 0, z_n)
\end{aligned}$$

all belong to the space of solutions W. Moreover, the vectors $Y_{r+1}, \ldots, Y_n$ are linearly independent. To see this suppose $c_{r+1}Y_{r+1} + \cdots + c_nY_n = 0$. Then $c_{r+1}z_{r+1} = 0 = c_{r+2}z_{r+2} = \cdots = c_nz_n$. However, each of the scalars $z_{r+1}, \ldots, z_n$ is different from zero. Therefore $c_{r+1} = \cdots = c_n = 0$. Next we assert that the vectors $Y_{r+1}, \ldots, Y_n$ form a basis for the space of solutions W. To prove this we must verify that if $X = (x_1, \ldots, x_n)$ and $x_1C_1 + \cdots + x_nC_n = 0$, then there exist appropriate scalars $c_{r+1}, \ldots, c_n$ so that

$$X = c_{r+1}Y_{r+1} + \cdots + c_nY_n.$$

Note first that if $X \in W$ and it is known that $x_{r+1} = \cdots = x_n = 0$, then $x_1 = \cdots = x_r = 0$ since the vectors $C_1, \ldots, C_r$ are linearly independent. Now let $X = (x_1, \ldots, x_n) \in W$. Then

$$X - \frac{x_{r+1}}{z_{r+1}}Y_{r+1} - \cdots - \frac{x_n}{z_n}Y_n \in W$$

since W is a subspace. Moreover if

$$U = (u_1, \ldots, u_n) = X - \frac{x_{r+1}}{z_{r+1}}Y_{r+1} - \cdots - \frac{x_n}{z_n}Y_n,$$

then $u_{r+1} = \cdots = u_n = 0$. Hence by the observation we have just made $U = 0$, and

$$X = \frac{x_{r+1}}{z_{r+1}}Y_{r+1} + \cdots + \frac{x_n}{z_n}Y_n.$$

This completes the proof.

The assertion that the vectors (13) on p. 70 form a basis for the space of solutions to (11) p. 70 now is a direct corollary of Theorem 2.13.

The following very important theorem on the rank of an $m \times n$ matrix also is a consequence of Theorem 2.13.

THEOREM 2.14 *Let r be the rank of an $m \times n$ matrix*

(4) $$\begin{pmatrix} a_{11} & \cdots & a_{1n} \\ \vdots & & \vdots \\ a_{m1} & \cdots & a_{mn} \end{pmatrix}.$$

Then r is the dimension of the space spanned by the rows of the matrix.

Proof. Consider the associated homogeneous linear equations

(5) $$\begin{aligned} a_{11}x_1 + \cdots + a_{1n}x_n &= 0 \\ &\vdots \\ a_{m1}x_1 + \cdots + a_{mn}x_n &= 0 \end{aligned}$$

Let W be the space of solutions to (5). By Theorem 2.13 this space has dimension $n-r$. On the other hand let V be the space spanned by the rows of the matrix (4) and assume dim $V = k$. Then by Theorem 2.9 there exists a basis $\{S_1, \ldots, S_k\}$ for V consisting of vectors which are in echelon form. If $S_j = (0, \ldots, 0, a'_{jl_j}, \ldots, a'_{jn})$ where a'_{jl_j} is the first nonzero entry in S_j for $j = 1, \ldots, k$, then we know from the discussion on p. 69 that the system

(6)
$$\begin{aligned} a'_{1l_1} x_{l_1} + \cdots\cdots\cdots + a'_{1n} x_n &= 0 \\ &\vdots \\ a'_{kl_k} x_{l_k} + \cdots + a'_{kn} x_n &= 0 \end{aligned}$$

has exactly the same solutions space as (5). However, the rank of the matrix

$$\begin{pmatrix} a'_{1l_1} & \cdots & & a'_{1n} \\ 0 & & & \\ \vdots & & & \\ 0 \cdots 0 a'_{kl_k} & \cdots & & a'_{kn} \end{pmatrix}$$

must equal k since we have already seen that for echelon matrices the dimension of the space spanned by the rows equals the dimension of the space spanned by the columns (p. 70). Hence dim $W = n - k$. Therefore $n - k = n - r$, or $k = r$.

The rank of a matrix is sometimes called the *column rank* of the matrix since this is the dimension of the space spanned by the columns. If we define the *row rank* of a matrix to be the dimension of the space spanned by the rows, then Theorem 2.14 asserts that the row and column ranks for a matrix are always the same.

EXERCISE

1. Let V be the space spanned by the rows of a matrix

$$\begin{pmatrix} a_{11} & \cdots & a_{1n} \\ \vdots & & \vdots \\ a_{m1} & \cdots & a_{mn} \end{pmatrix}$$

and let W be the space of solutions to the associated homogeneous equations

$$\begin{aligned} a_{11} x_1 + \cdots + a_{1n} x_n &= 0 \\ &\vdots \\ a_{m1} x_1 + \cdots + a_{mn} x_n &= 0 \end{aligned}$$

Show that $V_n(R) = V \oplus W$. (See p. 64 for the definition of $V \oplus W$.)

CHAPTER THREE

MATRICES AND LINEAR TRANSFORMATIONS

3.1 Matrices

Our study of linear equations in the last chapter was greatly facilitated when we focused attention on the coefficient matrix and the augmented matrix of the system. We wish now to investigate properties of matrices in some detail.

Definition. *An $m \times n$ array*

(1)
$$\begin{pmatrix} a_{11} & \cdots & a_{1n} \\ \vdots & & \vdots \\ a_{m1} & \cdots & a_{mn} \end{pmatrix}$$

of real numbers is called a matrix. *The first index m denotes the* number of rows *of the matrix, the second n denotes the* number of columns. *The real numbers a_{ij} are called the* coefficients *or* elements *of the matrix. The totality of all $m \times n$ matrices, will be denoted by $R_{m,n}$. The set $R_{m,n}$ is often called "the set of all $m \times n$ matrices with real coefficients."*

We shall use capital Latin letters to denote matrices, and we shall often use the compressed notation $A = (a_{ij})$ in place of

$$A = \begin{pmatrix} a_{11} & \cdots & a_{1n} \\ \vdots & & \vdots \\ a_{m1} & \cdots & a_{mn} \end{pmatrix}.$$

Two matrices $A = (a_{ij})$ and $B = (b_{ij})$ are equal if and only if $a_{ij} = b_{ij}$ for $i = 1, \ldots, m; j = 1, \ldots, n$.

If $A = (a_{ij})$ and $B = (b_{ij})$ both belong to $R_{m,n}$, then we may define the sum $A + B$ by the formula

$$A + B = (a_{ij} + b_{ij}) = \begin{pmatrix} a_{11} + b_{11} & \cdots & a_{1n} + b_{1n} \\ \vdots & & \vdots \\ a_{m1} + b_{m1} & \cdots & a_{mn} + b_{mn} \end{pmatrix}$$

Scalar multiplication of matrices by real numbers is defined by the formula

$$aA = (aa_{ij}) = \begin{pmatrix} aa_{11}, & \ldots, & aa_{1n} \\ \vdots & & \vdots \\ aa_{m1}, & \ldots, & aa_{mn} \end{pmatrix}.$$

Example 1. In $R_{2,3}$ if $A = \begin{pmatrix} 1 & 0 & -1 \\ 2 & 1 & 3 \end{pmatrix}$ and $B = \begin{pmatrix} 0 & -1 & 3 \\ 1 & 0 & -1 \end{pmatrix}$, then

$$A + B = \begin{pmatrix} 1 & -1 & 2 \\ 3 & 1 & 2 \end{pmatrix} \quad \text{and} \quad 2A = \begin{pmatrix} 2 & 0 & -2 \\ 4 & 2 & 6 \end{pmatrix}.$$

We leave it as an exercise for the student to verify that with this definition of addition and scalar multiplication $R_{m,n}$ is a vector space over the field of real numbers.

To ascertain the dimension of $R_{m,n}$ we introduce unit coordinate matrices I_{jk}. Each coefficient of I_{jk} equals zero except the jkth which is 1. (See Figure 1.)

$$\text{jth row} \begin{pmatrix} 0 & \cdots & 0 & \cdots & 0 \\ \vdots & & \vdots & & \vdots \\ 0 & \cdots & 1 & \cdots & 0 \\ \vdots & & \vdots & & \\ 0 & & \cdots & & 0 \end{pmatrix} = I_{jk}$$

$\uparrow$ kth column

Fig. 1 Unit coordinate matrix I_{jk}.

It is clear that the set $\{I_{jk} : j = 1, \ldots, m, k = 1, \ldots, n\}$ is linearly independent, and if $A = (a_{jk})$, then

(2) $$A = \sum_{j=1}^{m} \sum_{k=1}^{n} a_{jk} I_{jk}.$$

If care is not taken, matrix notation can get cumbersome. We shall make extensive use of the summation symbol Σ and also suppress indices and limits of summation if it is clear from the context what these indices and summation limits are. Thus we would abbreviate (2) by

$$A = \sum a_{jk} I_{jk}.$$

When this is done it is understood that the sum is to be taken over all possible values of the indices.

We now examine further the connection between matrices and linear equations. Let A be a fixed matrix in $R_{m,n}$ and let $X = (x_1, \ldots, x_n)$ be a vector from $V_n(R)$, the space of n-tuples of elements from R. Then the equations

$$a_{11}x_1 + \cdots + a_{1n}x_n = y_1$$
$$\vdots \qquad \qquad \vdots \qquad \qquad \vdots$$
$$a_{m1}x_1 + \cdots + a_{mn}x_n = y_m$$

define a new vector $Y = (y_1, \ldots, y_m)$ in $V_m(R)$. Thus associated with the matrix A is a function with *domain* $V_n(R)$ and *range of values* in $V_m(R)$. We could denote this functional relationship by $Y = A(X)$, but instead it is more customary to drop the parenthesis and write

$$Y = AX.$$

If we wish to use an expanded notation, it is customary, for reasons that will be apparent shortly, to write the vectors X and Y vertically. Thus $Y = AX$

becomes

(3) $$\begin{pmatrix} y_1 \\ \vdots \\ y_m \end{pmatrix} = \begin{pmatrix} a_{11} & \cdots & a_{1n} \\ \vdots & & \vdots \\ a_{m1} & \cdots & a_{mn} \end{pmatrix} \begin{pmatrix} x_1 \\ \vdots \\ x_n \end{pmatrix}.$$

Each entry y_j in the vector Y is defined by the formula

(4) $$y_j = \sum_{k=1}^{n} a_{jk} x_k.$$

One way of remembering this formula is to observe that each row of the $m \times n$ matrix A is a vector from $V_n(R)$, and y_j is just the inner product of the jth row of A with the vector X. Thus

$$\begin{pmatrix} 2 & -1 & 3 \\ -1 & 0 & 1 \end{pmatrix} \begin{pmatrix} 1 \\ -1 \\ 1 \end{pmatrix} = \begin{pmatrix} 6 \\ 0 \end{pmatrix}$$

and

$$\begin{pmatrix} -1 & 1 \\ 0 & 1 \\ 2 & 1 \end{pmatrix} \begin{pmatrix} 2 \\ 3 \end{pmatrix} = \begin{pmatrix} 1 \\ 3 \\ 7 \end{pmatrix}.$$

The function with domain $V_n(R)$ and range in $V_m(R)$ defined by the $m \times n$ matrix A is called a *transformation or mapping* from $V_n(R)$ to $V_m(R)$. We note two important properties possessed by this transformation. First if $X, Z \in V_n(R)$ and $Y = AX$ and $W = AZ$, then for each $j = 1, \ldots, m$

$$y_j = \sum_{k=1}^{n} a_{jk} x_k$$

and

$$w_j = \sum_{k=1}^{n} a_{jk} z_k.$$

where $Y = (y_1, \ldots, y_m)$ and $W = (w_1, \ldots, w_m)$. Hence adding we have

$$y_j + w_j = \sum_{k=1}^{n} a_{jk}(x_k + z_k)$$

or

$$AX + AZ = Y + W = A(X + Z).$$

Thus the transformation induced by A preserves vector addition. Also if $a \in R$ and

$$Z = aX,$$

then letting $W = AZ$ we have

$$w_j = \sum_{k=1}^{n} a_{jk} a x_k$$

$$= a \sum_{k=1}^{n} a_{jk} x_k = a y_j$$

where $Y = AX$. Thus

$$A(aX) = a(AX),$$

and the transformation induced by A preserves scalar multiplication.

MATRICES 79

Definition. *A transformation from a vector space V to a vector space W which satisfies*
$$T(X+Y) = TX + TY$$
and
$$T(aX) = aTX$$
for all $X, Y \in V$ *and* $a \in R$ *is called a* linear transformation *from V to W*.

The content of the preceding discussion can be summarized in the following theorem.

THEOREM 3.1 *Let A be an* $m \times n$ *matrix with real coefficients. The transformation* $Y = AX$ *defined by (3) from* $V_n(R)$ *to* $V_m(R)$ *is a linear transformation from* $V_n(R)$ *to* $V_m(R)$.

This theorem has an extremely important converse which we shall prove later in this chapter. Namely, if V and W are finite-dimensional vector spaces with dimension n, m, respectively, then each linear transformation from V to W can be "represented" by an $m \times n$ matrix. Just what is meant by "represented" will be made clear in due course.

The connection between the linear transformation induced by the matrix A and the problem of finding solutions to the linear equations

(5)
$$\begin{aligned} a_{11}x_1 + \cdots + a_{1n}x_n &= y_1 \\ &\vdots \\ a_{m1}x_1 + \cdots + a_{mn}x_n &= y_m \end{aligned}$$

can be made explicit. Equations (5) will have a solution $X = (x_1, \ldots, x_n)$ if and only if the vector $Y = (y_1, \ldots, y_m)$ lies in the range of the transformation induced by the matrix A.

EXERCISES

Determine $Y = AX$ for the following choices of matrices A and vectors X.

1. $A = \begin{pmatrix} -1 & 0 & 1 \\ 2 & -1 & 1 \end{pmatrix}, X = \begin{pmatrix} 1 \\ 0 \\ -1 \end{pmatrix}$.

2. $A = \begin{pmatrix} 2 & -1 & 1 & 1 \\ 4 & -2 & -1 & 1 \\ 0 & 1 & -1 & 2 \end{pmatrix}, X = \begin{pmatrix} 1 \\ 0 \\ -1 \\ 0 \end{pmatrix}$.

3. $A = \begin{pmatrix} -1 & 0 & 1 & -1 \\ 1 & -1 & 1 & 1 \\ 1 & 1 & -1 & 1 \\ 0 & 1 & 1 & 0 \end{pmatrix}, X = \begin{pmatrix} 1 \\ -1 \\ 1 \\ 2 \end{pmatrix}$.

If $A \in R_{m,n}$ and $X \in V_n(R)$, then $Y = AX \in V_m(R)$. If $B \in R_{p,m}$, then we may compute $Z = BY = BAX$. The vector $Z \in V_p(F)$. Compute BAX for the following choices of B, A, X.

4. $B = \begin{pmatrix} 1 & -1 \\ -1 & 1 \end{pmatrix}, A = \begin{pmatrix} -1 & 1 & 0 \\ 0 & -1 & 1 \end{pmatrix}, X = \begin{pmatrix} 1 \\ -1 \\ 1 \end{pmatrix}$.

5. $B = \begin{pmatrix} 2 & -1 & 1 \\ 0 & 1 & -1 \\ 1 & 1 & 0 \end{pmatrix}, A = \begin{pmatrix} -1 & 0 \\ 1 & 1 \\ 2 & -1 \end{pmatrix}, X = \begin{pmatrix} -1 \\ 1 \end{pmatrix}$.

80   MATRICES AND LINEAR TRANSFORMATIONS

6. If $A = \begin{pmatrix} -1 & 1 \\ 2 & 1 \end{pmatrix}$, $B = \begin{pmatrix} 1 & 0 \\ -1 & 2 \end{pmatrix}$, $X = \begin{pmatrix} 1 \\ 1 \end{pmatrix}$, compute ABX, BAX, and AAX.

7. Show that the space of all $m \times n$ matrices is a vector space over the field of real numbers.

8. If V is the vector space of all polynomials, show that the transformations

$$(Df)(x) = f'(x) \quad \text{(differentiation)}$$

$$(Sf)(x) = \int_0^x f(t)\, dt \quad \text{(integration)}$$

are linear transformations from V into V.

9. If $V = C[0, 1]$, the space of continuous functions on $[0, 1]$, do the transformations D and S of Exercise 8 define linear transformations from V into V?

10. Let T be a linear transformation from a vector space V into a vector space W. Let S be a linear transformation from W to a third vector space U. Then for $X \in V$,

$$Z = S(T(X)) \in U.$$

Verify that this composition of S and T is a linear transformation from V to U.

11. If $A = \begin{pmatrix} 1 & -1 & 1 \\ 1 & 1 & -1 \\ 1 & -1 & 0 \end{pmatrix}$ and $Y = \begin{pmatrix} 3 \\ 3 \\ 2 \end{pmatrix}$, find a vector $X \in V_3(R)$ such that $AX = Y$ if this is possible.

12. If $A = \begin{pmatrix} 1 & -1 & 1 \\ 0 & 1 & -1 \\ 1 & 0 & -1 \\ 2 & -1 & 1 \end{pmatrix}$ and $Y = \begin{pmatrix} 1 \\ 1 \\ 1 \\ 3 \end{pmatrix}$, find a vector $X \in V_3(R)$ such that $AX = Y$ if this is possible.

13. If $A \in R_{m,n}$ where $m < n$ show that there always must exist a nonzero vector $X \in V_n(R)$ such that $AX = 0$.

14. Let a_{jk}, x_j, y_k where $j = 1, \ldots, m$ and $k = 1, \ldots, n$, be real numbers. Let

$$c_j = \sum_{k=1}^{n} a_{jk} y_k,$$

and

$$x = \sum_{j=1}^{m} c_j x_j.$$

Also set

$$b_k = \sum_{j=1}^{m} x_j a_{jk},$$

and

$$y = \sum_{k=1}^{n} b_k y_k.$$

Show that $x = y$. (Which of the axioms of arithmetic justify this result?) We may write this result as

$$\sum_{j=1}^{m} \left(x_j \sum_{k=1}^{n} a_{jk} y_k \right) = \sum_{k=1}^{n} \left(\sum_{j=1}^{m} x_j a_{jk} \right) y_k.$$

Each of these expressions is a double summation, and the exercise states that the order of summation in a double sum may be interchanged without affecting the result. This is a simple but extremely useful fact. Since the order of summation is immaterial we may abbreviate the double sum by

$$\sum_{j,k} x_j a_{jk} y_k.$$

3.2 Matrix Multiplication

If A is a matrix from $R_{m,n}$ and B is a matrix from $R_{p,m}$ then in a natural way we may define the product BA of the two matrices.

Definition. If $A = (a_{lk}) \in R_{m,n}$ and $B = (b_{jl}) \in R_{p,m}$, then the matrix $C = (c_{jk}) \in R_{p,m}$, where $c_{jk}, j = 1, \ldots, p, k = 1, \ldots, n$ is defined by the formula

$$\text{(1)} \qquad c_{jk} = \sum_{l=1}^{m} b_{jl} a_{lk},$$

is called the *product of B and A* and we write $C = BA$.

To remember this product rule note that if $A \in R_{m,n}$ and $B \in R_{p,m}$ then the *columns* of A and the *rows* of B are vectors from $V_m(R)$. The matrix element c_{jk} is then the *inner product of the jth row of B with the kth column of A*. See Figure 2.

$$\text{jth row of } B \to \begin{pmatrix} \cdot & \cdot & \cdot \\ b_{j1}, & \ldots, & b_{jm} \\ \cdot & \cdot & \cdot \end{pmatrix} \begin{pmatrix} \cdot & a_{1k} & \cdot \\ \cdot & \vdots & \cdot \\ \cdot & a_{mk} & \cdot \end{pmatrix}$$

$$kth \text{ column of } A$$

$$(c_{jk}) = \left(\sum_{l=1}^{m} b_{jl} a_{lk} \right)$$

Fig. 2 Matrix product rule.

Example 1. If $A = \begin{pmatrix} 1 & -1 \\ 0 & 2 \end{pmatrix} \in R_{2,2}$ and $B = \begin{pmatrix} 0 & 1 \\ -1 & 2 \\ 1 & -1 \end{pmatrix} \in R_{3,2}$, compute BA.

Using the definition above we obtain

$$BA = \begin{pmatrix} 0 & 1 \\ -1 & 2 \\ 1 & -1 \end{pmatrix} \begin{pmatrix} 1 & -1 \\ 0 & 2 \end{pmatrix} = \begin{pmatrix} 0 & 2 \\ -1 & 5 \\ 1 & -3 \end{pmatrix}$$

It must be emphasized that in order to define the product BA the number of columns of B must equal the number of rows of A. If this is not the case, the product BA is not defined.

Example 2. If $A = \begin{pmatrix} 1 & -1 \\ 0 & 1 \end{pmatrix}$ and $B = \begin{pmatrix} 1 & 0 & 1 \\ -1 & 1 & 0 \end{pmatrix}$, then

$$AB = \begin{pmatrix} 1 & -1 \\ 0 & 1 \end{pmatrix} \begin{pmatrix} 1 & 0 & 1 \\ -1 & 1 & 0 \end{pmatrix} = \begin{pmatrix} 2 & -1 & 1 \\ -1 & 1 & 0 \end{pmatrix}.$$

However the product BA is not defined since the number of columns of B is not equal to the number of rows of A.

Even if AB and BA are both defined, the products may be different. For example if $A = \begin{pmatrix} 1 & 0 \\ 1 & 1 \end{pmatrix}$ and $B = \begin{pmatrix} 1 & 1 \\ 0 & 1 \end{pmatrix}$, then $AB = \begin{pmatrix} 1 & 0 \\ 1 & 1 \end{pmatrix} \begin{pmatrix} 1 & 1 \\ 0 & 1 \end{pmatrix} = \begin{pmatrix} 1 & 1 \\ 1 & 2 \end{pmatrix}$ and $BA = \begin{pmatrix} 1 & 1 \\ 0 & 1 \end{pmatrix} \begin{pmatrix} 1 & 0 \\ 1 & 1 \end{pmatrix} = \begin{pmatrix} 2 & 1 \\ 1 & 1 \end{pmatrix}$.

Thus $AB \neq BA$, and matrix multiplication is noncommutative.

Matrix multiplication is associative, however, as we shall show next. This fact follows from an observation we have alread used; namely, that the order of summation in a double sum is immaterial. This result in turn depends on the associativity of multiplication in the field of real numbers as well as the distributive law.

THEOREM 3.2 Let $A \in R_{p,m}$, $B \in R_{m,n}$ and $C \in R_{n,r}$, then
$$A(BC) = (AB)C.$$

Proof. We verify this by a direct calculation. If $A = (a_{ij})$, $B = (b_{jk})$ and $C = (c_{kl})$, then let
$$BC = (d_{jl})$$
and
$$AB = (e_{ik})$$
where
$$d_{jl} = \sum_{k=1}^{n} b_{jk} c_{kl}$$
and
$$e_{ik} = \sum_{j=1}^{m} a_{ij} b_{jk}.$$

Therefore, if $A(BC) = (f_{il})$ and $(AB)C = (g_{il})$, then for each pair of indices i, l

$$f_{il} = \sum_{j=1}^{m} a_{ij} d_{jl} = \sum_{j=1}^{m} a_{ij} \sum_{k=1}^{n} b_{jk} c_{kl}$$

$$= \sum_{j=1}^{m} \sum_{k=1}^{n} a_{ij}(b_{jk} c_{kl})$$

$$= \sum_{j=1}^{m} \sum_{k=1}^{n} (a_{ij} b_{jk}) c_{kl}$$

$$= \sum_{k=1}^{n} \sum_{j=1}^{m} a_{ij} b_{jk} c_{kl}$$

$$= \sum_{k=1}^{n} e_{ik} c_{kl} = g_{il}.$$

This completes the proof.

In a similar fashion it may be verified that the distributive law holds for matrix multiplication. Namely, if $A, B \in R_{m,n}$ and $C, D \in R_{n,p}$, then
$$A(C+D) = AC + AD$$
and
$$(A+B)C = AC + BC.$$

Also matrix multiplication "associates" with scalar multiplication. That is, if $A \in R_{m,n}$ and $B \in R_{n,p}$ and $a \in R$, then
$$a(AB) = (aA)B = A(aB).$$

The verification of this fact is left to the exercises.

The student should also note that if vectors $X = (x_1, \ldots, x_n) \in V_n(R)$ are written as $n \times 1$ matrices, i.e. $X = \begin{pmatrix} x_1 \\ \vdots \\ x_n \end{pmatrix}$, and $A = (a_{ij})$ is an $m \times n$ matrix, then the transformation $Y = AX$ from $V_n(R)$ to $V_m(R)$ is represented by the product of the matrices A and X. Thus

$$\begin{pmatrix} y_1 \\ \vdots \\ y_m \end{pmatrix} = \begin{pmatrix} a_{11} & \cdots & a_{1n} \\ \vdots & & \vdots \\ a_{m1} & \cdots & a_{mn} \end{pmatrix} \begin{pmatrix} x_1 \\ \vdots \\ x_n \end{pmatrix}$$

which is the notation we have been using. If A, B both belong to $R_{m,n}$ the product is not defined unless $n = m$. Matrices with the same number of rows and columns are called *square matrices*. The number of rows or columns in a square matrix is called the *order* of the matrix. Note that the product of two $n \times n$ matrices is again an $n \times n$ matrix. Hence $R_{n,n}$ is closed under matrix multiplication. In particular if $A \in R_{n,n}$, successive powers $A^2, A^3, \ldots, A^k, \ldots$ of the matrix can be defined. That the symbol $A^k = \underbrace{A \cdots A}_{k}$ is independent of the grouping of the factors follows from the associativity of matrix multiplication.

There is an identity element with respect to matrix multiplication in $R_{n,n}$. Indeed if

$$I = \begin{pmatrix} 1 & 0 & \cdots & 0 \\ 0 & 1 & & \cdot \\ \vdots & & \ddots & \vdots \\ 0 & \cdot & \cdots & 1 \end{pmatrix}$$

and $A \in R_{n,n}$, then

$$IA = AI = A.$$

However, division by matrices is not always possible. That is, for a given $n \times n$ matrix $A \neq 0$ it is not in general possible to find an $n \times n$ matrix B so that

$$AB = I.$$

Before giving an example let us note that if there exists a nonzero matrix $A \in R_{n,n}$ satisfying $A^2 = 0$, then it is impossible to find a matrix $B \in R_{n,n}$ such that

(2) $$AB = I.$$

For if (2) holds, then multiplying both sides by A we have

$$A(AB) = A \cdot I = A.$$

But

$$A(AB) = (AA)B = 0 \cdot B = 0.$$

therefore $A = 0$ which is a contradiction.

Now it is easily checked that if

$$A = \begin{pmatrix} 0 & \cdots & 0 & 1 \\ 0 & \cdots & 0 & 0 \\ \vdots & & \vdots & \vdots \\ 0 & \cdots & 0 & 0 \end{pmatrix},$$

then $A^2 = 0$. Hence there is no matrix $B \in R_{n,n}$ satisfying $AB = I$. More generally the above argument shows that if A is a nonzero matrix in $R_{n,n}$ and there exists a nonzero $B \in R_{n,n}$ satisfying $BA = 0$, then there does not exist a matrix C satisfying $AC = I$.

Whenever there exists a matrix B satisfying

(3) $$AB = BA = I$$

the matrix A is said to be *invertible*. It is not hard to check that if there exists a matrix B satisfying (3), there exists only one. This matrix B is called the inverse of A and we write $B = A^{-1}$. A fundamental problem that we shall consider in some detail is to determine necessary and sufficient conditions on the matrix A for it to be invertible. We summarize the properties of matrix multiplication in $R_{n,n}$ in the following theorem.

THEOREM 3.3 *The space $R_{n,n}$ of $n \times n$ matrices is closed under matrix multiplication, and for each $A, B, C \in R_{n,n}$ and $a \in R$ matrix multiplication satisfies the following properties.*
  (i) $A(BC) = (AB)C$ (*associative law*)
  (ii) $a(AB) = (aA)B = A(aB)$
  (iii) $A(B+C) = AB + AC$ (*distributive laws*)
      $(A+B)C = AC + BC$
  (iv) *There is an identity element I with respect to multiplication, that is*

$$AI = IA = A$$

for each $A \in R_{n,n}$.

We have already remarked that $R_{n,n}$ is a vector space over the field of real numbers. A vector space which is closed under a multiplication operation satisfying (i), (ii) and (iii) of Theorem 3.3 is called an *algebra* over the field of real numbers. Thus Theorem 3.3 states that the spaces of $n \times n$ matrices, $R_{n,n}$ is an algebra over the field of real numbers.

EXERCISES

1. Let $A = \begin{pmatrix} 1 & 2 \\ -1 & 1 \end{pmatrix}$, $B = \begin{pmatrix} 2 & 1 \\ -1 & 2 \\ 1 & 0 \end{pmatrix}$, $C = \begin{pmatrix} 1 & -1 & 1 \\ 0 & 1 & 2 \end{pmatrix}$, $D = \begin{pmatrix} 1 & -1 & 1 \\ 0 & 1 & 2 \\ -1 & 1 & 0 \end{pmatrix}$.

   Compute the following matrix products if they are defined. If the product is undefined specify the reason.
   (a) A^2
   (b) D^2
   (c) B^2
   (d) ACB
   (e) ABC
   (f) CBD
   (g) BCD.

2. If $A = \begin{pmatrix} -1 & 1 \\ 0 & 1 \end{pmatrix}$ and $B = \begin{pmatrix} 1 & 1 \\ 1 & 1 \end{pmatrix}$ compute

   (a) $AB - BA$
   (b) $A^2 - B^2$
   (c) $(A+B)(A-B)$.

3. If $A = \begin{pmatrix} -1 & 0 & 1 \\ 1 & -1 & 2 \\ 1 & 1 & 0 \end{pmatrix}$ and $B = \begin{pmatrix} 2 & -1 & 0 \\ 1 & 0 & -1 \\ 1 & 1 & 1 \end{pmatrix}$ compute

   (a) AB
   (b) BA
   (c) $A^2 - B^2$.

4. Verify that matrix multiplication is distributive with respect to addition of matrices [property (iii) of Theorem 3.3].
5. Verify that matrix multiplication satisfies property (ii) of Theorem 3.3.
6. Let $A \in R_{n,n}$. Show that if A has an inverse, then this inverse is unique.
7. Let $A \in R_{n,n}$. Show that if $AX = 0$ for some nonzero $X \in V_n(R)$ then A is not invertible. (Examine the argument on p. 83.)
8. Let $A \in R_{n,n}$. Show that if there are vectors $X, Y \in V_n(R)$ such that $X \neq Y$ and $AX = AY$, then A is not invertible. (Use Exercise 7.)
9. Let $A \in R_{n,n}$ and assume $A^k = 0$ for some positive integer k. Show that A is not invertible.
10. Let $A_1, \ldots, A_k$, be matrices in $R_{n,n}$. Show that $\{A_1, \ldots, A_k\}$ is linearly dependent if $k > n^2$.
11. Let $A \in R_{n,n}$. Show that there exist scalars $a_0, \ldots, a_{n^2}$ such that

$$a_0 I + a_1 A + a_2 A^2 + \cdots + a_{n^2} A^{n^2} = 0.$$

Thus a matrix $A \in R_{n,n}$ satisfies a polynomial equation of degree $\leq n^2$. (Use Exercise 10.)

12. Define a multiplication in $C[a,b]$, the vector space of continuous functions on the interval $[a,b]$, so that with this multiplication $C[a,b]$ is an algebra over the field of real numbers.

3.3 Linear Transformations

Let V and W be vector spaces and let T be a transformation from V to W. That is, for each $X \in V$, $T(X)$ is a unique vector in W. Then we have noted that T is called *linear* if for each $X, Y \in V$ and $a \in R$

$$T(X+Y) = T(X) + T(Y)$$

and
$$T(aX) = aT(X).$$

Combining these two statements we see that if T is a linear transformation and Z is a linear combination of the vectors X and Y, that is $Z = aX + bY$, then

(1) $$T(Z) = T(aX + bY) = aT(X) + bT(Y).$$

Conversely if the transformation satisfies

$$T(aX + bY) = aT(X) + bT(Y)$$

for each pair of vectors X, Y and scalars a, b, then T is linear. We may extend (1) by induction to arbitrary finite sums. That is, if $X = \Sigma a_i X_i$ and T is linear, then

$$T(X) = T\left(\sum a_i X_i\right) = \sum a_i T(X_i).$$

If $\{X_1, \ldots, X_n\}$ is a basis for V, then each $X \in V$ can be written uniquely as $X = \Sigma a_k X_k$, and applying the transformation T we have $T(X) = \Sigma a_k T(X_k)$. Thus a linear transformation T is completely determined by what it does to the basis vectors $X_1, \ldots, X_n$.

We have already noted that if A is an $m \times n$ matrix and $X \in V_n(R)$, then

$$Y = AX$$

defines a linear transformation from $V_n(R)$ to $V_m(R)$.
Let us list other examples of linear transformations.

Example 1. Let P_n be the space of polynomials of degree n. Then the differentiation operator

$$D(p)(x) = p'(x)$$

is a linear transformation from P_n to P_{n-1}. Clearly the derivative of a polynomial of degree n is a polynomial of degree $n-1$. That differentiation satisfies (1) is a familiar theorem from calculus.

Example 2. The integration operator

$$S(p)(x) = \int_a^x p(t)\, dt$$

is a linear transformation from P_n to P_{n+1}. We leave the verification of this as an exercise.

Example 3. In any vector space the identity transformation,

$$I(X) = X$$

and the zero transformation

$$T(X) = 0 \quad \text{for each } X$$

are linear transformations.

Example 4. Scalar multiplication is a linear transformation in a vector space. However, translation by a fixed vector Y is not linear unless $Y = 0$. We leave it as an exercise to verify this fact, i.e.,

$$T(X) = X + Y$$

is not linear unless $Y = 0$.

Example 5. In $V_2(R)$ the transformation T which rotates each vector through an angle θ is a linear transformation. (See Figure 3.)
We leave it as an exercise to fashion a geometric argument to prove that a rotation is a linear transformation.

Example 6. In $V_2(R)$ if l is a line through the origin, then the reflection of each vector through this line defines a linear transformation T in $V_2(R)$. See Figure 4. The reflected vector $T(X)$ has the same length as X and the angles that X and $T(X)$ make with the line are equal.

Example 7. Suppose the finite-dimensional vector space V is the direct sum of subspaces U and W. Thus $U + W = V$ and $U \cap W = \{0\}$. Then each vector $X \in V$ can be written uniquely as a sum $X = Y + Z$ where $Y \in U$ and $Z \in W$. (Exercise

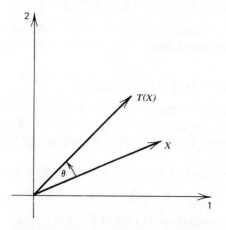

Fig. 3 Rotation by an angle θ.

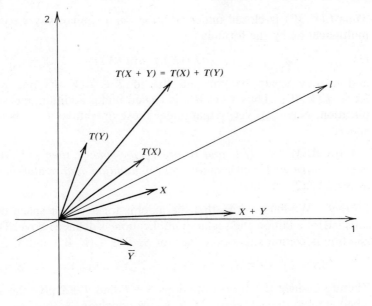

Fig. 4 Reflection through the line l.

20, p. 65.) If we write $Y = P(X)$, then it is an easy exercise to verify that P is a linear transformation from V to U. This transformation P is called the *projection of V onto U along the subspace W*.

Example 8. Let Y be a fixed vector in $V_3(R)$. Then for each $X \in V_3(R)$

$$T(X) = Y \times X$$

defines a linear transformation from $V_3(R)$ to itself. This may be easily verified by checking the axioms for the vector product on p. 37.

Next we investigate operations on linear transformations. If V and W are vector spaces, we shall denote the set of all linear transformations from V to W by $L(V, W)$. If it happens that $V = W$ then we abbreviate $L(V, V)$ by $L(V)$. Then an element $T \in L(V)$ is called a linear transformation in the vector space V.

If $S, T \in L(V, W)$, then we may define the *sum* of the transformations by the formula

(2) $$(S + T)(X) = S(X) + T(X).$$

$S + T$ is clearly a transformation from V to W. We check that $S + T$ is also *linear*. Indeed if $X, Y \in V$, then

$$\begin{aligned}
(S+T)(X+Y) &= S(X+Y) + T(X+Y) & \text{(definition of sums)} \\
&= S(X) + S(Y) + T(X) + T(Y) & \text{(since S and T are linear)} \\
&= S(X) + T(X) + S(Y) + T(Y) & \text{(Why?)} \\
&= (S+T)(X) + (S+T)(Y) & \text{(Why?)}
\end{aligned}$$

Therefore $S + T$ preserves addition of vectors. To show $S + T$ preserves scalar multiplication note that

$$\begin{aligned}
(S+T)(aX) &= S(aX) + T(aX) & \text{(by definition)} \\
&= aS(X) + aT(X) & \text{(Why?)} \\
&= a(S(X) + T(X)) & \text{(Why?)} \\
&= a(S+T)(X) & \text{(Why?)}
\end{aligned}$$

Thus $L(V, W)$ is closed under addition. In a similar way we define scalar multiplication by the formula

(3) $$(aT)(X) = a(T(X)),$$

and it may easily be checked that if $T \in L(V, W)$ and $a \in R$ then $aT \in L(V, W)$. Thus $L(V, W)$ is closed under addition and scalar multiplication. Next, we verify that under these operations $L(V, W)$ is a vector space.

THEOREM 3.4 *If V and W are vector spaces, then $L(V, W)$ is also a vector space under the operations of addition and scalar multiplication defined by (2) and (3).*

Proof. We must verify that the axioms for a vector space on p. 52 are satisfied for addition and scalar multiplication as defined above. To show that addition is commutative note that for $S, T \in L(V, W)$ and $X \in V$,

$$(S+T)(X) = S(X) + T(X) = T(X) + S(X) = (T+S)(X).$$

Therefore since the transformations $S+T$ and $T+S$ give the same result when applied to each vector $X \in V$, we conclude

$$S + T = T + S.$$

In a similar way we may check that addition is associative. There is a "zero transformation." Namely,

$$0(X) = 0$$

for all $X \in V$. Also each transformation T has an additive inverse $-T$ defined by the formula $(-T)(X) = -T(X)$. The remaining axioms for a vector space may be easily checked. We leave this verification as an exercise.

If T is a transformation from V to W, then the *domain* of T is the entire space V and the *range* of T is the set of all vectors Y such that

$$TX = Y$$

for some $X \in V$. Whenever T is linear, the range of T is a *subspace of W*, since if $TX_1 = Y_1$ and $TX_2 = Y_2$, then

$$T(a_1 X_1 + a_2 X_2) = a_1 Y_1 + a_2 Y_2.$$

Whenever the range of the transformation is all of the space W, we say T *maps V onto W*.

If T is a transformation from the vector space V to a vector space W and if S is a transformation from W to a third space U, then we may form the *composition* or *product P* of the transformations T and S by the formula

(4) $$P(X) = S(T(X)).$$

The transformation P takes the space V into U, and if both T and S are linear then it may be easily verified that P is also linear. We write this product $P = ST$. If the spaces V, W, and U are all the same, then (4) defines a multiplication under which $L(V)$ is closed and it may be easily verified that the axioms (i), (ii), and (iii) on p. 84 are satisfied. Hence $L(V)$ is an algebra over the real number field. We state this result as a theorem.

THEOREM 3.5 *If V is a real vector space, then $L(V)$ equipped with the multiplication defined by (4) is an algebra over the field of real numbers.*

LINEAR TRANSFORMATIONS

A transformation T is said to be *one-to-one* (written 1:1) if whenever $X \neq Y$, it follows that $T(X) \neq T(Y)$. Equivalently if $T(X) = T(Y)$, then it must follow that $X = Y$. When T is linear, this latter property may be given a very simple characterization. In this case $T(X) = T(Y)$ implies $0 = T(X) - T(Y) = T(X - Y)$. Hence T will be 1:1, if and only if, whenever $T(X) = 0$, it follows that $X = 0$. For a linear transformation defined on the vector space V the set of all vectors X such that $TX = 0$ is again a subspace, called the *null space* of the transformation. Thus for a linear transformation T from V to W the range of T is a subspace of W, and the null space of T is a subspace of V. Whenever the dimension of V is finite we have a very important formula connecting the dimensions of the range and null space of T.

THEOREM 3.6 *Let $T \in L(V, W)$ and assume $\dim V < \infty$. If U is the range of T and N is the null space of T, then*

(5) $$\dim V = \dim U + \dim N.$$

Proof. Let $n = \dim V$ and $m = \dim N$. If $X_1, \ldots, X_m$ is a basis for N, then we know by Corollary 2 of Theorem 2.4 that we may extend $\{X_1, \ldots, X_m\}$ to be a basis for V. Thus we may adjoin vectors $X_{m+1}, \ldots, X_n$ to the set $\{X_1, \ldots, X_m\}$ so that $\{X_1, \ldots, X_m, X_{m+1}, \ldots, X_n\}$ is linearly independent and spans V. To prove the theorem it suffices to show that $T(X_{m+1}), \ldots, T(X_n)$ is a basis for the range U. First we show that these vectors span U. If $Y \in U$, then $Y = T(X)$ where $X \in V$. But since $\{X_1, \ldots, X_n\}$ is a basis for V,

$$X = \sum_{i=1}^{n} a_i X_i.$$

Therefore

$$Y = T(X) = T\left(\sum_{i=1}^{n} a_i X_i\right)$$

$$= \sum_{i=1}^{n} a_i T(X_i) \qquad \text{since } T \text{ is linear}$$

$$= \sum_{i=m+1}^{n} a_i T(X_i)$$

since each of the vectors $X_1, \ldots, X_m$ belongs to the null space of T, and hence $T(X_1) = T(X_2) = \cdots = T(X_m) = 0$. Next we verify that the vectors $T(X_{m+1}), \ldots, T(X_n)$ are linearly independent. If $\sum_{i=m+1}^{n} a_i T(X_i) = 0$, then applying the linearity of T we have

$$T\left(\sum_{i=m+1}^{n} a_i X_i\right) = 0.$$

Thus $\sum_{m+1}^{n} a_i X$ belongs to the null space of T. Consequently

(6) $$\sum_{i=m+1}^{n} a_i X_i = \sum_{i=1}^{m} a_i X_i$$

for suitable scalars $a_1, \ldots, a_m$ since $\{X_1, \ldots, X_m\}$ is a basic for the null space of T. However, $\{X_1, \ldots, X_m, \ldots, X_n\}$ is linearly independent. Hence all of the scalars a_i in (6) must be identically zero. Thus $T(X_{m+1}), \ldots, T(X_n)$ are linearly independent and the proof is complete.

If A is a matrix in $R_{m,n}$, then we have seen that A defines a linear transformation from $V_n(R)$ to $V_m(R)$ by the formula $Y = AX$. The null space of this transformation is identical with the solution space of the equations

$$a_{11}x_1 + \cdots + a_{1n}x_n = 0$$
$$\vdots \qquad \qquad \vdots$$
$$a_{m1}x_1 + \cdots + a_{mn}x_n = 0$$

COROLLARY 1. *Let $A \in R_{m,n}$, then the rank of the matrix A equals the dimension of the range of the transformation from $V_n(R)$ to $V_m(R)$ defined by the matrix A.*

Proof. We know from the corollary to Theorem 2.13 that if $r =$ rank of the matrix A, then

(7) $$\dim \{X \in V_n(R) : AX = 0\} = n - r.$$

Therefore by the above theorem, rank $A =$ dim range of A.

COROLLARY 2. *If $\dim V$ is finite and there exists a one-to-one linear transformation T of the vector space V onto the vector space W, then*

$$\dim V = \dim W.$$

Proof. This follows from formula (7) since W is the range of T and the null space of T consists of the zero subspace.

This result has an important converse.

THEOREM 3.7 *If V and W are finite-dimensional vector spaces and $\dim V = \dim W$, then there exists a one-to-one linear transformation T mapping V onto W.*

Proof. Let $X_1, \ldots, X_n$ be a basis for V and $Y_1, \ldots, Y_n$ be a basis for W. For each X_i define the transformation T by the formula $T(X_i) = Y_i, i = 1, \ldots, n$. For arbitrary $X \in V$, X can be written *uniquely* as a linear combination of the vectors $X_1, \ldots, X_n$. If we write $X = \sum_{i=1}^{n} a_i X_i$, then we define

(8) $$T(X) = \sum_{i=1}^{n} a_i Y_i.$$

It now may be easily checked that the transformation T defined by (8) is linear, one-to-one, and maps V onto W.

A one-to-one linear transformation of one vector space V onto another W, is called a *linear isomorphism* and we say that V and W are *linearly isomorphic*. This means that as far as the linear structure of the spaces is concerned the two spaces are indistinguishable. Thus any linear phenomenon in one space can be reproduced in the other by means of the isomorphism T. The two previous results may be summarized in the following.

THEOREM 3.8 *Two finite-dimensional vector spaces are linearly isomorphic if and only if they have the same dimension.*

We leave it as an exercise to verify that if V has dimension n and $\{X_1, \ldots, X_n\}$ is a basis for V then the transformation $T(X) = (a_1, \ldots, a_n)$ where $X \in V$ and $X = \sum_{i=1}^{n} a_i X_i$ defines an isomorphism of V onto $V_n(R)$.

EXERCISES

In the following exercises P_n denotes the space of polynomials of degree $\leq n$. Primes, such as f', f'', denote first and second derivatives, etc.

1. Verify that $T(f) = f''$ is a linear transformation in P_n.
2. Verify that $T(f) = f'' + f$ defines a linear transformation in P_n.
3. For what values of the real number c is the transformation $T(f) = f' + c$ a linear transformation in P_n.
4. Verify that $T(f)(x) = (xf)'$ defines a linear transformation in P_n.
5. Verify that the integration operator $S(p)(x) = \int_a^x p(t)\,dt$ is a linear transformation from P_n to P_{n+1}.
6. Give a geometric argument to show that a rotation is a linear transformation in $V_2(R)$.
7. Give a geometric argument to show that a reflection is a linear transformation in $V_2(R)$.
8. If the vector space V is a direct sum of the subspaces U and W, show that the projection of V onto U along W is a linear transformation. (See example 7, p. 86)
9. If $Y \in V_3(R)$, verify that $T(X) = Y \times X$ defines a linear transformation in $V_3(R)$.
10. Verify that $T(f)(x) = \int_2^x tf(t)\,dt$, defines a linear transformation from P to P_{n+2}.
11. Determine the range and null space of the transformation $T(f) = f''$ in the space P_n. Is T one-to-one? Onto?
12. Determine the dimension of the null space and range of the transformation $T(f) = \int_1^x f(t)\,dt$ as a transformation from P_n to P_{n+1}. Is T one-to-one? Onto?
13. Determine the dimension of the range and null space of the transformation $T(f)(x) = (xf)'$ as a transformation in P_n Is T one-to-one? Onto?
14. Are reflections and rotations one-to-one transformations in $V_2(R)$? Do they map onto?
15. The matrix $\begin{pmatrix} 2 & 1 & 3 \\ 1 & 0 & 1 \\ 3 & 1 & 4 \end{pmatrix}$ defines a linear transformation in $V_3(R)$. What is the dimension of the range and the dimension of the null space of this transformation?
16. The matrix
$$A = \begin{pmatrix} 1 & 1 & 0 & -1 \\ 1 & 0 & 1 & -1 \\ 1 & 1 & -1 & 1 \\ 0 & 2 & 0 & 1 \end{pmatrix}$$
defines a linear transformation in $V_4(R)$ What is the dimension of the range and the dimension of the null space of A.
17. Answer Exercise 16 if
$$A = \begin{pmatrix} 1 & 2 & -1 & 3 \\ 2 & -1 & 4 & 1 \\ 3 & 0 & 9 & -1 \\ 0 & 3 & -6 & 5 \end{pmatrix}$$
18. Determine the range and the null space in $V_3(R)$ of the transformation $T(X) = I_3 \times X$ where $I_3 = (0, 0, 1)$.
19. Characterize geometrically the range and null space in $V_3(R)$ of the transformation $T(X) = Y \times X$ where Y is a fixed nonzero vector in $V_3(R)$.
20. Is $T(f) = f'$ a linear transformation in the space $C[0, 1]$ of all continuous functions on the unit interval? Explain.

92   MATRICES AND LINEAR TRANSFORMATIONS

21. Complete the proof of Theorem 3.4.
22. If V is an n-dimensional vector space, show that V is linearly isomorphic with $V_n(R)$. (If $\{X_1, \ldots, X_n\}$ is a basis for V and $X = \sum_{i=1}^n a_i X_i$, show that the transformation $T(X) = (a_1, \ldots, a_n)$ is a one-to-one linear transformation of V onto $V_n(R)$.
23. Verify that the product of two linear transformations is again a linear transformation.
24. Verify that the space $L(V)$ is an algebra over the field of real numbers.

3.4 The Matrix of a Linear Transformation

Let V and W be finite-dimensional vector spaces and let T be a linear transformation from V to W. If $\{X_1, \ldots, X_n\}$ is a basis for V and $\{Y_1, \ldots, Y_m\}$ is a basis for W, then we wish to associate an $m \times n$ matrix A_T with the transformation T. To see how this should be done we proceed as follows. We know by Theorem 2.8 that each $X \in V$ can be written uniquely as

(1) $$X = \sum_{i=1}^n a_i X_i$$

and each $Y \in W$ can be written uniquely as

(2) $$Y = \sum_{i=1}^m b_i Y_i.$$

The question we wish to answer is the following. Does there exist a matrix $A_T = (a_{ij}) \in R_{m,n}$ such that whenever $T(X) = Y$ it follows that

(3) $$\begin{pmatrix} b_1 \\ \vdots \\ b_m \end{pmatrix} = \begin{pmatrix} a_{11} & \cdots & a_{1n} \\ \vdots & & \vdots \\ a_{m1} & \cdots & a_{mn} \end{pmatrix} \begin{pmatrix} a_1 \\ \vdots \\ a_n \end{pmatrix}$$

where $(a_1, \ldots, a_n)$ and $(b_1, \ldots, b_m)$ are the coefficients appearing in (1) and (2), respectively. If such a matrix A_T exists, it clearly depends on the choice of the bases $\{X_1, \ldots, X_n\}$ and $\{Y_1, \ldots, Y_m\}$. Now for each j, $j = 1, \ldots, n$, the vector $T(X_j)$ is a linear combination of the vectors $Y_1, \ldots, Y_m$). Hence we may write

(4) $$T(X_j) = \sum_{i=1}^m a_{ij} Y_i.$$

The matrix A_T must be an $m \times n$ matrix, so one possibility is to define

$$A_T = (a_{ij})$$

where the coefficients a_{ij} are defined by (4). We now must verify that if $T(X) = Y$ and X and Y satisfy (1) and (2), respectively, then (3) holds. Equation (3) just says that for each i

$$b_i = \sum_{j=1}^m a_{ij} a_j.$$

However, if $T(X) = Y$ and $X = \sum_{j=1}^n a_j X_j$ and $Y = \sum_{i=1}^m b_i Y_i$, then using the linearity of T we have

$$T(X) = T\left(\sum_{j=1}^n a_j X_j\right)$$

$$= \sum_{j=1}^n a_j T(X_j)$$

$$= \sum_{j=1}^{n} a_j \sum_{i=1}^{m} a_{ij} Y_i$$
$$= \sum_{i=1}^{m} \sum_{j=1}^{n} a_{ij} a_j Y_i$$
$$= \sum_{i=1}^{m} b_i Y_i.$$

But since $T(X)$ has a unique representation as a linear combination of the vectors Y_i, we must have

$$b_i = \sum_{j=1}^{n} a_{ij} a_j \qquad i = 1, \ldots, m.$$

which verifies (3).

Conversely suppose $A \in R_{m,n}$. Let $(b_1, \ldots, b_m) \in V_m(R)$ and $(a_1, \ldots, a_n) \in V_n(R)$ be defined by (3). Then it can be easily checked that the transformation T taking $X = \sum_{j=1}^{n} a_j X_j$ into $Y = \sum_{i=1}^{m} b_i Y_i$ is a linear transformation from V into W. The matrix A_T defined by (3) is called the *matrix of the transformation T with respect to the bases* $\{X_1, \ldots, X_n\}$ and $\{Y_1, \ldots, Y_m\}$.

For example if V is a three-dimensional space with a basis $\{X_1, X_2, X_3\}$ and W is a two-dimensional space with basis $\{Y_1, Y_2\}$ and the linear transformation T satisfies

$$T(X_1) = 2Y_1 - Y_2$$
$$T(X_2) = 3Y_1 + 4Y_2$$
$$T(X_3) = Y_2,$$

then the matrix of the transformation T is the matrix

$$A_T = \begin{pmatrix} 2 & 3 & 0 \\ -1 & 4 & 1 \end{pmatrix}.$$

The matrix of the transformation T must be a 2×3 matrix so A_T is *not* the matrix

$$\begin{pmatrix} 2 & -1 \\ 3 & 4 \\ 0 & 1 \end{pmatrix}$$

which might be expected at first glance.

To best remember the formula for the matrix A_T one need only recall that the *j*th *column* of A_T is the coefficient vector $(a_{1j}, \ldots, a_{mj})$ where

$$T(X_j) = \sum_{i=1}^{m} a_{ij} Y_i.$$

In the following examples note that a linear transformation is completely defined by its action on the basis vectors of a vector space.

Example 1. Define a linear transformation T from $V_2(R)$ to $V_3(R)$ by the formulas
$$T(I_1) = (1, 1, 1)$$
$$T(I_2) = (1, -1, 1)$$

where $I_1 = (1, 0)$ and $I_2 = (0, 1)$. What is the matrix of T with respect to the basis $\{I_1, I_2\}$ in $V_2(R)$ and $J_1 = (1, 0, 0)$, $J_2 = (0, 1, 0)$, $J_3 = (0, 0, 1)$ in $V_3(R)$? Clearly

$$T(I_1) = J_1 + J_2 + J_3$$

and

$$T(I_2) = J_1 - J_2 + J_3.$$

Therefore the matrix $A_T = \begin{pmatrix} 1 & 1 \\ 1 & -1 \\ 1 & 1 \end{pmatrix}$. Furthermore if $X = x_1 I_1 + x_2 I_2$ and then
$$T(X) = y_1 J_1 + y_2 J_2 + y_3 J_3$$
$$\begin{pmatrix} y_1 \\ y_2 \\ y_3 \end{pmatrix} = \begin{pmatrix} 1 & 1 \\ 1 & -1 \\ 1 & 1 \end{pmatrix} \begin{pmatrix} x_1 \\ x_2 \end{pmatrix}.$$

Now $K_1 = (1, 1)$, $K_2 = (1, -1)$ also form a basis for $V_2(R)$. What is the matrix of T with respect to the basis $\{K_1, K_2\}$ in $V_2(R)$ and $\{J_1, J_2, J_3\}$ in $V_3(R)$? Now we must write $T(K_1)$ and $T(K_2)$ as linear combinations of the basis vectors $\{J_1, J_2, J_3\}$ in $V_3(R)$. Since $K_1 = I_1 + I_2$ and $K_2 = I_1 - I_2$ we have

$$\begin{aligned} T(K_1) &= T(I_1 + I_2) = T(I_1) + T(I_2) \\ &= (1, 1, 1) + (1, -1, 1) = (2, 0, 2) \\ &= 2J_1 + 2J_3. \end{aligned}$$

Also
$$\begin{aligned} T(K_2) &= T(I_1 - I_2) = T(I_1) - T(I_2) \\ &= (1, 1, 1) - (1, -1, 1) = (0, 2, 0) \\ &= 2J_2. \end{aligned}$$

Therefore the matrix A_T with respect to the bases $\{K_1, K_2\}$ in $V_2(R)$ and (J_1, J_2, J_3) in $V_3(R)$ is $A_T = \begin{pmatrix} 2 & 0 \\ 0 & 2 \\ 2 & 0 \end{pmatrix}$. Furthermore if the vector $X \in V_2(R)$ is written $X = z_1 K_1 + z_2 K_2$ and we write $T(X) = y_1 J_1 + y_2 J_2 + y_3 J_3$ then

$$\begin{pmatrix} y_1 \\ y_2 \\ y_3 \end{pmatrix} = \begin{pmatrix} 2 & 0 \\ 0 & 2 \\ 2 & 0 \end{pmatrix} \begin{pmatrix} z_1 \\ z_2 \end{pmatrix}.$$

The vectors $L_1 = (1, 1, 1)$, $L_2 = (0, 1, 1)$ and $L_3 = (0, 0, 1)$ also form a basis for $V_3(R)$. To compute the matrix of T with respect to the basis $\{K_1, K_2\}$ in $V_2(R)$ and $\{L_1, L_2, L_3\}$ in $V_3(R)$ we must compute the scalars a_{jk} for which

$$T(K_k) = \sum_{j=1}^{3} a_{jk} L_j.$$

But
$$T(K_1) = T(I_1 + I_2) = (2, 0, 2) = 2L_1 - 2L_2 + 2L_3$$
and
$$T(K_2) = T(I_1 - I_2) = (0, 2, 0) = 0L_1 + 2L_2 - 2L_3.$$

Therefore the matrix of T with respect to these bases is

$$\begin{pmatrix} 2 & 0 \\ -2 & 2 \\ 2 & -2 \end{pmatrix}.$$

As in the previous cases if $X = z_1 K_1 + z_2 K_2$ and we write $T(X) = w_1 L_1 + w_2 L_2 + w_3 L_3$ then

$$\begin{pmatrix} w_1 \\ w_2 \\ w_3 \end{pmatrix} = \begin{pmatrix} 2 & 0 \\ -2 & 2 \\ 2 & -2 \end{pmatrix} \begin{pmatrix} z_1 \\ z_2 \end{pmatrix}.$$

If a linear transformation T maps V into V and one basis $\{X_1, \ldots, X_n\}$ is used for V then the matrix A_T is called the *matrix of the transformation with respect to the basis* $\{X_1, \ldots, X_n\}$.

THE MATRIX OF A LINEAR TRANSFORMATION

Example 2. The vectors $X_1 = (1, 1, 1)$, $X_2 = (0, -1, 2)$, $X_3 = (0, 0, 3)$ are in echelon form, hence are linearly independent. If $T(X_1) = (1, 0, 1)$, $T(X_2) = (1, 1, 0)$ and $T(X_3) = (0, 1, 1)$, this defines a linear transformation from $V_3(R)$ to $V_3(R)$. What is the matrix of this transformation with respect to the canonical basis $I_1 = (1, 0, 0)$, $I_2 = (0, 1, 0)$, $I_3 = (0, 0, 1)$? We must first write I_1, I_2, I_3 as linear combinations of the vectors X_1, X_2, X_3 to determine $T(I_1)$, $T(I_2)$, $T(I_3)$. Clearly

$$I_1 = X_1 + X_2 - X_3$$
$$I_2 = -X_2 + \frac{2}{3}X_3$$
$$I_3 = \frac{1}{3}X_3.$$

Therefore

$$T(I_1) = T(X_1 + X_2 - X_3) = (1, 0, 1) + (1, 1, 0) - (0, 1, 1) = (2, 0, 0) = 2I_1$$

$$T(I_2) = T\left(-X_2 + \frac{2}{3}X_3\right) = -(1, 1, 0) + \frac{2}{3}(0, 1, 1) = \left(-1, -\frac{1}{3}, \frac{2}{3}\right)$$

$$= -I_1 - \frac{1}{3}I_2 + \frac{2}{3}I_3$$

and

$$T(I_3) = T\left(\frac{1}{3}X_3\right) = \frac{1}{3}(0, 1, 1) = \frac{1}{3}I_2 + \frac{1}{3}I_3.$$

Now the matrix $A_T = (a_{ij})$ of T with respect to the basis I_1, I_2, I_3 is defined by the formula

$$T(I_j) = \sum_{i=1}^{3} a_{ij}I_i.$$

Therefore from the above three equations

$$A_T = \begin{pmatrix} 2 & -1 & 0 \\ 0 & -1/3 & 1/3 \\ 0 & 2/3 & 1/3 \end{pmatrix}.$$

Example 3. Let P_n be the vector space of polynomials of degree $\leq n$. What is the matrix of the differentiation operator $D(p) = p'$ with respect to the basis $1, x, x^2, \ldots, x^n$? Now

$$D(x^k) = kx^{k-1} \qquad k = 1, 2, \ldots, n.$$

Therefore the matrix for D is

$$\begin{pmatrix} 0 & 1 & 0 & 0 & \cdots & 0 \\ 0 & 0 & 2 & 0 & & 0 \\ . & . & & 0 & 3 & \vdots \\ . & . & & & . & 0 \\ . & . & & & . & n \\ 0 & 0 & 0 & 0 & & 0 \end{pmatrix}.$$

Hence if $p(x) = a_0 + a_1x + \cdots + a_nx^n$ and $D(p)(x) = p'(x) = b_0 + b_1x + \cdots + b_nx^n$, then

$$\begin{pmatrix} b_0 \\ b_1 \\ b_2 \\ \vdots \\ \vdots \\ b_n \end{pmatrix} = \begin{pmatrix} 0 & 1 & 0 & . & . & . & . & 0 \\ 0 & 0 & 2 & 0 & . & . & . & 0 \\ 0 & 0 & 0 & 3 & 0 & . & . & 0 \\ \vdots & & & & & . & . & \\ & 0 & . & . & . & . & 0 & n \\ 0 & 0 & . & . & & & & 0 \end{pmatrix} \begin{pmatrix} a_0 \\ a_1 \\ a_2 \\ \vdots \\ \vdots \\ a_n \end{pmatrix}$$

Example 4. The integration operator

$$S(p) = \int_0^x p(t)\, dt,$$

maps P_n into P_{n+1}. If $p_k(x) = x^k$, then

$$S(p_k)(x) = \int_0^x p_k(t)\, dt = \frac{1}{k+1} x^{k+1} = \frac{1}{k+1} p_{k+1}(x).$$

Therefore the matrix for S with respect to the bases consisting of powers of x in both P_n and P_{n+1} is

$$n+2 \text{ rows } \left\{ \begin{pmatrix} 0 & 0 & \cdots & 0 \\ 1 & 0 & & 0 \\ 0 & 1/2 & & \vdots \\ \cdot & 0 & & \\ \cdot & \cdot & & 0 \\ 0 & 0 & & 1/(n+1) \end{pmatrix} \right.$$

$$\underbrace{}_{n+1 \text{ columns}}$$

Example 5. We saw in the previous section that a rotation T through an angle θ defines a linear transformation of $V_2(R)$ onto itself. The matrix of T with respect to the canonical basis is easy to determine. Assuming that the rotation is counterclockwise, then $T(I_1) = (\cos\theta, \sin\theta)$ and $T(I_2) = (-\sin\theta, \cos\theta)$ as is evident from Figure 5.

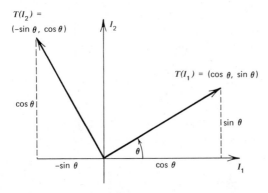

Fig. 5 Rotation through an angle θ.

The matrix A_T of the rotation is then given by

$$A_T = \begin{pmatrix} \cos\theta & -\sin\theta \\ \sin\theta & \cos\theta \end{pmatrix}.$$

A counterclockwise rotation in $V_2(R)$ is usually called a *positive* rotation. A clockwise rotation is a *negative* rotation. Thus the matrix of a positive rotation of 30° with respect to the canonical basis in $V_2(R)$ is

$$\begin{pmatrix} \sqrt{3}/2 & -1/2 \\ 1/2 & \sqrt{3}/2 \end{pmatrix}$$

Example 6. Another linear transformation in $V_2(R)$ is reflection through a line l passing through the origin. (See Figure 6.) To compute the matrix of a reflection with respect to the basis I_1, I_2 we need only compute $T(I_1)$ and $T(I_2)$. If the line l is the first coordinate axis, then $T(I_1) = I_1$ and $T(I_2) = -I_2$. Therefore the matrix is

THE MATRIX OF A LINEAR TRANSFORMATION   97

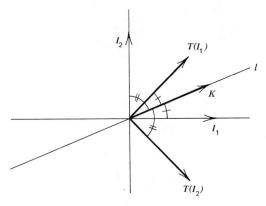

Fig. 6 Reflection through a line *l*.

$\begin{pmatrix} 1 & 0 \\ 0 & -1 \end{pmatrix}$. If the line l is the line $x_2 = x_1$, then $T(I_1) = I_2$ and $T(I_2) = I_1$. Hence the matrix is $\begin{pmatrix} 0 & 1 \\ 1 & 0 \end{pmatrix}$. In the general case if the line l has equation $a_1 x_1 + a_2 x_2 = 0$, then to determine $T(I_1)$ and $T(I_2)$ we use two geometric facts about reflections. First, $T(I_1)$ and $T(I_2)$ both have unit length since a reflection preserves the lengths of vectors. Second, if $T(X)$ is obtained from X by reflecting through the line l, then X and $T(X)$ make equal angles with the line l. If we notice that $K = (-a_2, a_1)$ is a vector parallel to l, then X and $T(X)$ must make equal angles with K. Phrasing this in terms of inner products we have

$$(X, K) = (T(X), K).$$

Using these two facts it is a simple calculation to write $T(I_1)$ and $T(I_2)$ as a linear combination of I_1 and I_2. The resulting matrix A for the reflection T is given by

$$A = \frac{1}{a_1^2 + a_2^2} \begin{pmatrix} a_2^2 - a_1^2 & -2a_1 a_2 \\ -2a_1 a_2 & -(a_2^2 - a_1^2) \end{pmatrix}$$

The verification of this formula we leave as an exercise (Exercise 26).

Example 7. Let l be a line through the origin in $V_3(R)$. Then a rotation T through an angle θ about the line l is a linear transformation of $V_3(R)$ onto itself. (See Figure 7.)

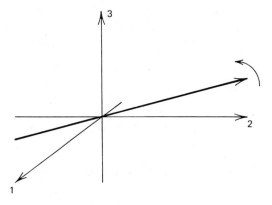

Fig. 7 Rotation in $V_3(R)$.

Again to determine the matrix of the rotation T with respect to the canonical basis we need only consider the effect of T on the basis vectors I_1, I_2, I_3. For example, if T is a rotation of 90° about I_3 in the counterclockwise sense as in Figure 8,

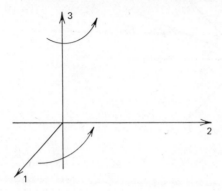

Fig. 8 Positive rotation about I_3.

we have $T(I_1) = I_2$, $T(I_2) = -I_1$, and $T(I_3) = I_3$. Therefore the matrix of T with respect to the basis I_1, I_2, I_3 is $\begin{pmatrix} 0 & -1 & 0 \\ 1 & 0 & 0 \\ 0 & 0 & 1 \end{pmatrix}$.

When representing the vectors of $V_3(R)$ as directed line segments we have agreed to represent I_1, I_2, I_3 as a right handed coordinate system. (See Figure 9.)

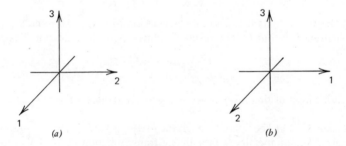

Fig. 9 (a) Right-handed coordinate system. (b) Left-handed coordinated system.

Therefore a positive rotation about a directed line l will be taken to be a right-handed rotation. That is, if the thumb of the right hand is directed positively along the line l, then the positive direction of rotation is toward the ends of the fingers of the clasped right hand. (See Figure 10.)

Fig. 10 (a) Positive rotation in right-handed system. (b) Positive rotation in a left-handed system.

To establish a positive direction along a line it is convenient to specify a vector X parallel to the line and in the appropriate direction.

THE MATRIX OF A LINEAR TRANSFORMATION

Example 8. Determine the matrix of the positive 120° rotation of $V_3(R)$ about the vector $X = (1, 1, 1)$. Thus we are rotating about the line $x_1 = x_2 = x_3$. The positive direction has been determined by the vector $X = (1, 1, 1)$. If T denotes the rotation, then $T(I_1) = I_2$, $T(I_2) = I_3$ and $T(I_3) = I_1$. Therefore the matrix is

$$\begin{pmatrix} 0 & 0 & 1 \\ 1 & 0 & 0 \\ 0 & 1 & 0 \end{pmatrix}.$$

EXERCISES

In $V_n(R)$ the basis consisting of $I_1 = (1, 0, \ldots, 0), \ldots, I_n = (0, \ldots, 0, 1)$ will be called the *canonical basis*.

1. Determine the matrix of the linear transformation T from $V_3(R)$ to $V_2(R)$ with respect to the canonical bases in both spaces if T is defined by $T(1, 0, 0) = (0, 1)$; $T((0, 1, 0)) = (1, 1)$; and $T((0, 0, 1)) = (1, 0)$. If $X = (3, 2, -1)$ determine $T(X)$.

2. If $T((1, 1, 1)) = (1, 1)$, $T((0, 1, -1)) = (0, -1)$, and $T((0, 0, 1)) = (1, 0)$, determine the matrix of the linear transformation T with respect to the canonical bases in $V_3(R)$ and $V_2(R)$. If $X = (-1, 1, 2)$, determine $T(X)$.

3. If $T((1, 1)) = (0, 2, -1, 1)$ and $T((1, -1)) = (1, 1, 0, -1)$ determine the matrix of the linear transformation T with respect to the canonical bases in $V_2(R)$ and $V_4(R)$. If $X = (2, -1)$ determine $T(X)$.

4. If $T((1, -1, 1)) = (-1, 2, 0)$, $T((0, 2, 1)) = (0, -1, 2)$, and $T((0, 0, -1)) = (1, 0, 1)$ determine the matrix of the linear transformation T with respect to the canonical basis in $V_3(R)$. If $X = (1, 1, 1)$ determine $T(X)$.

5. If $T((1, 0, 0)) = (1, 0)$, $T((0, 1, 0)) = (1, 1)$, and $T((0, 0, 1)) = (0, 1)$ determine the matrix of the linear transformation T with respect to the basis $J_1 = (1, 1, 1)$, $J_2 = (1, -1, 0)$, $J_3 = (2, 0, 0)$ in $V_3(R)$ and the basis $K_1 = (1, 1)$, $K_2 = (1, 0)$ in $V_2(R)$. If $X = 2J_1 - J_2 + J_3$, determine scalars y_1, y_2 so that $T(X) = y_1 K_1 + y_2 K_2$.

6. If $T(1, 0, 0) = (2, 1, 0)$, $T(0, 1, 0) = (0, -1, 1)$, $T(0, 0, 1) = (-1, 1, 1)$, determine the matrix of the linear transformation T with respect to the basis $L_1 = (1, 1, 1)$, $L_2 = (0, 1, 1)$, $L_3 = (0, 0, 1)$. If $X = 2L_1 - L_2 + L_3$ determine scalars y_1, y_2, y_3 so that $T(X) = y_1 L_1 + y_2 L_2 + y_3 L_3$.

7. Determine the matrices of rotations of 90°, 180°, 270° in $V_2(R)$ with respect to the canonical basis.

8. Determine the matrices of rotations by $-45°$, $-90°$ in $V_2(R)$ with respect to the canonical basis.

9. Let T be reflection about the line $x_1 + x_2 = 0$ in $V_2(R)$. What is the matrix of T with respect to the canonical basis? What is the matrix of T with respect to the basis consisting of $(1, -1)$ and $(1, 1)$?

10. Let T be a reflection in $V_2(R)$ about the line $a_1 x_1 + a_2 x_2 = 0$. Determine a basis $\{X_1, X_2\}$ for $V_2(R)$ so that the matrix of T with respect to this basis is the matrix
$$\begin{pmatrix} 1 & 0 \\ 0 & -1 \end{pmatrix}.$$

11. Determine the matrices of the following positive rotations in $V_3(R)$. The basis is the canonical one.
    (a) Rotation of 90° about I_1.
    (b) Rotation of 270° about $-I_2$.
    (c) Rotation of 120° about $(1, 1, 1)$.

12. Determine the matrices of the following reflections in $V_3(R)$. Use the canonical basis.
    (a) Reflection through the plane $x_3 = 0$.
    (b) Reflection through the plane $x_1 = x_2$.
    (c) Reflection through the plane $x_3 + x_2 = 0$.

In the following exercises P_n denotes the space of polynomials of degree $\leq n$. The canonical basis will consist of the powers of x.

13. Determine the matrix of the transformation $T(f) = f''$ in P_n with respect to the canonical basis.

14. Determine the matrix of the transformation $T(f) = f'' + f$ in P_n with respect to the canonical basis.

15. Determine the matrix of the transformation T in P_n with respect to the canonical basis if for each real number x, $T(f)(x) = (xf)'$.

16. The transformation $T(f)(x) = \int_0^x 2f(t)\, dt$ is a linear transformation from P_n to P_{n+1}. Determine the matrix of T with respect to the canonical bases in each space.

17. The transformation $T(f)(x) = \int_1^x f(t)\, dt$ is a linear transformation from P_n to P_{n+1}. Determine the matrix of T with respect to the canonical bases in each space.

18. The transformation $T(f)(x) = \int_2^x tf(t)\, dt$ is a linear transformation from P_n to P_{n+2}. Determine the matrix of T with respect to the canonical bases in each space.

19. In the space spanned by $\{1, \cos x, \sin x\}$ determine the matrix of the transformation $T(f) = f'' + f$ with respect to these functions as basis vectors.

20. The functions $\{\sin x, \cos x, x \sin x, x \cos x\}$ are linearly independent and hence span a four-dimensional vector space V. Verify that the differentiation operators $D(f) = f'$, $D^2(f) = f''$ are linear transformations in V. What is the matrix of D, D^2 with respect to the above basis of V?

21. In P_4 let $D(f) = f'$ and $(Tf)(x) = xf'(x)$. With respect to the canonical basis, determine the matrix for DT, TD and $DT - TD$.

22. In $V_3(R)$ determine the matrices of the linear transformations $T_1(X) = I_1 \times X$, $T_2(X) = I_2 \times X$, $T_3(X) = I_3 \times X$ with respect to the canonical basis.

23. If $Y = (y_1, y_2, y_3)$, determine the matrix of the transformation in $V_3(R)$ defined by $T(X) = Y \times X$. Use the canonical basis.

24. Determine the matrix of the projection of $V_3(R)$ onto the x_1, x_2 plane along the vector I_3.

25. Let $V_4(R)$ be the direct sum of the two-dimensional subspaces V and W. Explain how to choose a basis in $V_4(R)$ so that the matrix of the projection of $V_4(R)$ onto V along W with respect to this basis is given by

$$\begin{pmatrix} 1 & 0 & 0 & 0 \\ 0 & 1 & 0 & 0 \\ 0 & 0 & 0 & 0 \\ 0 & 0 & 0 & 0 \end{pmatrix}.$$

26. Show that the matrix of a reflection in $V_2(R)$ through the line $a_1 x_1 + a_2 x_2 = 0$ is given by

$$\frac{1}{a_1^2 + a_2^2}\begin{pmatrix} a_2^2 - a_1^2 & -2a_1 a_2 \\ -2a_1 a_2 & -(a_2^2 - a_1^2) \end{pmatrix}$$

if the canonical basis is used.

27. Let V and W be n- and m-dimensional vector spaces, respectively, over the field of real numbers. Show that $L(V, W)$ is linearly isomorphic to $R_{m,n}$. (Choose bases in V and W and determine the matrix representation for each $T \in L(V, W)$.)

28. Let $E = \{I_1, \ldots, I_n\}$, $F = \{J_1, \ldots, J_n\}$ be two bases for an n-dimensional vector space V. Then

(1) $$J_l = \sum_{k=1}^n b_{kl} I_k \qquad l = 1, \ldots, n$$

for suitable scalars b_{kl}. Let T be a linear transformation from V to an m-dimensional vector space W with a basis $G = \{K_1, \ldots, K_m\}$. If $A = (a_{jk})$ is the matrix of T with respect to the bases E and G, show that the matrix of T with respect to the bases F and G is given by AB where B is the $n \times n$ matrix defined by (1).

29. Let $T((1, 0, 0)) = (1, 1)$, $T((0, 1, 0)) = (1, -1)$, $T((0, 0, 1)) = (0, 1)$ define a linear transformation from $V_3(R)$ to $V_2(R)$. Using Exercise 28 write the matrix of T with respect to the basis $(1, 1, 1)$, $(1, -1, 0)$, $(1, 1, 0)$ in $V_3(R)$ and $(1, 0)$, $(0, 1)$ in $V_2(R)$ as a product of two matrices. Compute this product.

3.5 The Space $L(V)$

Let V be a vector space over the field of real numbers R. We shall abbreviate $L(V, V)$, the space of all linear transformations of V into itself by $L(V)$. If $\dim V = n < \infty$ and $X_1, \ldots, X_n$ is a basis for V, then we have seen that we may associate to each $T \in L(V)$ an $n \times n$ matrix $A = (a_{jk})$ defined by the formulas

$$(1) \qquad T(X_k) = \sum_{j=1}^{n} a_{jk} X_j \qquad k = 1, \ldots, n.$$

If U is a second linear transformation in $L(V)$ with matrix B, then it is easily checked that the matrix for the sum of the transformations, $T + U$, is the sum of the matrices, $A + B$. In addition to the sum of the transformations we may define the product TU by the formula

$$TU(X) = T(U(X)).$$

Our next task is to verify that the matrix of the transformation TU is just the matrix product AB.

THEOREM 3.9 *Let $\{X_1, \ldots, X_n\}$ be a basis for the finite dimensional vector space V. Let $T, U \in L(V)$. If A and B are the matrices of T and U, respectively, with respect to the basis $\{X_1, \ldots, X_n\}$, then AB is the matrix of the transformation TU.*

Proof. If $A = (a_{jk})$ and $B = (b_{kl})$, then by definition of the matrix of a linear transformation

$$U(X_l) = \sum_{k=1}^{n} b_{kl} X_k.$$

Applying T to this equation we obtain

$$T(U(X_l)) = T\left(\sum_{k=1}^{n} b_{kl} X_k\right) = \sum_{k=1}^{n} b_{kl} T(X_k),$$

the latter equality holding since T is linear. But

$$T(X_k) = \sum_{j=1}^{n} a_{jk} X_j$$

Hence

$$(2) \qquad T(U(X_l)) = \sum_{k=1}^{n} b_{kl} \sum_{j=1}^{n} a_{jk} X_j = \sum_{j=1}^{n} \left(\sum_{k=1}^{n} a_{jk} b_{kl}\right) X_j.$$

But if

$$c_{jl} = \sum_{k=1}^{n} a_{jk} b_{kl}$$

and C is the matrix (c_{jl}), then $C = AB$. Moreover (2) asserts that C is the matrix of the transformation TU which was to be proved.

If now V, W, Z are finite-dimensional vector spaces and $T \in L(V, W)$ and $U \in L(W, Z)$, then the product UT belongs to $L(V, Z)$. The same argument as the above now yields the fact that the matrix of the transformation UT is the product of the matrix of U and the matrix of T. We state the result as our next theorem, and leave the proof as an exercise.

THEOREM 3.10 *Let V, W, Z be finite-dimensional vector spaces and let $T \in L(V, W)$ and $U \in L(W, Z)$. Fix a basis in each of the spaces V, W, Z and let A be the matrix of the transformation T, and B be the matrix of the transformation U. Then the matrix of the transformation $UT \in L(V, Z)$ is the product BA.*

Returning to the case of one vector space V, if we denote by A_T the matrix of the transformation $T \in L(V)$, then the correspondence $T \to A_T$ defines a one-to-one mapping i of $L(V)$ onto $R_{n,n}$. If we write $A_T = i(T)$, then it is easily checked that the transformation i is linear. Hence the vector spaces $L(V)$ and $R_{n,n}$ are linearly isomorphic (Exercise 27, p. 100). Theorem 3.9 asserts that i is *multiplicative*, that is, $i(TU) = i(T)i(U)$ for each pair T, $U \in L(V)$. When a linear isomorphism i between two algebras, such as $L(V)$ and $R_{n,n}$, preserves multiplication we say that i is an *algebraic isomorphism of $L(V)$ onto $R_{n,n}$*. Just as a linear isomorphism between vector spaces allows us to identify the linear structure of the two vector spaces, the algebraic isomorphism i between $L(V)$ and $R_{n,n}$ allows us to identify the algebraic structure of these two algebras. Thus any phenomenon involving only multiplication, addition, or scalar multiplication in one algebra has its exact counterpart in the other.

Definition. *If $\mathcal{A}, \mathcal{B}$ are algebras over the field of real numbers and there exists a one-to-one linear mapping i of $\mathcal{A}$ onto $\mathcal{B}$ which preserves multiplication, that is,*

$$i(AB) = i(A) \cdot i(B)$$

for each $A, B \in \mathcal{A}$ then $\mathcal{A}$ and $\mathcal{B}$ are said to be algebraically isomorphic.

We may formalize our discussion in the next theorem.

THEOREM 3.11 *If V is a vector space of dimension n over the field of real numbers, then $L(V)$ and $R_{n,n}$ are algebraically isomorphic.*

The space $L(V)$ has an identity transformation I satisfying $I(X) = X$ for each $X \in V$. The matrix of this transformation with respect to any basis in V is the identity matrix

$$I = \begin{pmatrix} 1 & 0 & \cdots & & 0 \\ 0 & 1 & 0 & \cdots & \cdot \\ \cdot & & 1 & & \cdot \\ \cdot & & & \ddots & \cdot \\ 0 & & \cdots & & 1 \end{pmatrix}.$$

A fundamental problem is to determine for arbitrary vector spaces V which transformations T have inverses with respect to the multiplication in $L(V)$.

Definition. *A linear transformation T in $L(V)$ is said to be* invertible *if there exists a transformation $U \in L(V)$ satisfying*

(3) $$TU = UT = 1.$$

If there exists one transformation U satisfying (3), it is easily shown there is only one (Exercise 23). We call this transformation U the inverse of T and write $U = T^{-1}$. The following theorem gives conditions on the transformation T which insure that T is invertible.

THEOREM 3.12 *Let V be a vector space over the field of real numbers. For a linear transformation $T \in L(V)$ the following statements are equivalent.*
  (i) *T is invertible.*
  (ii) *T is one-to-one and T is onto (i.e., the range of T is all of V).*

Proof. We first assume T is invertible and prove T is one-to-one and onto. To prove T one-to-one suppose $X, Y \in V$ and $X \neq Y$. We assert $T(X) \neq T(Y)$. For if $T(X) = T(Y)$, then $T^{-1}T(X) = T^{-1}T(Y)$. But $T^{-1}T(X) = X$ and $T^{-1}T(Y) = Y$. Hence $X = Y$ which is impossible. To show T is onto we must show that for each $Y \in V$ there is an $X \in V$ such that $T(X) = Y$. But for Y given if we define $X = T^{-1}(Y)$, then $T(X) = TT^{-1}(Y) = Y$. Thus T is onto.

To prove the converse of the theorem we must show that if T is one-to-one and onto, then T is invertible. These two conditions imply that for each $Y \in V$ there is precisely one X such that $T(X) = Y$. Thus we may define a transformation U of V into itself by the formula $U(Y) = X$ whenever $T(X) = Y$.

Thus if $T(X) = Y$, $U(T(X)) = U(Y) = X$. This implies that UT is the identity transformation. However, if $U(Y) = X$, then $T(U(Y)) = T(X) = Y$ since T is assumed to be one-to-one. Therefore $UT = TU = I$ and (3) is satisfied. To complete the proof that $U = T^{-1}$, we need only show that U is linear.

To show linearity we must verify

$$U(a_1 Y_1 + a_2 Y_2) = a_1 U(Y_1) + a_2 U(Y_2),$$

for each $a_1, a_2 \in R$ and $Y_1, Y_2 \in V$. However if $U(Y_1) = X_1$ and $U(Y_2) = X_2$, then $T(X_1) = Y_1$ and $T(X_2) = Y_2$. Also $T(a_1 X_1 + a_2 X_2) = a_1 Y_1 + a_2 Y_2$ since T is linear. But by the definition of U this says that

$$U(a_1 Y_1 + a_2 Y_2) = a_1 X_1 + a_2 X_2 = a_1 U(Y_1) + a_2 U(Y_2)$$

which is the required statement of linearity.

If the dimension of V is finite, then we have a sharper result.

THEOREM 3.13 *If $\dim V = n < \infty$ and $T \in L(V)$, then T is one to one if and only if T is onto.*

Proof. We must show that the range of T is all of V if and only if the null space of T consists just of the zero vector. But by Theorem 3.6 we know that $n = \dim (\text{range of } T) + \dim (\text{null space of } T)$. But range of T equals V if and only if $\dim (\text{range of } T) = n$. This holds if and only if $\dim (\text{null space of } T) = 0$. The latter is equivalent to the statement that the null space of T consists of just the zero vector.

Combining these two results we have the following

THEOREM 3.14 *If the dimension of V is finite and $T \in L(V)$, then the following are equivalent.*

(i) T is invertible.
(ii) T is one-to-one.
(iii) T is onto.

An example which shows that the last two results are false for infinite-dimensional spaces is easy to construct.

Example 1. Let P be the vector space of all polynomials. If

$$D(p)(x) = p'(x) \quad \text{and} \quad S(p)(x) = \int_0^x p(t)\, dt$$

then both D and S are linear transformations in V. It is easily seen that the differentiation operator D maps P onto P, but D is not one-to-one. The integration operator S is one-to-one but is not onto. The proofs of these facts are left as exercises. Also by the fundamental theorem of the calculus

$$DS(p) = p$$

for each polynomial p. Hence $DS = I$. However, it is not true that $SD = I$, since $SD(p) = p$ for a polynomial p if and only if the polynomial has zero constant term.

We turn next to characterizing the invertibility of the transformation T in terms of the matrix A_T of the transformation defined by (1). Since the correspondence $T \to A_T$ defines an algebraic isomorphism of $L(V)$ onto $R_{n,n}$, T will be invertible if and only if the matrix A_T is invertible. But the matrix $A_T = (a_{ij})$ defines a linear transformation from $V_n(R)$ to itself by the formula

$$Y = \begin{pmatrix} y_1 \\ \vdots \\ y_n \end{pmatrix} = \begin{pmatrix} a_{11} & \cdots & a_{1n} \\ \vdots & & \vdots \\ a_{n1} & \cdots & a_{nn} \end{pmatrix} \begin{pmatrix} x_1 \\ \vdots \\ x_n \end{pmatrix} = A_T X.$$

By Theorem 3.14, A_T is invertible if and only if this transformation is one-to-one. However, we know from the corollary to Theorem 2.13 that for an arbitrary $n \times n$ matrix A of rank r the dimension of the set of solutions to the equation $AX = 0$ is precisely $n - r$. The transformation induced in $V_n(R)$ by A_T is one-to-one if and only if

$$A_T X = 0 \text{ implies } X = 0.$$

That is, the set of solutions to $A_T X = 0$ consists just of the zero subspace. Equivalently, the rank of A_T must equal n. We summarize this result in the next theorem.

THEOREM 3.15 *Let V be an n-dimensional vector space. The transformation T in $L(V)$ is invertible if and only if the matrix A_T of this transformation is invertible. This occurs if and only if the rank of A_T equals n.*

The actual computation of the inverse A^{-1} of an invertible matrix is facilitated by the following result which follows from Theorem 3.14.

THEOREM 3.16 *If $A \in R_{n,n}$, then A is invertible if there exists a matrix B such that $BA = I$ or $AB = I$. In either case $B = A^{-1}$.*

Proof. If $BA = I$, then the matrix A must induce a $1:1$ transformation in $V_n(R)$. For if $AX = 0$ for some $X \in V_n(R)$, then $0 = BAX = IX$. But if A is $1:1$, then by 3.14, A is invertible and multiplying the equation $BA = I$ by the matrix A^{-1} yields

$$BAA^{-1} = A^{-1} \quad \text{or} \quad B = A^{-1}.$$

If on the other hand $AB = I$ then the range of the transformation induced by A must be all of $V_n(R)$ (Why?). Thus A is onto, and by 3.14, A is inevitible. Multiplying the equation $AB = I$ *on the left by* A^{-1}, we see that $B = A^{-1}$.

Next we shall develop a technique for computing the inverses for a special class of $n \times n$ matrices. These are the so-called triangular matrices. If we call the vector $(a_{11}, a_{22}, \ldots, a_{nn})$ the main diagonal of the matrix A, then A is called *lower triangular* if all the entries above the main diagonal are zero. If all the entries below the main diagonal are zero, then the matrix is called *upper triangular*. (See Figure 11.)

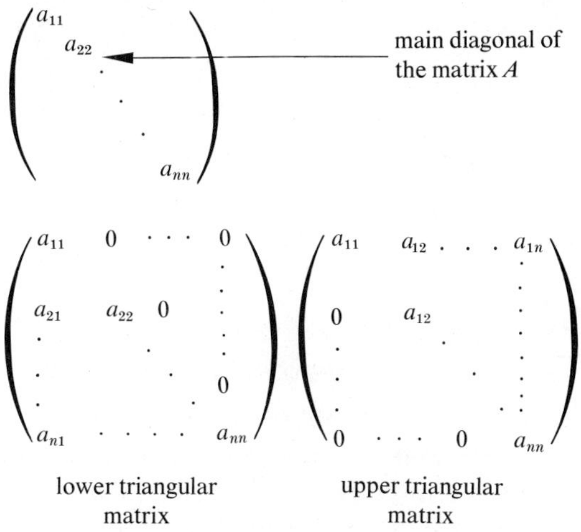

$$\begin{pmatrix} a_{11} & & & \\ & a_{22} & & \\ & & \ddots & \\ & & & a_{nn} \end{pmatrix} \longleftarrow \text{main diagonal of the matrix } A$$

$$\begin{pmatrix} a_{11} & 0 & \cdots & 0 \\ a_{21} & a_{22} & 0 & \\ \vdots & & \ddots & 0 \\ a_{n1} & \cdots & & a_{nn} \end{pmatrix} \quad \begin{pmatrix} a_{11} & a_{12} & \cdots & a_{1n} \\ 0 & a_{12} & & \vdots \\ \vdots & & \ddots & \\ 0 & \cdots & 0 & a_{nn} \end{pmatrix}$$

lower triangular matrix ⟍ upper triangular matrix

Fig. 11

Let us try to construct the inverse for a lower triangular matrix. By Theorem 3.16 it suffices to determine a matrix

$$X = \begin{pmatrix} x_{11} & \cdots & x_{1n} \\ \vdots & & \vdots \\ x_{n1} & \cdots & x_{nn} \end{pmatrix}$$

satisfying

(4)
$$\begin{pmatrix} a_{11} & 0 & \cdots & 0 \\ \vdots & & \ddots & 0 \\ a_{n1} & \cdots & & a_{nn} \end{pmatrix} \begin{pmatrix} x_{11} & \cdots & x_{1n} \\ \vdots & & \vdots \\ x_{n1} & \cdots & x_{nn} \end{pmatrix} = \begin{pmatrix} 1 & 0 & \cdots & 0 \\ 0 & 1 & & \vdots \\ \vdots & & \ddots & \\ 0 & \cdots & & 1 \end{pmatrix}.$$

In order for the matrix equation (4) to have a solution two conditions must be satisfied. These follow immediately from the definition of matrix multiplication. Find, if we compute the inner product of the first row of A with the first column of X we conclude that $a_{11} \neq 0$ and $x_{11} = 1/a_{11}$. Computing the inner product of the first row of A with the remaining columns of X we conclude that $x_{12} = x_{13} = \cdots = x_{1n} = 0$. Continuing inductively we conclude that if a lower triangular matrix A is invertible, then all the entries on the main diagonal must be nonzero. Furthermore the inverse A^{-1} is also a lower triangular matrix and the entries on the main diagonal are just the numbers

$a_{11}^{-1}, \ldots, a_{nn}^{-1}$. Thus the matrix equation (4) becomes

$$(5) \quad \begin{pmatrix} a_{11} & 0 & & 0 \\ a_{21} & a_{22} & & \\ \vdots & & \ddots & 0 \\ a_{n1} & \cdots & & a_{nn} \end{pmatrix} \begin{pmatrix} a_{11}^{-1} & 0 & & 0 \\ x_{21} & a_{22}^{-1} & & \\ \vdots & & \ddots & 0 \\ x_{n1} & \cdots & & a_{nn}^{-1} \end{pmatrix} = \begin{pmatrix} 1 & 0 & \cdots & 0 \\ 0 & 1 & & \\ \vdots & & \ddots & \\ 0 & & \cdots & 1 \end{pmatrix}.$$

On the other hand if the main diagonal terms of A are all nonzero, then the columns of A are independent. Hence the rank of $A = n$, and A is invertible by Theorem 3.15. This means that the remaining unknowns x_{ij} in the matrix equation (5) may always be computed. To see how to do this write

$$A^{-1} = \begin{pmatrix} a_{11}^{-1} & 0 & & 0 \\ x_{21} & a_{22}^{-1} & & 0 \\ \vdots & & \ddots & \\ x_{n1} & \cdots & & a_{nn}^{-1} \end{pmatrix}$$

and let $X_1, \ldots, X_n$ be the columns of A^{-1}. Denote the rows of A by $A_1, \ldots, A_n$. If we compute the inner product (A_2, X_1) we get

$$a_{21}a_{11}^{-1} + a_{22}x_{21} = (A_2, X_1) = 0.$$

If we solve this equation for x_{21} and substitute the resulting number back in the matrix X, we may compute x_{31} by solving the equation $(A_3, X_1) = 0$. Continuing we determine all the entries for the first column of X. This process is then continued for the remaining columns of X. This method for determining the inverse of a lower triangular matrix is called the method of *back substitution*. To see what is going on let us consider an example.

Example 2. Determine A^{-1} if

$$A = \begin{pmatrix} 1 & 0 & 0 \\ 1 & 1 & 0 \\ 1 & 1 & 1 \end{pmatrix}.$$

The matrix A is invertible since there are no zeroes on the main diagonal. By (5) we must solve the matrix equation

$$A^{-1} = \begin{pmatrix} 1 & 0 & 0 \\ 1 & 1 & 0 \\ 1 & 1 & 1 \end{pmatrix} \begin{pmatrix} 1 & 0 & 0 \\ x_{21} & 1 & 0 \\ x_{31} & x_{32} & 1 \end{pmatrix} = \begin{pmatrix} 1 & 0 & 0 \\ 0 & 1 & 0 \\ 0 & 0 & 1 \end{pmatrix}.$$

To determine x_{21} compute the inner product of the second row of A with the first column of A^{-1}. Setting this result equal to zero yields

$$1 + x_{21} = 0 \quad \text{or} \quad x_{21} = -1.$$

Next, compute the inner product of the third row of A with the first column of A^{-1}. This yields $1 + x_{21} + x_{31} = 0$. Since $x_{21} = -1$ we conclude $x_{31} = 0$. Continuing for the second column of X we conclude that $x_{32} = -1$. Hence

$$A^{-1} = \begin{pmatrix} 1 & 0 & 0 \\ -1 & 1 & 0 \\ 0 & -1 & 1 \end{pmatrix}.$$

If the matrix A is upper triangular, then A has an inverse if and only if there are no zeroes on the main diagonal. Furthermore the inverse of an upper triangular matrix is again a matrix of the same type. To compute A^{-1} we must solve the matrix equation

$$A^{-1} = \begin{pmatrix} a_{11} & \cdots & a_{1n} \\ 0 & a_{22} & \cdots & a_{2n} \\ \vdots & & \ddots & \vdots \\ 0 & \cdots & & a_{2n} \end{pmatrix} \begin{pmatrix} a_{11}^{-1} & x_{12} & \cdots & x_{1n} \\ 0 & a_{22}^{-1} & \cdots & x_{2n} \\ \vdots & & \ddots & \vdots \\ 0 & 0 & \cdots & a_{nn}^{-1} \end{pmatrix} = \begin{pmatrix} 1 & 0 & \cdots & 0 \\ 0 & 1 & & \\ \vdots & & \ddots & \\ 0 & & \cdots & 1 \end{pmatrix}.$$

We compute the x_{ij} by back substitution, except that we begin at the bottom of each column of A^{-1} and proceed up. We shall consider the problem of computing the inverses of general $n \times n$ matrices in the next section.

EXERCISES

Determine the inverses of the following matrices if they exist.

1. $\begin{pmatrix} 1 & 0 \\ 0 & 2 \end{pmatrix}$

2. $\begin{pmatrix} 1 & 0 \\ -1 & 0 \end{pmatrix}$

3. $\begin{pmatrix} 2 & 0 \\ 1 & -1 \end{pmatrix}$

4. $\begin{pmatrix} 1 & -1 \\ 0 & 1 \end{pmatrix}$

5. $\begin{pmatrix} 1 & 0 & 0 \\ 0 & 2 & 0 \\ 0 & 0 & 3 \end{pmatrix}$

6. $\begin{pmatrix} 1 & -1 & 1 \\ 0 & 0 & 1 \\ 0 & 0 & 1 \end{pmatrix}$

7. $\begin{pmatrix} 2 & 1 & 0 \\ 0 & 1 & -1 \\ 0 & 0 & 1 \end{pmatrix}$

8. $\begin{pmatrix} 1 & 0 & 0 \\ -1 & 2 & 0 \\ 1 & 2 & 3 \end{pmatrix}$

9. $\begin{pmatrix} 1 & 0 & 0 \\ 0 & -1 & 0 \\ 1 & 2 & 1 \end{pmatrix}$

10. $\begin{pmatrix} 1 & 0 & 1 \\ 0 & 1 & 0 \\ 0 & 0 & -1 \end{pmatrix}$

11. $\begin{pmatrix} 1 & 0 & 0 & 0 \\ -1 & 1 & 0 & 0 \\ 1 & -1 & 0 & 0 \\ 1 & 2 & 1 & -1 \end{pmatrix}$

12. $\begin{pmatrix} 1 & 0 & 1 & 0 \\ 0 & -1 & 0 & -1 \\ 0 & 0 & 1 & 0 \\ 0 & 0 & 0 & -1 \end{pmatrix}$

13. $\begin{pmatrix} 2 & 0 & 0 & 0 \\ 0 & -1 & 0 & 0 \\ 0 & 0 & 1 & 0 \\ 1 & 0 & 0 & 2 \end{pmatrix}$

14. $\begin{pmatrix} 1 & -1 & 1 & -1 \\ 0 & 1 & -1 & 1 \\ 0 & 0 & -1 & 1 \\ 0 & 0 & 0 & -1 \end{pmatrix}$

15. In P_n, the polynomials of degree $\leq n$, define the linear transformation T by the formula
$$Tf = f' + f.$$

(a) Verify that T is an invertible linear transformation by examining an appropriate matrix for T.
(b) Determine the matrix of T^{-1} in P_3 with respect to the basis consisting of powers of x.
(c) Find a solution in P_3 to the differential equation $f' + f = x^2$. Is this solution unique?

16. Show that $T(f) = f'' + f$ is an invertible transformation in P_n. Find a solution in P_n to the differential equation
$$f'' + f = x.$$
Is this solution unique?

17. For what values of a does the transformation
$$Tf = f' + af$$
define a one-to-one linear transformation in P_n?

18. Let $D(f) = f'$ and $S(f)(x) = \int_0^x f(t)\,dt$ be the differentiation and integration operators in the space of all polynomials P. Verify that D is onto but not one-to-one, and verify that S is one-to-one but not onto.

19. Let V be a vector space and assume $T \in L(V)$. If there exists a transformation, $U \in L(V)$ satisfying $UT = I$, show that T is one-to-one. If there exists a linear transformation U satisfying $TU = I$, show that T is onto.

20. Prove that an upper triangular $n \times n$ matrix A is invertible if and only if all the coefficients on the main diagonal are nonzero. If A is invertible, what must the diagonal elements of A^{-1} be? (Adapt the argument in the text for lower triangular matrices.)

21. By examining an appropriate matrix for the transformation
$$Tf = f'' + af' + bf$$
in P_n, prove that T is invertible if and only if $b \neq 0$.

22. Consider the set of n linear equations in n unknowns
$$a_{11}x_1 + \cdots + a_{1n}x_n = y_1$$
$$\vdots \qquad\qquad \vdots$$
$$a_{n1}x_1 + \cdots + a_{nn}x_n = y_n.$$
Prove that to *each* vector $Y = (y_1, \ldots, y_n)$ there is a solution $X = (x_1, \ldots, x_n)$ if and only if to *no* vector Y is there more than one solution $X = (x_1, \ldots, x_n)$.

23. Let V be a vector space and $T \in L(V)$. If there exists $U \in L(V)$ satisfying $TU = UT = I$, show there can be only one such transformation.

3.6 Invertible Matrices

Let A be an invertible $n \times n$ matrix. If $Y \in V_n(R)$ is specified and it is desired to solve the equation $Y = AX$ for the vector X, then as we have seen the solution X is given by
$$X = A^{-1}Y.$$
If this equation must be solved for many choices of the vector Y, then it is desirable to compute the matrix A^{-1}. To develop a technique for this let us formalize the "elimination" technique for linear equations discussed in Section 2.7. There it was shown that for an arbitrary matrix A there exists a matrix B, the rows of which are in echelon form, such that
$$AX = 0 \quad \text{if and only if} \quad BX = 0.$$
Furthermore B was obtained from A by performing certain operations on the rows of A.

Definition. *A matrix, the rows of which are in echelon form, is called an echelon matrix.*

An $n \times n$ echelon matrix is an upper triangular matrix, but not every upper triangular matrix is an echelon matrix. We saw in the last section that a triangular matrix was invertible if and only if all terms on the main diagonal are nonzero. This is, of course, true for echelon matrices as well. However,

INVERTIBLE MATRICES

if an $n \times n$ echelon matrix has a zero in the ith position on the main diagonal, then zero will be in the jth position for each $j > i$.

To formalize our discussion of the row operations on an $m \times n$ matrix we need some terminology.

Definition. *Let A be an $m \times n$ matrix and let $R_1, \ldots, R_m$ be the rows of A. Each of the following operations on the rows of A is called an* elementary row operation *on A.*

1. *Replace R_i by aR_i where a is a nonzero scalar.*
2. *Interchange two rows of A.*
3. *Replace R_j by $R_j + aR_i$ where $j \neq i$.*

If the matrix B can be obtained from A by a finite sequence of elementary row operations, then we say B is *row equivalent* to A.

The following results summarize the content of Section 2.7.

THEOREM 3.17 *If A is an $m \times n$ matrix, then A is row equivalent to an echelon matrix. If A and B are row equivalent and $X \in V_n(R)$ then*

$$AX = 0 \text{ if and only if } BX = 0.$$

Next we wish to show that each elementary row operation on an $m \times n$ matrix A may be accomplished by multiplying A on the left by an invertible $m \times n$ matrix.

To see what is going on let us first perform the elementary row operations on the identity matrix.

1. Replacing the ith row $R_i = (0, \ldots, 1, 0, \ldots, 0)$ by $aR_i = (0, \ldots, a, 0, \ldots, 0)$ yields

$$D_i(a) = \begin{pmatrix} 1 & & & & & 0 \\ & \ddots & & & & \\ & & a & & & \\ & & & 1 & & \\ & & & & \ddots & \\ 0 & & & & & 1 \end{pmatrix} i\text{th row}.$$

Clearly $D_i(a)$ is invertible. Indeed $D_i(a) \cdot D_i(1/a) = I$.

2. If we interchange the ith and jth rows of I we obtain

$$E_{ij} = \begin{pmatrix} 1 & & & & & & 0 \\ & \ddots & & & & & \\ & & 0 & & 1 & & \\ & & & 1 & & & \\ & & & & \ddots & & \\ & & 1 & & 0 & & \\ & & & & & 1 & \\ & & & & & & \ddots \\ 0 & & & & & & 1 \end{pmatrix} \begin{matrix} i\text{th row} \\ \\ \\ j\text{th row} \end{matrix}$$

E_{ij} is also invertible. Indeed $E_{ij}^2 = I$.

110   MATRICES AND LINEAR TRANSFORMATIONS

3. If we multiply the ith row of I by a and add it to the jth row we obtain

$$F_{ij}(a) = \begin{pmatrix} 1 & & & & & 0 \\ & \ddots & & & & \\ & & 1 & & & \\ & & & \ddots & & \\ & & a & & 1 & \\ & & & & & \ddots \\ 0 & & & & & 1 \end{pmatrix} \begin{matrix} \\ \\ i\text{th row} \\ \\ j\text{th row} \\ \\ \end{matrix}$$

Again $F_{ij}(a)$ is invertible with inverse

$$F_{ij}(-a) = \begin{pmatrix} 1 & & & & & 0 \\ & \ddots & & & & \\ & & 1 & & & \\ & & & \ddots & & \\ & & -a & & 1 & \\ & & & & & \ddots \\ 0 & & & & & 1 \end{pmatrix} \begin{matrix} \\ \\ i\text{th row} \\ \\ j\text{th row} \\ \\ \end{matrix}$$

The matrices above obtained from the identity matrix I by performing an elementary row operation on I are called *elementary* matrices. It may now be immediately checked that each row operation on an $m \times n$ matrix A may be achieved by multiplying on the left by the corresponding $m \times m$ elementary matrix. The verification of these facts we leave as exercises.

THEOREM 3.18 *If A and B are $m \times n$ matrices and B is row equivalent to A, then there exists an invertible $m \times m$ matrix C such that $B = CA$.*

Proof. If B is row equivalent to A, then B may be obtained from A by a finite sequence of elementary row operations. However, by the previous discussion an elementary row operation on an $m \times n$ matrix A may be accomplished by multiplying A on the left by an elementary matrix which is necessarily invertible. Therefore, if B is row equivalent to A, there exists a finite sequence $E_1, \ldots, E_m$ of elementary matrices such that $B = E_m \cdots E_1 A$. Letting $C = E_m \cdots E_1$ we have $B = CA$. The matrix C is invertible since $C^{-1} = E_1^{-1} \cdots E_m^{-1}$.

Example 1. Let $A = \begin{pmatrix} 1 & -1 & 1 \\ -1 & 1 & 2 \end{pmatrix}$. If we add the first row of A to the second row obtaining $B = \begin{pmatrix} 1 & -1 & 1 \\ 0 & 0 & 3 \end{pmatrix}$, the matrix B is in echelon form and is row equivalent to A. The elementary matrix corresponding to this elementary row operation is $F_{12}(1) = \begin{pmatrix} 1 & 0 \\ 1 & 0 \end{pmatrix}$. Clearly $\begin{pmatrix} 1 & -1 & 1 \\ 0 & 0 & 3 \end{pmatrix} = \begin{pmatrix} 1 & 0 \\ 1 & 1 \end{pmatrix}\begin{pmatrix} 1 & -1 & 1 \\ -1 & 1 & 2 \end{pmatrix}$. B in turn is row equivalent to $C = \begin{pmatrix} 1 & -1 & 1 \\ 0 & 0 & 1 \end{pmatrix}$ since C is obtained by multiplying the second row of B by $\frac{1}{3}$. The elementary matrix for this row operation is in the diagonal matrix

INVERTIBLE MATRICES 111

$$D_2(\tfrac{1}{3}) = \begin{pmatrix} 1 & 0 \\ 0 & \tfrac{1}{3} \end{pmatrix}$$

and

$$\begin{pmatrix} 1 & -1 & 1 \\ 0 & 0 & 1 \end{pmatrix} = \begin{pmatrix} 1 & 0 \\ 0 & \tfrac{1}{3} \end{pmatrix} \begin{pmatrix} 1 & -1 & 1 \\ 0 & 0 & 3 \end{pmatrix}.$$

Example 2. Find an echelon matrix row equivalent to $A = \begin{pmatrix} 0 & 1 & 2 \\ 1 & -1 & 1 \\ 1 & 1 & 1 \end{pmatrix}$. The elementary row operations necessary are:

1. Interchange first and second row obtaining $\begin{pmatrix} 1 & -1 & 1 \\ 0 & 1 & 2 \\ 1 & 1 & 1 \end{pmatrix}$.

2. Subtract the first row from the third obtaining $\begin{pmatrix} 1 & -1 & 1 \\ 0 & 1 & 2 \\ 0 & 2 & 0 \end{pmatrix}$.

3. Multiply the second row by -2 and add it to the third obtaining $\begin{pmatrix} 1 & -1 & 1 \\ 0 & 1 & 2 \\ 0 & 0 & -4 \end{pmatrix}$

which is an echelon matrix row equivalent to A.

The elementary matrices for each of these row operations are

$$E_{12} = \begin{pmatrix} 0 & 1 & 0 \\ 1 & 0 & 0 \\ 0 & 0 & 1 \end{pmatrix}, \quad F_{13}(-1) = \begin{pmatrix} 1 & 0 & 0 \\ 0 & 1 & 0 \\ -1 & 0 & 1 \end{pmatrix}, \quad F_{23}(-2) = \begin{pmatrix} 1 & 0 & 0 \\ 0 & 1 & 0 \\ 0 & -2 & 1 \end{pmatrix},$$

respectively. Clearly

$$\begin{pmatrix} 1 & -1 & 1 \\ 0 & 1 & 2 \\ 1 & 1 & 1 \end{pmatrix} = \begin{pmatrix} 0 & 1 & 0 \\ 1 & 0 & 0 \\ 0 & 0 & 1 \end{pmatrix} \begin{pmatrix} 0 & 1 & 2 \\ 1 & -1 & 1 \\ 1 & 1 & 1 \end{pmatrix},$$

$$\begin{pmatrix} 1 & -1 & 1 \\ 0 & 1 & 2 \\ 0 & 2 & 0 \end{pmatrix} = \begin{pmatrix} 1 & 0 & 0 \\ 0 & 1 & 0 \\ -1 & 0 & 1 \end{pmatrix} \begin{pmatrix} 1 & -1 & 1 \\ 0 & 1 & 2 \\ 1 & 1 & 1 \end{pmatrix},$$

and

$$\begin{pmatrix} 1 & -1 & 1 \\ 0 & 1 & 2 \\ 0 & 0 & -4 \end{pmatrix} = \begin{pmatrix} 1 & 0 & 0 \\ 0 & 1 & 0 \\ 0 & -2 & 1 \end{pmatrix} \begin{pmatrix} 1 & -1 & 1 \\ 0 & 1 & 2 \\ 0 & 2 & 0 \end{pmatrix}.$$

Writing out the matrix product corresponding to each row operation is somewhat cumbersome. A more compressed notation is to use an arrow with the appropriate elementary matrix above it. Thus, the three row operations above would be written

$$\begin{pmatrix} 0 & 1 & 2 \\ 1 & -1 & 1 \\ 1 & 1 & 1 \end{pmatrix} \xrightarrow{E_{12}} \begin{pmatrix} 1 & -1 & 1 \\ 0 & 1 & 2 \\ 1 & 1 & 1 \end{pmatrix}$$

$$\xrightarrow{F_{13}(-1)} \begin{pmatrix} 1 & -1 & 1 \\ 0 & 1 & 2 \\ 0 & 2 & 0 \end{pmatrix} \xrightarrow{F_{23}(-2)} \begin{pmatrix} 1 & -1 & 1 \\ 0 & 1 & 2 \\ 0 & 0 & -4 \end{pmatrix}$$

For square matrices the technique of the proof of 3.18 yields the following useful corollary.

COROLLARY. *If A is an $n \times n$ matrix then A can be written as a product.*

$$A = E_m \cdots E_1 B$$

where $E_1, \ldots, E_m$ are elementary matrices and B is upper triangular.

We now consider the problem of determining the inverse of an invertible $n \times n$ matrix.

THEOREM 3.19 *An $n \times n$ matrix A is invertible if and only if A is row equivalent to the identity matrix.*

Proof. The matrix A is row equivalent to the identity if and only if there exists a sequence $E_1, \ldots, E_m$ of elementary matrices such that $E_m \cdots E_1 A = I$. Hence if A is row equivalent to the identity, A must be invertible by Theorem 3.16 and indeed $A^{-1} = E_m \cdots E_1$. For the converse observe first that an invertible echelon matrix B is clearly row equivalent to the identity since B has no zeros on its main diagonal. Now any $n \times n$ matrix A is row equivalent to an echelon matrix B. Furthermore B is invertible if and only if A is. Therefore if A is invertible, there is an invertible echelon matrix B row equivalent to A. Since B is row equivalent to the identity so is A.

We may rephrase 3.19 with the equivalent statement than an $n \times n$ matrix A is invertible if and only if A is the product of elementary matrices.

Theorem 3.19 contains a computational recipe for determining the inverse of an invertible matrix A. We need only determine the elementary row operations on A which reduce A to the identity I. If $E_1, \ldots, E_m$ are the elementary matrices corresponding to these row operations, then

$$A^{-1} = E_m \cdots E_1.$$

Hence if we perform this sequence of elementary row operations beginning with the identity matrix, then the end result will be the inverse A^{-1}. We illustrate this with an example.

Example 2. Show that $A = \begin{pmatrix} 1 & 1 \\ 2 & 1 \end{pmatrix}$ is invertible and compute its inverse. To show A is row equivalent to the identity we perform the following elementary row operations

$$\begin{pmatrix} 1 & 1 \\ 2 & 1 \end{pmatrix} \xrightarrow{F_{12}(-2)} \begin{pmatrix} 1 & 1 \\ 0 & -1 \end{pmatrix} \xrightarrow{F_{21}(1)} \begin{pmatrix} 1 & 0 \\ 0 & -1 \end{pmatrix} \xrightarrow{D_2(-1)} \begin{pmatrix} 1 & 0 \\ 0 & 1 \end{pmatrix}.$$

The inverse of A is the product

$$A^{-1} = D_2(-1) \cdot F_{21}(1) \cdot F_{12}(-2)$$

$$= \begin{pmatrix} 1 & 0 \\ 0 & -1 \end{pmatrix}\begin{pmatrix} 1 & 1 \\ 0 & 1 \end{pmatrix}\begin{pmatrix} 1 & 0 \\ -2 & 1 \end{pmatrix} = \begin{pmatrix} 1 & 1 \\ 0 & -1 \end{pmatrix}\begin{pmatrix} 1 & 0 \\ -2 & 1 \end{pmatrix} = \begin{pmatrix} -1 & 1 \\ 2 & -1 \end{pmatrix}.$$

Performing these row operations beginning with the identity yields

$$\begin{pmatrix} 1 & 0 \\ 0 & 1 \end{pmatrix} \xrightarrow{F_{12}(-2)} \begin{pmatrix} 1 & 0 \\ -2 & 1 \end{pmatrix} \xrightarrow{F_{21}(1)} \begin{pmatrix} -1 & 1 \\ -2 & 1 \end{pmatrix} \xrightarrow{D_2(-1)} \begin{pmatrix} -1 & 1 \\ 2 & -1 \end{pmatrix} = A^{-1}.$$

We may minimize our computation if we combine the two sequences of row operations. If we write the matrices A and I together as a 2×4 matrix using the notation

$$(A|I) = \begin{pmatrix} 1 & 2 & | & 1 & 0 \\ 2 & 1 & | & 0 & 1 \end{pmatrix}$$

then performing the above sequence of row operations on this 2×4 matrix we obtain
$$(I|A^{-1}) = \begin{pmatrix} 1 & 0 & -1 & 1 \\ 0 & 1 & 2 & -1 \end{pmatrix}.$$
Indeed
$$(A|I) = \begin{pmatrix} 1 & 1 & 1 & 0 \\ 2 & 1 & 0 & 1 \end{pmatrix} \xrightarrow{F_{12}(-2)} \begin{pmatrix} 1 & 1 & 1 & 0 \\ 0 & -1 & -2 & 1 \end{pmatrix} \xrightarrow{F_{21}(1)}$$

$$\begin{pmatrix} 1 & 0 & -1 & 1 \\ 0 & -1 & -2 & 1 \end{pmatrix} \xrightarrow{D_2(-1)} \begin{pmatrix} 1 & 0 & -1 & 1 \\ 0 & 1 & 2 & -1 \end{pmatrix} = (I|A^{-1}).$$

Example 3. Determine if $A = \begin{pmatrix} 1 & -1 & 1 \\ 0 & 0 & 1 \\ 1 & 1 & -1 \end{pmatrix}$ is invertible and if so compute its

inverse. We first perform the row operations which reduce A to an echelon matrix. Beginning with the matrix $(A|I)$ we obtain

$$(A|I) = \begin{pmatrix} 1 & -1 & 1 & 1 & 0 & 0 \\ 0 & 0 & 1 & 0 & 1 & 0 \\ 1 & 1 & -1 & 0 & 0 & 1 \end{pmatrix} \xrightarrow{F_{13}(-1)} \begin{pmatrix} 1 & -1 & 1 & 1 & 0 & 0 \\ 0 & 0 & 1 & 0 & 1 & 0 \\ 0 & 2 & -2 & -1 & 0 & 1 \end{pmatrix}$$

$$\xrightarrow{E_{23}} \begin{pmatrix} 1 & -1 & 1 & 1 & 0 & 0 \\ 0 & 2 & -2 & -1 & 0 & 1 \\ 0 & 0 & 1 & 0 & 1 & 0 \end{pmatrix}.$$

Since A has been reduced by elementary row operations to the echelon matrix $\begin{pmatrix} 1 & -1 & 1 \\ 0 & 2 & -2 \\ 0 & 0 & 1 \end{pmatrix}$ which has no zeroes on the main diagonal, the matrix A is invertible.

We construct the remaining elementary row operations which reduce A to the identity.

$$\begin{pmatrix} 1 & -1 & 1 & 1 & 0 & 0 \\ 0 & 2 & -2 & -1 & 0 & 1 \\ 0 & 0 & 1 & 0 & 1 & 0 \end{pmatrix} \xrightarrow{F_{21}(\frac{1}{2})} \begin{pmatrix} 1 & 0 & 0 & \frac{1}{2} & 0 & \frac{1}{2} \\ 0 & 2 & -2 & -1 & 0 & 1 \\ 0 & 0 & 1 & 0 & 1 & 0 \end{pmatrix}$$

$$\xrightarrow{D_2(\frac{1}{2})} \begin{pmatrix} 1 & 0 & 0 & \frac{1}{2} & 0 & \frac{1}{2} \\ 0 & 1 & -1 & -\frac{1}{2} & 0 & \frac{1}{2} \\ 0 & 0 & 1 & 0 & 1 & 0 \end{pmatrix} \xrightarrow{F_{32}(1)} \begin{pmatrix} 1 & 0 & 0 & \frac{1}{2} & 0 & \frac{1}{2} \\ 0 & 1 & 0 & -\frac{1}{2} & 0 & \frac{1}{2} \\ 0 & 0 & 1 & 0 & 1 & 0 \end{pmatrix}$$
$$= (I|A^{-1}).$$

Hence
$$A^{-1} = \begin{pmatrix} \frac{1}{2} & 0 & \frac{1}{2} \\ -\frac{1}{2} & 1 & \frac{1}{2} \\ 0 & 1 & 0 \end{pmatrix}$$
and we are through.

EXERCISES

Reduce each of the following matrices to an echelon matrix by performing a sequence of elementary row operations.

1. $\begin{pmatrix} 1 & 1 \\ 1 & -1 \end{pmatrix}$

2. $\begin{pmatrix} 2 & -1 & 1 \\ 1 & 0 & 2 \end{pmatrix}$

3. $\begin{pmatrix} 1 & -1 \\ 1 & 0 \\ 2 & 1 \end{pmatrix}$

4. $\begin{pmatrix} 2 & -1 & 1 \\ 1 & 0 & 1 \\ -1 & 1 & 1 \end{pmatrix}$

5. $\begin{pmatrix} 1 & 1 & -1 \\ 0 & 1 & 0 \\ -1 & 1 & 0 \\ 2 & 1 & 1 \end{pmatrix}$

Determine if the following $n \times n$ matrices are invertible. If so compute the inverse.

6. $\begin{pmatrix} 1 & 2 \\ -1 & 1 \end{pmatrix}$

7. $\begin{pmatrix} 2 & 1 \\ -2 & -1 \end{pmatrix}$

8. $\begin{pmatrix} 1 & 1 \\ -1 & 1 \end{pmatrix}$

9. $\begin{pmatrix} 1 & 1 \\ -1 & -1 \end{pmatrix}$

10. $\begin{pmatrix} 1 & -1 & 1 \\ 2 & 1 & 3 \\ 1 & 0 & 1 \end{pmatrix}$

11. $\begin{pmatrix} 1 & -1 & 1 \\ 0 & 1 & 0 \\ 1 & 0 & 1 \end{pmatrix}$

12. $\begin{pmatrix} 1 & 0 & 1 \\ 2 & -1 & 1 \\ 1 & -1 & 0 \end{pmatrix}$

13. $\begin{pmatrix} 0 & 1 & 0 \\ 1 & 0 & 0 \\ 0 & 0 & 1 \end{pmatrix}$

14. $\begin{pmatrix} 1 & 0 & -1 \\ 1 & -1 & 0 \\ -1 & 0 & 1 \end{pmatrix}$

15. $\begin{pmatrix} 1 & 0 & 0 & 1 \\ -1 & 1 & 0 & 1 \\ 2 & 0 & 1 & 0 \\ 1 & 1 & 1 & 1 \end{pmatrix}$

16. $\begin{pmatrix} 1 & -1 & 1 & 1 \\ 1 & 0 & 1 & 0 \\ 1 & 1 & 0 & -1 \\ 1 & 2 & 0 & -2 \end{pmatrix}$

17. $\begin{pmatrix} 0 & 0 & 0 & 1 \\ 1 & 0 & 0 & 0 \\ 0 & 1 & 0 & 0 \\ 0 & 0 & 1 & 0 \end{pmatrix}$

18. Show that each elementary row operation on an $m \times n$ matrix can be accomplished by multiplying on the left by the corresponding elementary matrix.

19. A *diagonal* $m \times n$ matrix is a matrix with all terms *off* the main diagonal equal to zero. Prove or disprove the following statement. Each $m \times n$ matrix is row equivalent to a diagonal matrix. (The terms $a_{11}, a_{22}, \ldots$, form the main diagonal of an $m \times n$ matrix.)

20. Show that the invertible $n \times n$ matrices are closed under matrix multiplication. Show that the inverse of an invertible matrix is again invertible.

21. If A and B are $n \times n$ matrices, show that AB is invertible if and only if both A and B are invertible.

22. Let $T(f)(x) = xf''(x) + f(x)$. Show that T is an invertible linear transformation in P_n, the space of polynomials of degree $\leq n$. Compute the matrix for T^{-1} with respect to the canonical basis in P_3. In P_3 solve the differential equation

$$xy'' + y = x^2 + x.$$

Is this solution unique?

23. An $n \times n$ matrix with one entry in each row and each column equal to one and all other entries equal to zero is called a *permutation matrix*. For example

$$\begin{pmatrix} 0 & 1 \\ 1 & 0 \end{pmatrix}, \quad \begin{pmatrix} 0 & 1 & 0 \\ 0 & 0 & 1 \\ 1 & 0 & 0 \end{pmatrix}, \quad \text{and} \quad \begin{pmatrix} 0 & 1 & 0 & 0 \\ 1 & 0 & 0 & 0 \\ 0 & 0 & 0 & 1 \\ 0 & 0 & 1 & 0 \end{pmatrix}$$

are permutation matrices. Show that each permutation matrix is invertible, and furthermore the inverse of a permutation matrix is again a permutation matrix. Show also that the set of permutation matrices is closed under matrix multiplication.

24. Show that if A is an invertible matrix, then A is the product of elementary matrices.

25. Show that if A and B are $m \times n$ matrices and $B = CA$ where C is an invertible $m \times m$ matrix, then A and B are row equivalent.

3.7 Change of Basis Matrices

In this section we shall consider the following problem.

Suppose $\{X_1, \ldots, X_n\}$ is a basis for the vector space V, and suppose the matrix of a linear transformation $T \in L(V)$ with respect to this basis is given by A_T. If now $\{Y_1, \ldots, Y_n\}$ is another basis for V, what is the matrix of T with respect to this second basis?

To begin the discussion we note that changing the basis from $\{X_1, \ldots, X_n\}$ to $\{Y_1, \ldots, Y_n\}$ defines an invertible matrix $C = (c_{jk})$ in $R_{n,n}$. The matrix (c_{jk}) is defined by the formula

$$\text{(1)} \qquad Y_k = \sum_{j=1}^{n} c_{jk} X_j.$$

To show this matrix (c_{jk}) is invertible and to compute its inverse we reverse the process and write X_l as a linear combination of the basis vectors Y_k. Then

$$\text{(2)} \qquad X_l = \sum_{k=1}^{n} d_{kl} Y_k$$

which defines an $n \times n$ matrix $D = (d_{kl})$. Substituting (1) in (2) we have

$$\text{(3)} \qquad X_l = \sum_{j=1}^{n} \sum_{k=1}^{n} c_{jk} d_{kl} X_j.$$

This equation in turn defines a matrix $E = (e_{jl})$ whose entries are

$$\text{(4)} \qquad e_{jl} = \sum_{k=1}^{n} c_{jk} d_{kl},$$

and indeed $E = CD$. Combining (3) and (4) we obtain

$$X_l = \sum_{j=1}^{n} e_{jl} X_j \qquad \text{for } l = 1, \ldots, n.$$

But since $\{X_1, \ldots, X_n\}$ is a basis, X_l can be written uniquely as a linear combination of the vectors $\{X_1, \ldots, X_n\}$. Hence $e_{jl} = 0$ if $j \neq l$ and $e_{ll} = 1$. This is just the statement that

$$CD = E = I$$

where I is the identity matrix. By Theorem 3.16 this implies that the matrix C is invertible and $C^{-1} = D$.

Definition. If $\{X_1, \ldots, X_n\}$ and $\{Y_1, \ldots, Y_n\}$ are bases for the vector space V, then the matrix $C = (c_{jk})$ defined by

$$\text{(1)} \qquad Y_k = \sum_{j=1}^{n} c_{jk} X_j$$

is called the matrix of the change of basis from the basis $\{X_j\}$ to the basis $\{Y_k\}$.

The reason for the terminology is that the matrix C defined by (1) is the same as the matrix of the linear transformation which takes X_k into Y_k for $k = 1, \ldots, n$. Hence we often speak of transforming one basis onto another by a reflection, rotation, etc.

Example 1. In $V_3(R)$ the change of basis matrix from the canonical basis $\{I_k\}$ to $J_1 = (1, 1, 1)$, $J_2 = (1, -1, 0)$, $J_3 = (2, 0, 0)$ is

$$\begin{pmatrix} 1 & 1 & 2 \\ 1 & -1 & 0 \\ 1 & 0 & 0 \end{pmatrix}.$$

In general in $V_n(R)$ the change of basis matrix from the canonical basis $\{I_k\}$ to a basis whose elements are $J_k = (a_{1k}, \ldots, a_{nk})$ has its *columns* equal to the basis vectors $\{J_1, \ldots, J_n\}$.

If $\{X_j\}$ and $\{Y_k\}$ are bases for the vector space V and we write a fixed vector X as a linear combination of these basis elements, that is,

(5) $$X = \sum_{j=1}^{n} p_j X_j$$

and

(6) $$X = \sum_{k=1}^{n} q_k Y_k$$

then (1) enables us to determine the scalars $\{p_j\}$ in terms of the scalars $\{q_k\}$. Indeed substituting (1) in (6) we obtain

$$X = \sum_{k=1}^{n} q_k \sum_{j=1}^{n} c_{jk} X_j$$

$$= \sum_{j=1}^{n} \sum_{k=1}^{n} c_{jk} q_k X_j.$$

Since X may be written uniquely as a linear combination of the vectors X_j, we conclude that for each j

(7) $$p_j = \sum_{k=1}^{n} c_{jk} q_k.$$

If $P = \begin{pmatrix} p_1 \\ \vdots \\ p_n \end{pmatrix}$ and $Q = \begin{pmatrix} q_1 \\ \vdots \\ q_n \end{pmatrix}$ are the coefficient vectors in (5) and (6), then (7) implies that $P = CQ$.

Example 2. If $J_1 = (1, -1, 1)$, $J_2 = (2, 1, 0)$ and $J_3 = (1, 0, 0)$, then the change of basis matrix from the canonical basis to $\{J_1, J_2, J_3\}$ is given by

$$\begin{pmatrix} 1 & 2 & 1 \\ -1 & 1 & 0 \\ 1 & 0 & 0 \end{pmatrix}.$$

If $X = J_1 - J_2 + 2J_3$, then to write $X = p_1 I + p_2 I_2 + p_3 I_3$ we compute

$$\begin{pmatrix} p_1 \\ p_2 \\ p_3 \end{pmatrix} = \begin{pmatrix} 1 & 2 & 1 \\ -1 & 1 & 0 \\ 1 & 0 & 0 \end{pmatrix} \begin{pmatrix} 1 \\ -1 \\ 2 \end{pmatrix} = \begin{pmatrix} 1 \\ -2 \\ 1 \end{pmatrix}$$

Hence $X = I_1 - 2I_2 + I_3 = (1, -2, 1)$.

Next we compute the matrix of a linear transformation T under a change of basis.

THEOREM 3.20. *Let $T \in L(V)$ and assume A is the matrix of T with respect to the basis $\{X_1, \ldots, X_n\}$ and B is the matrix of T with respect to the basis $\{Y_1, \ldots, Y_n\}$. If C is the matrix of the change of basis from $\{X_1, \ldots, X_n\}$ to $\{Y_1, \ldots, Y_n\}$, then*

$$B = C^{-1}AC.$$

Proof. We prove the result by a direct calculation. Let $C = (c_{jk})$ be the change of basis matrix from $\{X_j\}$ to $\{Y_k\}$ and let $C^{-1} = (d_{ml})$ be its inverse. If $A = (a_{lj})$ is the matrix of T with respect to the basis $\{X_j\}$, then the matrix $B = (b_{mk})$ of the transformation T with respect to the basis $\{Y_k\}$ is defined by

$$T(Y_k) = \sum_{m=1}^{n} b_{mk} Y_m.$$

But

(8) $$Y_k = \sum_{j=1}^{n} c_{jk} X_j.$$

Applying T to both sides of (8) yields

$$T(Y_k) = \sum_{j=1}^{n} c_{jk} T(X_j)$$

$$= \sum_{j=1}^{n} c_{jk} \sum_{l=1}^{n} a_{lj} X_l \text{ (by definition of the matrix } A = (a_{lj}))$$

$$= \sum_{j=1}^{n} c_{jk} \sum_{l=1}^{n} a_{lj} \sum_{m=1}^{n} d_{ml} Y_m \text{ (Why?)}$$

$$= \sum_{m=1}^{n} \left(\sum_{j=1}^{n} \sum_{l=1}^{n} d_{ml} a_{lj} c_{jk} \right) Y_m = \sum_{k=1}^{n} b_{mk} Y_m.$$

Since $\{Y_1, \ldots, Y_n\}$ is a basis, we infer that for each choice of the indices m, k

$$b_{mk} = \sum_{j} \sum_{l} (d_{ml} a_{lj} c_{jk}).$$

However, the right hand side of this expression is just the mkth entry in the matrix $C^{-1}AC$. Therefore $B = C^{-1}AC$.

Example 3. If the matrix of a transformation T in $V_2(R)$ is $\begin{pmatrix} 1 & 1 \\ 1 & 2 \end{pmatrix}$ with respect to the canonical basis $I_1 = (1, 0), I_2 = (0, 1)$, what is the matrix of T with respect to the basis $J_1 = (1, 1), J_2 = (1, -1)$? The change of basis matrix from the basis $\{I_k\}$ to the basis $\{I_k\}$ is $C = \begin{pmatrix} 1 & 1 \\ 1 & -1 \end{pmatrix}$. To compute C^{-1} we reduce $(C|I)$ to $(I|C^{-1})$ by elementary row operations. Indeed

$$(C|I) = \begin{pmatrix} 1 & 1 & | & 1 & 0 \\ 1 & -1 & | & 0 & 1 \end{pmatrix} \xrightarrow{F_{12}(-1)} \begin{pmatrix} 1 & 1 & | & 1 & 0 \\ 0 & -2 & | & -1 & 1 \end{pmatrix} \xrightarrow{D_2(\frac{1}{2})}$$

$$\begin{pmatrix} 1 & 1 & | & 1 & 0 \\ 0 & 1 & | & \frac{1}{2} & -\frac{1}{2} \end{pmatrix} \xrightarrow{F_{2,1}(-1)} \begin{pmatrix} 1 & 0 & | & \frac{1}{2} & \frac{1}{2} \\ 0 & 1 & | & \frac{1}{2} & -\frac{1}{2} \end{pmatrix} = (I|C^{-1})$$

Therefore the change of basis matrix from $\{J_1, J_2\}$ to $\{I_1, I_2\}$ is

$$C^{-1} = \begin{pmatrix} \frac{1}{2} & \frac{1}{2} \\ \frac{1}{2} & -\frac{1}{2} \end{pmatrix}.$$

Consequently matrix B of the transformation T with respect to the basis $\{J_1, J_2\}$ is

$$B = \begin{pmatrix} \frac{1}{2} & \frac{1}{2} \\ \frac{1}{2} & -\frac{1}{2} \end{pmatrix} \begin{pmatrix} 1 & 1 \\ 1 & 2 \end{pmatrix} \begin{pmatrix} 1 & 1 \\ 1 & -1 \end{pmatrix}$$

$$= \frac{1}{2}\begin{pmatrix} 1 & 1 \\ 1 & -1 \end{pmatrix} \begin{pmatrix} 2 & 0 \\ 3 & -1 \end{pmatrix}$$

$$= \frac{1}{2}\begin{pmatrix} 5 & -1 \\ -1 & 1 \end{pmatrix} = \begin{pmatrix} \frac{5}{2} & -\frac{1}{2} \\ -\frac{1}{2} & \frac{1}{2} \end{pmatrix}.$$

EXERCISES

1. In $V_2(R)$ compute the change of basis matrix from the canonical basis $\{I_1, I_2\}$ to the basis $\{J_1, J_2\}$ and compute the change of basis matrix from $\{J_1, J_2\}$ to the canonical basis if

   (a) $J_1 = (-1, 0), J_2 = (1, 1)$
   (b) $J_1 = (1, -1), J_2 = (0, 2)$
   (c) $J_1 = (2, 1), J_2 = (1, 2)$
   (d) $J_1 = (1, -1), J_2 = (1, 1)$.

2. For each of the bases $\{J_1, J_2\}$ of Exercise 1 if $X = 3J_1 - 2J_2$ determine scalars x_1, x_2 so that $X = x_1 I_1 + x_2 I_2$. If $X = 2I_1 + I_2$ determine scalars $\{y_1, y_2\}$ so that $X = y_1 J_1 + y_2 J_2$.

3. In $V_3(R)$ compute the change of basis matrix from the canonical basis $\{I_1, I_2, I_3\}$ to the basis $\{J_1, J_2, J_3\}$ and compute the change of basis matrix from $\{J_1, J_2, J_3\}$ to the canonical basis, if

   (a) $J_1 = (1, -1, 1), J_2 = (0, 1, 1), J_3 = (0, 0, 2)$
   (b) $J_1 = (2, 0, 0), J_2 = (1, -1, 0), J_3 = (2, 1, -1)$
   (c) $J_1 = (2, 1, -1), J_2 = (1, 0, 1), J_3 = (1, 1, 1)$
   (d) $J_1 = (1, 0, -1), J_2 = (1, 1, 1), J_3 = (0, -1, 1)$.

4. For each of the bases $\{J_1, J_2, J_3\}$ of Exercise 3 if $X = 2J_1 + J_2 - J_3$ determine scalars x_1, x_2, x_3 so that $X = x_1 I_1 + x_2 I_2 + x_3 I_3$.

5. In P_3, the space of polynomials of degree ≤ 3, compute the change of basis matrix from the basis $\{1, x, x^2, x^3\}$ to the basis $\{1, 1+x, 1+x+x^2, 1+x+x^2+x^3\}$. Compute the change of basis matrix from the second basis to the first.

6. In P_3 compute the change of basis matrix from $\{1, 1+x, (1+x)^2, (1+x)^3\}$ to $\{1, x, x^2, x^3\}$.

7. Let V be the space spanned by $\{\cos^2 x, \sin^2 x\}$. $\{\cos 2x, 1\}$ is also a basis for this space. Compute the change of basis matrix from the first to the second basis and from the second to the first.

8. In $V_2(R)$ a linear transformation has matrix $\begin{pmatrix} 1 & 2 \\ -1 & 1 \end{pmatrix}$ with respect to the canonical basis. Determine the matrix of this transformation with respect to the basis $\{J_1, J_2\}$ if

   (a) $J_1 = (1, -1), J_2 = (0, 1)$
   (b) $J_1 = (2, 0), J_2 = (1, 2)$
   (c) $J_1 = (2, 1), J_2 = (1, 2)$
   (d) $J_1 = (1, -1), J_2 = (1, 1)$.

9. In $V_3(R)$ a linear transformation has matrix $\begin{pmatrix} 1 & 0 & -1 \\ -1 & 1 & 0 \\ 0 & 0 & 1 \end{pmatrix}$ with respect to the canonical basis. Determine the matrix of this transformation with respect to the basis $\{J_1, J_2, J_3\}$ if

   (a) $J_1 = (1, 0, 1), J_2 = (0, 1, 1), J_3 = (0, 0, -1)$

(b) $J_1 = (-1, 0, 0), J_2 = (1, -1, 0), J_3 = (1, 1, 1)$
(c) $J_1 = (2, 1, -1), J_2 = (1, 0, 1), J_3 = (1, 1, 1)$
(d) $J_1 = (1, 0, -1), J_2 = (1, 1, 1), J_3 = (0, -1, 1)$.

10. Determine the matrix of the differentiation operator in P_3 with respect to the basis $\{1, 1 + x, 1 + x + x^2, 1 + x + x^2 + x^3\}$.

11. Let T be a reflection in $V_2(R)$ through the line $a_1x_1 + a_2x_2 = 0$.
    (a) Determine a basis $\{J_1, J_2\}$ for $V_2(R)$ so that the matrix for T with respect to this basis is $\begin{pmatrix} 1 & 0 \\ 0 & -1 \end{pmatrix}$.
    (b) Using part (a) show that the matrix of T with respect to the canonical basis is given by
$$A = \frac{1}{a_1^2 + a_2^2}\begin{pmatrix} a_2^2 - a_1^2 & -2a_1a_2 \\ -2a_1a_2 & -(a_2^2 - a_1^2) \end{pmatrix}.$$

3.8 Similarity of Matrices; Eigenvalues and Eigenvectors

If T is a reflection in $V_2(R)$ through a line l, then we have seen that the matrix for T with respect to the canonical basis in $V_2(R)$ is given by

(1) $$A = \frac{1}{a_1^2 + a_2^2}\begin{pmatrix} a_2^2 - a_1^2 & -2a_1a_2 \\ -2a_1a_2 & -(a_2^2 - a_1^2) \end{pmatrix}$$

where $a_1x_1 + a_2x_2 = 0$ is the equation of the line l. However, if we choose a basis $\{J_1, J_2\}$ for V_2 consisting of a vector J_1 parallel to l and a vector J perpendicular to l then the matrix of the reflection T with respect to this basis is the very simple diagonal matrix $\begin{pmatrix} 1 & 0 \\ 0 & -1 \end{pmatrix}$ See Figure 12. This is clear since $T(J_1) = J_1$ and $T(J_2) = -J_2$.

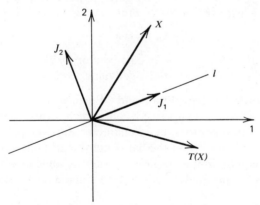

Fig. 12 Matrix of reflection T through the line l is $\begin{pmatrix} 1 & 0 \\ 0 & -1 \end{pmatrix}$ if $\{J_1, J_2\}$ is the basis.

We may infer from Theorem 3.20 of the last section that in this case there is an invertible change of basis matrix C such that
$$C^{-1}AC = \begin{pmatrix} 1 & 0 \\ 0 & -1 \end{pmatrix}.$$

Indeed C is just the matrix of the rotation of the basis $\{I_1, I_2\}$ into the basis $\{J_1, J_2\}$.

If T now is a linear transformation in an n-dimensional vector space V, and A is the matrix of T with respect to some basis then we know from

Theorem 3.20 that the matrix B will be the matrix for T with respect to another basis in V if and only if there exists an invertible matrix C satisfying

(2) $$C^{-1}AC = B.$$

This property of matrices is very important, and we give it a name.

Definition. *The $n \times n$ matrices A and B are called* similar *if there exists an invertible $n \times n$ matrix C such that*

$$B = C^{-1}AC.$$

An important problem in this regard is to determine which matrices A are similar to diagonal matrices. The discussion above shows that this is, for example, a property of reflection matrices. We are not in a position to solve this question completely in this section, but we shall develop some preliminary material.

If a linear transformation $T \in L(V)$ has a diagonal matrix

$$D = \begin{pmatrix} a_1 & & 0 \\ & \ddots & \\ 0 & & a_n \end{pmatrix}$$

with respect to some basis $\{Y_1, \ldots, Y_n\}$, then it follows immediately from the definition of the matrix of a linear transformation, that

$$T(Y_k) = a_k Y_k$$

for each of the vectors Y_k. We need some notation to describe this situation.

Definition. *Let V be a vector space and suppose $T \in L(V)$. If X is a nonzero vector in V with the property that*

(3) $$T(X) = \lambda X$$

for some scalar λ, then X is called an eigenvector *for the transformation T. The scalar λ is called the associated* eigenvalue.

The German words "Eigenvektor" and "Eigenwert" have literal translations "own vector" and "own value." Since such terms do not represent idiomatic English, the expressions "characteristic vector" or "proper vector" are often used. We prefer the half-English half-German expressions which are more common in the literature. Also it is somewhat traditional to denote eigenvalues by the small Greek letter λ (lambda). We shall adhere to that terminology.

If A is an $n \times n$ matrix and X is a nonzero vector in $V_n(R)$ with the property that

(4) $$AX = \lambda X$$

for some scalar λ, then X is an eigenvector for the matrix A and λ is the associated eigenvalue. We may interpret this definition geometrically in the following way. If $\lambda = 0$, then X is an eigenvector associated with the eigenvalue 0 if $X \neq 0$, and $AX = 0$. Thus a nonzero vector in the null space of A is an eigenvector associated with the eigenvalue 0. If $\lambda \neq 0$, then $AX = \lambda X$ just says that AX is parallel to the vector X. Notice that if X is an eigenvector associated with an eigenvalue λ, then X is not uniquely determined since if $AX = \lambda X$ then $A(aX) = \lambda(aX)$ for any scalar a. Hence for $a \neq 0$, aX is an eigenvector associated with λ.

We note also that if A is any matrix for a transformation $T \in L(V)$, then λ is an eigenvalue for T if and only if λ is an eigenvalue for the matrix A. The easiest way to see that is to observe that by definition the number λ is an eigenvalue of T if and only if the transformation $T - \lambda I$ is not one-to-one. This will occur if and only if $T - \lambda I$ is *not* invertible. This happens if and only if the matrix $A - \lambda I$ is not invertible. The latter is equivalent to the statement that for some nonzero $X \in V_n(R)$.

$$(A - \lambda I)X = 0$$

or

$$AX = \lambda X.$$

The above observation shows also that if A and B are similar $n \times n$ matrices, that is $B = C^{-1}AC$ for some invertible matrix C, then A and B have the same eigenvalues. This follows since A and B are similar if and only if the two matrices represent the same linear transformation with respect to different bases. An alternative verification of this fact is contained in Exercise 13.

We are now in a position to give a necessary and sufficient condition for a matrix A to be similar to a diagonal matrix Λ.

THEOREM 3.21. *The $n \times n$ matrix A is similar to a diagonal matrix if and only if there is a basis for $V_n(R)$ consisting of eigenvectors for A.*

Proof. Let $X_1, \ldots, X_n$ be n eigenvectors for the matrix A and let $\lambda_1, \ldots, \lambda_n$ be associated eigenvalues. Let C be the $n \times n$ matrix having the vectors $X_1, \ldots, X_n$ as columns. Set Λ equal to the diagonal matrix

$$\begin{pmatrix} \lambda_1 & & 0 \\ & \ddots & \\ 0 & & \lambda_n \end{pmatrix}.$$

It follows immediately from the definition of matrix multiplication that the columns of the matrix AC are precisely the vectors $AX_1, \ldots, AX_n$. Furthermore the columns of the matrix $C\Lambda$ are precisely the vectors $\lambda_1 X_1, \ldots, \lambda_n X_n$. (The student should convince himself that these statements are correct.) Therefore $AX_k = \lambda_k X_k$, $k = 1, \ldots, n$, if and only if

(5) $$AC = C\Lambda.$$

Now the matrix C has an inverse if and only if the rank of $C = n$, or equivalently if the columns of C are linearly independent. This will be the case if and only if $\{X_1, \ldots, X_n\}$ is a basis for $V_n(R)$. Thus if $\{X_1, \ldots, X_n\}$ is a basis we may multiply (5) on the left by C^{-1} obtaining

(6) $$C^{-1}AC = \Lambda.$$

Thus we have proved that A is similar to Λ.

For the converse, if (6) holds then by multiplying this equation on the left by C we obtain (5). This in turn implies that the columns of C are linearly independent eigenvectors for A. Hence $V_n(R)$ has a basis consisting of eigenvectors for A and the proof is complete.

We note that if T is a reflection in $V_2(R)$ through the line l then the eigenvalues of T are $+1$ and -1. Corresponding eigenvectors are vectors J_1, parallel to l, and J_2 perpendicular to l.

The following condition on the eigenvalues of a linear transformation T is helpful in constructing a basis for the space V consisting of eigenvectors.

THEOREM 3.22. *Let $T \in L(V)$ and let $\lambda_1, \ldots, \lambda_k$ be distinct eigenvalues for the transformation T. If for each j, X_j is an eigenvector associated with the eigenvalue λ_j, then the set $\{X_1, \ldots, X_k\}$ is linearly independent.*

Proof. We argue by induction on the number k. If $k = 1$, the result is certainly true since an eigenvector by definition is nonzero. So assume $X_1, \ldots, X_{k+1}$ are eigenvectors associated with the distinct eigenvalues $\lambda_1, \ldots, \lambda_{k+1}$ and assume any subset of k of the vectors $\{X_1, \ldots, X_{k+1}\}$ is linearly independent. To prove $\{X_1, \ldots, X_{k+1}\}$ is linearly independent assume

(7) $$c_1 X_1 + \cdots + c_{k+1} X_{k+1} = 0.$$

We must show $c_1 = c_2 = \cdots = c_{k+1} = 0$. Applying T to equation (5) and using the fact that each vector X_j is an eigenvector, we have

(8) $$\lambda_1 c_1 X_1 + \cdots + \lambda_{k+1} c_{k+1} X_{k+1} = 0.$$

Multiplying (7) by λ_1 and subtracting from (8) yields

$$c_2 (\lambda_2 - \lambda_1) X_2 + \cdots + c_{k+1}(\lambda_{k+1} - \lambda_1) X_{k+1} = 0.$$

Since $\lambda_j - \lambda_1 \neq 0$, $j = 2, \ldots, k+1$ and $\{X_2, \ldots, X_{k+1}\}$ is linearly independent, it follows that $c_2 = \cdots = c_{k+1} = 0$. Since $X_1 \neq 0$, this implies that $c_1 = 0$. Hence $\{X_1, \ldots, X_{k+1}\}$ is linearly independent. An application of the principle of induction completes the proof.

COROLLARY. *If $\dim V = n$ and T has n distinct eigenvalues $\lambda_1, \ldots, \lambda_n$, then there exists a basis for V consisting of eigenvectors for T. Moreover, each matrix for T is similar to the diagonal matrix*

$$D = \begin{pmatrix} \lambda_1 & & & 0 \\ & \lambda_2 & & \\ & & \ddots & \\ 0 & & & \lambda_n \end{pmatrix}.$$

In order to determine if an $n \times n$ matrix A is similar to a diagonal matrix we need some technique for determining the eigenvalues of the matrix A. The fundamental tool in this regard is the notion of the *determinant* of the matrix A. The definition and properties of this concept we develop in the next chapter.

EXERCISES
1. If the linear transformation T is the reflection in $V_2(R)$ through the line with equation $x_1 - x_2 = 0$, determine a basis for $V_2(R)$ consisting of eigenvectors for T.
2. Answer Exercise 1 if the line l has equation $2x_1 + x_2 = 0$.
3. If the linear transformation T is the reflection in $V_3(R)$ through the plane $x_3 = 0$, determine a basis for $V_3(R)$ consisting of eigenvectors for T. What are the eigenvalues for T? Determine the matrix for T with respect to the chosen basis.
4. Answer Exercise 3 if the plane has equation $x_1 + x_2 + x_3 = 0$.
5. Let the n-dimensional space V be the direct sum of the sub-spaces W_1 and W_2. If T is the projection of V onto W_1 along W_2, show that a basis for V may be chosen so that the matrix of T with respect to this basis is a diagonal matrix. What are the eigenvalues for T?

6. Let A be a nonzero $n \times n$ matrix. If 0 is the only eigenvalue for A, show that A is not similar to a diagonal matrix.
7. If A is an $n \times n$ matrix satisfying $A^2 = 0$, show that zero is the only eigenvalue for A.
8. Show that $\begin{pmatrix} 0 & 1 \\ 0 & 0 \end{pmatrix}$ is not similar to a diagonal matrix.
9. Show that if $T \in L(V)$ and $T^m = 0$ for some positive integer m, then 0 is the only eigenvalue for T.
10. Show that if A is a nonzero $n \times n$ matrix such that $A^k = 0$ for some positive integer k, then A is not similar to a diagonal matrix.
11. Let $D(f) = f'$ be the differentiation transformation in P_n. Can a basis be constructed for P_n so that the matrix for D is a diagonal matrix? Explain. (What are the eigenvalues for D?)
12. If $T \in L(V)$ and dim $V = n$, show that T can have at most n distinct eigenvalues.
13. Let A and B be similar $n \times n$ matrices. Show that λ is an eigenvalue for A if and only if λ is an eigenvalue for B. (If $B = C^{-1}AC$ and if X is an eigenvector for A associated with the eigenvalue λ, show that $Y = C^{-1}X$ is an eigenvector for B associated with the eigenvalue λ. Why must Y be different from zero?)
14. If A is an $n \times n$ matrix and λ is a scalar, show that the set of vectors X satisfying $AX = \lambda X$ is a subspace of $V_n(R)$. If this subspace is not the zero space, it is called the *eigenspace* associated with the eigenvalue λ.

CHAPTER FOUR

DETERMINANTS

4.1 Definition of a Determinant Function

In Chapter 1 a 2×2 determinant was defined by the formula

$$\begin{vmatrix} a_{11} & a_{12} \\ a_{21} & a_{22} \end{vmatrix} = a_{11}a_{22} - a_{21}a_{12}.$$

To discuss the cross product of two vectors in $V_3(R)$ we extended this notion to a 3×3 determinant. To discuss determinants in general we shall change our point of view slightly. First we shall regard the determinant as a real-valued function defined on the set of all $n \times n$ matrices. In place of vertical lines we shall use the capital letter D to denote the determinant function. Thus if $A = \begin{pmatrix} a_{11} & a_{12} \\ a_{21} & a_{22} \end{pmatrix}$, then we write $D(A) = D\begin{pmatrix} a_{11} & a_{12} \\ a_{21} & a_{22} \end{pmatrix}$
$= a_{11}a_{22} - a_{21}a_{12}.$

Before trying to define the determinant of an $n \times n$ matrix, let us investigate some of the properties of the determinant considered as a function defined on the set of 2×2 matrices. We ask first if D is a linear function. If so $D(aA) = aD(A)$ for each scalar a and 2×2 matrix A. However, if $I = \begin{pmatrix} 1 & 0 \\ 0 & 1 \end{pmatrix}$, then $D(I) = 1$ but $D(2I) = 4$. Hence D is not linear.

If we change our point of view slightly, we may establish a certain linearity property for the function D. First if $R_1 = (a_{11}, a_{12})$, and $R_2 = (a_{21}, a_{22})$ denote the rows of the matrix A, then the set of all 2×2 matrices A may be identified with the set of all pairs of vectors R_1, R_2 from $V_2(R)$. Accordingly let us write

(1) $$D(A) = D(R_1, R_2).$$

If we fix the second variable R_2 in (1) then we claim that D is linear in the first variable R_1. To verify this we must prove that if $R_1 = (a_{11}, a_{12})$ and $S_1 = (b_{11}, b_{12})$ and a and b are scalars, then

$$D(aR_1 + bS_1, R_2) = aD(R_1, R_2) + bD(S_1, R_2).$$

But

$$D(aR_1 + bS_1, R_2) = D\begin{pmatrix} aa_{11} + bb_{11} & aa_{12} + bb_{12} \\ a_{21} & a_{22} \end{pmatrix}$$

$$= (aa_{11} + bb_{11})a_{22} - a_{21}(aa_{12} + bb_{12})$$

$$= aa_{11}a_{22} - aa_{21}a_{12} + bb_{11}a_{22} - ba_{21}b_{12}$$

$$= aD\begin{pmatrix} a_{11} & a_{12} \\ a_{21} & a_{22} \end{pmatrix} + bD\begin{pmatrix} b_{11} & b_{12} \\ a_{21} & a_{22} \end{pmatrix}$$

$$= aD(R_1, R_2) + bD(S_1, R_2).$$

In a similar fashion we may verify that $D(R_1, R_2)$ is linear in the second variable when the first is held fixed.

Secondly we note that if the rows of A are identical then $D(A) = 0$. This is clear since if $A = \begin{pmatrix} a & b \\ a & b \end{pmatrix}$ then $D(A) = D(R_1, R_1) = ab - ab = 0$. Lastly if $I = \begin{pmatrix} 1 & 0 \\ 0 & 1 \end{pmatrix}$ then $D(I) = 1$.

Before proceeding let us note that the determinant function D is the only function defined on the set 2×2 matrices with these properties. Suppose that F is a real-valued function defined on the set of 2×2 matrices A, or equivalently the set of pairs of vectors R_1, R_2 from $V_2(R)$, which satisfies
  (i) $F(aR_1 + bS_1, R_2) = aF(R_1, R_2) + bF(S_1, R_2)$
  (ii) $F(R_1, aR_2 + bS_2) = aF(R_1, R_2) + bF(R_1, S_2)$
  (iii) $F(R_1, R_2) = 0$ whenever $R_1 = R_2$
  (iv) $F(I_1, I_2) = 1$
where $I_1 = (1, 0)$, $I_2 = (0, 1)$. Then we assert that the function F is identical with the determinant function D. To verify that this is the case is easy. Note first that by (iii)

$$F(I_1 + I_2, I_1 + I_2) = 0.$$

But applying (i) and (ii) we have

$$\begin{aligned}
0 = F(I_1 + I_2, I_1 + I_2) &= F(I_1, I_1 + I_2) + F(I_2, I_1 + I_2) \\
&= F(I_1, I_1) + F(I_1, I_2) + F(I_2, I_1) + F(I_2, I_2) \\
&= F(I_1, I_2) + F(I_2, I_1)
\end{aligned}$$

by (iii). Therefore $F(I_2, I_1) = -F(I_1, I_2) = -1$. Now let $A = \begin{pmatrix} a & b \\ c & d \end{pmatrix}$. (For simplicity we drop the subscript notation.) Then

$$R_1 = (a, b) = aI_1 + bI_2$$

and

$$R_2 = (c, d) = cI_1 + dI_2.$$

Applying (i) and (ii) we have

$$\begin{aligned}
F(A) = F(R_1, R_2) &= F(aI_1 + bI_2, cI_1 + dI_2) \\
&= aF(I_1, cI_1 + dI_2) + bF(I_2, cI_1 + dI_2) && \text{(by (i))} \\
&= acF(I_1, I_1) + adF(I_1, I_2) + bcF(I_2, I_1) + bdF(I_2, I_2) && \text{(by (ii))} \\
&= adF(I_1, I_2) + bcF(I_2, I_1) && \text{(by (iii))} \\
&= ad - bc.
\end{aligned}$$

Thus we have shown that if we define the determinant D of a 2×2 matrix A by the formula $D(A) = a_{11}a_{22} - a_{21}a_{12}$ and regard $D(A) = D(R_1, R_2)$ as a function of pairs of vectors R_1, R_2, then D satisfies properties (i)–(iv). Conversely if $F(A) = F(R_1, R_2)$ is any function satisfying (i)–(iv) then we must have

$$F(A) = a_{11}a_{22} - a_{21}a_{12} = D(A).$$

We wish now to generalize these ideas to $n \times n$ matrices.

Suppose F is a real-valued function defined on $R_{n,n}$. Thus for each $n \times n$ matrix A, $F(A)$ is a specific real number. Now if we denote the rows of A by $R_1, \ldots, R_n$, then each vector $R_k \in V_n(R)$, and we may write $F(A) = F(R_1, \ldots, R_n)$. In this way we may regard F as a function of n variables $R_1, \ldots, R_n$ each variable ranging over the vector space $V_n(R)$. Thus if

$X_1, \ldots, X_n$ are n vectors from $V_n(R)$, then we may form the matrix B whose rows are $X_1, \ldots, X_n$ and indeed

$$F(B) = F(X_1, \ldots, X_n).$$

In this way we may regard any function of n variables where each variable ranges over $V_n(R)$ as a function defined on the $n \times n$ matrices and conversely. In particular $F(aX_1 + bY_1, X_2, \ldots, X_n)$ is the value of the function F at the matrix B whose first row is the vector $aX_1 + bY_1$, and whose remaining rows are the vectors $X_2, \ldots, X_n$.

Definition. Let $R_1, \ldots, R_n$ be the rows of the matrix A in $R_{n,n}$. Let F be a real-valued function defined on $R_{n,n}$. Then F is called n-linear if $F(A) = F(R_1, \ldots, R_n)$ is linear in each variable R_k when the others are held fixed. That is, for each $k, k = 1, \ldots, n$, if $R_1, \ldots, R_k, S_k, R_{k+1}, \ldots, R_n$ are vectors in $V_n(R)$, and $a, b \in R$, then

$$F(R_1, \ldots, aR_k + bS_k, \ldots, R_n)$$
$$= aF(R_1, \ldots, R_k, \ldots, R_n) + bF(R_1, \ldots, S_k, \ldots, R_n).$$

The n-linear function F is called alternating if

$$F(R_1, \ldots, R_n) = 0$$

whenever two of the vectors R_i, R_j, $i \neq j$ are identical.

Thus if two rows of the matrix A are identical, an alternating n-linear function F satisfies $F(A) = 0$. The reason the term alternating is used for this property is because of the following result.

PROPOSITION 4.1 Let F be an alternating n-linear function defined on $R_{n,n}$. Let A' be the $n \times n$ matrix obtained from A by interchanging two rows. Then

$$F(A) = -F(A').$$

Proof. The argument is exactly the same as the one used to prove that $F(I_2, I_1) = -F(I_1, I_2)$ on p. 125 where F is a 2-linear alternating function defined on the set of 2×2 matrices. If $R_1, \ldots, R_n$ are the rows of the $n \times n$ matrix A, let B be the matrix whose ith and jth rows are $R_i + R_j$, and whose other rows are the corresponding rows of A. Then since F is alternating, $F(B) = 0$. But since F is n-linear as well as alternating, $F(B) = F(A) + F(A')$. (See the argument on p. 125.) Consequently

$$F(A) = -F(A'),$$

and the proposition is proved.

Definition. An alternating n-linear function F defined on the set of $n \times n$ matrices which satisfies $F(I) = 1$ is called a determinant function.

Examples of n-linear functions are easy to construct. Indeed

$$F(A) = a_{11} \cdot a_{22} \cdots a_{nn}$$

where a_{kk} are the diagonal elements of the matrix A is easily seen to be one such function. However, considered as a function defined on all of $R_{n,n}$ this function is not alternating. On the other hand, if we set $F(A) = 0$ for each $A \in R_{n,n}$ then this defines an alternating n-linear function on $R_{n,n}$. But it is by no means obvious that there are any nonzero alternating n-linear functions.

DEFINITION OF A DETERMINANT FUNCTION

Our problem is to show that indeed nonzero alternating n-linear functions defined on the set of $n \times n$ matrices exist. Moreover there is only one such function with the property that $F(I) = 1$. This function is what we call the *determinant* of an $n \times n$ matrix. We summarize the problem in the following theorem. Its proof will occupy us for some time.

THEOREM 4.2 *There exists a unique alternating n-linear real-valued function defined on $R_{n,n}$ satisfying*

$$D(I) = 1.$$

EXERCISES

1. Let $X, Y \in V_2(R)$ and let (X, Y) be the inner product of the vectors X and Y. Verify that (X, Y) is a 2-linear function defined on $R_{2,2}$. Is (X, Y) alternating?
2. Let $A = (a_{ij})$ be a 3×3 matrix. Verify that $F(A) = a_{11} \cdot a_{22} \cdot a_{33}$ is a 3-linear function defined on $R_{3,3}$. Is F alternating?
3. Let $\tilde{R}_{3,3}$ be the set of upper triangular 3×3 matrices. Verify that

$$F(A) = a_{11} \cdot a_{22} \cdot a_{33}$$

   is a determinant function defined on this set. Is Proposition 4.1 meaningful in this context?
4. Verify that if F is a determinant function defined on $\tilde{R}_{3,3}$ (Exercise 3), then for each such matrix A we must have

$$F(A) = a_{11}a_{22}a_{33}.$$

   [Let I_1, I_2, I_3 be the unit coordinate vectors in $V_3(R)$. Write

$$F(A) = (a_{11}I_1 + a_{12}I_2 + a_{13}I_3, a_{22}I_2 + a_{23}I_3, a_{33}I_3).$$

   Now use the properties of a determinant function to show that $F(A)$ must equal $a_{11}a_{22}a_{33}$.]
5. Let $\{k_1, k_2, k_3\}$ be a permutation (rearrangement) of the set $\{1, 2, 3\}$. Show that

$$F(A) = a_{1k_1}a_{2k_2}a_{3k_3}$$

   defines a 3-linear function with domain the set of 3×3 matrices.
6. If $D_i(a)$, E_{ij}, $F_{ij}(a)$ are the elementary 3×3 matrices and F is a determinant function defined on $R_{3,3}$, compute

$$F(D_i(a)), F(E_{ij}), F(F_{ij}(a)).$$

7. Let D be the determinant function defined on $\tilde{R}_{3,3}$ (see Exercises 3 and 4). Show that $D(A \cdot B) = D(A) \cdot D(B)$ whenever $A, B \in \tilde{R}_{3,3}$.
8. If D is the determinant function defined on $\tilde{R}_{3,3}$ show that $D(A) = 0$ if and only if the rows of A are linearly dependent.
9. If X_1, X_2, X_3 are vectors in $V_3(R)$ show that the scalar triple product $(X_1, X_2 \times X_3)$, defines a determinant function on $R_{3,3}$ (see Section 1.8).
10. Show that the set of all n-linear functions defined on $R_{n,n}$ is a vector space over the field of real numbers.
11. Show that $D(A) = a_{11} \cdot a_{22} \cdots a_{nn}$ is a determinant function defined on $\tilde{R}_{n,n}$ (the set of upper triangular $n \times n$ matrices).
12. Verify that the determinant function D defined on $\tilde{R}_{n,n}$ is unique. (Generalize the argument of Exercise 4.)
13. Let $\{k_1, \ldots, k_n\}$ be a permutation of the set $\{1, 2, 3, \ldots, n\}$. Verify that $D(A) = a_{1k_1}a_{2k_2} \cdots a_{nk_n}$ is an n-linear function defined on $\tilde{R}_{n,n}$.
14. Let D be the determinant function defined on $\tilde{R}_{n,n}$. Show that $D(A) \neq 0$ if and only if A is invertible. (Use Exercise 11.)

4.2 Existence of a Determinant Function

To show the existence of a determinant function defined on $R_{n,n}$ we need the following notation. If $A = (a_{ij})$ is an $n \times n$ matrix, let A_{ij} be the $n-1 \times n-1$ matrix obtained from A by striking out the ith row and the jth column. For example, if

$$A = \begin{pmatrix} 2 & -1 & 0 \\ 1 & 2 & 3 \\ -1 & 1 & -2 \end{pmatrix},$$

then

$$A_{11} = \begin{pmatrix} 2 & 3 \\ 1 & -2 \end{pmatrix}, \quad A_{31} = \begin{pmatrix} -1 & 0 \\ 2 & 3 \end{pmatrix}$$

and

$$A_{22} = \begin{pmatrix} 2 & 0 \\ -1 & -2 \end{pmatrix}.$$

The result we want is the following.

THEOREM 4.3 *There exists an n-linear alternating function D defined on $R_{n,n}$ which satisfies $D(I) = 1$.*

Proof. The argument proceeds by induction on n. We know by the results of Section 4.1 that the result is true for $n = 2$. To see how to proceed in general let us define a determinant function for $n = 3$. Let

$$\begin{pmatrix} a_{11} \\ a_{21} \\ a_{31} \end{pmatrix}$$

be the first column of A, and let A_{11}, A_{21}, A_{31} be the matrices obtained by striking out the first row and first column of A, second row and first column; and third row and first column. Then if D is the determinant function for 2×2 matrices, let us check that the formula

(1) $\qquad D_3(A) = a_{11}D(A_{11}) - a_{21}D(A_{21}) + a_{31}D(A_{31})$

defines a determinant function on the set of 3×3 matrices. To emphasize the order of the matrices involved we shall for the moment use the notation D_2 for the determinant of a 2×2 matrix and D_3 for the determinant of a 3×3.

If $R_k = (a_{k1}, a_{k2}, a_{k3})$ is the kth row of the matrix A let $R'_k = (a_{k2}, a_{k3})$ be the vector in $V_2(R)$ obtained by dropping the first component of R_k. Then the rows of the 2×2 submatrices A_{11}, A_{21}, A_{31} are two of the vectors R'_1, R'_2, R'_3. Furthermore (1) becomes

(2) $\qquad D_3(R_1, R_2, R_3) = a_{11}D_2(R'_2, R'_3) - a_{21}D_2(R'_1, R'_3) + a_{31}D_2(R'_1, R'_2).$

Let us check first that $D_3(R_1, R_2, R_3)$ is linear in each variable. For definiteness we consider the first variable. If $S_1 = (b_{11}, b_{12}, b_{13})$ and a and b are scalars, then we must show

$$D_3(aR_1 + bS_1, R_2, R_3) = aD_3(R_1, R_2, R_3) + bD_3(S_1, R_2, R_3).$$

But from (2)

$D_3(aR_1 + bS_1, R_2, R_3) = (aa_{11} + bb_{11})D_2(R'_2, R'_3) - a_{21}D_2(aR'_1 + bS'_1, R'_3)$

(3) $\qquad\qquad\qquad + a_{31}D_2(aR'_1 + bS'_1, R'_2).$

Now using the fact that D_2 is linear in the first variable and collecting terms in (3) we conclude that

$$D_3(aR_1+bS_1, R_2, R_3) = (aa_{11}+bb_{11})D_2(R'_2, R'_3) - a_{21}(aD_2(R'_1, R'_3)+$$
$$bD_2(S'_1, R'_3)) + a_{31}(aD_2(R'_1, R'_2)+bD_2(S'_1, R'_3))$$
$$= a[a_{11}D_2(R'_2, R'_3) - a_{21}D_2(R'_1, R'_3) + a_{31}D_2(R'_1, R'_2)]$$
$$+ b[b_{11}D_2(S'_2, R'_3) - a_{21}D_2(S'_1, R'_2) + a_{31}D_2(S'_1, R'_2)]$$
$$= aD_3(R_1, R_2, R_3) + bD(S_1, R_2, R_3)$$

which is the desired linearity. In the same way we may verify that D_3 is linear in the second and third variables when the others are fixed.

To show that D_3 is alternating assume first that $R_1 = R_2$. Then $a_{11} = a_{21}$ and using the fact that D_2 is alternating we conclude that

$$D_3(R_1, R_1, R_3) = a_{11}D_2(R'_1, R'_3) - a_{21}D_2(R'_1, R'_3) + a_{31}D_2(R'_1, R'_1)$$
$$= a_{11}D_2(R'_1, R'_3) - a_{11}D_2(R'_1, R'_3) = 0.$$

Similarly if $R_1 = R_3$, then $a_{11} = a_{31}$ and

$$D_3(R_1, R_2, R_1) = a_{11}D_2(R'_2, R'_1) - a_{21}D_2(R'_1, R'_1) + a_{31}D_2(R'_1, R'_2)$$
$$= -a_{11}D_2(R'_1, R'_2) + a_{11}D_2(R'_1, R'_2) = 0$$

since $D_2(R'_1, R'_1) = 0$ and $D_2(R'_2, R'_1) = -D_2(R'_1, R'_2)$. The case $R_2 = R_3$ proceeds similarly.

Lastly we assert $D_3(I) = 1$. But if I_1, I_2, I_3 are the unit coordinate vectors in $V_3(R)$ then

$$D_3(I) = D_3(I_1, I_2, I_3) = 1D_2(I'_2, I'_3) - 0D_2(I'_1, I'_3) + 0D_2(I'_1, I'_2) = 1.$$

For $n \times n$ matrices A we assume that a determinant function D_{n-1} has been inductively defined on the $n-1 \times n-1$ matrices. Then if A_{i1} is the submatrix of A obtained by striking out the ith row and first column, we define

$$D_n(A) = a_{11}D_{n-1}(A_{11}) - a_{21}D_{n-1}(A_{21}) + D_{31}D_{n-1}(A_{31}) - \cdots$$
$$+ (-1)^{n+1}a_{n1}D_{n-1}(A_{n1}).$$

Using the summation sign this becomes

(4) $$D_n(A) = \sum_{i=1}^{n} (-1)^{i+1}a_{i1}D_{n-1}(A_{i1}).$$

We now must check that D_n is n-linear, alternating, and $D_n(I) = 1$. The verification is similar to the case $n = 3$ and we omit it.

In what follows we shall drop the subscript from D_n, since the order of the matrices is clear. Then (4) becomes

(5) $$D(A) = \sum_{i=1}^{n} (-1)^{i+1}a_{i1}D(A_{i1}).$$

Example 1.

If $A = \begin{pmatrix} 1 & 0 & 1 \\ 2 & -1 & 1 \\ 3 & 0 & 1 \end{pmatrix}$ compute $D(A)$. By (5)

$$D(A) = D(A_{11}) - 2D(A_{21}) + 3D(A_{31}) = D\begin{pmatrix} -1 & 1 \\ 0 & 1 \end{pmatrix} - 2D\begin{pmatrix} 0 & 1 \\ 0 & 1 \end{pmatrix} + 3D\begin{pmatrix} 0 & 1 \\ -1 & 1 \end{pmatrix}$$
$$= -1 + 0 + 3 = 2.$$

Example 2. Compute $D(A)$ if

$$A = \begin{pmatrix} 2 & -1 & 1 & 0 \\ 1 & 0 & 1 & 0 \\ 3 & 0 & 0 & 1 \\ 2 & 1 & -1 & 1 \end{pmatrix}.$$

By (5) $D(A) = 2D(A_{11}) - D(A_{21}) + 3D(A_{31}) - 2D(A_{41})$. Using (5) to compute $D(A_{i1})$ for the 3×3 submatrices A_{i1} we see that $D(A_{11}) = 1$, $D(A_{21}) = 0$, $D(A_{31}) = -1$, and $D(A_{41}) = -1$. Hence $D(A) = 2 - 3 + 2 = 1$.

Formula (5) has many generalizations. To discuss these we first need some terminology. The number, $D(A_{ij})$, is called the i,jth *minor* of the matrix and $(-1)^{i+j}D(A_{ij})$ is the i,jth *cofactor*. For example if

$$A = \begin{pmatrix} 2 & -1 & 0 \\ 1 & 2 & 3 \\ -1 & 1 & -2 \end{pmatrix}$$

is the matrix considered at the beginning of the section, $D(A_{12}) = 1$, $D(A_{22}) = -4$, and $D(A_{32}) = 6$ are the 1, 2; 2, 2; and 3, 2 minors, respectively. The corresponding cofactors are $(-1)^{1+2}D(A_{12}) = -1$, $(-1)^{2+2}D(A_{22}) = -4$, and $(-1)^{3+2}D(A_{32}) = -6$, respectively. Formula (5) says then that $D(A)$ is the inner product of the first column of A with a vector $((-1)^{1+1}D(A_{11}), \ldots, (-1)^{n+1}D(A_{n1}))$ whose components are the cofactors of the corresponding entries in the first column of the matrix A. As we shall see presently this result is true for any row or column of A. Indeed if $S = (a_1, \ldots, a_n)$ is any row or column of A and $T = (b_1, \ldots, b_n)$ is the vector whose entries are the cofactors corresponding to the entries of S, then $D(A)$ is the inner product of S and T. Expressed formally we have that for each $j = 1, \ldots, n$

(6) $$D(A) = \sum_{i=1}^{n} (-1)^{i+j} a_{ij} D(A_{ij}).$$

Also for each $i = 1, 2, \ldots, n$

(7) $$D(A) = \sum_{j=1}^{n} (-1)^{i+j} a_{ij} D(A_{ij}).$$

Each of the sums (6) is an inner product of a *column* of A with the vector whose entries are the associated cofactors. Each of the sums (7) is an inner product of a *row* of A with the corresponding cofactor vector. Formulas (6) and (7) are called the *expansion of a determinant by minors*. Formula (6) is the column formula or column expansion of $D(A)$; formula (7) is the corresponding row formula. We shall verify these formulas in Section 4.4. For the following calculations we shall assume the validity of both these formulas. The signs occurring in each of these formulas associated with the term $a_{ij}D(A_{ij})$ can be remembered easily if one remembers that they alternate like squares in a checkerboard:

$$\begin{pmatrix} + & - & + & - & + & \cdot & \cdot & \cdot \\ - & + & - & + & \cdot & \cdot & \cdot & \\ + & - & + & \cdot & \cdot & \cdot & & \\ - & + & \cdot & \cdot & \cdot & & & \\ + & \cdot & \cdot & \cdot & & & & \\ \cdot & \cdot & \cdot & & & & & \\ \cdot & \cdot & & & & & & \\ \cdot & & & & & & & \end{pmatrix}$$

EXISTENCE OF A DETERMINANT FUNCTION

Example 3. Compute $D(A)$ if
$$A = \begin{pmatrix} 1 & 1 & 0 \\ 3 & 2 & 0 \\ -1 & 1 & 2 \end{pmatrix}$$

Using the column formula (6) for the third column we have
$$D(A) = +2 \cdot D(A_{33}) = 2(-1) = -2.$$

Example 4. Compute $D(A)$ if
$$A = \begin{pmatrix} 1 & -1 & 1 & 3 \\ -1 & 0 & 2 & 0 \\ 1 & 1 & 1 & 2 \\ 1 & 0 & 0 & 2 \end{pmatrix}.$$

To take advantage of the zeros in the fourth row we should use the row formula (7) for $i = 4$. Then
$$D(A) = -1 \cdot D(A_{41}) + 2D(A_{44}).$$

Expanding $D(A_{41})$ about its middle or second row and $D(A_{44})$ about its second column we have
$$D(A_{41}) = 2D \begin{pmatrix} -1 & 3 \\ 1 & 2 \end{pmatrix} = -10,$$
and
$$D(A_{44}) = 1D \begin{pmatrix} -1 & 2 \\ 1 & 1 \end{pmatrix} - 1D \begin{pmatrix} 1 & 1 \\ -1 & 2 \end{pmatrix}$$
$$= -6.$$

Therefore $D(A) = 10 - 12 = -2$.

EXERCISES

Compute $D(A)$ for the following matrices A using an appropriate row of column expansion [formulas (6) or (7)].

1. $\begin{pmatrix} 1 & 2 & 1 \\ 2 & 1 & -1 \\ -1 & 1 & 0 \end{pmatrix}$.

2. $\begin{pmatrix} 1 & 1 & 2 \\ 0 & 1 & 1 \\ -1 & 0 & 1 \end{pmatrix}$.

3. $\begin{pmatrix} 0 & 1 & -1 \\ 1 & 0 & 2 \\ 0 & -1 & 1 \end{pmatrix}$.

4. $\begin{pmatrix} 1 & -1 & 1 \\ 2 & 0 & 1 \\ 1 & -1 & 2 \end{pmatrix}$.

5. $\begin{pmatrix} 1 & -1 & 0 \\ 0 & 1 & 2 \\ 1 & 1 & -1 \end{pmatrix}$.

6. $\begin{pmatrix} 1 & -2 & 1 & 0 \\ 2 & 0 & -3 & 0 \\ 0 & 2 & 1 & 1 \\ 1 & 0 & 0 & 1 \end{pmatrix}$.

7. $\begin{pmatrix} 1 & 0 & -1 & 0 \\ 0 & 2 & -1 & 1 \\ -1 & 1 & 0 & 2 \\ 0 & 0 & -1 & 3 \end{pmatrix}$

8. $\begin{pmatrix} 1 & 2 & 0 & 0 \\ -1 & 0 & 0 & -1 \\ 0 & 0 & -4 & 2 \\ 0 & -2 & 1 & 0 \end{pmatrix}$.

9. $\begin{pmatrix} 1 & 0 & 0 & 0 & -1 \\ 0 & 2 & 1 & 0 & 0 \\ 0 & 0 & 3 & -2 & 0 \\ -1 & 0 & 0 & 2 & 0 \\ 0 & 1 & 0 & 0 & -3 \end{pmatrix}$

For the following matrices A form the matrix $A - xI$. Determine the values of x which satisfy $D(A - xI) = 0$.

10. $A = \begin{pmatrix} 1 & 0 & 0 \\ 1 & 1 & 0 \\ 0 & 0 & 2 \end{pmatrix}$.

11. $A = \begin{pmatrix} 2 & 2 & 1 \\ 2 & 2 & 1 \\ 0 & 0 & 1 \end{pmatrix}$.

12. $A = \begin{pmatrix} 1 & 2 & 1 \\ -1 & 1 & 1 \\ 0 & 3 & 2 \end{pmatrix}$.

13. $A = \begin{pmatrix} 1 & 0 & -1 & 2 \\ -1 & 1 & 0 & 1 \\ 0 & 1 & -1 & 3 \\ -2 & 1 & 1 & -1 \end{pmatrix}$.

14. If A is a 3×3 matrix show that for $j = 2, 3, \sum_{i=1}^{3} (-1)^{i+j} a_{ij} D(A_{ij})$ defines an alternating, 3-linear function on $R_{3,3}$ which takes on the value 1 at the identity matrix I.

15. If A is an upper triangular matrix show that $D(A)$ is the product of the diagonal elements.

16. If A is an upper triangular matrix, show that A is invertible if and only if $D(A) \neq 0$.

4.3 Properties of Determinant Functions; Uniqueness

In the previous section we showed that there exists one determinant function which we have denoted by D. It was defined inductively by the formula $D(A) = \sum_{i=1}^{n} (-1)^{i+1} a_{i1} D(A_{i1})$, where A_{i1} is the submatrix obtained from the matrix A by striking out the ith row and the first column. In this section we wish to show that D is the only determinant function defined on $R_{n,n}$. To do this we shall need to develop some properties of determinant functions. We first consider upper triangular matrices. It follows immediately from the definition of D above that if A is upper triangular, then $D(A) = a_{11} a_2 \cdots a_{nn}$. We wish to show next that $D(A) = a_{11} \cdots a_{nn}$ is the only alternating n-linear function defined on the upper triangular matrices satisfying $D(I) = 1$.

THEOREM 4.4 *Let $\tilde{R}_{n,n}$ be the set of upper triangular matrices. If $A = (a_{ij}) \in \tilde{R}_{n,n}$, then the function*

(1) $$D(A) = a_{11} a_{22} \cdots a_{nn}$$

is the only alternating n-linear function on $\tilde{R}_{n,n}$ satisfying $D(I) = 1$.

Proof. Let $R_1, \ldots, R_n$ be the rows of the upper triangular matrix

$$A = \begin{pmatrix} a_{11} & a_{12} & \cdots & a_{1n} \\ 0 & a_{22} & \cdots & a_{2n} \\ 0 & 0 & a_{33} & \cdots & a_{3n} \\ \vdots & & & & \\ 0 & & \cdots & & a_{nn} \end{pmatrix}.$$

To show that the function defined by (1) is the only determinant function we proceed exactly as in the 2×2 case which was done in Section 4.1. If $I_1, \ldots, I_n$ are the unit coordinate vectors in $V_n(R)$, then

$$R_1 = (a_{11}, \ldots, a_{1n}) = a_{11} I_1 + \cdots + a_{1n} I_n$$
$$R_2 = (0, a_{22}, \ldots, a_{2n}) = a_{22} I_2 + \cdots + a_{2n} I_n$$
$$\vdots$$
$$R_n = (0, \ldots, 0, a_{nn}) = a_{nn} I_n.$$

Let $F(A) = F(R_1, \ldots, R_n)$ be an alternating n-linear function defined on

$\tilde{R}_{n,n}$ satisfying $F(I) = 1$. We assert $F(A) = a_{11} \cdots a_{nn}$. Exploiting the n-linearity of F beginning with the nth variable we have

$$F(A) = F(R_1, \ldots, R_n) = a_{nn} F(R_1, \ldots, R_{n-1}, I_n).$$

But $R_{n-1} = a_{n-1,n-1} I_{n-1} + a_{n-1,n} I_n$. Hence since F is alternating and linear in the $n-1$ variable we have

$$a_{nn} F(R_1, \ldots, R_{n-1}, I_n) = a_{nn} [a_{n-1,n-1} F(R_1, \ldots, R_{n-2}, I_{n-1}, I_n)$$
$$+ a_{n-1,n} F(R_1, \ldots, R_{n-2}, I_n, I_n)]$$
$$= a_{nn} a_{n-1,n-1} F(R_1, \ldots, R_{n-2}, I_{n-1}, I_n).$$

since $F(R_1, \ldots, R_{n-2}, I_n, I_n) = 0$. Continuing inductively we conclude that

$$F(A) = a_{nn} \cdots a_{11} F(I_1, \ldots, I_n).$$

Since $F(I) = F(I_1, \ldots, I_n) = 1$ by hypothesis, it follows that

$$F(A) = a_{11} \cdots a_{nn} = D(A).$$

The proof is now complete.

COROLLARY. *If A is upper triangular, then $D(A) \neq 0$ if and only if A is invertible.*

Proof. If A is upper triangular, then A is invertible if and only if there are no zeros on the main diagonal of A. This will occur if and only if

$$D(A) = a_{11} \cdots a_{nn} \neq 0.$$

We now turn our attention to general $n \times n$ matrices. We wish to investigate what happens to $F(A)$ for a determinant function F if an elementary row operation is performed on the matrix A. Let $R_1, \ldots, R_n$ be the rows of A. For the moment we shall use the notation $A = \begin{pmatrix} R_1 \\ \vdots \\ R_n \end{pmatrix}$. First, if the ith row is multiplied by a nonzero scalar a, then

(2)
$$A = \begin{pmatrix} R_1 \\ \vdots \\ R_i \\ \vdots \\ R_n \end{pmatrix} \longrightarrow \begin{pmatrix} R_1 \\ \vdots \\ aR_i \\ \vdots \\ R_n \end{pmatrix} = B$$

By the n-linearity of a determinant function F we have

$$F(B) = aF(A).$$

In particular if

$$I = \begin{pmatrix} I_1 \\ \vdots \\ I_n \end{pmatrix},$$

134 DETERMINANTS

then the elementary matrix

$$D_i(a) = \begin{pmatrix} I_1 \\ \vdots \\ aI_i \\ \vdots \\ I_n \end{pmatrix}$$

and $F(D_i(a)) = a$. We know from our discussion of elementary row operations in Section 3.6 that if B is obtained from A by an elementary row operation, then B is obtained from A by multiplying on the left by the corresponding elementary matrix. Hence for the case considered above [formula (2)], $B = D_i(a)A$. If F is a determinant function then since $F(D_i(a)) = a$, and $F(B) = aD(A)$, we conclude that

(3) $$F(D_i(a)A) = F(D_i(a))F(A).$$

Let us investigate to what extent the product formula (3) holds for the other elementary matrices E_{ij} and $F_{ij}(a)$.

If we interchange two rows of A, then

(4) $$A = \begin{pmatrix} R_1 \\ \vdots \\ R_i \\ \vdots \\ R_j \\ \vdots \\ R_n \end{pmatrix} \longrightarrow \begin{pmatrix} R_1 \\ \vdots \\ R_j \\ \vdots \\ R_i \\ \vdots \\ R_n \end{pmatrix} = B.$$

But by Proposition 4.1 if F is a determinant function

$$F(B) = -F(A).$$

In particular for the elementary matrix

$$E_{ij} = \begin{pmatrix} I_1 \\ \vdots \\ I_j \\ \vdots \\ I_i \\ \vdots \\ I_n \end{pmatrix} \begin{matrix} \\ \\ (i\text{th row}) \\ \\ (j\text{th row}) \\ \\ \end{matrix}$$

$$F(E_{ij}) = -1.$$

Hence if B is given by (4), then $B = E_{ij}A$ and

(5) $$F(E_{ij}A) = -1F(A) = F(E_{ij})F(A).$$

PROPERTIES OF DETERMINANT FUNCTIONS

Lastly if aR_i is added to R_j, then

(6) $$A = \begin{pmatrix} R_1 \\ \vdots \\ R_n \end{pmatrix} \longrightarrow \begin{pmatrix} R_1 \\ \vdots \\ aR_i + R_j \\ \vdots \\ R_n \end{pmatrix} = B,$$

and $B = F_{ij}(a)A$ where

$$F_{ij}(a) = \begin{pmatrix} I_1 \\ \vdots \\ aI_i + I_j \\ \vdots \\ I_n \end{pmatrix} \quad j\text{th row}.$$

Applying the n-linearity of the function F we conclude from (5) that

$$F(B) = F\begin{pmatrix} R_1 \\ \vdots \\ R_i \\ \vdots \\ aR_i \\ \vdots \\ R_n \end{pmatrix} + F\begin{pmatrix} R_1 \\ \vdots \\ R_i \\ \vdots \\ R_j \\ \vdots \\ R_n \end{pmatrix} = F\begin{pmatrix} R_1 \\ \vdots \\ R_i \\ \vdots \\ aR_i \\ \vdots \\ R_n \end{pmatrix} + F(A).$$

However,

$$F\begin{pmatrix} R_1 \\ \vdots \\ R_i \\ \vdots \\ aR_i \\ \vdots \\ R_n \end{pmatrix} = aF\begin{pmatrix} R_1 \\ \vdots \\ R_i \\ \vdots \\ R_i \\ \vdots \\ R_n \end{pmatrix} = 0$$

since F is alternating. Therefore $F(A) = F(B)$. Since $B = F_{ij}(a)A$ and $F(F_{ij}(a)) = 1$ we conclude that

(7) $$F(F_{ij}(a)A) = F(F_{ij}(a))F(A)$$

Combining formulas (3), (5), and (7) we see that if E is an elementary $n \times n$ matrix and F is a determinant function, then for each $n \times n$ matrix A

(8) $$F(EA) = F(E)F(A).$$

Furthermore $F(D_i(a)) = a$, $F(E_{ij}) = -1$, and $F(F_{ij}(a)) = 1$. The next result summarizes what we have proved about determinant functions.

PROPOSITION 4.5 *Let $A \in R_{n,n}$ and let E be one of the elementary matrices $F_i(a)$, E_{ij}, $F_{ij}(a)$. If F is a determinant function then*

(9) $$F(EA) = F(E)F(A)$$

and $F(D_i(a)) = a$, $F(E_{ij}) = -1$, and $F(F_{ij}(a)) = 1$.

We are now in a position to establish the uniqueness of the determinant function.

THEOREM 4.6 *If A is an $n \times n$ matrix, then*

$$D(A) = \sum_{i=1}^{n} (-1)^{i+1} a_{i1} D(A_{i1})$$

is the only determinant function defined on $R_{n,n}$.

Proof. Let F be a determinant function defined on $R_{n,n}$. If B is upper triangular, then we know from Theorem 4.4 that $F(A) = D(A)$. If E is an elementary matrix, then it follows from Theorem 4.5 that $F(E) = D(E)$. However, we know from the corollary to Theorem 3.18 on p. 112 that an arbitrary matrix A can be written

$$A = E_1 \cdots E_m B$$

where E_i, $i = 1, \ldots, m$, are elementary matrices and B is upper triangular. Extending formula (8) by induction we conclude that

$$\begin{aligned} F(A) &= F(E_1 \cdots E_m B) \\ &= F(E_1) F(E_2 \cdots E_m B) \\ &= F(E_1) \cdots F(E_m) F(B) \\ &= D(E_1) \cdots D(E_m) D(B) \\ &= D(E_1 \cdots E_m B) = D(A) \end{aligned}$$

which proves the desired uniqueness.

From now on we shall refer to the function $D(A)$ as *the* determinant of the matrix A. Theorem 4.6 then asserts that this function D is the only real-valued function defined on $R_{n,n}$ which is alternating, n-linear, and satisfies $D(I) = 1$.

Formula (8) is a special case of an important product formula for determinants. Namely, if $A, B \in R_{n,n}$, then

(10) $$D(AB) = D(A)D(B).$$

We prove this result in several stages. First we verify the following.

PROPOSITION 4.7 *If $A, B \in R_{n,n}$ and A is invertible, then*

$$D(AB) = D(A)D(B).$$

Proof. The argument is the same as a portion of the proof of 4.6. We know by Theorem 3.19 that if A is invertible then A is the product of elementary matrices $E_1, \ldots, E_m$. Therefore $AB = E_1 \cdots E_m B$. By 4.5

$$\begin{aligned} D(AB) &= D(E_1 \cdots E_m B) = D(E_1) D(E_2 \cdots E_m B) \\ &= D(E_1) D(E_2) D(E_3 \cdots E_m B). \end{aligned}$$

Continuing by induction we have

$$D(AB) = D(E_1) \cdots D(E_m) D(B).$$

Again applying 4.5 and induction we infer that
$$D(E_1) \cdots D(E_m) = D(E_1 \cdots E_m) = D(A).$$
Combining these two statements we have the desired formula
$$D(AB) = D(A) \cdot D(B) \text{ if } A \text{ is invertible.}$$

To establish formula (10) for arbitrary $n \times n$ matrices we need one further preliminary result that is important in its own right.

PROPOSITION 4.8 *If A is an $n \times n$ matrix, then $D(A) \neq 0$ if and only if A is invertible.*

Proof. Again let us write $A = E_1 \cdots E_m B$ where E_i are elementary matrices and B is upper triangular. By 4.6 and induction
$$D(A) = D(E_1) \cdots D(E_m) D(B).$$
Since $D(E_i) \neq 0$, this implies that $D(A) \neq 0$ if and only if $D(B) \neq 0$. Moreover, by the corollary to Theorem 4.4, $D(B) \neq 0$ if and only if B is invertible. Since A is invertible if and only if B is, we have proved the theorem.

We now prove (10).

THEOREM 4.9 *If $A, B \in R_{n,n}$ then*

(10) $$D(AB) = D(A)D(B).$$

Proof. If A is invertible, we have already proved the result. If A is not invertible, note first that AB is not invertible either. To see this, observe that if $(AB)^{-1}$ exists then
$$AB(AB)^{-1} = I.$$
Thus if we set $C = B(AB)^{-1}$, we have $AC = I$. This implies by Theorem 3.16 that A is invertible, contrary to our assumption.

Now A not invertible implies $D(A) = 0$. Since AB is not invertible, $D(AB) = 0$. Thus
$$0 = D(AB) = 0 \cdot D(B) = D(A)D(B)$$
and the proof is complete.

The use of elementary row operations and the corresponding elementary matrices provides us with an alternative method for computing the value of the determinant of a matrix A.

Example 1. Compute $D(A)$ if
$$A = \begin{pmatrix} 1 & -1 & 1 \\ 2 & 1 & 0 \\ 0 & 1 & 2 \end{pmatrix}.$$

Reducing A to an upper triangular matrix by a sequence of elementary row operations we have

$$A = \begin{pmatrix} 1 & -1 & 1 \\ 2 & 1 & 0 \\ 0 & 1 & 2 \end{pmatrix} \xrightarrow{F_{12(-2)}} \begin{pmatrix} 1 & -1 & 1 \\ 0 & 3 & -2 \\ 0 & 1 & 2 \end{pmatrix} \xrightarrow{E_{23}} \begin{pmatrix} 1 & -1 & 1 \\ 0 & 1 & 2 \\ 0 & 3 & -2 \end{pmatrix}$$

$$\xrightarrow{F_{23(-3)}} \begin{pmatrix} 1 & -1 & 1 \\ 0 & 1 & 2 \\ 0 & 0 & -8 \end{pmatrix} = B.$$

However, each row operation performed on a matrix A may be accomplished by multiplying A on the left by the corresponding elementary matrix. Hence we conclude that B is given by the following matrix product

$$B = F_{23}(-3)E_{23}F_{12}(-2)A.$$

Therefore, using the product formula (10) we have

$$-8 = D(B) = D(F_{23}(-3))D(E_{23})D(F_{12}(-2))D(A)$$
$$= 1 \cdot (-1) \cdot 1 \cdot D(A).$$

Hence $D(A) = 8$.

Example 2. Compute the real numbers λ for which $A - \lambda I$ has no inverse if

$$A = \begin{pmatrix} 1 & -1 & 0 \\ 0 & 1 & 1 \\ 0 & 0 & 2 \end{pmatrix}.$$

Since a matrix is invertible if and only if its determinant is nonzero we must compute those λ for which

$$D(A - \lambda I) = 0.$$

But

$$A - \lambda I = \begin{pmatrix} 1-\lambda & -1 & 0 \\ 0 & 1-\lambda & 1 \\ 0 & 0 & 2-\lambda \end{pmatrix}.$$

Since this matrix is upper triangular, its determinant is the product of the diagonal elements. Hence

$$D(A - \lambda I) = (1 - \lambda)^2(2 - \lambda).$$

$D(A - \lambda I) = 0$ if and only if $\lambda = 1, 2$.

Observe that the matrix $A - \lambda I$ has no inverse if and only if the transformation $A - \lambda I$ is not one-to-one. This means that for some nonzero vector X, $(A - \lambda I)X = 0$ or $AX = \lambda X$. This says that λ is an eigenvalue of the matrix A. Thus it follows from 4.8 that λ is an eigenvalue of the matrix A if and only if $D(A - \lambda I) = 0$. We state this important result as a theorem.

THEOREM 4.10 *The scalar λ is an eigenvalue for the $n \times n$ matrix A if and only if*

$$D(A - \lambda I) = 0.$$

Example 3. Determine the eigenvalues of the matrix

$$A = \begin{pmatrix} 1 & -1 & 1 \\ 2 & 1 & 0 \\ 3 & 0 & 1 \end{pmatrix}.$$

We must compute those values of x for which

$$D(A - xI) = D\begin{pmatrix} 1-x & -1 & 1 \\ 2 & 1-x & 0 \\ 3 & 0 & 1-x \end{pmatrix} = 0.$$

Expanding this determinant along the third row we see that

$$D(A - xI) = 3(x - 1) + (1 - x)[(1 - x)^2 + 2]$$
$$= (1 - x)[(1 - x)^2 - 1] = x(1 - x)(x - 2).$$

Hence the eigenvalues are 0, 1, 2.

It follows from the definition of D, that if A is an $n \times n$ matrix, then $D(A - xI)$ is a polynomial of degree n. This polynomial is called the *charac-*

teristic polynomial* of the matrix A. Since $D(A-xI)$ has degree n, the equation

$$D(A-xI) = 0$$

can have at most n solutions. This is an alternative proof of Exercise 12, p. 123.

EXERCISES

Compute the determinants of the following matrices by first reducing the matrix to an upper triangular matrix.

1. $A = \begin{pmatrix} -1 & 0 & 1 \\ 2 & 1 & 1 \\ 0 & 1 & 1 \end{pmatrix}$

2. $A = \begin{pmatrix} 1 & -1 & 1 \\ 2 & 1 & 0 \\ 1 & 1 & 1 \end{pmatrix}$

3. $A = \begin{pmatrix} 1 & 1 & -1 \\ 0 & 2 & 1 \\ 1 & 0 & -1 \end{pmatrix}$

4. $A = \begin{pmatrix} 2 & 1 & 0 \\ 0 & 1 & 2 \\ 1 & 0 & 1 \end{pmatrix}$

5. $A = \begin{pmatrix} 1 & 0 & -1 & 1 \\ 1 & 1 & 0 & 1 \\ 0 & 1 & -1 & 0 \\ 1 & 1 & -1 & 1 \end{pmatrix}$

6. $A = \begin{pmatrix} 1 & 0 & 0 & 1 \\ 0 & -1 & 1 & 0 \\ 1 & 2 & -1 & 0 \\ 1 & 0 & -1 & 1 \end{pmatrix}$

7. $A = \begin{pmatrix} 0 & 0 & 1 & 0 & 0 \\ 1 & 0 & 0 & 0 & 0 \\ 0 & 1 & 0 & 0 & 0 \\ 0 & 0 & 0 & 0 & 1 \\ 0 & 0 & 0 & 1 & 0 \end{pmatrix}$

8. $A = \begin{pmatrix} 0 & 0 & 0 & 1 & 0 \\ 1 & 0 & 0 & 0 & 0 \\ 0 & 0 & 1 & 0 & 0 \\ 0 & 0 & 0 & 0 & 1 \\ 0 & 1 & 0 & 0 & 0 \end{pmatrix}$

9. $A = \begin{pmatrix} 2 & 0 & -1 & 1 \\ 0 & 1 & 0 & 1 \\ 0 & 1 & 1 & 0 \\ 1 & 0 & 1 & -1 \end{pmatrix}$

10. Compute the eigenvalues of the following matrices. Evaluate the appropriate determinants by any convenient method.

(a) $\begin{pmatrix} 1 & 1 \\ 0 & -1 \end{pmatrix}$

(b) $\begin{pmatrix} 1 & 1 \\ 1 & -1 \end{pmatrix}$

(c) $\begin{pmatrix} 1 & 2 \\ 2 & 1 \end{pmatrix}$

(d) $\begin{pmatrix} 1 & 0 & -1 \\ 0 & -1 & 1 \\ 0 & 0 & 2 \end{pmatrix}$

(e) $\begin{pmatrix} -1 & 0 & 0 \\ 1 & 1 & 0 \\ 2 & -1 & 3 \end{pmatrix}$

(f) $\begin{pmatrix} 1 & 0 & 1 \\ 0 & 1 & 0 \\ 1 & 0 & 1 \end{pmatrix}$

(g) $\begin{pmatrix} 1 & 1 & 1 \\ 1 & 2 & 0 \\ 1 & 0 & 2 \end{pmatrix}$

(h) $\begin{pmatrix} 1 & 0 & 1 \\ 0 & 1 & 3 \\ 1 & 3 & -2 \end{pmatrix}$

11. A linear transformation T in $V_2(R)$ is defined by requiring that $T(I_1) = I_1 - I_2$ and $T(I_2) = -I_1 + I_2$, where $\{I_1, I_2\}$ is the canonical basis in $V_2(R)$. Determine the eigenvalues of this transformation.

12. Let I_1, I_2, I_3 be the canonical basis in $V_3(R)$. If the linear transformation T satisfies

$$T(I_1) = I_1 + I_2 + I_3$$
$$T(I_2) = -I_2 - 2I_3$$

and

$$T(I_3) = I_1 - I_3$$

determine the eigenvalues of T.

13. Find the eigenvalues of the linear transformation T in $V_3(R)$ if
$$T(I_1) = I_1 - 3I_3$$
$$T(I_2) = -2I_2$$
$$T(I_3) = -3I_1 + I_3.$$

Determine the eigenvalues of the following linear transformations defined in P_2, the polynomials of degree ≤ 2.

14. $T(f)(x) = f(x) + f''(x)$.
15. $T(f) = 2f(x) + xf'(x)$.
16. $T(f) = x^2 f''(x) + 2xf'(x) + f(x)$.
17. Let $x_1, x_2, \ldots, x_{n-1}$ be specified real numbers. For $k = 2, \ldots, n$ define the matrix

$$A_k(x) = \begin{pmatrix} 1 & x_1 & x_1^2 & \cdots & x_1^{k-1} \\ 1 & x_2 & x_2^2 & \cdots & x_2^{k-1} \\ \vdots & & & & \vdots \\ 1 & x_{k-1} & x_{k-1}^2 & \cdots & x_{k-1}^{k-1} \\ 1 & x & x^2 & \cdots & x^{k-1} \end{pmatrix}.$$

Prove that for each k

$$D(A_k(x)) = D(A_{k-1}(x_{k-1}))(x - x_1) \cdots (x - x_{k-1}).$$

(Use the fact that a polynomial $p(x)$ of degree n with roots $a_1, \ldots, a_n$ can be written uniquely as

$$p(x) = C(x - a_1) \cdots (x - a_n)$$

where C is a constant.)

18. Using the result of Exercise 17 show that

$$D\begin{pmatrix} 1 & x_1 & x_1^2 & \cdots & x_1^{n-1} \\ 1 & x_2 & x_2^2 & \cdots & x_2^{n-1} \\ \vdots & & & & \\ 1 & x_n & x_n^2 & \cdots & x_n^{n-1} \end{pmatrix} = \prod_{\substack{k,j=1 \\ k>j}}^{n} (x_k - x_j).$$

The symbol Π stands for product, so the right-hand side of the above expression equals the product of all terms $(x_k - x_j)$ where $k, j = 1, 2, \ldots, n$ and $k > j$.

19. Deduce from Exercise 18 that the functions $1, x, \ldots, x^{n-1}$ are linearly independent. (Suppose $c_0 + c_1 x + \cdots + c_{n-1} x^{n-1} = 0$ for $x = x_1, \ldots, x_n$. Conclude from 18 that $c_0 = c_1 = \cdots = c_{n-1} = 0$.)

4.4 The Transpose of a Matrix; Permutation Matrices

To establish the expansion formulas (6) and (7) on p. 130 for the determinant of a matrix A we need to introduce the notion of the transpose of a matrix A.

Definition. *If $A = (a_{ij})$ is an $n \times n$ matrix, then the matrix $A^t = (a_{ji})$ is called the* **transpose** *of A.*

The matrix A^t is obtained from A by reflecting A about the main diagonal. Thus the rows of A are the columns of A^t and vice versa.

Example 1. If

$$A = \begin{pmatrix} 2 & 1 \\ 0 & 1 \end{pmatrix} \quad \text{then} \quad A^t = \begin{pmatrix} 2 & 0 \\ 1 & 1 \end{pmatrix}.$$

If
$$A = \begin{pmatrix} 1 & 0 & 2 \\ -1 & 1 & 1 \\ 0 & 1 & 0 \end{pmatrix} \quad \text{then} \quad A^t = \begin{pmatrix} 1 & -1 & 0 \\ 0 & 1 & 1 \\ 2 & 1 & 0 \end{pmatrix}.$$

Let us note some elementary properties of the transposition operation.

1. If A, B are $n \times n$ matrices and a is a scalar, then
$$(A+B)^t = A^t + B^t$$
$$(aA)^t = aA^t$$
$$A^{tt} = A.$$

The first two of these properties imply that the correspondence $A \to A^t$ defines a linear transformation from $R_{n,n}$ to $R_{n,n}$. Indeed $A \to A^t$ is one-to-one transformation of $R_{n,n}$ onto $R_{n,n}$.

2. If A and B are $n \times n$ matrices, then
$$(AB)^t = B^t A^t.$$

To see this let $R_1, \ldots, R_n$ be the rows of A and $C_1, \ldots, C_n$ be the columns of B. By the definition of the transpose, the vectors $R_1, \ldots, R_n$ are the columns of A^t and the vectors $C_1, \ldots, C_n$ are the rows of B^t. By definition of the matrix product if
$$D = (d_{ij}) = B^t A^t,$$
then d_{ij} is the inner product of the vectors C_i and R_j. Thus
$$d_{ij} = (C_i, R_j).$$
On the other hand if
$$C = (c_{ij}) = (AB)^t$$
then
$$c_{ij} = (R_j, C_i).$$
Since $(R_j, C_i) = (C_i, R_j)$, $c_{ij} = d_{ij}$ for each i and j and the result follows.

3. If A is an invertible $n \times n$ matrix, then A^t is invertible and
$$(A^t)^{-1} = (A^{-1})^t.$$

This follows directly from (2) since $AA^{-1} = I$ implies
$$(A^{-1})^t A^t = I^t = I.$$

4. If A is an $n \times n$ matrix, then Rank A = Rank A^t. This follows since the rank of A which is defined to be the dimension of the space spanned by the columns of A is also the dimension of the space spanned by the rows (Theorem 2.14, p. 74).

5. A diagonal $n \times n$ matrix has the property that $A = A^t$. Any $n \times n$ matrix with this property is said to be *symmetric*.

Next we wish to show that the determinant of a matrix A is the same as the determinant of the transpose of A.

THEOREM 4.11 *If A is an $n \times n$ matrix, and A^t is its transpose, then*
$$D(A) = D(A^t).$$

Proof. If A is not invertible, neither is A^t, and in both cases
$$D(A) = 0 = D(A^t).$$

If A is invertible then by Theorem 3.19, A is the product of elementary matrices $D_i(a)$, E_{ij}, and $F_{ij}(a)$.

For an elementary matrix of the first kind we have

$$D_i(a) = D_i(a)^t.$$

Therefore

$$D(D_i(a)) = D(D_i(a)^t).$$

For the second kind we have also

$$E_{ij}^t = E_{ij}.$$

Hence

$$D(E_{ij}^t) = D(E_{ij}).$$

Lastly

$$F_{ij}^t(a) = F_{ji}(a).$$

But

$$D(F_{ij}(a)) = 1.$$

Therefore

$$D(F_{ij}^t(a)) = D(F_{ij}(a)).$$

Hence if E is an elementary matrix

$$D(E) = D(E^t).$$

Writing A as a product of elementary matrices E_i we have

$$A = E_1 \cdots E_m.$$

Therefore

$$A^t = E_m{}^t \cdots E_1{}^t.$$

Applying Theorem 4.9 we have

$$D(A) = D(E_1) \cdots D(E_m)$$

and

$$D(A^t) = D(E_m{}^t) \cdots D(E_1{}^t)$$
$$= D(E_1{}^t) \cdots D(E_m{}^t)$$
$$= D(E_1) \cdots D(E_m).$$

Hence $D(A) = D(A^t)$, and we are done.

COROLLARY. *If the matrix A' is obtained from A by interchanging two columns then $D(A') = -D(A)$.*

Proof. We know that the sign of the determinant of a matrix is reversed if two rows are interchanged. Applying Theorem 4.11 we conclude the same result if two columns are interchanged.

We may use this corollary to give a simple proof of the column formula (6), p. 130. Namely,

(6) $$D(A) = \sum_{i=1}^{n} (-1)^{i+j} a_{ij} D(A_{ij}) \quad \text{for } j = 1, 2, \ldots, n$$

where $A = (a_{ij})$ and A_{ij} is the submatrix of A obtained by striking out the ith row and jth column. We already have the formula for $j = 1$. Let us establish it for $j = 2$. To do this let A' be the matrix obtained from A by interchanging the first and second columns. If a'_{ij} are the entries and A'_{ij} the corresponding submatrices of A', then for each i $a'_{i1} = a_{i2}$ and $A'_{i1} = A_{i2}$ since

$$A' = \begin{pmatrix} a_{12} & a_{11} & \cdots & a_{1n} \\ a_{22} & a_{21} & & \cdot \\ \vdots & \vdots & & \cdot \\ \vdots & \vdots & & \cdot \\ a_{n2} & a_{n1} & \cdots & a_{nn} \end{pmatrix}.$$

Hence

$$D(A') = \sum_{i=1}^{n} (-1)^{i+1} a'_{i1} D(A'_{i1})$$

$$= \sum_{i=1}^{n} (-1)^{i+1} a_{i2} D(A_{i2}).$$

But by the corollary $D(A') = -D(A)$. Hence

$$D(A) = -D(A') = -\sum_{i=1}^{n} (-1)^{i+1} a_{i2} D(A_{i2})$$

$$= \sum_{i=1}^{n} (-1)^{i+2} a_{i2} D(A_{i2})$$

which is the column formula (6) for $j = 2$. The verification for $j = 3, \ldots, n$ proceeds exactly the same.

The row formulas (7), p. 130, namely,

(7) $$D(A) = \sum_{j=1}^{n} (-1)^{i+j} a_{ij} D(A_{ij}), \quad i = 1, \ldots, n,$$

now follow directly from the column formulas and the fact that $D(A) = D(A^t)$.

The row and column expansion formulas (6) and (7) are important theoretical expressions for the determinant of the matrix A, but they are only practical formulas for evaluating $D(A)$ when the order of A is small or when the entries of A are largely zero. To evaluate a determinant of large order by using one of the formulas (6) or (7) requires many more multiplication and addition operations than are necessary to reduce A to an upper triangular matrix. Indeed to evaluate $D(A)$ by (6) or (7) requires $n! \cdot n$ multiplications and additions, whereas to evaluate $D(A)$ using elementary row operations as described in Section 4.3 requires less than $\frac{2}{3}n^3$ multiplications and additions.

We finish this section with the derivation of a formula for $D(A)$ which often serves as the definition. First we note some facts about permutation matrices. A *permutation matrix* P is an $n \times n$ matrix which can be obtained from the identity matrix by a sequence of row interchanges. Thus a permutation matrix P has precisely one nonzero entry equal to one in each row and column. Indeed P is the product of row interchange matrices E_{ij}. Since $D(E_{ij}) = -1$, this implies that $D(P) = \pm 1$. If $D(P) = 1$, then P is the product of an even number of matrices E_{ij}, and P is called an *even permutation matrix*. If $D(P) = -1$, then P is the product of an odd number of matrices E_{ij}, and P is called an *odd permutation matrix*. It is natural then to call the determinant of a permutation matrix its *sign*.

Definition. *The function* sgn (P), *read "sign P," is defined on the set of permutation matrices by the formula*

$$\text{sgn }(P) = 1 \text{ if } P \text{ is even}$$
$$= -1 \text{ if } P \text{ is odd}.$$

Associated with a permutation matrix is a permutation of the set of integers $\{1, 2, \ldots, n\}$. Now a permutation of a set of integers is just a one-to-one function p which transforms the set onto itself. Thus $p(1) = 2$, $p(2) = 3$, $p(3) = 1$ defines a permutation of the set $\{1, 2, 3\}$. The correspondence between an $n \times n$ permutation matrix and a permutation p of $\{1, 2, \ldots, n\}$ is obtained by observing that if the nonzero entries of a permutation matrix occur in the $1, p(1); 2, p(2); \ldots n, p(n)$ position then the function p is the desired permutation of $\{1, 2, \ldots, n\}$.

THEOREM 4.12 *If D is the determinant function defined on the set of $n \times n$ matrices $R_{n,n}$, then for each $A = (a_{ij}) \in R_{n,n}$*

(1) $$D(A) = \sum_P a_{1p(1)} \cdots a_{np(n)} \operatorname{sgn}(P)$$

where the sum is taken over all permutation matrices P in $R_{n,n}$. The function p is the permutation of $\{1, \ldots, n\}$ corresponding to the permutation matrix P.

Proof. Let $R_1, \ldots, R_n$ be the rows of A. Then for each i we may write $R_i = a_{i1}I_1 + \cdots + a_{in}I_n$ where I_k are the unit coordinate vectors in $V_n(R)$. Hence we may expand

$$D(A) = D(R_1, \ldots, R_n) = D(a_{11}I_1 + \cdots + a_{1n}I_n, \ldots, a_{n1}I_1 + \cdots + a_{nn}I_n)$$

by exploiting the n-linearity of the function D to obtain a sum of the form

$$D(A) = \sum_{j,k,\ldots} a_{1j}a_{2k} \cdots a_{nm} D(I_j, I_k, \ldots, I_m).$$

Now $D(I_j, I_k, \ldots, I_m) = 0$ unless the vectors $I_j, I_k, \ldots, I_m$ are all distinct. If they are distinct, then we label them $I_{p(1)}, \ldots, I_{p(n)}$. This defines a permutation p of $\{1, \ldots, n\}$ and $I_{p(1)}, \ldots, I_{p(n)}$ are the rows of the associated permutation matrix P. The corresponding term in the sum above is just

$$a_{1p(1)} \cdots a_{np(n)} D(I_{p(1)}, \ldots, I_{p(n)})$$
$$= a_{1p(1)} \cdots a_{np(n)} D(P) = a_{1p(1)} \cdots a_{np(n)} \operatorname{sgn}(P).$$

Therefore

$$D(A) = \sum_P a_{1p(1)} \cdots a_{np(n)} \operatorname{sgn}(P)$$

as required.

In order to use formula (1) as the definition of the determinant function one must prove independently the fact that permutation matrices are either the product of an even number of matrices E_{ij} or an odd number, but not both. The approach we have taken avoids the necessity of doing that.

EXERCISES

Using Theorem 4.11 establish the following facts about determinants.
1. If one column of an $n \times n$ matrix is multiplied by a scalar a and is added to another column, then the determinant of the matrix is unchanged.
2. If two columns of the $n \times n$ matrix A are the same, show that $D(A) = 0$.
3. If the $n \times n$ matrix B is obtained from A by multiplying one column of A by a scalar a, show that $D(B) = aD(A)$.
4. Show that the function D is n-linear considered as a function of the columns of the $n \times n$ matrices A.
5. The transpose of an $m \times n$ matrix is defined exactly as for a square matrix. If A is an $m \times n$ matrix and B is an $n \times p$ matrix show that $(AB)^t = B^t A^t$.

FURTHER PROPERTIES OF DETERMINANT FUNCTION

The following series of exercises establishes the connection between permutations of the set $\{1, 2, \ldots, n\}$ and permutation matrices.

6. If $Z_n = \{1, 2, \ldots, n\}$, then a permutation p of Z_n is a one-to-one function mapping Z_n onto itself. Let π_n be the set of all permutations of Z_n. For $p, q \in \pi_n$ let $p \cdot q$ be the composition of p and q. That is

$$(p \cdot q)(k) = p(q(k)) \qquad k = 1, \ldots, n.$$

Prove that π_n is closed under the multiplication $p \cdot q$. Prove that this multiplication is associative. Prove also that there is an identity with respect to this multiplication and that every permutation p has an inverse p^{-1}.

7. To each permutation $p \in \pi_n$ there corresponds a permutation matrix P with nonzero entries in the $(1, p(1)), \ldots, (n, p(n))$ positions. If $\iota(p) = P$ denotes this correspondence, show that ι is a one-to-one function mapping π_n onto the set of permutation matrices of order n. Show further that ι reverses multiplication, that is, if $\iota(p) = P$ and $\iota(q) = Q$, then $\iota(p \cdot q) = Q \cdot P$ where $Q \cdot P$ is the matrix product.

8. Let $\iota(p) = P$. Show that the nonzero entries of P^{-1} are in the $(1, p^{-1}(1)), \ldots, (n, p^{-1}(n))$ position.

9. A transposition p_k is a permutation of Z_n leaving each element of Z_n fixed except $k, k+1$. Furthermore p_k interchanges these two numbers, that is

$$p_k(k) = k+1 \qquad \text{and} \qquad p_k(k+1) = k.$$

(Note: $p_n(n) = 1$, $p_n(1) = n$, and $p_n(k) = k$, $k = 2, 3, \ldots, n-1$.) Therefore there are exactly n transpositions. Prove that each permutation p is a product of transpositions. (You may prove this directly or you may use Exercise 7 and known facts about permutation matrices.)

10. Prove that a permutation p is a product of an even number of transpositions or an odd number but not both. (Use Exercise 9 and known facts about determinants.)

4.5 Further Properties of the Determinant Function

We know that an $n \times n$ matrix A is invertible if and only if $D(A) \neq 0$. The row and column formulas established in the previous section enable us to write a formula for A^{-1} if A is invertible. First we need some terminology.

Definition. If $A = (a_{ij})$ is an $n \times n$ matrix, let $c_{ij} = (-1)^{i+j} D(A_{ij})$ be the i, jth cofactor of A. Then the matrix $C = (c_{ij})$ is called the **cofactor matrix of** A. The transpose C^t of the cofactor matrix is called the **adjoint of** A. We shall denote the adjoint of A by A^*.

Thus if $A = (a_{ij})$ then

$$A^* = ((-1)^{i+j} D(A_{ji})).$$

For example if

$$A = \begin{pmatrix} 1 & 2 \\ 0 & 3 \end{pmatrix},$$

the cofactor matrix $C = \begin{pmatrix} 3 & 0 \\ -2 & 1 \end{pmatrix}$ and $A^* = \begin{pmatrix} 3 & -2 \\ 0 & 1 \end{pmatrix}$. If

$$A = \begin{pmatrix} 1 & 0 & -1 \\ 1 & -1 & 0 \\ 0 & 1 & 1 \end{pmatrix},$$

the cofactor matrix

$$C = \begin{pmatrix} -1 & -1 & 1 \\ -1 & 1 & -1 \\ -1 & -1 & -1 \end{pmatrix}$$

and

$$A^* = C^t = \begin{pmatrix} -1 & -1 & -1 \\ -1 & 1 & -1 \\ 1 & -1 & -1 \end{pmatrix}.$$

THEOREM 4.13 *If A is an $n \times n$ matrix, then*

$$AA^* = D(A) \cdot I.$$

Thus if A is invertible,

$$A^{-1} = \frac{1}{D(A)} A^*.$$

Proof. If $R_1, \ldots, R_n$ are the rows of A and $C_1, \ldots, C_n$ are the columns of A^*, then the i,jth entry of AA^* is (R_i, C_j). But $R_i = (a_{i1}, \ldots, a_{in})$ and

$$C_j = \begin{pmatrix} (-1)^{j+1} D(A_{j1}) \\ \vdots \\ (-1)^{j+n} D(A_{jn}) \end{pmatrix}$$

Therefore the inner product

$$(R_i, C_j) = \sum_{k=1}^{n} (-1)^{j+k} a_{ik} D(A_{jk}).$$

We must show

$$(R_i, C_j) = D(A) \quad \text{if } i = j$$
$$= 0 \quad \text{otherwise.}$$

Now if $i = j$, then by the row formula (7) of Section 4.2

$$D(A) = \sum_{k=1}^{n} (-1)^{i+k} a_{ik} D(A_{ik}).$$

To show that for $i \neq j$

(1) $$\sum_{k=1}^{n} (-1)^{j+k} a_{ik} D(A_{jk}) = 0$$

we again apply formula (7). To establish (1) observe that in the row formula $D(A) = \sum_{k=1}^{n} (-1)^{j+k} a_{jk} D(A_{jk})$ if we replace the jth row $(a_{j1}, \ldots, a_{jn})$ by a vector $(y_1, \ldots, y_n)$ then the resulting sum $\sum_{k=1}^{n} (-1)^{j+k} y_k D(A_{jk})$ is the determinant of the matrix A' obtained from A by replacing the jth row $(a_{j1}, \ldots, a_{jn})$ by $(y_1, \ldots, y_n)$. Hence the sum

$$\sum_{k=1}^{n} (-1)^{j+k} a_{ik} D(A_{jk})$$

is the determinant of the matrix A' obtained from A by replacing the jth row $(a_{j1}, \ldots, a_{jn})$ by the ith row $(a_{i1}, \ldots, a_{in})$. Since the ith and jth rows of A' are now identical we have

$$D(A') = \sum_{k=1}^{n} (-1)^{j+k} a_{ik} D(A_{jk}) = 0$$

when $i \neq j$. This establishes (1).

As a consequence of this result we may derive Cramer's rule for solving n equations in n unknowns

FURTHER PROPERTIES OF DETERMINANT FUNCTION

(2)
$$a_{11}x_1 + \cdots + a_{1n}x_n = y_1$$
$$\vdots \qquad \vdots \qquad \vdots$$
$$a_{n1}x_1 + \cdots + a_{nn}x_n = y_n$$

If A is the coefficient matrix of this system and

$$X = \begin{pmatrix} x_1 \\ \vdots \\ x_n \end{pmatrix}, \quad Y = \begin{pmatrix} y_1 \\ \vdots \\ y_n \end{pmatrix},$$

then
$$AX = Y.$$

Assuming A to be invertible we have

$$X = A^{-1}Y = \frac{1}{D(A)} A^*Y.$$

Now the ith component of the vector A^*Y is the inner product of the ith row of A^* with the vector Y. Since the ith row of A^* is

$((-1)^{i+1}D(A_{1i}), \ldots (-1)^{i+n}D(A_{ni}))$ we have for $i = 1, \ldots, n$.

$$x_i = \frac{1}{D(A)} \sum_{k=1}^{n} (-1)^{k+i} y_k D(A_{ki}).$$

Now, however, we observe that since

$$D(A) = \sum_{k=1}^{n} (-1)^{k+i} a_{ki} D(A_{ki})$$

we may infer that the sum

$$\sum_{k=1}^{n} (-1)^{k+i} y_k D(A_{ki})$$

is the determinant of the matrix A' obtained from A by replacing the ith column $\begin{pmatrix} a_{1i} \\ \vdots \\ a_{ni} \end{pmatrix}$ by the vector $\begin{pmatrix} y_1 \\ \vdots \\ y_n \end{pmatrix}$. Consequently for $i = 1, \ldots, n$

$$x_i = \frac{D\begin{pmatrix} a_{11} \cdots y_1 \cdots a_{1n} \\ \vdots \qquad \vdots \qquad \vdots \\ a_{n1} \cdots y_n \cdots a_{nn} \end{pmatrix}}{D\begin{pmatrix} a_{11} \cdots a_{1n} \\ \vdots \qquad \vdots \\ a_{n1} \cdots a_{nn} \end{pmatrix}}.$$

(with the ith column indicated)

These formulas are known as *Cramer's rule* for solving the linear equations (2). Because of the large number of arithmetic operations involved these formulas are only practical as methods for computing $x_1, \ldots, x_n$ when n is

DETERMINANTS

small. When $n \geq 4$ the elimination techniques of Chapter 2 are preferable unless most of the entries of A are zero. Also there may be solutions to the equations $AX = Y$ when A has no inverse. In this case Cramer's rule is not applicable.

Example 1. Solve the following system of equations by Cramer's rule

(3)
$$\begin{aligned} x_1 + x_2 &= 1 \\ -x_2 + x_4 &= 0 \\ -x_1 + x_3 &= 0 \\ -x_3 + x_4 &= 1. \end{aligned}$$

The coefficient matrix

$$A = \begin{pmatrix} 1 & 1 & 0 & 0 \\ 0 & -1 & 0 & 1 \\ -1 & 0 & 1 & 0 \\ 0 & 0 & -1 & 1 \end{pmatrix} \quad \text{and} \quad D(A) = -2.$$

Therefore A is invertible and we may apply Cramer's rule. Replacing the first column of A by $B = \begin{pmatrix} 1 \\ 0 \\ 0 \\ 1 \end{pmatrix}$ yields

$$x_1 = \frac{D\begin{pmatrix} 1 & 1 & 0 & 0 \\ 0 & -1 & 0 & 1 \\ 0 & 0 & 1 & 0 \\ 1 & 0 & -1 & 1 \end{pmatrix}}{D(A)} = 0.$$

Replacing the second column by $B = \begin{pmatrix} 1 \\ 0 \\ 0 \\ 1 \end{pmatrix}$ yields

$$x_2 = \frac{D\begin{pmatrix} 1 & 1 & 0 & 0 \\ 0 & 0 & 0 & 1 \\ -1 & 0 & 1 & 0 \\ 0 & 1 & -1 & 0 \end{pmatrix}}{D(A)} = \frac{-2}{-2} = 1.$$

Continuing the process or substituting back in equations (3) we get $x_3 = 0$ and $x_4 = 1$.

EXERCISES

Compute the inverses of the following matrices using Theorem 4.13.

1. $\begin{pmatrix} 2 & -1 \\ 1 & 1 \end{pmatrix}$.

2. $\begin{pmatrix} 1 & 2 \\ -1 & 1 \end{pmatrix}$.

3. $\begin{pmatrix} 3 & 1 \\ 0 & 1 \end{pmatrix}$.

4. $\begin{pmatrix} 1 & -1 & 1 \\ 2 & 1 & 0 \\ 0 & 1 & 1 \end{pmatrix}$.

5. $\begin{pmatrix} 1 & 0 & 1 \\ -1 & 1 & 0 \\ 0 & 1 & 2 \end{pmatrix}$.

6. $\begin{pmatrix} 0 & 1 & -1 \\ 1 & 0 & 2 \\ 2 & 1 & 0 \end{pmatrix}$.

7. $\begin{pmatrix} 1 & 0 & 0 & -1 \\ -1 & 1 & 0 & 0 \\ 0 & -1 & 1 & 0 \\ 0 & 0 & 1 & 1 \end{pmatrix}$.

8. $\begin{pmatrix} 0 & 0 & 1 & 0 \\ 1 & 0 & 0 & 0 \\ 0 & 1 & 0 & 0 \\ 0 & 0 & 0 & 1 \end{pmatrix}$.

Solve the following sets of linear equations by Cramer's rule if the technique is applicable.

9. $x_1 - x_2 + x_3 = 1$
   $2x_1 + x_2 - x_3 = 0$
   $x_1 + x_2 - x_3 = 1.$

10. $x_1 - x_3 = 0$
    $x_2 + x_3 = 1$
    $x_1 + x_2 + x_3 = -1.$

11. $x_1 + x_2 - x_3 = 1$
    $x_2 - x_3 + x_4 = -1$
    $-x_1 + x_3 - x_4 = 1$
    $x_1 - x_2 + x_4 = 0.$

12. $x_1 - x_3 = 0$
    $x_2 + x_4 = 1$
    $x_1 - x_2 = -1$
    $x_3 + x_4 = 0.$

13. $2x_1 - x_2 + x_3 = 1$
    $x_1 + x_2 - x_3 = 0$
    $4x_1 + x_2 - x_3 = 2.$

14. $x_1 - x_2 + x_3 = 1$
    $2x_1 - x_3 = 0$
    $2x_2 + x_3 = -1.$

15. Let $y_0, \ldots, y_n$ be real numbers and let $x_0, \ldots, x_n$ be real numbers which are assumed to be distinct. Prove that there exists a unique polynomial $p(x) = a_0 + a_1 x + \cdots + a_n x^n$ satisfying
$$p(x_k) = y_k \quad k = 0, 1, \ldots, n.$$
(This is Lagrange's interpolation theorem. Use Exercise 18, p. 140.)

16. Determine the unique polynomial of degree 2 satisfying
$$p(1) = -1, \quad p(2) = 1, \quad p(3) = 4.$$

17. Derive general formulas for the coefficients of the polynomial in Exercise 15.

18. Let $x_0, y_0, y_1, \ldots, y_n$ be given real numbers. Show that there exists a unique polynomial $p(x)$ of degree n satisfying
$$p(x_0) = y_0$$
$$p'(x_0) = y_1$$
$$\vdots$$
$$p^{(n)}(x_0) = y_n.$$

4.6 Eigenvalues and Applications to Differential Equations

If A is an $n \times n$ matrix, we know by Theorem 4.10 that the eigenvalues of A are precisely the roots of the nth-degree polynomial

(1) $$D(A - xI) = 0.$$

Moreover if there are exactly n distinct eigenvalues $\lambda_1, \ldots, \lambda_n$, then by Theorem 3.22, $V_n(R)$ has a basis consisting of eigenvectors $\{X_1, \ldots, X_n\}$ associated with the eigenvalues $\{\lambda_1, \ldots, \lambda_n\}$. If P is the change of basis matrix from the canonical basis to the basis $\{X_1, \ldots, X_n\}$, then

$$P^{-1}AP = \begin{pmatrix} \lambda_1 & & 0 \\ & \ddots & \\ 0 & & \lambda_n \end{pmatrix} = \Lambda$$

since for each i, $AX_i = \lambda_i X_i$. Hence A is similar to the diagonal matrix Λ and to determine P we need only observe that the columns of P are the eigenvectors $\{X_1, \ldots, X_n\}$. It is important to note that if the eigenvalues $\lambda_1, \ldots, \lambda_n$ of A are not all distinct, then A is not in general similar to a diagonal matrix.

Example 1. Show that $A = \begin{pmatrix} 2 & 2 \\ 1 & 3 \end{pmatrix}$ is similar to a diagonal matrix and find a

matrix P such that
$$\begin{pmatrix} \lambda_1 & 0 \\ 0 & \lambda_2 \end{pmatrix} = P^{-1}AP$$
where λ_1, λ_2 are the eigenvalues for A. Now
$$A - xI = \begin{pmatrix} 2-x & 2 \\ 1 & 3-x \end{pmatrix}$$
$$D(A - xI) = (2-x)(3-x) - 2 = x^2 - 5x + 4 = (x-4)(x-1).$$
Therefore 4, 1 are the eigenvalues for A. Associated eigenvectors are nonzero solutions of the equations
$$(A - 4I)X = \begin{pmatrix} -2 & 2 \\ 1 & -1 \end{pmatrix}\begin{pmatrix} x_1 \\ x_2 \end{pmatrix} = 0$$
and
$$(A - I)X = \begin{pmatrix} 1 & 2 \\ 1 & 2 \end{pmatrix}\begin{pmatrix} x_1 \\ x_2 \end{pmatrix} = 0.$$
Such vectors are $\begin{pmatrix} 1 \\ 1 \end{pmatrix}$ and $\begin{pmatrix} 2 \\ -1 \end{pmatrix}$, respectively. Hence we may take $P = \begin{pmatrix} 1 & 2 \\ 1 & -1 \end{pmatrix}$. Then
$$P^{-1} = -\frac{1}{3}\begin{pmatrix} -1 & -2 \\ -1 & 1 \end{pmatrix} = \frac{1}{3}\begin{pmatrix} 1 & 2 \\ 1 & -1 \end{pmatrix},$$
and
$$P^{-1}AP = \begin{pmatrix} 4 & 0 \\ 0 & 1 \end{pmatrix}.$$

Now the polynomial $D(A-xI)$ is of degree n and hence may have no more than n real roots. However, it may have less. For example if $A = \begin{pmatrix} 0 & -1 \\ 1 & 0 \end{pmatrix}$ then $D(A-xI) = D\begin{pmatrix} -x & -1 \\ 1 & -x \end{pmatrix} = x^2 + 1$. In the field of real numbers, the equation $x^2 + 1 = 0$ has no solution. Hence the matrix $A = \begin{pmatrix} 0 & -1 \\ 1 & 0 \end{pmatrix}$ has no real eigenvalues. If, however, the field of scalars is the complex number field instead of the real number field we can say more. Then since $D(A-xI) = (x-i)(x+i)$ where $i^2 = -1$, we see that $D(A-xI) = 0$ if and only if $x = \pm i$. Hence $\pm i$ are complex eigenvalues for the matrix A.

In general for an $n \times n$ matrix if the polynomial $D(A-xI)$ has n distinct complex roots $\{\lambda_1, \ldots, \lambda_n\}$, then A is similar to the complex diagonal matrix
$$\begin{pmatrix} \lambda_1 & & 0 \\ & \ddots & \\ 0 & & \lambda_n \end{pmatrix}.$$ Now, however, the columns of the change of basis matrix P are eigenvectors for A with complex components.

Example 2. To determine a 2×2 matrix P satisfying $P^{-1}AP = \begin{pmatrix} i & 0 \\ 0 & -i \end{pmatrix}$ where $A = \begin{pmatrix} 0 & -1 \\ 1 & 0 \end{pmatrix}$ we must solve the equations
$$(A - iI)X = \begin{pmatrix} -i & -1 \\ 1 & -i \end{pmatrix}\begin{pmatrix} x_1 \\ x_2 \end{pmatrix} = 0$$
and
$$(A + iI)X = \begin{pmatrix} i & -1 \\ 1 & i \end{pmatrix}\begin{pmatrix} x_1 \\ x_2 \end{pmatrix} = 0.$$
These eigenvectors (x_1, x_2) are $(1, -i)$ and $(1, i)$, respectively. Hence $P = \begin{pmatrix} 1 & 1 \\ -i & i \end{pmatrix}$. The inverse P^{-1} may be computed in the usual way even though now the scalars are complex numbers. Indeed

$$P^{-1} = \frac{1}{2i}\begin{pmatrix} i & -1 \\ i & 1 \end{pmatrix}.$$

If some of the roots of $D(A-xI)$ are repeated, then A may or may not be similar to a diagonal matrix. To go into these matters here, however, is beyond the scope of the present treatment.

We illustrate the wide applicability of these notions by considering a problem in linear differential equations. Recall that a function $y = y(t)$ is a solution of the first-order homogeneous linear differential equation

$$y' = ky = 0$$

if and only if

$$y = ce^{kt}$$

where c is a constant. The scalar k may be either real or complex. We consider first the problem of finding n functions $y_1(t), \ldots, y_n(t)$ which are to satisfy the following system of differential equations

(2)
$$\begin{aligned} y_1'(t) &= a_{11}y_1(t) + \cdots + a_{1n}y_n(t) \\ y_2'(t) &= a_{21}y_2(t) + \cdots + a_{2n}y_n(t) \\ &\vdots \\ y_n'(t) &= a_{n1}y_n(t) + \cdots + a_{nn}y_n(t). \end{aligned}$$

These equations are to hold, of course, for all t in some interval I. Define the vector function $Y(t)$ by the formula

$$Y(t) = (y_1(t), \ldots, y_n(t)).$$

Then

$$Y'(t) = (y_1'(t), \ldots, y_n'(t)).$$

If we write $Y(t)$ and $Y'(t)$ as column vectors and let A be the coefficient matrix of (2), then we may rewrite (2) in the equivalent vector form

$$\begin{pmatrix} y_1'(t) \\ \vdots \\ y_n'(t) \end{pmatrix} = \begin{pmatrix} a_{11} & \cdots & a_{1n} \\ \vdots & & \vdots \\ a_{n1} & \cdots & a_{nn} \end{pmatrix} \begin{pmatrix} y_1(t) \\ \vdots \\ y_n(t) \end{pmatrix}$$

or

(3) $$Y'(t) = AY(t).$$

Suppose now that A has n distinct eigenvalues $\lambda_1, \ldots, \lambda_n$. Then A is similar to the diagonal matrix

$$\Lambda = \begin{pmatrix} \lambda_1 & & 0 \\ & \ddots & \\ 0 & & \lambda_n \end{pmatrix}$$

If $P^{-1}AP = \Lambda$, define the vector function $Z(t)$ by the formula.

$$Z(t) = P^{-1}Y(t).$$

Then

(4) $$Y(t) = PZ(t)$$

and differentiating we get

(5) $$Y'(t) = PZ'(t)$$

since P is a constant matrix. On the other hand, combining (3), (4) and (5) we get
$$Y'(t) = AY(t) = APZ(t)$$
and
(6) $$PZ'(t) = APZ(t).$$

Multiplying (6) by P^{-1} yields
$$Z'(t) = P^{-1}APZ(t) = \Lambda Z(t).$$

This latter vector differential equation consists of just the n first-order scalar differential equations
$$z_1'(t) = \lambda_1 z_1(t)$$
$$z_2'(t) = \lambda_2 z_2(t)$$
$$\vdots \qquad \vdots$$
$$z_n'(t) = \lambda_n z_n(t).$$

Each solution of these scalar equations has the form
$$z_k(t) = c_k e^{\lambda_k t}.$$
where c_k is an arbitrary constant. Hence each solution vector $Y(t)$ to the original system (2) is given by
$$Y(t) = PZ(t) = P \begin{pmatrix} c_1 e^{\lambda_1 t} \\ \vdots \\ c_n e^{\lambda_n t} \end{pmatrix}.$$

To specify the constants $c_1, \ldots, c_n$ one need only specify initial conditions $y_1(t_0), \ldots, y_n(t_0)$. For if
$$Y(t_0) = \begin{pmatrix} y_1(t_0) \\ \vdots \\ y_n(t_0) \end{pmatrix}$$
then
$$Y(t_0) = P \cdot \begin{pmatrix} c_1 e^{\lambda_1 t_0} \\ \vdots \\ c_n e^{\lambda_n t_0} \end{pmatrix} \quad \text{or} \quad \begin{pmatrix} c_1 e^{\lambda_1 t_0} \\ \vdots \\ c_n e^{\lambda_n t_0} \end{pmatrix} = P^{-1} Y(t_0).$$

Hence to determine the constant vector $\begin{pmatrix} c_1 \\ \vdots \\ c_n \end{pmatrix}$ we may either solve the equations $Y(t_0) = P \begin{pmatrix} c_1 e^{\lambda_1 t_0} \\ \vdots \\ c_n e^{\lambda_n t_0} \end{pmatrix}$ or if the inverse P^{-1} is known, we use the relation $\begin{pmatrix} c_1 e^{\lambda_1 t_0} \\ \vdots \\ c_n e^{\lambda_n t_0} \end{pmatrix} = P^{-1} Y(t_0).$

EIGENVALUES

Example 3. Determine functions $y_1(t)$, $y_2(t)$ satisfying
$$y_1'(t) = 2y_1(t) + 2y_2(t)$$
$$y_2'(t) = y_1(t) + 3y_2(t)$$

and $y_1(0) = 1$ and $y_2(0) = 0$. We have already seen on p. 150 that the matrix $\begin{pmatrix} 2 & 2 \\ 1 & 3 \end{pmatrix}$ has eigenvalues 4, 1 with associated eigenvectors $\begin{pmatrix} 1 \\ 1 \end{pmatrix}$ and $\begin{pmatrix} 2 \\ -1 \end{pmatrix}$. Therefore $P = \begin{pmatrix} 1 & 2 \\ 1 & -1 \end{pmatrix}$ and $P^{-1} = 1/3 \begin{pmatrix} 1 & 2 \\ 1 & -1 \end{pmatrix}$. Now

$$\begin{pmatrix} z_1'(t) \\ z_2'(t) \end{pmatrix} = \begin{pmatrix} 4z_1(t) \\ z_2(t) \end{pmatrix}$$

implies $z_1(t) = c_1 e^{4t}$ and $z_2(t) = c_2 e^t$. However

$$Z(0) = \begin{pmatrix} z_1(0) \\ z_2(0) \end{pmatrix} = \begin{pmatrix} c_1 \\ c_2 \end{pmatrix} = P^{-1} \begin{pmatrix} y_1(0) \\ y_2(0) \end{pmatrix} = \frac{1}{3} \begin{pmatrix} 1 & 2 \\ 1 & -1 \end{pmatrix} \begin{pmatrix} 1 \\ 0 \end{pmatrix} = \begin{pmatrix} \frac{1}{3} \\ \frac{1}{3} \end{pmatrix}.$$

Therefore

$$Y(t) = PZ(t) = \begin{pmatrix} 1 & 2 \\ 1 & -1 \end{pmatrix} \begin{pmatrix} \frac{1}{3} e^{4t} \\ \frac{1}{3} e^t \end{pmatrix} = \frac{1}{3} \begin{pmatrix} e^{4t} + 2e^t \\ e^{4t} - e^t \end{pmatrix}.$$

Consequently

$$y_1(t) = \frac{1}{3}(e^{4t} + 2e^t)$$

and

$$y_2(t) = \frac{1}{3}(e^{4t} - e^t)$$

are the desired functions. Moreover these functions are unique. To determine $\begin{pmatrix} c_1 \\ c_2 \end{pmatrix}$ without knowing P^{-1} we observe that since $P = \begin{pmatrix} 1 & 2 \\ 1 & -1 \end{pmatrix}$

$$\begin{pmatrix} 1 & 2 \\ 1 & -1 \end{pmatrix} \begin{pmatrix} c_1 \\ c_2 \end{pmatrix} = P \begin{pmatrix} c_1 \\ c_2 \end{pmatrix} = Y(0) = \begin{pmatrix} 1 \\ 0 \end{pmatrix}$$

or
$$c_1 + 2c_2 = 1$$
$$c_1 - c_2 = 0.$$

Hence $c_1 = c_2 = \frac{1}{3}$.

We summarize these results in the following theorem.

THEOREM 4.14 *Let A be an $n \times n$ matrix with n distinct eigenvalues $\lambda_1, \ldots, \lambda_n$. Then the vector differential equation*

$$Y'(t) = AY(t)$$

with initial conditions

$$Y(t_0) = Y_0$$

has a unique solution

(7) $$Y(t) = P \begin{pmatrix} c_1 e^{\lambda_1 t} \\ \vdots \\ c_n e^{\lambda_n t} \end{pmatrix}$$

where P is a matrix whose columns are eigenvectors associated with the eigenvalues $\lambda_1, \ldots, \lambda_n$. The constants $c_1, \ldots, c_n$ are uniquely deter-

mined by requiring that

$$P^{-1}Y(t_0) = \begin{pmatrix} c_1 e^{\lambda_1 t_0} \\ \vdots \\ c_n e^{\lambda_n t_0} \end{pmatrix}.$$

EXERCISES

For the following matrices A compute the eigenvalues of A and determine matrices P, P^{-1} such that $P^{-1}AP$ is a diagonal matrix.

1. $\begin{pmatrix} 3 & 2 \\ 1 & 2 \end{pmatrix}.$

2. $\begin{pmatrix} 1 & 0 \\ -1 & 0 \end{pmatrix}.$

3. $\begin{pmatrix} -1 & 2 \\ 3 & 4 \end{pmatrix}.$

4. $\begin{pmatrix} 0 & 1 \\ -6 & -5 \end{pmatrix}.$

5. $\begin{pmatrix} 1 & -1 & 1 \\ 2 & 1 & 0 \\ 3 & 0 & 1 \end{pmatrix}.$

6. $\begin{pmatrix} 0 & 1 & 0 \\ 0 & 0 & 1 \\ 0 & -3 & 4 \end{pmatrix}.$

7. $\begin{pmatrix} 0 & 1 & 0 \\ 0 & 0 & 1 \\ 0 & -2 & 3 \end{pmatrix}.$

Certain of the eigenvalues for the following matrices A may be complex. Determine matrices P, P^{-1} with complex entries so that $P^{-1}AP$ is a diagonal matrix.

8. $\begin{pmatrix} 0 & -2 \\ 2 & 0 \end{pmatrix}.$

9. $\begin{pmatrix} 0 & 1 \\ 5 & 4 \end{pmatrix}.$

10. $\begin{pmatrix} 0 & 1 & 0 \\ 0 & 0 & 1 \\ 0 & -1 & 0 \end{pmatrix}.$

11. Show that an odd order square matrix with real coefficients always has at least one real eigenvalue.

12. Show that the matrix

$$\begin{pmatrix} 0 & 1 & 0 \\ 0 & 0 & 1 \\ -a_2 & -a_1 & -a_2 \end{pmatrix}$$

is similar to a diagonal matrix if the polynomial $x^3 + a_2 x^2 + a_1 x + a_0$ has no multiple roots. (Take the field of scalars to be the field of complex numbers.)

13. Generalize Exercise 12 to the matrix

$$\begin{pmatrix} 0 & 1 & 0 & \cdots & 0 \\ & 0 & 1 & & \vdots \\ \vdots & & & & 0 \\ 0 & & \cdots & & 1 \\ -a_0 & -a_1 & \cdots & & -a_n \end{pmatrix}.$$

Determine the solutions $(y_1(t), y_2(t))$ to the following first-order systems of differential equations which satisfy the given initial conditions.

14. $y_1'(t) = y_1(t) + 2y_2(t)$
    $y_2'(t) = 2y_1(t) + y_2(t)$
    $y_1(0) = 1, y_2(0) = -1.$

15. $y_1'(t) = 3y_1(t) + y_2(t)$
    $y_2'(t) = y_1(t) + 3y_2(t)$
    $y_1(0) = 1, y_2(0) = 0.$

16. $y_1'(t) = y_2(t)$
    $y_2'(t) = 3y_1(t) - 2y_2(t)$
    $y_1(0) = 0, y_2(0) = 1.$

Solve the following systems of differential equations subject to the given initial conditions. We suppress the variable t.

17. $y_1' = y_1 - y_2 + y_3$
    $y_2' = 2y_1 + y_2$
    $y_3' = 3y_1 + y_3$
    $y_1(0) = 1$, $y_2(0) = -1$, $y_3(0) = 0$.

18. $y_1' = -y_1 + 3y_2$.
    $y_2' = 3y_1 - y_2$
    $y_3' = -2y_1 - 2y_2 + 6y_3$
    $y_1(0) = 1$, $y_2(0) = 0$, $y_3(0) = 1$.

4.7* Further Applications to Differential Equations

As a second application of the problem of finding eigenvalues for matrices we shall consider the problem of finding all solutions to the nth-order homogeneous linear differential equation.

(1) $$y^{(n)}(t) + a_{n-1} y^{(n-1)}(t) + \cdots + a_0 y(t) = 0.$$

To solve (1) we must construct a system of first-order equations

(2) $$\begin{aligned} y_1'(t) &= a_{11} y_1(t) + \cdots + a_{1n} y_n(t) \\ &\vdots \\ y_n'(t) &= a_{n1} y_n(t) + \cdots + a_{nn} y_n(t) \end{aligned}$$

such that each solution of (1) is a solution of (2) and conversely. We do this by the following stratagem. Let $y_1(t) = y(t)$ be the unknown function which is a solution to (1). Define

$$\begin{aligned} y_2(t) &= y_1'(t) \\ y_3(t) &= y_2'(t) = y_1''(t) \\ y_4(t) &= y_3'(t) = y_1'''(t) \\ &\vdots \\ y_n(t) &= y_{n-1}'(t) = y_1^{(n-1)}(t). \end{aligned}$$

Then since $y_1(t)$ is a solution to (1), if we substitute the functions $y_2, y_3 \cdots$ for $y_1', y_1'', \ldots$ we have

$$\begin{aligned} y_n'(t) &= -a_0 y_1(t) - a_1 y_1''(t) - \cdots - a_{n-1} y_1^{(n-1)}(t) \\ &= -a_0 y_1(t) - a_1 y_2(t) - \cdots - a_{n-1} y_{n-1}(t). \end{aligned}$$

Thus we have constructed the following system of first-order differential equations

$$\begin{aligned} y_1'(t) &= y_2(t) \\ y_2'(t) &= y_3(t) \\ &\vdots \\ y_{n-1}'(t) &= y_n'(t) \\ y_n'(t) &= -a_0 y_1(t) - \cdots - a_{n-1} y_n(t). \end{aligned}$$

If

$$Y(t) = \begin{pmatrix} y_1(t) \\ \vdots \\ y_n(t) \end{pmatrix},$$

156 DETERMINANTS

this system of first-order equations may be written

(3) $$Y'(t) = AY(t)$$

where

(4) $$A = \begin{pmatrix} 0 & 1 & 0 & \cdots & 0 \\ 0 & 0 & 1 & 0 & \cdots & 0 \\ \vdots & & & & & \\ 0 & & \cdots & & 0 & 1 \\ -a_0 & -a_1 & & \cdots & & -a_{n-1} \end{pmatrix}.$$

Clearly from the construction $y_1(t)$ is a solution of (1) if and only if

$$Y(t) = \begin{pmatrix} y_1(t) \\ y_2(t) \\ \vdots \\ y_n(t) \end{pmatrix} = \begin{pmatrix} y_1(t) \\ y_1'(t) \\ \vdots \\ y_1^{(n-1)}(t) \end{pmatrix}$$

is a solution to (3). We may now apply Theorem 4.14 whenever the matrix A can be shown to have n distinct eigenvalues. However

$$D(A - xI) = D\begin{pmatrix} -x & 1 & 0 & \cdots & 0 \\ 0 & -x & 1 & & \vdots \\ \vdots & & & & \\ & & & & 1 \\ -a_0 & -a_1 & \cdots & & -(x+a_{n-1}) \end{pmatrix}.$$

Expanding this determinant along the first column we see that

$$D(A - xI) = -xD\begin{pmatrix} -x & 1 & 0 & \cdots & 0 \\ \vdots & & & & \\ & & & & 1 \\ -a_1 & & \cdots & & -(x+a_{n-1}) \end{pmatrix} + (-1)^n a_0$$

$$= -x\left[-xD\begin{pmatrix} -x & 1 & 0 & \cdots & 0 \\ \vdots & & & & \\ & & & & 1 \\ -a_2 & & \cdots & & -(x+a_{n-1}) \end{pmatrix} + (-1)^{n-1} a_1 \right] + (-1)^n a_0$$

which by induction is easily seen to be equal to

$$(-1)^n[a_0 + a_1 x + \cdots + a_{n-1} x^{n-1} + x^n].$$

Thus A will have n distinct eigenvalues if and only if the polynomial

$$a_0 + a_1 x + \cdots + a_{n-1} x^{n-1} + x^n$$

has n distinct roots $\lambda_1, \ldots, \lambda_n$. These roots are the eigenvalues of A.

If $X_1, \ldots, X_n$ are corresponding eigenvectors and P is the matrix whose columns are the vectors $X_1, \ldots, X_n$, then since

$$Y(t) = \begin{pmatrix} y_1(t) \\ \vdots \\ y_n(t) \end{pmatrix}$$

satisfies (3) we may infer from the discussion p. 152 that

(5) $$Y(t) = \begin{pmatrix} y_1(t) \\ \vdots \\ y_n(t) \end{pmatrix} = P \begin{pmatrix} c_1 e^{\lambda_1 t} \\ \vdots \\ c_n e^{\lambda_n t} \end{pmatrix}.$$

Hence $y(t)$ is a solution to (1) if and only if

(6) $$y(t) = y_i(t) = p_{11} c_1 e^{\lambda_1 t} + p_{12} e^{\lambda_2 t} + \cdots + p_{1n} c_n e^{\lambda_n t}$$

where $(p_{11}, \ldots, p_{1n})$ is the first row of the matrix P and $(c_1, \ldots, c_n)$ is an arbitrary constant vector. Next we claim that each linear combination of the functions $e^{\lambda_1 t}, \ldots, e^{\lambda_n t}$ is a solution of (1). This will follow from the fact that each function $y(t)$ defined by (6) is a solution to (1) provided that $p_{1i} \neq 0$ for $i = 1, \ldots, n$. The problem of showing that this is always the case we leave as an exercise. The functions $e^{\lambda_1 t}, \ldots, e^{\lambda_n t}$ are called *fundamental solutions* to equation (1) and indeed they span the space of all solutions. To show that they form a basis for the space of solutions to (1) we need only show that they are linearly independent. This is not hard and we leave it as an exercise. Thus the space of solutions to (1) has dimension n. To determine the constants $(c_1, \ldots, c_n)$ in (5) we need only solve the equations

(7) $$\begin{pmatrix} c_1 e^{\lambda_1 t_0} \\ \vdots \\ c_n e^{\lambda_n t_0} \end{pmatrix} = P^{-1} Y(t_0).$$

These constants $c_1, \ldots, c_n$ can be determined without computing the matrices P and P^{-1} in the following way. We construct n linear equations in the unknowns $c_1, \ldots, c_n$ by differentiating the function

$$y(t) = c_1 e^{\lambda_1 t} + \cdots + c_n e^{\lambda_n t}$$

$n-1$ times and evaluating the successive derivatives of $y(t)$ at $t = t_0$. We obtain as a result the set of equations

$$c_1 e^{\lambda_1 t_0} + \cdots + c_n e^{\lambda_n t_0} = y(t_0)$$
$$\lambda_1 c_1 e^{\lambda_1 t_0} + \cdots + \lambda_n c_n e^{\lambda_n t_0} = y'(t_0)$$
$$\vdots \qquad \qquad \vdots$$
$$\lambda_1^{n-1} c_1 e^{\lambda_1 t_0} + \cdots + \lambda_n^{n-1} c_n e^{\lambda_n t_0} = y^{(n-1)}(t_0).$$

If $t_0 = 0$, these equations reduce to

(8) $$\begin{aligned} c_1 + \cdots + c_n &= y(0) \\ \lambda_1 c_1 + \cdots + \lambda_n c_n &= y'(0) \\ &\vdots \\ \lambda_1^{n-1} c_1 + \cdots + \lambda_n^{n-1} c_n &= y^{(n-1)}(0). \end{aligned}$$

Often if $t_0 = 0$ it may be more convenient to solve (8) directly for the coefficients $c_1, \ldots, c_n$ rather than to compute the matrix P^{-1} and to determine the coefficients from (7).

Example 1. Find the solution $y(t)$ to $y'' - 2y' - 3y = 0$ satisfying $y(0) = 1$ and $y'(0) = 0$. For this differential equation the matrix A in formula (4) is given by

$$A = \begin{pmatrix} 0 & 1 \\ 3 & 2 \end{pmatrix}.$$

Moreover $D(A - xI) = x^2 - 2x - 3 = (x-3)(x+1)$. Hence the eigenvalues of A are 3 and -1. Corresponding eigenvectors are $\begin{pmatrix} 1 \\ 3 \end{pmatrix}$ and $\begin{pmatrix} 1 \\ -1 \end{pmatrix}$. Consequently $P = \begin{pmatrix} 1 & 1 \\ 3 & -1 \end{pmatrix}$ and

$$P^{-1} = -\frac{1}{4}\begin{pmatrix} -1 & -1 \\ -3 & 1 \end{pmatrix} = \frac{1}{4}\begin{pmatrix} 1 & 1 \\ 3 & -1 \end{pmatrix}.$$

Now

$$P^{-1}Y(0) = P^{-1}\begin{pmatrix} y_1(0) \\ y_1'(0) \end{pmatrix} = P^{-1}\begin{pmatrix} 1 \\ 0 \end{pmatrix} = \begin{pmatrix} \frac{1}{4} \\ \frac{3}{4} \end{pmatrix}.$$

Since $P^{-1}Y(t) = \begin{pmatrix} c_1 e^{3t} \\ c_2 e^{-t} \end{pmatrix}$, we have $c_1 = \frac{1}{4}$, $c_2 = \frac{3}{4}$. Therefore

$$Y(t) = P\begin{pmatrix} \frac{1}{4}e^{3t} \\ \frac{3}{4}e^{-t} \end{pmatrix}$$

and the solution $y(t)$ which is given by the first component in the vector $Y(t)$ is just

$$y(t) = \frac{e^{3t}}{4} + \frac{3}{4}e^{-t}.$$

To determine the solution without computing the matrices P and P^{-1} observe that $y(t) = c_1 e^{3t} + c_2 e^{-t}$. Hence $y'(t) = 3c_1 e^{3t} - c_2 e^{-t}$. Evaluating $y(t)$ and $y'(t)$ for $t = 0$, we obtain

$$c_1 + c_2 = y(0) = 1$$
$$3c_1 - c_2 = y'(0) = 0.$$

Solving these equations we obtain $c_1 = \frac{1}{4}$ and $c_2 = \frac{3}{4}$. Hence $y(t) = \frac{1}{4}e^{3t} + \frac{3}{4}e^{-t}$ as before.

Example 2. Find the solution to $y'' + 3y' + \frac{5}{2}y = 0$ satisfying $y(0) = 0$ and $y'(0) = 1$. Proceeding as before the coefficient matrix

$$A = \begin{pmatrix} 0 & 1 \\ -\frac{5}{2} & -3 \end{pmatrix}$$

and $D(A - xI) = x^2 + 3x + \frac{5}{2}$. Now this polynomial has distinct complex roots $x = -(3+i)/2$ and $x = -(3-i)/2$. The associated eigenvectors are

$$\begin{pmatrix} 1 \\ \frac{3-i}{2} \end{pmatrix} \text{ and } \begin{pmatrix} 1 \\ \frac{3+i}{2} \end{pmatrix}.$$

Consequently

$$P = \begin{pmatrix} 1 & 1 \\ \frac{3-i}{2} & \frac{3+i}{2} \end{pmatrix}$$

and

$$P^{-1} = \frac{1}{i}\begin{pmatrix} \frac{3+i}{2} & -1 \\ \frac{-3+i}{2} & 1 \end{pmatrix}.$$

The fundamental solutions are $e^{-(3+i)t/2}$ and $e^{-(3-i)t/2}$. To determine the coefficients c_1, c_2 in (8) we note that

$$P^{-1}\begin{pmatrix}0\\1\end{pmatrix} = \begin{pmatrix}c_1\\c_2\end{pmatrix}$$

implies $c_1 = -1/i$ and $c_2 = 1/i$. Therefore

(9) $$y(t) = \frac{1}{i}\left(-e^{-(3+i)t/2} + e^{-(3-i)t/2}\right).$$

To remove the complex exponentials we note that $\cos t = (e^{it} + e^{-it})/2$ and $\sin t = (e^{it} - e^{-it})/2i$ or $e^{it} = \cos t + i \sin t$. Substituting the latter in (9) we obtain

$$y(t) = \frac{1}{i}e^{-3t/2}[-e^{it/2} + e^{-it/2}]$$

$$= \frac{1}{i}e^{-3t/2}\left[-\left(\cos\frac{t}{2} + i\sin\frac{t}{2}\right) + \left(\cos\frac{t}{2} - i\sin\frac{t}{2}\right)\right]$$

$$= -2e^{-3t/2}\sin\frac{t}{2}.$$

EXERCISES

Solve the following homogeneous linear differential equations subject to the given initial conditions.

1. $y'' - y = 0, y(0) = 1, y'(0) = -1.$
2. $y'' - 3y' - 4y = 0, y(0) = 1, y'(0) = 0.$
3. $y'' + 2y' + 2y = 0, y(0) = 0, y'(0) = 1.$
4. $y''' + 5y'' + 4y' = 0.$
   $y(0) = 1, y'(0) = 0, y''(0) = 1.$

Determine the space of all solutions to the following differential equations.

5. $y'' - y = 0.$
6. $y'' - y' - 2y = 0.$
7. $y''' + 2y'' - 3y' = 0.$
8. $y^{\text{iv}} - 5y'' + 4y = 0.$
9. Let λ be an eigenvalue of the $n \times n$ matrix

$$A = \begin{pmatrix} 0 & 1 & & & 0 \\ 0 & 0 & 1 & & \\ \vdots & & & & \\ & & & & 1 \\ -a_0 & -a_1 & \cdots & & -a_{n-1} \end{pmatrix}.$$

If $X = (x_1, \ldots, x_n)$ is a corresponding eigenvector show that $x_1 \neq 0$. (Examine the equation $AX = \lambda X$.)

10. Show that if the numbers $\lambda_1, \ldots, \lambda_n$ are all distinct, then the functions

$$e^{\lambda_1 t}, \ldots, e^{\lambda_n t}$$

are linearly independent. (Suppose $c_1 e^{\lambda_1 t} + c_2 e^{\lambda_2 t} + \cdots + c_n e^{\lambda_n t} = 0$. Differentiate this equation $n - 1$ times, with respect to t and examine the n equations in the n unknowns $c_1, \ldots, c_n$ which result if $t = 0$. Show that these equations have only the trivial solution. You will need the result of Exercise 18 on p. 140.)

CHAPTER FIVE

SYMMETRIC MATRICES AND QUADRATIC FORMS

5.1 Orthogonal Transformations

In Chapter 4 we considered the problem of determining which $n \times n$ matrices A were similar to diagonal matrices. We showed for example that this was the case if A had n distinct eigenvalues. In this chapter we wish to show that if A is symmetric, that is, $A = A^t$, then A is similar to a diagonal matrix. First, however, we need to discuss some preliminary material. We must determine the structure of those linear transformations in $V_n(R)$ which preserve the inner product.

Definition. *Let T be a linear transformation in $V_n(R)$ then T is said to be orthogonal if for each pair of vectors $X, Y \in V_n(R)$*

$$(T(X), T(Y)) = (X, Y).$$

An equivalent condition for the linear transformation T to be orthogonal is that T preserve the length of each vector X. The proof of this fact is easy, and we give it next.

THEOREM 5.1 *Let T be a linear transformation in $V_n(R)$. Then T is an orthogonal transformation if and only if for each $X \in V_n(R)$*

$$|TX| = |X|.$$

Proof. If T is orthogonal, then $|TX|^2 = (TX, TX) = (X, X) = |X|^2$, and T preserves lengths. For the converse observe that if $|TX| = |X|$, then $(T(X), T(X)) = (X, X)$. Therefore for each $X, Y \in V_n(R)$

$$(T(X+Y), T(X+Y)) = (X+Y, X+Y).$$

But

$$(T(X+Y), T(X+Y)) = (T(X), T(X)) + 2(T(X), T(Y)) + (T(Y), T(Y))$$

and

$$(X+Y, X+Y) = (X, X) + 2(X, Y) + (Y, Y).$$

Since $(T(X), T(X)) = (X, X)$ and $(T(Y), T(Y)) = (Y, Y)$, it follows from the two equations above that

$$(T(X), T(Y)) = (X, Y),$$

and T is orthogonal.

The most familiar example of an orthogonal transformation is a *rotation* in $V_2(R)$. That a rotation R_θ is orthogonal follows immediately from Theorem 5.1 since for each $X \in V_2(R)$, $|R_\theta X| = |X|$. Let us compute the matrix of R_θ with respect to the canonical basis. By definition we must write $R_\theta(I_1)$,

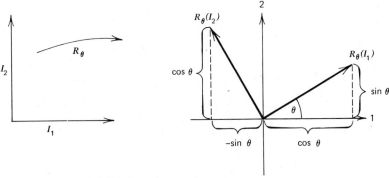

Fig. 1 Rotation of $V_2(R)$ through an angle θ.

$R_\theta(I_2)$ as a linear combination of the basis vectors I_1, I_2. Clearly from Figure 1

$$R_\theta(I_1) = \cos\theta I_1 + \sin\theta I_2$$

and

$$R_\theta(I_2) = -\sin\theta I_1 + \cos\theta I_2.$$

Therefore the matrix A_θ of the rotation R_θ with respect to the canonical basis is given by

$$A_\theta = \begin{pmatrix} \cos\theta & -\sin\theta \\ \sin\theta & \cos\theta \end{pmatrix}.$$

It may be immediately verified that for each $X, Y \in V_2(R)$, $(A_\theta X, A_\theta Y) = (X, Y)$ which gives another proof that a rotation is orthogonal. The details of this calculation we leave to the exercises.

Rotations are not the only orthogonal transformations in $V_2(R)$, however. Reflections also are orthogonal. To see this suppose S is a reflection of $V_2(R)$ about the line l through the origin. Then by definition of a reflection, X and $S(X)$ have equal lengths and make equal angles with the line l. The first of these statements implies by Theorem 5.1 that S is orthogonal.

The matrix of a reflection is extremely simple if we choose the correct bases $\{J_1, J_2\}$ for $V_2(R)$. Choose J_1 to be parallel to l and to have unit length and choose J_2 to be perpendicular to l and also to have unit length. See

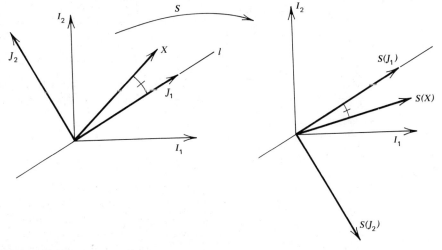

Fig. 2 Reflection through a line l.

Figure 2. Then
$$S(J_1) = J_1$$
and
$$S(J_2) = -J_2.$$

Therefore the matrix of S with respect to the basis J_1, J_2 is $\begin{pmatrix} 1 & 0 \\ 0 & -1 \end{pmatrix}$.

Both reflections and rotations are invertible transformations. The inverse of $\begin{pmatrix} 1 & 0 \\ 0 & -1 \end{pmatrix}$ is $\begin{pmatrix} 1 & 0 \\ 0 & -1 \end{pmatrix}$, and the inverse for the matrix

$$A_\theta = \begin{pmatrix} \cos\theta & -\sin\theta \\ \sin\theta & \cos\theta \end{pmatrix}$$

is clearly obtained by rotating through an angle of $-\theta$. Hence

$$A_{-\theta} = A_\theta^{-1} = \begin{pmatrix} \cos\theta & \sin\theta \\ -\sin\theta & \cos\theta \end{pmatrix} = A_\theta^t.$$

In both cases we see that the inverse of the matrix is given by its transpose.

Before giving further properties of orthogonal transformations in $V_n(R)$ we need some terminology. We have defined a set of vectors $\{X_1, \ldots, X_k\}$ in $V_n(R)$ to be orthogonal if $(X_i, X_j) = 0$ for $i \neq j$. If in addition these vectors have unit length, i.e. $(X_i, X_i) = 1$ for each i, then the set $\{X_1, \ldots, X_k\}$ is said to be *orthonormal*. A basis for $V_n(R)$ consisting of an orthonormal set of vectors is called an *orthonormal basis*. If $\{X_1, \ldots, X_n\}$ is an orthonormal basis for $V_n(R)$ and for $X \in V_n(R)$ we write $X = x_1 X_1 + \cdots + x_n X_n$, then the coefficients x_k in this equation may be very easily computed. Indeed, taking the inner product of this equation with the vectors $X_1, \ldots, X_n$ in turn, it follows that $x_k = (X, X_k)$ since $(X_k, X_k) = 1$ and $(X_j, X_k) = 0$ if $j \neq k$. The condition that a linear transformation be orthogonal can be interpreted very simply if an orthonormal basis is chosen for $V_n(R)$.

THEOREM 5.2 Let $\{X_1, \ldots, X_n\}$ be an orthonormal basis for $V_n(R)$ and let $T \in L(V_n(R))$. Then T is orthogonal if and only if $\{T(X_1), \ldots, T(X_n)\}$ is an orthonormal basis for $V_n(R)$.

Proof. If T is orthogonal, then the requirement that $(X, Y) = (T(X), T(Y))$ implies that $\{T(X_1), \ldots, T(X_n)\}$ must be an orthonormal set if $\{X_1, \ldots, X_n\}$ is. Since $\{T(X_1), \ldots, T(X_n)\}$ is linearly independent and the dimension of $V_n(R)$ is known to be n, $\{T(X_1), \ldots, T(X_n)\}$ must be a basis for $V_n(R)$. Conversely if $\{T(X_1), \ldots, T(X_n)\}$ is an orthonormal basis, we must show that $(X, Y) = (T(X), T(Y))$ for each $X, Y \in V_n(R)$. But if $X = \Sigma x_k X_k$ and $Y = \Sigma y_k X_k$, then $T(X) = \Sigma x_k T(X_k)$ and $T(Y) = \Sigma y_k T(X_k)$. Since $\{X_1, \ldots, X_n\}$ and $\{T(X_1), \ldots, T(X_n)\}$ are orthonormal, we have

$$(X, Y) = \left(\Sigma x_k X_k, \Sigma y_k X_k \right) = \Sigma x_k y_k$$

and

$$(T(X), T(Y)) = \left(\Sigma x_k T(X_k), \Sigma y_k T(X_k) \right) = \Sigma x_k y_k.$$

Therefore $(X, Y) = (T(X), T(Y))$, and we are done.

Since an orthogonal transformation T preserves the lengths of vectors, T must be one-to-one. Hence by Theorem 3.14 we know that T is invertible. Hence if A is the matrix of T with respect to some basis in $V_n(R)$, A is an invertible matrix. If the basis is chosen to be orthonormal, then the inverse for A is just the transpose A^t. This is our next result.

THEOREM 5.3 *Let $\{X_1, \ldots, X_n\}$ be an orthonormal basis for $V_n(R)$. For $T \in L(V_n(R))$ let A be the matrix of T with respect to this basis. Then T is orthogonal if and only if the matrix A satisfies*

(1) $$A^t A = I.$$

Proof. If $A = (a_{jk})$ is the matrix of T with respect to the basis $\{X_1, \ldots, X_n\}$, we have by definition

$$T(X_k) = \sum_{j=1}^{n} a_{jk} X_j \qquad k = 1, \ldots, n.$$

By Theorem 5.2 we know that T is orthogonal if and only if $\{T(X_1), \ldots, T(X_n)\}$ is an orthonormal set. However, since $\{X_1, \ldots, X_n\}$ is orthonormal, this implies that for each k and l

$$(T(X_k), T(X_l)) = \left(\sum_j a_{jk} X_j, \sum_j a_{jl} X_j\right) = \sum_j a_{jk} a_{jl}.$$

Hence T is orthogonal if and only if

(2) $$\sum_j a_{jk} a_{jl} = 1 \quad \text{if } k = l$$
$$= 0 \quad \text{if } k \neq l.$$

But if $C_k = \begin{pmatrix} a_{1k} \\ \vdots \\ a_{nk} \end{pmatrix}$ are the columns of the matrix A, then (2) just expresses the fact that

(3) $$(C_k, C_l) = 1 \quad \text{if } k = l$$
$$= 0 \quad \text{if } k \neq l.$$

However, the columns of A are the rows of A^t. Hence (3) holds if and only if

$$A^t A = I.$$

A matrix A satisfying the identity $A^t A = I$ is called an *orthogonal* matrix. Clearly a matrix A is orthogonal if and only if the columns of A form an orthonormal set in $V_n(R)$. Since $A^t A = I$ if and only if $AA^t = I$, a matrix A is orthogonal if and only if the rows of A are an orthonormal set.

COROLLARY. *If A is an orthogonal matrix, then $D(A) = \pm 1$.*

Proof. For any matrix, $D(A) = D(A^t)$. Since A is orthogonal, $A^t A = 1$. Hence $1 = D(A^t A) = D(A^t)D(A) = [D(A)]^2$. This implies $D(A) = \pm 1$.

Next we shall determine all orthogonal transformations in $V_2(R)$. Let $A = \begin{pmatrix} a & c \\ b & d \end{pmatrix}$ be the matrix of such a transformation with respect to some orthonormal basis. By Theorem 5.3 the columns of A are orthonormal. Hence

$$ac + bd = 0$$
$$a^2 + b^2 = 1$$
$$c^2 + d^2 = 1.$$

Let us fix the first column of A and determine c and d. Since the vector $(a, b) \neq 0$, the first of the above equations implies that the vector $(c, d) = \lambda(-b, a)$ where λ is a real number. However, $c^2 + d^2 = \lambda^2(b^2 + a^2) = \lambda^2$. Since $c^2 + d^2 = 1$, $\lambda = \pm 1$. Hence there are two possibilities

$$A = \begin{pmatrix} a & -b \\ b & a \end{pmatrix} \quad \text{or} \quad A = \begin{pmatrix} a & b \\ b & -a \end{pmatrix}.$$

In the first case since $a^2 + b^2 = 1$, there is a unique angle θ, $0 \leq \theta < 2\pi$ such that $a = \cos\theta$ and $b = \sin\theta$. Hence

$$A = \begin{pmatrix} \cos\theta & -\sin\theta \\ \sin\theta & \cos\theta \end{pmatrix},$$

and A is the matrix of a rotation through an angle θ. In the second case we assert A is similar to $\begin{pmatrix} 1 & 0 \\ 0 & -1 \end{pmatrix}$. To see this compute

$$D(A - xI) = D\begin{pmatrix} a-x & b \\ b & -a-x \end{pmatrix} = x^2 - (a^2 + b^2) = x^2 - 1 = (x-1)(x+1).$$

Therefore A has two distinct eigenvalues, namely, ± 1. It now follows by the corollary to Theorem 3.22 that A is similar to the matrix $\begin{pmatrix} 1 & 0 \\ 0 & -1 \end{pmatrix}$. Let us compute the eigenvectors associated with the eigenvalues 1, and -1. First, if $(A - I)X = 0$ then

(4)
$$\begin{pmatrix} a-1 & b \\ b & -a-1 \end{pmatrix}\begin{pmatrix} x_1 \\ x_2 \end{pmatrix} = 0$$

and the vector $(x_1, x_2) = (b, 1-a)$ is easily seen to satisfy (4). Similarly if $X = (-b, a+1)$, then

$$(A + I)X = 0.$$

Hence $X_1 = (b, 1-a)$ and $X_2 = (-b, a+1)$ are eigenvectors associated with the eigenvalues 1 and -1, respectively. Notice that these eigenvectors are orthogonal. Since $AX_1 = X_1$ and $AX_2 = -X_2$, it is apparent that A is the matrix of a reflection about the line through the origin parallel to the vector X_1. It is easily seen that this line has equation $-bx_1 + (a+1)x_2 = 0$.

Thus we have shown that an orthogonal transformation T in $V_2(R)$ is either a rotation or a reflection. If A is the matrix of T with respect to an orthonormal basis, then T is a rotation if and only if $D(A) = 1$. The orthogonal transformation T is a reflection if and only if $D(A) = -1$.

EXERCISES

For the following 2×2 orthogonal matrices A determine if A defines a rotation or a reflection. If A is a rotation, determine the angle of rotation. If A is a reflection, determine a vector about which $V_2(R)$ is reflected.

1. $\dfrac{1}{\sqrt{2}}\begin{pmatrix} 1 & -1 \\ 1 & 1 \end{pmatrix}$.

2. $\dfrac{1}{\sqrt{2}}\begin{pmatrix} 1 & 1 \\ 1 & -1 \end{pmatrix}$.

3. $\dfrac{1}{5}\begin{pmatrix} 3 & 4 \\ -4 & 3 \end{pmatrix}$.

4. $\dfrac{1}{5}\begin{pmatrix} -3 & -4 \\ -4 & 3 \end{pmatrix}$.

5. $\dfrac{1}{\sqrt{5}}\begin{pmatrix} 2 & -1 \\ 1 & 2 \end{pmatrix}$.

6. $\dfrac{1}{-13}\begin{pmatrix} 5 & -12 \\ 12 & 5 \end{pmatrix}$.

7. $\dfrac{1}{\sqrt{5}}\begin{pmatrix} 1 & -2 \\ 2 & 1 \end{pmatrix}$.

8. $\dfrac{1}{\sqrt{13}}\begin{pmatrix} -2 & 3 \\ 3 & 2 \end{pmatrix}$.

Determine eigenvectors for the following reflection matrices.

9. $\dfrac{1}{\sqrt{5}}\begin{pmatrix} 2 & 1 \\ 1 & -2 \end{pmatrix}$.

10. $\dfrac{1}{13}\begin{pmatrix} -12 & 5 \\ 5 & 12 \end{pmatrix}$.

11. $\dfrac{1}{\sqrt{2}}\begin{pmatrix} -1 & 1 \\ 1 & 1 \end{pmatrix}$.

12. $\dfrac{1}{5}\begin{pmatrix} 4 & -3 \\ -3 & -4 \end{pmatrix}$.

13. Let $A = \begin{pmatrix} a & b \\ b & -a \end{pmatrix}$ where $a^2 + b^2 = 1$ be the matrix of a reflection through the line l. Determine the cosine of the angle this line makes with the positive x_1 axis.

14. Let A be a rotation in $V_2(R)$. Show that A is similar to a real diagonal matrix if and only if $A = \pm I$. (What are the eigenvalues of A?)

15. Let A be a rotation matrix in $V_2(R)$. Show by a direct calculation that $(AX, AY) = (X, Y)$ for each pair of vectors $X, Y \in V_2(R)$.

16. Let P and Q be 2×2 rotation matrices. Show that $PQ = QP$.

17. Let $\{J_1, J_2\}$ be an orthonormal basis in $V_2(R)$. Show that $\{J_1, J_2\}$ can be obtained from $\{I_1, I_2\}$ by either a rotation or a reflection of $V_2(R)$.

18. Let $\{J_1, J_2\}$ be an orthonormal basis for $V_2(R)$ obtained from the canonical basis $\{I_1, I_2\}$ by a rotation. If A is the matrix of a rotation R_θ in $V_2(R)$ with respect to the basis $\{I_1, I_2\}$, show that R_θ has the same matrix A with respect to the basis $\{J_1, J_2\}$.

19. If A is the matrix of a rotation R_θ in $V_2(R)$ with respect to an orthonormal basis $\{J_1, J_2\}$, show that A^t is the matrix of the rotation with respect to $\{J_2, J_1\}$.

20. Let G be a nonempty set which is closed under an associative operation. If there exists an identity for this operation (i.e. there is an element $e \in G$ such that $ex = xe = x$ for each $x \in G$) and if every $x \in G$ has an inverse (i.e. for each $x \in G$ there exists an element $x^{-1} \in G$ such that $xx^{-1} = x^{-1}x = e$) then G is called a *group*. Verify that the invertible $n \times n$ matrices form a group under the operation of matrix multiplication. Verify that the orthogonal matrices form a group under the same operation. The first of these groups is called the *full linear group*. The second is called the *orthogonal group*.

21. (See Exercise 20.) If G is a group, and S is a nonempty subset of G then S is called a subgroup if S is closed under the multiplication in G and if for each $x \in S$, it follows that $x^{-1} \in S$. Let G be the orthogonal group. If $S = \{A \in G : D(A) = 1\}$ show that S is a subgroup of G. Is $\{A \in G : D(A) = -1\}$ a subgroup of G?

5.2 Orthogonal Transformations in $V_3(R)$

In this section we shall determine all orthogonal transformations in $V_3(R)$. However, to analyze by a direct calculation the structure of a 3×3 matrix whose columns form an orthonormal set as we did for 2×2 matrices in the previous section would be quite complicated. We note first that the matrices

$$\begin{pmatrix} 1 & 0 & 0 \\ 0 & \cos\theta & -\sin\theta \\ 0 & \sin\theta & \cos\theta \end{pmatrix}, \quad \begin{pmatrix} -1 & 0 & 0 \\ 0 & 1 & 0 \\ 0 & 0 & 1 \end{pmatrix}$$

are certainly orthogonal. The first represents a rotation about the vector I_1. The second represents a reflection through the plane spanned by I_2 and I_3. It is also apparent that the matrix

$$\begin{pmatrix} -1 & 0 & 0 \\ 0 & \cos\theta & -\sin\theta \\ 0 & \sin\theta & \cos\theta \end{pmatrix}$$

which is the product of the first two is also orthogonal. We shall show that if T is an orthogonal transformation in $V_3(R)$, then there exists an orthonormal basis J_1, J_2, J_3 for $V_3(R)$ such that the transformation T is either a rotation about the vector J_1, a reflection through the plane spanned by J_2 and J_3, or a product of two such transformations. The matrix of T with respect to the basis $\{J_1, J_2, J_3\}$ will then have the form

$$\begin{pmatrix} \pm 1 & 0 & 0 \\ 0 & \cos\theta & -\sin\theta \\ 0 & \sin\theta & \cos\theta \end{pmatrix}.$$

This matrix is called the *canonical form* for the matrix of an orthogonal transformation in $V_3(R)$.

To establish these facts let A be the matrix of an orthogonal transformation T with respect to the canonical basis I_1, I_2, I_3. We assert that $+1$ or -1 is an eigenvalue of the orthogonal matrix A according as the determinant $D(A) = 1$ or $D(A) = -1$. We consider the case $D(A) = 1$, leaving the other as an exercise. To prove our assertion we must verify that $D(A-I) = 0$. Since A is orthogonal, $A^tA = I = AA^t$ and we know for any matrix A, $D(A) = D(A^t)$. Hence

$$D(A-I) = D((A-I)^t) = D(A^t - I) = D(A^t - A^tA)$$
$$= D(A^t(I-A)) = D(A^t)D(I-A).$$

Since $D(A^t) = D(A) = 1$ we conclude that

(1) $$D(A-I) = D(I-A).$$

Now observe that for 3×3 matrices B

$$D(-B) = (-1)^3 D(B) = -D(B).$$

Hence $D(I-A) = -D(A-I)$. Combining this with (1) we conclude that $D(A-I) = 0$. Hence $+1$ is an eigenvalue for A.

Now let J_1 be an eigenvector of unit length for the eigenvalue $\lambda = 1$. Let W be the plane perpendicular to J_1. Then $W = \{X \in V_3(R) : (J_1, X) = 0\}$. If $Y \in W$, then since A is orthogonal.

(2) $$0 = (J_1, Y) = (AJ_1, AY) = (J_1, AY).$$

($AJ_1 = J_1$ since J_1 is an eigenvector associated with the eigenvalue $+1$.) Equation (2) asserts that if $Y \in W$, then $AY \in W$. Hence if $\{J_2, J_3\}$ is an orthonormal basis for W we must have

$$AJ_1 = J_1$$
$$AJ_2 = aJ_2 + bJ_3$$
$$AJ_3 = cJ_2 + dJ_3.$$

Hence the matrix of the transformation T with respect to the basis J_1, J_2, J_3 has the form

$$\begin{pmatrix} 1 & 0 & 0 \\ 0 & a & c \\ 0 & b & d \end{pmatrix}.$$

The 2×2 submatrix $\begin{pmatrix} a & c \\ b & d \end{pmatrix}$ is, of course, orthogonal and has determinant equal to $+1$. Therefore by the discussion in Section 5.1

$$\begin{pmatrix} a & c \\ b & d \end{pmatrix} = \begin{pmatrix} \cos\theta & -\sin\theta \\ \sin\theta & \cos\theta \end{pmatrix}.$$

Hence if the matrix A of the orthogonal transformation T has determinant 1, the transformation T is a rotation about an eigenvector J_1 associated with the eigenvalue $\lambda = 1$. The matrix of this transformation with respect to the orthonormal base J_1, J_2, J_3 is then

ORTHOGONAL TRANSFORMATIONS IN $V_3(R)$

$$A = \begin{pmatrix} 1 & 0 & 0 \\ 0 & \cos\theta & -\sin\theta \\ 0 & \sin\theta & \cos\theta \end{pmatrix}.$$

If we interchange the order of J_2, and J_3 in this basis, the matrix A becomes A^t. To avoid this ambiguity we shall always assume that $\{J_1, J_2, J_3\}$ can be obtained from the canonical basis $\{I_1, I_2, I_3\}$ by a rotation. This will be the case if the orthogonal matrix with J_1, J_2, J_3 as columns has determinant equal to $+1$. This amounts to the requirement that $\{J_1, J_2, J_3\}$ forms a right-handed coordinate system. Note however that if we choose $-J_1$ as the eigenvector associated with $\lambda = 1$ and take $\{-J_1, J_3, J_2\}$, as our orthonormal basis, then the matrix for T is also A^t. Furthermore $\{-J_1, J_3, J_2\}$ is right-handed if and only if $\{J_1, J_2, J_3\}$ is. Thus to specify the canonical form for the matrix of an orthogonal transformation T we must specify the eigenvector J_1 associated with the eigenvalue $\lambda = 1$ as well as the sense of the orthonormal basis $\{J_1, J_2, J_3\}$.

We leave it as an exercise to show that if $D(A) = -1$, then a right-handed orthonormal basis J_1, J_2, J_3 can be chosen for $V_3(R)$ such that J_1 is an eigenvector associated with -1. Furthermore the matrix of the transformation with respect to this basis has the form

$$\begin{pmatrix} -1 & 0 & 0 \\ 0 & \cos\theta & -\sin\theta \\ 0 & \sin\theta & \cos\theta \end{pmatrix}.$$

If $\theta = 0$, this represents a reflection through the plane orthogonal to J_1. If $\theta \neq 0$, then this is the matrix of such a reflection followed by rotation about J_1. (see Figure 3.)

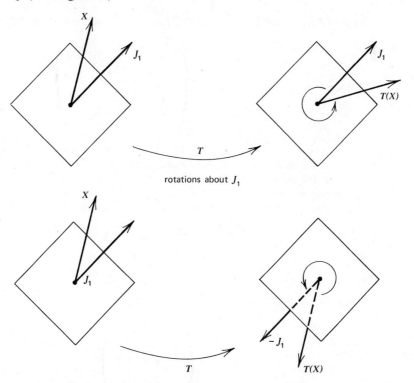

Fig. 3 A reflection through the plane perpendicular to J_1 followed by a rotation about J_1.

168 SYMMETRIC MATRICES AND QUADRATIC FORMS

To determine the canonical form of a given orthogonal transformation T, or equivalently an orthogonal matrix A, we need only to determine an eigenvector J_1 of unit length associated with the eigenvalue $+1$ if $D(A) = 1$ or associated with the eigenvalue -1 if $D(A) = -1$. Then if $\{J_2, J_3\}$ is an orthonormal basis for the plane perpendicular to J_1, we must solve the equations
$$AJ_2 = aJ_2 + bJ_3$$
and
$$AJ_3 = -bJ_2 + aJ_3.$$

Since $\{J_2, J_3\}$ is orthonormal, $a = (AJ_2, J_2)$ and $b = (AJ_2, J_3)$. Letting $a = \cos\theta$ and $b = \sin\theta$ the canonical form for A is then

$$\begin{pmatrix} 1 & 0 & 0 \\ 0 & \cos\theta & -\sin\theta \\ 0 & \sin\theta & \cos\theta \end{pmatrix} \quad \text{or} \quad \begin{pmatrix} -1 & 0 & 0 \\ 0 & \cos\theta & -\sin\theta \\ 0 & \sin\theta & \cos\theta \end{pmatrix}.$$

For definiteness we always assume that $\{J_1, J_2, J_3\}$ is right-handed. That is, $\{J_1, J_2, J_3\}$ can be obtained from $\{I_1, I_2, I_3\}$ by a rotation. This will be the case if the determinant of the matrix whose columns are J_1, J_2, J_3 has value $+1$.

Example 1. Determine the canonical form for the orthogonal matrices

$$A = \begin{pmatrix} \frac{1}{\sqrt{2}} & 0 & \frac{-1}{\sqrt{2}} \\ 0 & 1 & 0 \\ \frac{1}{\sqrt{2}} & 0 & \frac{1}{\sqrt{2}} \end{pmatrix} \quad \text{and} \quad A = \begin{pmatrix} \frac{1}{\sqrt{2}} & 0 & \frac{1}{\sqrt{2}} \\ 0 & 1 & 0 \\ \frac{1}{\sqrt{2}} & 0 & \frac{-1}{\sqrt{2}} \end{pmatrix}$$

In the first case the determinant has value $+1$. Hence A is the matrix of a rotation.

$$A - I = \begin{pmatrix} \frac{1}{\sqrt{2}} - 1 & 0 & \frac{-1}{\sqrt{2}} \\ 0 & 0 & 0 \\ \frac{1}{\sqrt{2}} & 0 & \frac{1}{\sqrt{2}} - 1 \end{pmatrix}$$

and $J_1 = I_2 = (0, 1, 0)$ is clearly an eigenvector associated with the eigenvalue $\lambda = 1$. If we let $J_2 = I_3 = (0, 0, 1)$ and $J_3 = I_1 = (1, 0, 0)$, $\{J_2, J_3\}$ is an orthonormal basis for the plane perpendicular to J_1 and $\{J_1, J_2, J_3\}$ is a right-handed system. The canonical form for the matrix A, or equivalently for the transformation defined by A, is just the matrix of the transformation with respect to the basis $\{J_1, J_2, J_3\}$. Since

$$AJ_2 = \left(\frac{-1}{\sqrt{2}}, 0, \frac{1}{\sqrt{2}}\right) = \frac{1}{\sqrt{2}}J_2 - \frac{1}{\sqrt{2}}J_3,$$

and

$$AJ_3 = \left(\frac{1}{\sqrt{2}}, 0, \frac{1}{\sqrt{2}}\right) = \frac{1}{\sqrt{2}}J_2 + \frac{1}{\sqrt{2}}J_3$$

this matrix is clearly

$$\begin{pmatrix} 1 & 0 & 0 \\ 0 & \frac{1}{\sqrt{2}} & \frac{1}{\sqrt{2}} \\ 0 & \frac{-1}{\sqrt{2}} & \frac{1}{\sqrt{2}} \end{pmatrix}.$$

Thus A defines a rotation of 135° about the vector J_1.
In the second case $D(A) = -1$ and

$$A + I = \begin{pmatrix} 1 + \frac{1}{\sqrt{2}} & 0 & \frac{+1}{\sqrt{2}} \\ 0 & 2 & 0 \\ \frac{1}{\sqrt{2}} & 0 & 1 - \frac{1}{\sqrt{2}} \end{pmatrix}.$$

An eigenvector associated with the eigenvalue $\lambda = -1$ is

$$\left(\frac{1}{\sqrt{2}}, 0, -\left(1 + \frac{1}{\sqrt{2}}\right)\right) = \frac{1}{\sqrt{2}}(1, 0, -1 - \sqrt{2}).$$

To get an eigenvector of unit length we may take

$$J_1 = \frac{1}{\sqrt{4 + 2\sqrt{2}}}(1, 0, -1 - \sqrt{2}).$$

The plane perpendicular to J_1 is spanned by $J_2 = (0, 1, 0)$ and

$$J_3 = \frac{1}{\sqrt{4 - 2\sqrt{2}}}(+1, 0, -1 + \sqrt{2}).$$

Furthermore
$$AJ_2 = J_2$$
and
$$AJ_3 = J_3.$$

Therefore with respect to the right-handed basis $\{J_1, J_2, J_3\}$, A has canonical form

$$\begin{pmatrix} -1 & 0 & 0 \\ 0 & 1 & 0 \\ 0 & 0 & 1 \end{pmatrix}.$$

Thus the transformation defined by A is a reflection through the plane spanned by J_2 and J_3.

EXERCISES

For the following 3×3 orthogonal matrices A determine if A is a rotation, or a reflection followed by a rotation. Determine the canonical form for A with respect to an appropriate right-handed coordinate system.

1. $\begin{pmatrix} 0 & 1 & 0 \\ 0 & 0 & 1 \\ 1 & 0 & 0 \end{pmatrix}.$

2. $\begin{pmatrix} 0 & 0 & 1 \\ 1 & 0 & 0 \\ 0 & 1 & 0 \end{pmatrix}.$

3. $\begin{pmatrix} 0 & 0 & 1 \\ 0 & 1 & 0 \\ 1 & 0 & 0 \end{pmatrix}.$

4. $\begin{pmatrix} 0 & 0 & 1 \\ 0 & -1 & 0 \\ 1 & 0 & 0 \end{pmatrix}.$

5. $\begin{pmatrix} 0 & 1 & 0 \\ \frac{3}{5} & 0 & -\frac{4}{5} \\ \frac{4}{5} & 0 & \frac{3}{5} \end{pmatrix}.$

6. $\begin{pmatrix} 0 & \frac{4}{5} & \frac{3}{5} \\ 0 & \frac{3}{5} & -\frac{4}{5} \\ 1 & 0 & 0 \end{pmatrix}.$

7. $\begin{pmatrix} \frac{2}{3} & \frac{1}{3} & \frac{2}{3} \\ -\frac{2}{3} & \frac{2}{3} & \frac{1}{3} \\ \frac{1}{3} & \frac{2}{3} & -\frac{2}{3} \end{pmatrix}.$

8. If A is a 3×3 orthogonal matrix, show that A is similar to a real diagonal matrix if and only if A has no complex eigenvalues (Examine the canonical form for A.)

9. Let A be an odd order orthogonal matrix. If $D(A) = +1$, show that $+1$ is an eigenvalue for A. (Examine the argument for the 3×3 case in the text.)

170   SYMMETRIC MATRICES AND QUADRATIC FORMS

10. Let A be an orthogonal matrix. If $D(A) = -1$, show that -1 is an eigenvalue for A.

11. Let $\{J_1, J_2, J_3\}$ be an orthonormal basis in $V_3(R)$. Show that $\{J_1, J_2, J_3\}$ can be obtained from $\{I_1, I_2, I_3\}$ by either a rotation or a rotation followed by a reflection. Convince yourself that in the first case the basis is right-handed, in the second, left-handed.

5.3 Symmetric Matrices

In this section we shall return to the question of determining which matrices are similar to diagonal matrices. One very important class for which this is true is the set of symmetric matrices. Recall that an $n \times n$ matrix is symmetric if $A^t = A$. The difficulty in showing that a symmetric matrix A is similar to a diagonal matrix is establishing that a symmetric matrix must have at least one real eigenvalue. Assuming this fact it is not hard to show that in fact all of the eigenvalues $\lambda_1, \ldots, \lambda_n$ of a symmetric $n \times n$ matrix A must be real, and moreover there exists an orthogonal matrix P satisfying

$$P^t A P = \begin{pmatrix} \lambda_1 & & 0 \\ & \ddots & \\ 0 & & \lambda_n \end{pmatrix}.$$

The columns $X_1, \ldots, X_n$ of P are an orthonormal set of eigenvectors associated with the eigenvalues $\lambda_1, \ldots, \lambda_n$.

The result we shall assume is the following.

THEOREM 5.4 *Let A be a symmetric $n \times n$ matrix, then A has at least one real eigenvalue.*

Surprisingly enough the proof of Theorem 5.4 involves methods from the calculus. A sketch of how the argument goes is contained in Exercises 14–16 at the end of the section. Assuming 5.4 we are in a position to prove our main result.

THEOREM 5.5 *Let A be a symmetric $n \times n$ matrix. Then there exists an orthonormal basis for $V_n(R)$ consisting of eigenvectors $\{X_1, \ldots, X_n\}$ for A. The corresponding eigenvalues $\lambda_1, \ldots, \lambda_n$ are all real numbers. Furthermore if P is the orthogonal matrix with columns $X_1, \ldots, X_n$ then*

(1) $$P^t A P = \begin{pmatrix} \lambda_1 & & 0 \\ & \ddots & \\ 0 & & \lambda_n \end{pmatrix}.$$

Proof. Observe first that if A is symmetric and P is orthogonal, then $P^t A P$ is symmetric. This is clear since

$$(P^t A P)^t = P^t A^t P^{tt} = P^t A P.$$

Furthermore $P^t A P$ has the same eigenvalues as A. To see this let $B = P^t A P$. Then

$$B - \lambda I = P^t A P - a P^t P = P^t (A - \lambda I) P.$$

Hence
$$D(B - \lambda I) = D(P^t (A - \lambda I) P)$$
$$= D(P^t) D(A - \lambda I) D(P)$$
$$= D(A - \lambda I)$$

since $D(P^t) = D(P)^{-1}$ for an orthogonal matrix P. Therefore $D(B - \lambda I) = 0$ if and only if $D(A - \lambda I) = 0$.

We prove the theorem by induction on n. For $n = 1$ there is nothing to prove. So assuming that the result holds for $n = k-1$, we must prove it for $n = k$. By Theorem 5.4 if A is a symmetric $k \times k$ matrix, then A has at least one real eigenvalue which we denote by λ_1. Let X_1 be an eigenvector of unit length associated with λ_1 and let $\{X_1, \ldots, X_k\}$ be an orthonormal basis for $V_k(R)$ containing X_1 as the initial vector. (Exercise 21 on p. 65 shows how to construct such a basis.) Let Q be the orthogonal matrix with columns $X_1, \ldots, X_k$, and let $B = Q^t A Q$. Then B is symmetric and has the same eigenvalues as A. Let us compute the first column of B. The first column of AQ is the vector AX_1. But since X_1 is the eigenvector associated with λ_1, we have $AX_1 = \lambda_1 X_1$. Now again by definition of the matrix product the first column of $Q^t A Q$ is the inner product of the rows of Q^t with the vector $\lambda_1 X_1$. However, the rows of Q^t are just the vectors $X_1, \ldots, X_k$ which are assumed to be orthonormal. Hence the first column of $Q^t A Q$ is the vector $\begin{pmatrix} \lambda_1 \\ 0 \\ \vdots \\ 0 \end{pmatrix}$. Since the matrix $Q^t A Q$ is symmetric we may infer from this that

$$Q^t A Q = \begin{pmatrix} \lambda_1 & 0 & \cdots & 0 \\ 0 & b_{22} & \cdots & b_{2k} \\ \vdots & \vdots & & \vdots \\ 0 & b_{k2} & \cdots & b_{kk} \end{pmatrix}.$$

Let B_{11} be the $k-1 \times k-1$ symmetric matrix

$$\begin{pmatrix} b_{22} & \cdots & b_{2k} \\ \vdots & & \vdots \\ b_{k2} & \cdots & b_{kk} \end{pmatrix}.$$

By our inductive assumption there is a $k-1 \times k-1$ orthogonal matrix S such that

$$S^t B_{11} S = \begin{pmatrix} \lambda_2 & & 0 \\ & \ddots & \\ 0 & & \lambda_k \end{pmatrix}$$

where $\lambda_2, \ldots, \lambda_k$ are the eigenvalues for B_{11}. Furthermore $\lambda_2, \ldots, \lambda_k$ are real numbers, although they may not all be distinct.

Now define the $k \times k$ matrix R by setting

$$R = \begin{pmatrix} 1 & 0 & \cdots & 0 \\ 0 & & & \\ \vdots & & S & \\ 0 & & & \end{pmatrix}.$$

Then R is orthogonal and setting $P = QR$ we observe that P is orthogonal.

Moreover

(1) $$P^t A P = \begin{pmatrix} \lambda_1 & & 0 \\ & \ddots & \\ 0 & & \lambda_k \end{pmatrix}$$

since

$$P^t A P = (QR)^t A Q R = R^t Q^t A Q R = R^t \begin{pmatrix} \lambda_1 & 0 \cdots 0 \\ 0 & \\ \vdots & B_{11} \\ 0 & \end{pmatrix} R = \begin{pmatrix} \lambda_1 & & 0 \\ & \ddots & \\ 0 & & \lambda_k \end{pmatrix}.$$

If $X_1, \ldots, X_k$ are the columns of P, then $\{X_1, \ldots, X_k\}$ is an orthonormal set and $AX_j = \lambda_j X_j$ for $j = 1, \ldots, k$. Hence $\{\lambda_1, \ldots, \lambda_k\}$ are the real eigenvalues for A and the proof is complete.

We remark also that the matrix P may always be taken to have determinant $+1$. For if $D(P) = -1$, then we may multiply the first column of P by -1, obtaining an orthogonal matrix $\tilde{P}$ satisfying $D(\tilde{P}) = 1$. Clearly

$$\tilde{P}^t A \tilde{P} = \begin{pmatrix} \lambda_1 & & 0 \\ & \ddots & \\ 0 & & \lambda_n \end{pmatrix}.$$

Example 1. Let $A = \begin{pmatrix} 1 & 1 \\ 1 & 1 \end{pmatrix}$. Determine an orthogonal matrix P such that $P^t A P = \begin{pmatrix} \lambda_1 & 0 \\ 0 & \lambda_2 \end{pmatrix}$ where λ_1, λ_2 are the eigenvalues of A. To determine the eigenvalues, compute

$$D(A - xI) = D\begin{pmatrix} 1 - x & 1 \\ 1 & 1 - x \end{pmatrix} = (1 - x)^2 - 1 = x(x - 2).$$

Hence the eigenvalues are 0 and 2. Associated eigenvectors are $(1, -1)$ and $(1, 1)$. These vectors are orthogonal and if we take $X_1 = 1/\sqrt{2}(1, -1)$ and $X_2 = 1/\sqrt{2}(1, 1)$ we have an orthonormal basis for $V_2(R)$ consisting of eigenvectors for A. The matrix

$$P = \frac{1}{\sqrt{2}} \begin{pmatrix} 1 & 1 \\ -1 & 1 \end{pmatrix}.$$

Example 2. Let $A = \begin{pmatrix} 1 & 0 & -3 \\ 0 & -2 & 0 \\ -3 & 0 & 0 \end{pmatrix}$. Determine the eigenvalues for A and an orthogonal matrix P such that $P^t A P$ is diagonal.

$$D(A - xI) = D\begin{pmatrix} 1 - x & 0 & -3 \\ 0 & -2 - x & 0 \\ -3 & 0 & 1 - x \end{pmatrix}$$
$$= -(1 - x)^2(2 + x) + 9(2 + x)$$
$$= (2 + x)[8 + 2x - x^2] = (2 + x)^2(4 - x).$$

Therefore the eigenvalues are 4 and -2. Clearly $(1, 0, -1)$ is an eigenvector associated with the eigenvalue 4. For the eigenvalue -2

$$A + 2I = \begin{pmatrix} 3 & 0 & -3 \\ 0 & 0 & 0 \\ -3 & 0 & 3 \end{pmatrix}.$$

The rank of this matrix is clearly one. Hence the dimension of the space of vectors X satisfying $(A + 2I)X = 0$ must equal two. An orthogonal basis for this space may be taken to be $(0, 1, 0)$ and $(1, 0, 1)$. The desired orthogonal basis for $V_3(R)$ consisting of eigenvectors for A is $1/\sqrt{2}(1, 0, -1)$, $(0, 1, 0)$, and $1/\sqrt{2}(1, 0, 1)$. Hence if

$$P = \begin{pmatrix} \frac{1}{\sqrt{2}} & 0 & \frac{1}{\sqrt{2}} \\ 0 & 1 & 0 \\ \frac{-1}{\sqrt{2}} & 0 & \frac{1}{\sqrt{2}} \end{pmatrix},$$

then

$$P^t A P = \begin{pmatrix} 4 & 0 & 0 \\ 0 & -2 & 0 \\ 0 & 0 & -2 \end{pmatrix}.$$

EXERCISES

Determine the eigenvalues for the following symmetric matrices. For each matrix A determine an orthogonal matrix P such that $P^t AP$ is diagonal.

1. $\begin{pmatrix} 1 & -1 \\ -1 & 1 \end{pmatrix}$.

2. $\begin{pmatrix} 0 & -1 \\ -1 & 0 \end{pmatrix}$.

3. $\begin{pmatrix} 4 & 3 \\ 3 & -4 \end{pmatrix}$.

4. $\begin{pmatrix} 6 & -2 \\ -2 & 9 \end{pmatrix}$.

5. $\begin{pmatrix} \frac{1}{2} & 0 & 0 \\ 0 & -1 & 1 \\ 0 & 1 & -1 \end{pmatrix}$.

6. $\begin{pmatrix} 5 & 4 & 2 \\ 4 & 5 & 2 \\ 2 & 2 & 2 \end{pmatrix}$.

7. $\begin{pmatrix} 1 & 2 & 0 \\ 2 & 2 & 2 \\ 0 & 2 & 3 \end{pmatrix}$.

8. $\begin{pmatrix} 16 & 0 & 12 \\ 0 & 25 & 0 \\ 12 & 0 & 9 \end{pmatrix}$.

9. $\begin{pmatrix} 0 & -1 & 0 & 0 \\ -1 & 0 & 0 & 0 \\ 0 & 0 & 0 & -1 \\ 0 & 0 & -1 & 0 \end{pmatrix}$.

10. If A is an $n \times n$ matrix and $X, Y \in V_n(R)$ show that

$$(AX, Y) = (X, A^t Y).$$

In particular if A is symmetric, then

$$(AX, Y) = (X, AY).$$

[If $A = (a_{ij})$, $B = (b_{ij})$ are $n \times n$ matrices and $X = (x_1, \ldots, x_n)$ $Y = (y_1, \ldots, y_n)$ are vectors from $V_n(R)$, then

$$(AX, Y) = \sum_i \left(\sum_j a_{ij} x_j \right) y_i = \sum_{i,j} a_{ij} x_j y_i$$

and

$$(X, BY) = \sum_i x_i \left(\sum_j b_{ij} y_j \right) = \sum_{i,j} b_{ij} x_i y_j.$$

Show that these two sums are equal if $B = A^t$.]

*11. An $n \times n$ matrix is called *positive definite* if A is symmetric and all the eigenvalues of A are positive. Show that if $A = B^t B$ where B is invertible, then A is positive definite. [If $A = B^t B$ then A is symmetric, and by Theorem 5.5 there is an orthogonal matrix P satisfying

$$P^t A P = P^t B^t B P = (BP)^t (BP) = \begin{pmatrix} \lambda_1 & & 0 \\ & \ddots & \\ 0 & & \lambda_n \end{pmatrix}$$

where $\lambda_1, \ldots, \lambda_n$ are the eigenvalues of A_0. Now, however if I_k is the kth coordinate vector in $V_n(R)$ it follows from Exercise 10 that

$$|BPI_k|^2 = (BPI_k, BPI_k) = (I_k, (BP)^t(BP)I_k) = (I_k, \lambda_k I_k) = \lambda_k.$$

From this conclude that $\lambda_k > 0$.]

*12. Show that if A is positive definite, then there exists an invertible matrix B such that $A = B^t B$. [If $P^t A P = \begin{pmatrix} \lambda_1 & & 0 \\ & \ddots & \\ 0 & & \lambda_n \end{pmatrix} = D$, consider $\sqrt{D} = \begin{pmatrix} \sqrt{\lambda_1} & & 0 \\ & \ddots & \\ 0 & & \sqrt{\lambda_n} \end{pmatrix}$.

How should B be defined?]

13. If $A = (a_{ij})$ is a positive definite $n \times n$ matrix, show that for all real numbers $x_1, \ldots, x_n$, $\sum_{i,j=1}^n a_{ij} x_i x_j \geq 0$. Furthermore $\sum_{i,j=1}^n a_{ij} x_i x_j = 0$ if and only if $x_1 = x_2 = \cdots = x_n = 0$. [Note that if $X = (x_1, \ldots, x_n)$, then

$$\sum_{i,j=1}^n a_{ij} x_i x_j = (X, AX).$$

Now use Theorem 5.5.]

The following exercises sketch a proof of the fact that a symmetric matrix must have at least one real eigenvalue.

*14. Show that the quadratic polynomial $p(x) = ax^2 + bx$ is nonnegative for all values of x if and only if $a \geq 0$ and $b = 0$. (Investigate the graph.)

*15. Let A be a symmetric $n \times n$ matrix and assume that for each $X \in V_n(R)$, $(AX, X) \geq 0$. Show that if $(AX_0, X_0) = 0$, then $AX_0 = 0$. [Let $Y_0 = AX_0$ and let x be a real number. Verify that

$$0 \leq (A(X_0 + xY_0), X_0 + xY_0) = 2x(AX_0, Y_0) + x^2(AY_0, Y_0).$$

If for all values of x this quadratic polynomial is nonnegative, conclude from Exercise 14 that $(AX_0, Y_0) = 0$, and hence $AX_0 = 0$.]

*16. To produce a real eigenvalue λ for the symmetric matrix A we assume that there exists a solution to the following extremal problem. Let

$$S = \{X \in V_n(R) : |X| = 1\}.$$

Consider the real-valued function $F(X) = (AX, X)$ defined on S. Now it can be shown that this function attains its maximum value on the set S. That is, there is a vector $X_0 \in S$ such that $(AX_0, X_0) \geq (AX, X)$ for each $X \in S$. In fact this theorem has the same structure as the theorem that asserts that a continuous function defined on a closed interval attains its maximum. Problems of this nature will be dealt with in Chapter 7. Show that if $\lambda = (AX_0, X_0) \geq (AX, X)$ for each $X \in S$, then λ is an eigenvalue for A and X_0 is an associated eigenvector. (Apply the result of Exercise 15 to the matrix $\lambda I - A$.)

5.4 Quadratic Forms

Let X and Y be vectors in $V_n(R)$, and consider the function

(1) $$F(X, Y) = \sum_{\substack{i=1 \\ j=1}}^n a_{ij} x_i y_j$$

where a_{ij} are real numbers and x_i, y_j are the components of the vectors X, and Y, respectively. Such a function is called a *real bilinear form* defined on $V_n(R)$. The bilinearity refers to the fact that F is linear in each of its variables X and Y.

If $A = (a_{ij})$ is the $n \times n$ matrix with coefficients a_{ij} then an easy calculation verifies that

(2) $$F(X,Y) = (X,AY) = (A^tX,Y).$$

See Exercise 10, p. 173. The function F is *symmetric*, that is, $F(X,Y) = F(Y,X)$ if and only if the matrix A is symmetric.

If in (1) we set $X = Y$, this defines a function

(3) $$Q(X) = \sum_{i,j=1}^{n} a_{ij}x_ix_j.$$

This function Q is called a *real quadratic form*, and if $A = (a_{ij})$ is the matrix of coefficients in (3), then it follows from (2) that

(4) $$Q(X) = (X, AX).$$

Now, however, without loss of generality we may assume that the matrix A is symmetric. To see this observe that

$$a_{ij}x_ix_j + a_{ji}x_jx_i = (a_{ij} + a_{ji})x_ix_j.$$

Hence if we replace a_{ij} and a_{ji} by $(a_{ij}+a_{ji})/2$, the value of Q is unchanged and the matrix A is now symmetric. This symmetric matrix A is called the *matrix of the quadratic form Q*.

Example 1. If $Q(X) = 2x_1^2 - x_1x_2 - 3x_2^2$, then the matrix of Q is $\begin{pmatrix} 2 & -\frac{1}{2} \\ -\frac{1}{2} & -3 \end{pmatrix}$. If $Q(X) = 3x_1^2 - x_2^2 + x_3^2 - 2x_1x_2 + x_2x_3$, then Q has matrix

$$\begin{pmatrix} 3 & -1 & 0 \\ -1 & -1 & \frac{1}{2} \\ 0 & \frac{1}{2} & 1 \end{pmatrix}.$$

When Q is written explicitly as in the above example, customarily only the one term $a_{ij}x_ix_j$ appears. Therefore if $i \neq j$, then the ijth entry of the matrix for Q is $a_{ij}/2$.

Let us observe next what happens to a quadratic form if we change to another orthonormal basis $J_1, \ldots, J_n$ in $V_n(R)$. Let $P = (p_{ij})$ be the orthogonal change of basis matrix from the canonical basis $I_1, \ldots, I_n$ to the basis $J_1, \ldots, J_n$. Then writing $X = \sum_k y_k J_k$ we have by the definition of P,

$$X = \sum_k y_k J_k = \sum_k y_k \sum_j p_{jk} I_j = \sum_j \sum_k p_{jk} y_k I_j = \sum_j x_j I_j.$$

Hence for each j, $x_j = \sum_k p_{jk} y_k$, or

(5) $$\begin{pmatrix} x_1 \\ \vdots \\ x_n \end{pmatrix} = \begin{pmatrix} p_{11} & \cdots & p_{1n} \\ \vdots & & \vdots \\ p_{n1} & \cdots & p_{nn} \end{pmatrix} \begin{pmatrix} y_1 \\ \vdots \\ y_n \end{pmatrix}.$$

Therefore identifying X with $\begin{pmatrix} x_1 \\ \vdots \\ x_n \end{pmatrix}$ and Y with $\begin{pmatrix} y_1 \\ \vdots \\ y_n \end{pmatrix}$ we have $X = PY$.

Substituting in (4) we have

(6) $$Q(X) = (X, AX) = (PY, APY) = (Y, P^tAPY).$$

Now we know by Theorem 5.5 that there is an orthogonal matrix P such that

$$P^tAP = \Lambda$$

176 SYMMETRIC MATRICES AND QUADRATIC FORMS

where $\Lambda = \begin{pmatrix} \lambda_1 & & 0 \\ & \ddots & \\ 0 & & \lambda_n \end{pmatrix}$ is the diagonal matrix of eigenvalues of A. Substituting this in (6) we obtain

(7) $$Q(X) = (X, AX) = (Y, \Lambda Y) = \sum_{i=1}^{n} \lambda_i y_i^2.$$

When $Q(X)$ is written as a linear combination of squares as in (7), we say that the *quadratic form Q has been reduced to diagonal form*. To perform this reduction, we need only to compute the eigenvalues of the matrix A. If the quadratic form Q is in diagonal form, that is $Q(X) = \sum_{i=1}^{n} \lambda_i y_i^2$, then to compute $Q(X)$ for a given value of $X = \sum x_k I_k$ we must determine the scalars $y_1, \ldots, y_n$ so that $X = \sum_{k=1}^{n} y_k J_k$. If P is the orthogonal matrix which diagonalizes A, then by (5)

$$\begin{pmatrix} y_1 \\ \vdots \\ x_n \end{pmatrix} = P \begin{pmatrix} y_1 \\ \vdots \\ y_n \end{pmatrix} \quad \text{or} \quad \begin{pmatrix} y_1 \\ \vdots \\ y_n \end{pmatrix} = P^t \begin{pmatrix} x_1 \\ \vdots \\ x_n \end{pmatrix}$$

since $P^t P = I$.

Let us use these techniques to investigate the geometric nature of the set S of vectors X satisfying

(8) $$Q(X) = k, \quad k \text{ a constant.}$$

This set S is called the *graph* of the quadratic equation (8).

We examine first the situation for $n = 2$. If $k \neq 0$, we may divide by it and consider the set S of vectors X satisfying

$$Q(X) = 1.$$

Let us assume first that Q is in diagonal form. Then $Q(X) = ax_1^2 + bx_1^2$, and the character of the graph of the equation $ax_1^2 + bx_2^2 = 1$ depends on the algebraic sign of a and b. We know from elementary analytic geometry that the following are the only possibilities.

(i) If $a > 0$ and $b > 0$, then S is an ellipse. (Figure 4.)

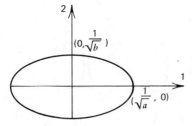

Fig. 4 Graph of $ax_1^2 + bx_2^2 = 1$; $a, b > 0$.

(ii) If $a > 0$ and $b < 0$, or if $a < 0$ and $b > 0$, then S is a hyperbola. (Figure 5.)
(iii) If $a > 0$ and $b = 0$ (or $a = 0$ and $b > 0$), then S is a pair of parallel lines. In the first instance we have the lines $x_1 = \pm (1/\sqrt{a})$. In the second we have $x_2 = \pm (1/\sqrt{b})$.
(iv) If $a \leq 0$ and $b \leq 0$, then S is void.

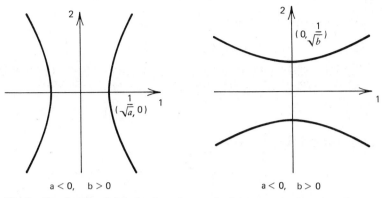

Fig. 5 Graphs of $ax_1^2 + bx_2^2 = 1$: a, b opposite in sign.

When $c = 0$ we have the following possibilities.

(v) If a and b are nonzero and of opposite sign, then S is a pair of lines intersecting at the origin. These lines have equation

$$x_2 = \pm \sqrt{\frac{|a|}{|b|}} x_1.$$

(vi) If a and b are nonzero and have the same sign, then S consists only of the origin.

(vii) If a or b is zero, then S is one of the coordinate axes.

Thus if S is not empty, S is a plane section of a cone or a pair of parallel lines. When the quadratic form Q is diagonal we say that the graph S is in *standard position*. For the general case let $Q(X) = ax_1^2 + bx_1x_2 + cx_2^2$. Then there exists a 2×2 orthogonal matrix P, indeed a rotation, satisfying

$$P^t \begin{pmatrix} a & \frac{b}{2} \\ \frac{b}{2} & c \end{pmatrix} P = \begin{pmatrix} \lambda_1 & 0 \\ 0 & \lambda_2 \end{pmatrix}.$$

where λ_1, λ_2 are the eigenvalues for the matrix of Q. Letting $X = P^t Y$ we have

$$Q(X) = ax_1^2 + bx_1x_2 + cx_2^2 = \lambda_1 y_1^2 + \lambda_2 y_2^2.$$

Thus the graph of the equation $Q(X) = k$ is a conic section or a pair of parallel lines. This graph may be put into standard position by a rotation of the coordinate axis. To determine the matrix P of the rotation we need only compute orthonormal eigenvectors associated with the eigenvalues of the matrix

$$\begin{pmatrix} a & \frac{b}{2} \\ \frac{b}{2} & c \end{pmatrix}.$$

By interchanging the order of these eigenvectors if necessary we assume $D(P) = 1$.

Example 2. Identify and sketch the graph of the quadratic equation

$$3x_1^2 - 2x_1x_2 + 3x_2^2 = 1.$$

The matrix A of the quadratic form is $\begin{pmatrix} 3 & -1 \\ -1 & 3 \end{pmatrix}$. Furthermore

$$D(A - xI) = (3 - x)^2 - 1 = (x - 2)(x - 4),$$

hence the eigenvalues of A are 2, 4. Eigenvectors associated with these eigenvalues are $(1, 1)$ and $(-1, 1)$, respectively. Therefore the rotation matrix

$$P = \frac{1}{\sqrt{2}} \begin{pmatrix} 1 & -1 \\ 1 & 1 \end{pmatrix}.$$

will diagonalize the quadratic form.

If $J_1 = 1/\sqrt{2}\,(1, 1)$ and $J_2 = 1/\sqrt{2}\,(-1, 1)$, then

$$Q(X) = 3x_1^2 - 2x_1 x_2 + 3x_2^2 = 2y_1^2 + 4y_2^2$$

where $X = x_1 I_1 + x_2 I_2 = y_1 J_1 + y_2 J_2$. The graph is an ellipse and its sketch is Figure 6.

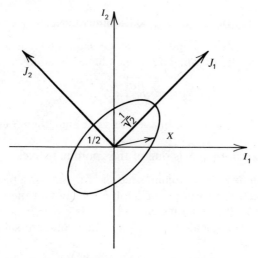

Fig. 6 Graph of $Q(X) = 3x_1^2 - 2x_1 x_2 + 3x_2^2$.

To determine the graph of the general quadratic equation

$$F(x_1, x_2) = ax_1^2 + bx_1 x_2 + cx_2^2 + dx_1 + ex_2 + f = 0$$

we first perform a rotation which diagonalizes the quadratic form $ax_1^2 + bx_1 x_2 + cx_2^2$. If P is the matrix of that rotation, and

$$\begin{pmatrix} y_1 \\ y_2 \end{pmatrix} = P^t \begin{pmatrix} x_1 \\ x_2 \end{pmatrix},$$

then

(9) $\qquad F(x_1, x_2) = a' y_1^2 + b' y_2^2 + c' y_1 + d' y_2 + e'.$

Next assuming $a' \neq 0$ and $b' \neq 0$ we complete the square in y_1 and y_2 in the above expression obtaining

$$F(x_1, x_2) = a'(y_1 - h)^2 + b'(y_2 - k)^2 + c''.$$

Therefore if we let $z_1 = y_1 - h$ and $z_2 = y_2 - k$, we have

$$F(x_1, x_2) = a' z_1^2 + b' z_2^2 + c''$$

and the graph of $F(x_1, x_2) = 0$ may be determined by examining the coefficients a', b', c'' as before. If in (9) $a' = 0$, then we complete the square in

y_2 and write
$$F(x_1, x_2) = c'(y_1 - h) + b'(y_2 - k)^2 = c'z_1 + b'z_2^2.$$

In this case the graph of $F(x_1, x_2) = 0$ is a parabola. The case $b' = 0$ is treated similarly.

Example 3. Identify and sketch the graph of the quadratic equation

(10) $$F(x_1, x_2) = x_1^2 + 4x_1x_2 + 4x_2^2 - \sqrt{5}x_1 + 8\sqrt{5}x_2 + \frac{5}{4} = 0.$$

The matrix of the quadratic form $x_1^2 + 4x_1x_2 + 4x_2^2$ is $\begin{pmatrix} 1 & 2 \\ 2 & 4 \end{pmatrix}$, and the eigenvalues of the matrix $\begin{pmatrix} 1 & 2 \\ 2 & 4 \end{pmatrix}$ are easily seen to be 0 and 5. Associated orthonormal eigenvectors are $J_1 = 1/\sqrt{5}(2, -1)$ and $J_2 = 1/\sqrt{5}(1, 2)$. Therefore

$$P = \frac{1}{\sqrt{5}} \begin{pmatrix} 2 & 1 \\ -1 & 2 \end{pmatrix}, \quad \text{and} \quad \begin{pmatrix} x_1 \\ x_2 \end{pmatrix} = P \begin{pmatrix} y_1 \\ y_2 \end{pmatrix} = \frac{1}{\sqrt{5}} \begin{pmatrix} 2y_1 + y_2 \\ -y_1 + 2y_2 \end{pmatrix}.$$

Therefore substituting $x_1 = 1/\sqrt{5}(2y_1 + y_2)$ and $x_2 = 1/\sqrt{5}(-y_1 + 2y_2)$ in (10) we obtain

$$F(x_1, x_2) = 5y_2^2 - (2y_1 + y_2) + 8(-y_1 + 2y_2) + \frac{5}{4} = 5y_2^2 - 10y_1 + 15y_2 + \frac{5}{4}.$$

Completing the square in y_2 we have

$$F(x_1, x_2) = 5\left(y_2^2 + 3y_2 + \frac{9}{4}\right) - \frac{45}{4} - 10y_1 + \frac{5}{4} = 5\left(y_2 + \frac{3}{2}\right)^2 - 10(y_1 + 1).$$

Therefore if $z_1 = y_1 + 1$ and $z_2 = y_2 + \frac{3}{2}$. $F(x_1, x_2) = 5z_2^2 - 10z_1$. Setting this equal to zero we have $z_1 = +\frac{1}{2}z_2^2$. The graph is a parabola, and the sketch is Figure 7.

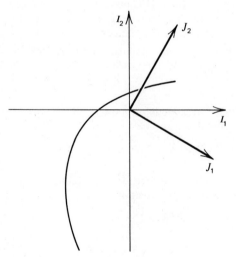

Fig. 7 Graph of equation (10).

EXERCISES

Determine the matrices of the following quadratic forms

1. $2x_1^2 - 3x_1x_2 + x_2^2$.
2. $x_1x_2 - 4x_2^2$.
3. $x_1^2 - x_1x_3 + x_2^2 - x_3^2$.
4. $3x_1^2 + x_1x_2 - x_2x_3 + x_2^2$.
5. $\sum_{i=1}^{3} \sum_{j=1}^{3} 2^{i+j} x_i x_j$.
6. $\sum_{i=1}^{4} \sum_{j=1}^{4} i \cdot j x_i x_j$.

Reduce each of the following quadratic forms $Q(X)$ to its diagonal form $\Sigma\, a_i y_i^2$. Determine the rotation matrix P which performs the diagonalization.

7. $4x_1 x_2$.
8. $x_1^2 + 4x_1 x_2 + 4x_2^2$.
9. $x_1^2 - 6x_1 x_2 + x_2^2$.
10. $-5x_1^2 - 20x_1 x_2 + 10x_2^2$.
11. $25x_1^2 + 48x_1 x_2 + 25x_2^2$.
12. $x_1^2 + 2x_1 x_2 - 2x_1 x_3 - 4x_2 x_3$.
13. $2x_1^2 + x_2^2 - 4x_1 x_2 - 4x_2 x_3$.
14. $3x_1^2 + x_2^2 + x_3^2 - 2x_1 x_2 + 2x_1 x_3 - 2x_2 x_3$.

Identify and sketch the graphs of the following quadratic equations.

15. $5x_1^2 + 6x_1 x_2 - 3x_2^2 = 3$.
16. $25x_1^2 - 30x_1 x_2 + 9x_2^2 = 5$.
17. $x_1 x_2 = 4$.
18. $x_1^2 + 4x_1 x_2 + 4x_2^2 = 4$.
19. $x_1^2 - 6x_1 x_2 + x_2^2 = 9$.
20. $5x_1^2 + 8x_1 x_2 + 5x_2^2 = 25$.
21. $16x_1^2 - 24x_1 x_2 + 9x_2^2 = 64$.
22. $3x_1^2 - 6x_1 x_2 - 5x_2^2 - 6x_1 + 22x_2 + 24 = 0$.
23. $9x_1^2 - 24x_1 x_2 + 16x_2^2 - 2x_1 - 14x_2 + 15 = 0$.
24. Let $Q(X) = ax_1^2 + bx_1 x_2 + cx_2^2$. Assuming that the graph of $Q(X) = 1$ is not empty, show that this graph is an ellipse if and only if $D(A) > 0$ where A is the matrix of Q. Show that the graph is a hyperbola if and only if $D(A) < 0$. What is the graph if $D(A) = 0$?
25. Let X, Y be vectors from $V_n(R)$ and let $F(X, Y)$ be a bilinear function defined on pairs of vectors from $V_n(R)$. Show that

$$F(X, Y) = \sum_{\substack{i=1 \\ j=1}}^{n} a_{ij} x_i y_j$$

for suitably defined scalars a_{ij}. [Examine $F(X, Y) = F(x_1 I_1 + \cdots + x_n I_n, y_1 I_1 + \cdots + y_n I_n)$]. Conclude that there exists an $n \times n$ matrix A such that $F(X, Y) = (X, AY)$ where (X, Y) is the inner product in $V_n(R)$.]
26. If in Exercise 25 the function F is symmetric, $(F(X, Y) = F(Y, X))$ show that A is symmetric. If in addition F is positive definite $(F(X, X) \geq 0$ and $F(X, X) = 0$ if and only if $X = 0)$, show that the matrix A is positive definite. (See Exercise 11, p. 173.)
27. Let Q be a real-valued function defined on $V_n(R)$. Show that Q defines a quadratic form if and only if

$$F(X, Y) = Q(X + Y) - Q(X) - Q(Y)$$

is bilinear.
28. Let $F(X, Y)$ be a symmetric, positive definite, bilinear function defined on $V_2(R)$. Show that the graph of the equation

$$F(X, X) = 1$$

is an ellipse.

5.5 Quadric Surfaces

Let $Q(x_1, x_2, x_3)$ be a quadratic polynomial in the three variables x_1, x_2, x_3. That is,

(1) $$Q(x_1, x_2, x_3) = ax_1^2 + bx_2^2 + cx_3^2 + dx_1 x_2 + ex_1 x_3 \\ + fx_2 x_3 + gx_1 + hx_2 + lx_3 + m$$

and we assume that not all of the coefficients a through f are zero. If the graph of the equation $Q(x_1, x_2, x_3) = 0$ is nonempty, it is called a *quadric surface*. If A is the matrix of the quadratic form $ax_1^2 + \cdots + fx_2 x_3$, then A is symmetric, and hence there exists an orthogonal matrix P, indeed a rotation, which reduces the quadratic form to a linear combination of squares $a'y_1^2 + b'y_2^2 + c'y_3^2$. The coefficients a', b', c' are, of course, just the eigenvalues of the matrix A. Letting

$$\begin{pmatrix} y_1 \\ y_2 \\ y_3 \end{pmatrix} = P^t \begin{pmatrix} x_1 \\ x_2 \\ x_3 \end{pmatrix}$$

we have

(2) $\quad Q(x_1, x_2, x_3) = a'y_1^2 + b'y_2^2 + c'y_3^2 + d'y_1 + e'y_2 + f'y_3 + g'.$

We may further simplify this polynomial by completing the square in the variables y_1, y_2, and y_3. If both y_1^2 and y_1 are present in (2), then the first-order term may be eliminated by completing the square in y_1 and then making substitution of the form $z_1 = y_1 - h$. Proceeding in a similar fashion for the variables y_2 and y_3 we reduce $Q(x_1, x_2, x_3)$ to a quadratic polynomial in z_1, z_2, z_3 where $z_1 = y_1 - k_1$, $z_2 = y_2 - k_2$, $z_3 = y_3 - k_3$. This polynomial in z_1, z_2, z_3 does not have any cross product terms and does not have both quadratic and first-order terms in any of the variables. The resulting quadric surface $Q(x_1, x_2, x_3) = 0$ is said to be in *standard position*. This simplification of the polynomial $Q(x_1, x_2, x_3)$ has been accomplished by first performing a rotation of the coordinate axes. Then the origin has been translated to the point (k_1, k_2, k_3). To describe the graph of $Q(x_1, x_2, x_3) = 0$ we may, therefore, restrict ourselves to those surfaces in standard position. As in the two-dimensional case the sign of the coefficients determines the geometric character of the surface. To sketch the surfaces we examine various plane cross sections. We shall distinguish three cases depending on the number of first-order terms present after the appropriate rotation and translation of the coordinate axes.

I. There are no first-order terms present. Then the quadratic equation takes the form

(3) $\quad\quad\quad\quad\quad\quad ax_1^2 + bx_2^2 + cx_3^2 = 1$

or

(4) $\quad\quad\quad\quad\quad\quad ax_1^2 + bx_2^2 + cx_3^2 = 0.$

We treat first the case when the right-hand side is not zero.

(i) If all the coefficients in (3) are positive, the graph is an *ellipsoid*. The intersection of the graph with each coordinate plane is an ellipse. The semimajor axes of these ellipses have lengths $1/\sqrt{a}$, $1/\sqrt{b}$, and $1/\sqrt{c}$, respectively. (Figure 8.)

If one or two of the coefficients a, b, c in (3) is negative, the graph is a hyperboloid. There are two possibilities. If one coefficient is negative we have for instance

$$ax_1^2 + bx_2^2 - cx_3^2 = 1 \quad\quad a, b, c > 0$$

and the graph is a *hyperboloid* of *one sheet*. (The number of sheets refers to the number of distinct surfaces present in the graph.) The intersection of the graph with the 1, 3 plane and the 2, 3 plane is a hyperbola, whereas the intersection with the 1, 2 plane is an ellipse. (Figure 9.)

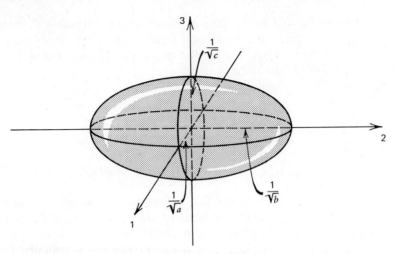

Fig. 8 Ellipsoid $ax_1^2 + bx_2^2 + cd_3^2 = 1$; $a, b, c > 0$.

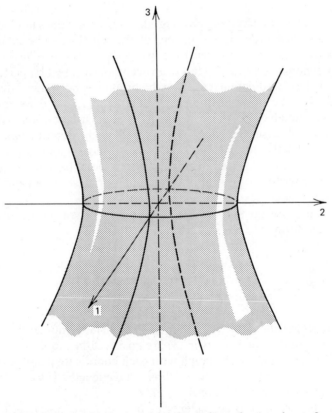

Fig. 9 Hyperboloid of one sheet $ax_1^2 + bx_2^2 - cx_3^2 = 1$; $a, b, c > 0$.

If two coefficients are negative, for instance,

$$-ax_1^2 - bx_2^2 + cx_3^2 = 1 \quad a, b, c > 0$$

then the graph is a *hyperboloid* of *two sheets*. The intersections with the 1, 3, and 2, 3 planes are hyperbolas. The intersection with the 1, 2 plane is void. However, if $1 - cx_3^2 < 0$ or, equivalently, if $|x_3| > 1/\sqrt{c}$, the intersection of the graph with the plane $x_3 =$ constant is an ellipse. (Figure 10.)

QUADRIC SURFACES 183

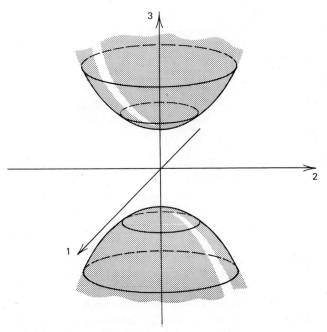

Fig. 10 Hyperboloid of two sheets $-ax_1^2 - bx_2^2 + cx_3^2 = 1; a, b, c > 0$.

If all coefficients a, b, c are negative then the graph is empty.

(iii) If one coefficient in (3) is zero then the graph is a cylinder. If we have, for example, $ax_1^2 + bx_2^2 = 1$ where $a, b > 0$, then the graph is an *elliptic cylinder*. If we have $ax_1^2 - bx_2^2 = 1$ then the graph is a *hyperbolic cylinder*. (Figure 11.)

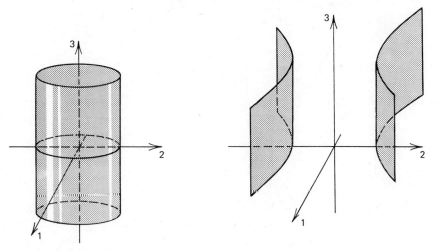

Fig. 11 Elliptic cylinder $\quad ax_1^2 + bx_2^2 = 1, a, b > 0$.
Hyperbolic cylinder $\quad ax_1^2 - bx_2^2 = 1, a, b > 0$.

(iv) If two coefficients in (3) are zero, then the graph is a *pair of parallel planes*. For example, see Figure 12.

Next we consider the case

(4) $$ax_1^2 + bx_2^2 + cx_3^2 = 0.$$

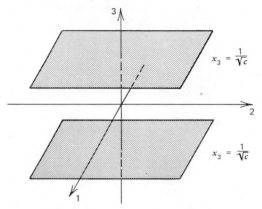

Fig. 12 Two parallel planes $\quad cx_3^2 = 1 \quad c > 0$.

If all three coefficients are positive (or negative), the graph is the single point at the origin. If $a, b \geq 0$ and $c < 0$, then we may divide by c and rewrite the equation in the form $ax_1^2 + bx_2^2 = x_3^2$. If both $a, b > 0$ then the graph is an *elliptic cone*. The intersection of the graph with planes parallel to the 1, 2 plane is always an ellipse, whereas the intersection of the graph with either the 1, 3 plane or the 2, 3 plane is a pair of lines intersecting at the origin. (Figure 13.) If a or b vanish, for instance, $a = 0$, then the graph is the pair of planes $x_3 = \pm \sqrt{b}x_2$.

II. The quadratic equation has one first-order term present. In that case

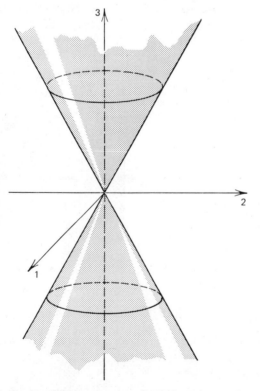

Fig. 13 Elliptic cone $\quad ax_1^2 + bx_2^2 = x_3^2 \quad a, b > 0$.

there is no constant term since we assume it has been eliminated by an appropriate translation of the origin. We consider

(5) $$ax_1^2 + bx_2^2 + cx_3 = 0.$$

Assuming $c \neq 0$ we write this in the form

$$ax_1^2 + bx_2^2 = x_3.$$

If neither a nor b vanishes, the intersection of the graph of this equation with the 1, 3 and 2, 3 coordinate planes is a parabola. If both a and b are positive or both are negative, then the intersection of the graph with the plane $x_3 =$ constant is either an ellipse or is empty. If a and b have opposite signs, this intersection is a hyperbola. In the former case the surface is called an *elliptic paraboloid*. In the second the surface is a *hyperbolic paraboloid*. Sketches of these surfaces are Figures 14 and 15. Note that the hyperbolic paraboloid is a saddle-shaped surface.

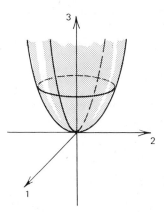

Fig. 14 Elliptic paraboloid $ax_1^2 + bx_2^2 = x_3$, $a, b > 0$.

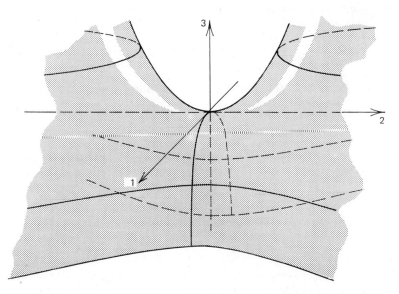

Fig. 15 Hyperbolic paraboloid $ax_1^2 + bx_2^2 = x_3$, $a, b > 0$.

186 SYMMETRIC MATRICES AND QUADRATIC FORMS

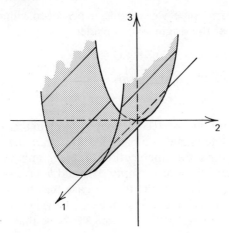

Fig. 16 Parabolic cylinder $x_3 = ax_2^2$.

If either a or b is zero, then the surface defined by (5) is a *parabolic cylinder*. (Figure 16.)

III. The quadratic equation has one quadratic and two first-order terms present. An example of this equation is

$$ax_1^2 + bx_2 + cx_3 = 0.$$

This may be reduced to the case where only one quadratic and one first-order term are present by a rotation about the first coordinate axis. Indeed if

$$x_1 = y_1$$
$$x_2 = \frac{1}{\sqrt{b^2 + c^2}}(by_2 - cy_3)$$
$$x_3 = \frac{1}{\sqrt{b^2 + c^2}}(cy_2 + by_3),$$

then $ax_1^2 + bx_2 + cx_3 = ay_1^2 + \sqrt{b^2 + c^2}\, y_2$. The graph of the equation is a parabolic cylinder. The matrix of the rotation is

$$\begin{pmatrix} 1 & 0 & 0 \\ 0 & b/\sqrt{b^2+c^2} & -c/\sqrt{b^2+c^2} \\ 0 & c/\sqrt{b^2+c^2} & b/\sqrt{b^2+c^2} \end{pmatrix}.$$

To identify the graph of the quadratic equation

$$Q(x_1, x_2, x_3) = ax_1^2 + bx_2^2 + cx_3^2 + dx_1x_2 + ex_1x_3 + fx_2x_3 + gx_1 + hx_2 + lx_3 + m$$
$$= 0$$

one must first determine the algebraic sign of the eigenvalues of the matrix of the quadratic form $ax_1^2 + \cdots + fx_2x_3$. This may be done without computing these eigenvalues by exploiting Descartes' rule of signs. This rule states that if a polynomial $p(x) = a_n x^n + \cdots + a_0$ of degree n is known to have n real roots, then it will have exactly as many positive roots as there are changes in sign of the nonzero coefficients $\{a_n, a_{n-1}, \ldots, a_0\}$. (For a proof see Uspensky, Theory of Equations, McGraw-Hill Book Co. pp. 121–126).

Example 1. Identify the surface $2x_1^2 - x_2^2 + 4x_1x_2 - 2x_2x_3 = 1$. The matrix of the quadratic form is

$$A = \begin{pmatrix} 2 & 2 & 0 \\ 2 & -1 & -1 \\ 0 & -1 & 0 \end{pmatrix}.$$

$$D(A - xI) = D\begin{pmatrix} 2-x & 2 & 0 \\ 2 & -1-x & -1 \\ 0 & -1 & -x \end{pmatrix} = (2-x)(x^2+x-1) - 2(-2x) = -x^3 + x^2 + 7x - 2.$$

For a 3×3 symmetric matrix A, the cubic equation $D(A - xI) = 0$ is known to have three real roots. (Why?) Hence we may use Descartes' rule of signs to determine the signs of the roots. In this example the coefficients of $D(A - xI) = -x^3 + x^2 + 7x - 2$ have two changes in sign.

Hence $D(A - xI)$ has two positive roots, and the surface is a hyperboloid of one sheet.

Example 2. Identify the surface

$$Q(x_1, x_2, x_3) = x_1^2 + 3x_3^2 + 2x_1x_2 + 4x_1x_3 + 2x_2x_3 = 1.$$

The matrix of the quadratic form is

$$A = \begin{pmatrix} 1 & 1 & 2 \\ 1 & 0 & 1 \\ 2 & 1 & 3 \end{pmatrix}.$$

and

$$D(A - xI) = (1-x)[x^2 - 3x - 1] - (1-x) + 2 + 4x$$
$$= -x^3 + 4x^2 + 3x = x(-x^2 + 4x + 3).$$

One root of this polynomial is zero. Since the coefficients of the polynomial $-x^2 + 4x + 3$ have one sign change, the polynomial has one positive and one negative root. Therefore A has one positive eigenvalue, one negative one, and one equal to zero. Therefore, under the appropriate rotation $Q(x_1, x_2, x_3) = 1$ becomes $ay_1^2 - by_2^2 = 1$ where $a, b > 0$. Therefore the surface is a hyperbolic cylinder.

Example 3. Identify the surface

$$x_1^2 + 2x_2^2 + 4x_3^2 - 2x_1 + 4x_2 - 1 = 0.$$

There are no cross product terms, so a translation alone will put the surface in standard position. Since all the coefficients of the quadratic terms are positive, the graph will be an ellipsoid if it is nonempty. Completing the square in the variables x_1 and x_2 we have

$$x_1^2 + 2x_2^2 + 4x_3^2 - 2x_1 + 4x_2 - 1 = (x_1^2 - 2x_1 + 1) + (2x_2^2 + 4x_2 + 2) + 4x_3^2 - 4$$
$$= (x_1 - 1)^2 + 2(x_2 + 1)^2 + 4x_3^2 - 4.$$

If $y_1 = x_1 - 1$, $y_2 = x_2 + 1$, $y_3 = x_3$, the equation becomes

$$y_1^2 + 2y_2^2 + 4y_3^2 = 4.$$

The graph is clearly an ellipsoid.

EXERCISES

Identify the graphs of the following quadratic equations. Determine the translation T (or rotation R) necessary to put the graph in standard position.

1. $x_1^2 - x_2^2 + x_3 - 1 = 0.$
2. $x_1^2 + x_2^2 - x_3^2 + 2x_1 - 4 = 0.$
3. $x_1^2 + 4x_2^2 + 2x_3^2 - 2x_1 + 4x_2 + 8 = 0.$
4. $x_1^2 + 2x_2^2 - 2x_1 + x_3 - 4 = 0.$
5. $x_2^2 - 3x_3^2 + 4x_1 - 6x_3 + 6 = 0.$
6. $2x_1^2 - x_2^2 - x_3^2 + 4x_1 - 2x_3 - 1 = 0.$
7. $x_3^2 - 4x_3 - 1 = 0.$

8. $x_2^2 - 3x_1 + 4x_3 = 0$.
9. $2x_1^2 - 4x_1 - x_2 + 3x_3 + 1 = 0$.
10. $x_1^2 + 2x_2^2 - 2x_1 + x_3 - 1 = 0$.

Identify the graphs of the following quadratic equations. It is not necessary to compute explicitly the eigenvalues of the matrices of the quadratic forms.

11. $x_1^2 - x_2^2 + x_3^2 - 4x_1x_2 = 1$.
12. $x_1^2 + x_2^2 + 2x_3^2 + 2x_1x_2 - 4x_2x_3 = 0$.
13. $2x_1^2 - x_2^2 - x_3^2 + 4x_1x_2 - 8x_1x_3 = 1$.
14. $x_1^2 + 2x_1x_2 - 2x_1x_3 - 4x_2x_3 = 1$.
15. $2x_1^2 + 2x_2^2 - x_3^2 + 8x_1x_2 - 4x_1x_3 - 4x_2x_3 = 1$.
16. $x_3^2 - x_1x_2 = 0$.
17. $2x_1^2 - x_2^2 + x_1x_3 = 1$.
18. $x_1^2 + 2x_2^2 + 2x_3^2 - 2x_1x_2 + 2x_1x_3 = 0$.
19. $x_1^2 + x_2^2 + 5x_3^2 - 2x_1x_3 + 4x_2x_3 = 1$.
20. $x_1^2 + x_2^2 + 4x_3^2 - 2x_1x_2 + 4x_1x_3 - 4x_2x_3 = 1$.

CHAPTER SIX

THE CALCULUS OF VECTOR FUNCTIONS

6.1 Vector Functions

In this chapter we shall study the structure of functions defined on subsets of $V_n(R)$ with values in $V_m(R)$. The simplest such functions are, of course, linear functions. These can be represented by matrices, and during the past chapters we have thoroughly investigated the structure of such functions. To deal with nonlinear functions we must utilize the techniques of the calculus. The central idea here is the notation of a limit.

The abstract notion of a function is, of course, familiar. Let S and T be nonempty sets and let f be a set of ordered pairs (s, t) where $s \in S$ and $t \in T$. Then f is a function if there do *not* exist ordered pairs $P = (s_1, t_1)$ and $Q = (s_2, t_2)$ in f with the same first coordinate but different second coordinates. Thus if f is a function and $P = (s, t)$ is a point in f, then P is the only point of f having first coordinate s. The set of *first* coordinates of points in f is called the *domain* of f. The set of *second* coordinates is called the *range* or *image* of f. The definition implies that for each s belonging to the domain of f there is one and only one point t in the range such that (s, t) is an ordered pair of f. It is customary to write this point $t = f(s)$. We then say that t is the *value* of the function at the point t. It is suggestive to think of a function as a transformation or mapping. See Figure 1. Hence we shall often use the terminology, "f maps S into T" to mean that the domain of the function f is all of S and the range of f is a subset of T.

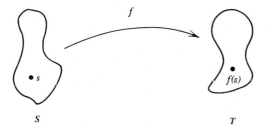

Fig. 1 Function f with domain in S and range in T.

We shall confine ourselves to functions defined on subsets of $V_n(R)$ with values in $V_m(R)$. Such a function is called a vector-valued function of a vector variable, or just a vector function. To emphasize the fact that the domain of f is a subset of $V_n(R)$ and the range is a subset of $V_m(R)$ we say that f is a function from $V_n(R)$ to $V_m(R)$. In the case that the domain of f is all of $V_n(R)$ and we wish to emphasize this fact, we say that f maps $V_n(R)$

into $V_m(R)$. Thus an $n \times m$ matrix defines a linear transformation, or a linear function, mapping $V_n(R)$ into $V_m(R)$. If the image of f is all of $V_m(R)$, then we say f is a function or mapping of $V_n(R)$ *onto* $V_m(R)$. Thus an invertible linear transformation in $V_n(R)$ maps $V_n(R)$ onto $V_n(R)$.

Now a vector $X \in V_n(R)$ can be written uniquely as a linear combination of the basis vectors $I_1, \ldots, I_n$. Indeed we commonly write

(1) $$X = (x_1, \ldots, x_n) = \sum x_k I_k.$$

So it is natural to write $f(x_1, \ldots, x_n)$ in place of $f(X)$, and call the function f a function of the n coordinates $x_1, \ldots, x_n$. Strictly speaking, $f(X)$ denotes the value of the function at the vector X, and f denotes the function. Just as for scalar functions it is awkward to always insist on this distinction. Hence we will often write "the function $Y = f(X)$," in place of "the function f." Also it is common usage to speak of a function defined on a set S in $V_n(R)$ as a function of n variables. The n variables, of course, represent the n coordinates $x_1, \ldots, x_n$ of a point $X \in S$.

In the future when we have occasion to consider a basis for $V_n(R)$ we shall always consider the canonical one. Hence we shall freely identify vectors X in $V_n(R)$ with the coefficient vector $(x_1, \ldots, x_n)$ in (1). Thus we would write $Y = f(X)$ in coordinate form as

$$(y_1, \ldots, y_m) = f(x_1, \ldots, x_n).$$

Since the coordinates $y_1, \ldots, y_m$ are uniquely determined by the vector Y, associated with the vector-valued function f, we have m scalar-valued functions $f_1, \ldots, f_m$ called the *coordinate functions of* f. These are defined by the requirements that

$$y_k = f_k(X) \quad \text{if} \quad Y = (y_1, \ldots, y_m) = f(X).$$

For example, if $X \in V_3(R)$ and $f(X) = (x_1 x_2 x_3, x_1 + x_2 + x_3)$, then f is a function defined on $V_3(R)$ with values in $V_2(R)$. The coordinate functions of f are

$$f_1(X) = x_1 x_2 x_3$$
$$f_2(X) = x_1 + x_2 + x_3.$$

When a vector function f is specified by a formula, such as for example

$$f(x_1, x_2) = \left(\frac{1}{x_1 - x_2}, \frac{1}{x_1 + x_2} \right),$$

then the domain of f is taken to be the set of vectors X in $V_n(R)$ for which $f(X)$ is a well-defined vector in $V_m(R)$. In the example above, this domain would be the set of vectors X lying off the two lines

$$x_1 = x_2$$

and
$$x_1 = -x_2.$$

If $f(X) = (\sqrt{x_1 + x_2}, \sqrt{x_1 - x_2})$, then this rule defines a vector $Y \in V_2(R)$ if and only if $x_1 + x_2 \geq 0$ and $x_1 - x_2 \geq 0$. This will occur if X lies in the quadrant above the line $x_1 + x_2 = 0$ and above the line $x_1 - x_2 = 0$. This is the shaded portion in Figure 2. Since $\sqrt{x} \geq 0$, f maps this quadrant into the set $T = \{Y = (y_1, y_2) : y_1 \geq 0 \text{ and } y_2 \geq 0\}$. This set is indeed the image of f since

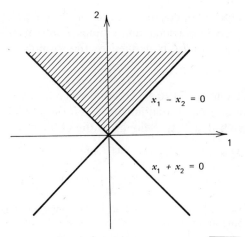

Fig. 2 Domain of the function $f(X) = (\sqrt{x_1 + x_2}, \sqrt{x_1 - x_2})$.

it may easily be checked that if $y_1 \geq 0$ and $y_2 \geq 0$, then x_1, x_2 may be determined such that
$$y_1 = \sqrt{x_1 + x_2}$$
and
$$y_2 = \sqrt{x_1 - x_2}.$$

Indeed, if $x_1 = (y_1^2 + y_2^2)/2$ and $x_2 = (y_1^2 - y_2^2)/2$, then $X = (x_1, x_2)$ is the required vector. Thus if $S = \{X = (x_1, x_2) : x_1 + x_2 \geq 0 \text{ and } x_1 - x_2 \geq 0\}$ then f maps S onto T. To determine the domain of a function is usually straightforward. Often this is not the case with the image.

Since $V_m(R)$ is closed under addition and scalar multiplication, these operations may be defined on functions f and g from $V_n(R)$ to $V_m(R)$. Indeed, we define $(af)(X) = a(f(X))$ and $(f+g)(X) = f(X) + g(X)$. The domain of af is the domain of f, and the domain of $f+g$ is the intersection of the domain of f and the domain of g. This is clear since $f(X) + g(X)$ only makes sense if X belongs to both the domain of f and the domain of g.

If f and g are scalar-valued functions, then since the real numbers are closed under multiplication, we may form the product of the functions f and g. Indeed,
$$(f \cdot g)(X) = f(X) \cdot g(X)$$
and the domain of $f \cdot g$ is the intersection of the domain of f and g.

It is reasonably clear that whenever $V_m(R)$ is closed under a binary operation "$\odot$," this operation may be applied to pairs of functions having their range in $V_m(R)$. An example of this is the vector product in $V_3(R)$. Recall that if $X, Y \in V_3(R)$, then we define

$$X \times Y = D \begin{pmatrix} I_1 & I_2 & I_3 \\ x_1 & x_2 & x_3 \\ y_1 & y_2 & y_3 \end{pmatrix}$$
$$= (x_2 y_3 - y_2 x_3) I_1 - (x_1 y_3 - y_1 x_3) I_2 + (x_1 y_2 - y_1 x_2) I_3.$$

Hence if f and g are functions from $V_n(R)$ to $V_3(R)$, then

$$(f \times g)(X) = f(X) \times g(X) = D \begin{pmatrix} I_1 & I_2 & I_3 \\ f_1(X) & f_2(X) & f_3(X) \\ g_1(X) & g_2(X) & g_3(X) \end{pmatrix}.$$

Perhaps the most important operation on functions is function composition. To recall, if f maps a set S into a set T and g maps T into a set U, then $g \circ f$, called the *composition of g and f*, is defined by the formula

$$(g \circ f)(s) = g(f(s)).$$

Thus if f is a function from $V_n(R)$ to $V_m(R)$ and g is a function from $V_m(R)$ to $V_p(R)$, the function $g \circ f$ is a function from $V_n(R)$ to $V_p(R)$ with domain the set of vectors X such that $f(X)$ belongs to the domain of g.

Example 1. If $g(x_1, x_2) = ((x_1 + x_2)^2, x_1 - x_2)$ and $f(x_1, x_2, x_3) = (e^{x_1+x_2}, 1/x_3)$

$$(g \circ f)(x_1, x_2, x_3) = \left(\left(e^{x_1+x_2} + \frac{1}{x_3}\right)^2, e^{x_1+x_2} - \frac{1}{x_3}\right).$$

Clearly the domain of f is $S = \{X = (x_1, x_2, x_3) : x_3 \neq 0\}$. Since the domain of g is all of $V_2(R)$, S is also the domain of $g \circ f$.

In general our practice will be to write vectors $X \in V_n(R)$ as row vectors. Thus $X = (x_1, \ldots, x_n)$. However, if A is an $n \times m$ matrix, then to indicate the linear transformation $Y = AX$ we write

$$Y = AX = \begin{pmatrix} a_{11} & \cdots & a_{1n} \\ \vdots & & \vdots \\ a_{m1} & \cdots & a_{mn} \end{pmatrix} \begin{pmatrix} x_1 \\ \vdots \\ x_n \end{pmatrix} = \begin{pmatrix} y_1 \\ \vdots \\ y_m \end{pmatrix}.$$

Hence if g is a function defined on $V_m(R)$

$$(g \circ A)(X) = g(AX) = g\left(\begin{pmatrix} a_{11} & \cdots & a_{1n} \\ \vdots & & \vdots \\ a_{m1} & \cdots & a_{mn} \end{pmatrix} \begin{pmatrix} x_1 \\ \vdots \\ x_n \end{pmatrix}\right) = g\begin{pmatrix} y_1 \\ \vdots \\ y_m \end{pmatrix} = g(Y).$$

Thus writing $g\begin{pmatrix} y_1 \\ \vdots \\ y_m \end{pmatrix}$ in place of $g(y_1, \ldots, y_m)$ is only a notational convenience. They both represent the vector $g(Y)$. The only point to remember is that if f is a linear function from $V_n(R)$ to $V_m(R)$ which is represented by the matrix $A = (a_{ij})$, then

$$f(X) = f(x_1, \ldots, x_n) = f\begin{pmatrix} x_1 \\ \vdots \\ x_n \end{pmatrix}$$

is actually computed by evaluating

$$\begin{pmatrix} a_{11} & \cdots & a_{1n} \\ \vdots & & \vdots \\ a_{m1} & \cdots & a_{mn} \end{pmatrix} \begin{pmatrix} x_1 \\ \vdots \\ x_n \end{pmatrix}.$$

Example 2. Let $g(x_1, x_2) = (x_1/x_2, x_2/x_1)$ and let A be the 2×2 matrix $\begin{pmatrix} 2 & 1 \\ -1 & 0 \end{pmatrix}$. Compute $(A + g)$, $A \circ g$ and $g \circ A$. Now

$$(A+g)(X) = \left(\begin{pmatrix} 2 & 1 \\ -1 & 0 \end{pmatrix} + g\right)\begin{pmatrix} x_1 \\ x_2 \end{pmatrix} = \begin{pmatrix} 2x_1 + x_2 \\ -x_1 \end{pmatrix} + \begin{pmatrix} \frac{x_1}{x_2} \\ \frac{x_2}{x_1} \end{pmatrix}$$

$$= \begin{pmatrix} 2x_1 + x_2 + \frac{x_1}{x_2} \\ -x_1 + \frac{x_2}{x_1} \end{pmatrix} = \left(2x_1 + x_2 + \frac{x_1}{x_2}, -x_1 + \frac{x_2}{x_1}\right).$$

Evaluating $A + g$ at the vector $(2, 1)$ we have $(A+g)(2,1) = (7,1)$. Next

$$(A \circ g)(X) = A(g(X)) = \begin{pmatrix} 2 & 1 \\ -1 & 0 \end{pmatrix}\begin{pmatrix} \frac{x_1}{x_2} \\ \frac{x_2}{x_1} \end{pmatrix} = \begin{pmatrix} \frac{2x_1}{x_2} + \frac{x_2}{x_1} \\ -\frac{x_1}{x_2} \end{pmatrix}$$

$$= \left(\frac{2x_1^2 + x_2^2}{x_1 x_2}, -\frac{x_1}{x_2}\right)$$

and

$$(g \circ A)(X) = g(A(X)) = g\left(\begin{pmatrix} 2 & 1 \\ -1 & 0 \end{pmatrix}\begin{pmatrix} x_1 \\ x_2 \end{pmatrix}\right) = g\begin{pmatrix} 2x_1 + x_2 \\ -x_1 \end{pmatrix} = g(2x_1 + x_2, -x_1)$$

$$= \left(\frac{2x_1 + x_2}{-x_1}, \frac{-x_1}{2x_1 + x_2}\right).$$

Evaluating these two functions at specific points we have $(A \circ g)(1,1) = (3,1)$ and $(g \circ A)(1,-1) = (-1,-1)$. Note that the domain of $A \circ g$ is $\{X = (x_1, x_2) : x_1 x_2 \neq 0\}$; whereas the domain of $g \circ A$ equals

$$\{X = (x_1, x_2) : x_1 \neq 0\} \cap \{X = (x_1, x_2) : 2x_1 + x_2 \neq 0\}.$$

Example 3. Let $f(x_1, x_2, x_3) = x_1 x_2 x_3$ and let L be the linear function from $V_3(R)$ to R defined by the 1×3 matrix $(1, 2, 3)$. Thus

$$L(X) = (1,2,3)\begin{pmatrix} x_1 \\ x_2 \\ x_3 \end{pmatrix} = x_1 + 2x_2 + 3x_3$$

and

$$(f + L)(X) = x_1 x_2 x_3 + x_1 + 2x_2 + 3x_3.$$

Indeed if $X = (1, -1, 2)$ then

$$(f + L)(1, -1, 2) = -2 + 1 - 2 + 6 = 7.$$

Note that the number $(f + L)(1, -1, 2)$ could just as well have been written

$$(f + L)\begin{pmatrix} 1 \\ -1 \\ 2 \end{pmatrix}.$$

EXERCISES

1. If $f(x_1, x_2) = (1/(x_1 + x_2), 1/(x_1 - x_2))$ and $g(x_1, x_2) = (x_1 + x_2, x_2^2)$ compute $(f - g)(1, 2)$, $(f + g)(3, 4)$, $(f \circ 3g)(1, 2)$, $(g \circ f)(0, 1)$.

2. If $f(x_1, x_2) = ((x_1 - x_2)^2, e^{x_1})$ and $A = \begin{pmatrix} 2 & 0 \\ -1 & 1 \end{pmatrix}$, compute

$(2f + A)(1, 0),$ $(f - 3A)(1, 2)$
$(A \circ f)(1, 1),$ $(f \circ (A^2))(0, 1).$

3. If $f(x_1, x_2, x_3) = (x_1 x_2, x_2 x_3, x_1 x_3)$ and $A = \begin{pmatrix} 2 & -1 \\ 1 & 0 \end{pmatrix}$ and $B = \begin{pmatrix} 1 & -1 & 0 \\ 0 & 2 & 1 \\ 1 & 0 & -1 \end{pmatrix}$

compute $(B \circ f)(X)$ and $(f \circ B)(X)$. Can $f \circ A$ and $A \circ f$ be computed?

4. If $f(x_1, x_2) = ((x_1 - x_2)/(x_1 + x_2), e^{x_1})$ and $g(x_1, x_2) = (x_1 x_2, (x_1 + x_2)^2)$ compute $f - g, f + g, 2f - g, f \circ g$, and $g \circ f$. Determine the coordinate functions and their domains.

5. If $f(x_1, x_2) = (\sqrt{x_1 x_2}, x_1 - x_2)$, $g(x_1, x_2) = (x_1 x_2, \sqrt{x_1 - x_2})$, compute $f - g$, $f + g, f - 3g, f \circ 2g$. Determine the coordinate functions and their domains.

6. Let $f(x) = (x, x^2)$, $A = (2, -1)$. For $X \in V_2(R)$ and $x \in R$ compute AX, $(f \circ A)(X)$, and $(A \circ f)(x)$. In particular compute these three quantities if $X = (0, 1), (-1, 1), (2, 1)$ and $x = 1, 2$.

7. Let $f(x) = (x, (x-1), (x-2)^2)$ and $A = (1, -1, 2)$. For $X \in V_3(R)$ compute AX, $(f \circ A)(X)$, $(A \circ f)(X)$. Evaluate these three functions at $X = (2, 0, 1)$.

8. Let $f(x) = (x^2, 1/x)$, $g(x_1, x_2) = (x_1 x_2, x_1/x_2)$, $h(x_1, x_2) = x_1 x_2$, $A = \begin{pmatrix} 2 & -1 \\ 0 & 1 \end{pmatrix}$, $B = (1, -1)$. Compute $(B \circ A \circ g)(X)$, $(f \circ B \circ A \circ A)(X)$, $(f \circ h \circ A)(X)$ and $(A \circ f \circ h \circ g)(X)$. In particular evaluate these quantities when $X = (1, -1)$.

9. Let f be a vector function with domain D. If $f_1, \ldots, f_n$ are the coordinate functions of f with domains $D_1, \ldots, D_n$, respectively, show that $D = D_1 \cap \cdots \cap D_n$.

10. Let W be the set of all functions mapping $V_n(R)$ into $V_m(R)$. Show that W is a vector space.

6.2 Limits and Continuity

We wish next to define for a function f from $V_n(R)$ to $V_m(R)$ the notion of the limit of $f(X)$ as X approaches X_0. We recall the appropriate definition for scalar functions. Let x_0 be a point of the open interval (a, b) and assume f is defined on (a, b) except possibly at the point x_0. Then $\lim_{x \to x_0} f(x) = l$ if for each $\epsilon > 0$ there exists a corresponding $\delta > 0$ such that whenever $0 < |x - x_0| < \delta$ it follows that $|f(x) - l| < \epsilon$. Thus no matter how small an interval J is chosen about l, a corresponding interval I may be chosen about x_0 so that if $x \in I$ and $x \neq x_0$ then $f(x) \in J$. Expressed in another way we may say that the values $f(x)$ will approximate the number l as closely as desired just so long as the numbers x are sufficiently close to x_0. Since the limit of $f(x)$ as x approaches x_0 is to be completely independent of the existence of the number $f(x_0)$, we only consider the values $f(x)$ for $x \neq x_0$.

To extend these ideas to vector functions we need only replace $|x - x_0|$ and $|f(x) - l|$ by $|X - X_0|$ and $|f(X) - L|$. The definition then extends verbatim. Before stating it explicitly, however, let us introduce some terminology concerning sets in $V_n(R)$.

If r is a positive number, the set $D_r(X) = \{Y \in V_n(R) : |Y - X| < r\}$ will be called *the open sphere of radius r about the vector X*. In place of sphere we shall often substitute the words disc or ball. If $r = 1$, $D_1(X)$ is called the *open unit ball* about X. The words disc, ball, and sphere are meant to be geometrically suggestive since in $V_2(R)$

$$D_r(X) = \{(y_1, y_2) : \sqrt{(y_1 - x_1)^2 + (y_2 - x_2)^2} < r\},$$

which is a disc of radius r about X. See Figure 3. In $V_3(R)$, $D_r(X)$ is a sphere of radius r about X. If $n = 1$ then $D_r(X)$ is just the open interval of length $2r$ centered at X.

Let S now be an arbitrary subset of $V_n(R)$. A vector or point $X \in S$ is called an *interior point of S* if for r sufficiently small

$$D_r(X) \subset S.$$

Thus in order for X to be an interior point of S the entire ball of radius r must

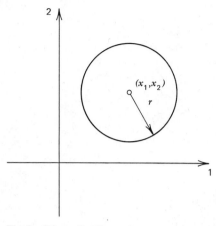

Fig. 3 Discs of radius r about (x_1, x_2).

be contained in S for r sufficiently small. If each point of S is an interior point of S, then S is said to be *open*.

For example, in $V_2(R)$ the square $T = \{(x_1, x_2) : |x_1| < 1 \text{ and } |x_2| < 1\}$ is open since for each $X \in T$, r can be chosen small enough so that $D_r(X)$ lies entirely in T. See Figure 4. If we consider $T_1 = \{(x_1, x_2) : |x_1| \leq 1 \text{ and } |x_2| \leq 1\}$ then T_1 is not open since points on the edge of T_1 are not interior points. For if $|x_1| = 1$ or $|x_2| = 1$ and $X = (x_1, x_2)$, then no matter how small the *positive* number r is chosen $D_r(X)$ always contains points not in T_1. See Figure 5.

The set of interior points of a set S is called the *interior* of S and is written S^0. Thus for the above example, the open square is the interior of $T_1{}^0$. We would then write $T = T_1{}^0$.

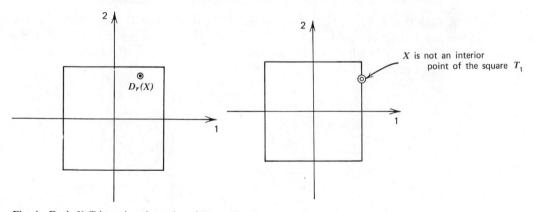

Fig. 4 Each $X \epsilon T$ is an interior point of T.   **Fig. 5**

In $V_3(R)$ the box

$$S = \{X = (x_1, x_2, x_3) : |x_1| < 1, |x_2| < 1 \text{ and } |x_3| < 1\}$$

is open. However, the projection of the box onto a coordinate plane, for example, $\{X \in S : x_3 = 0\}$, has no interior points. See Figure 6.

We now state the formal definition of a limit for vector functions.

Definition. *Let U be an open subset of $V_n(R)$ and let $X_0 \in U$. Assume*

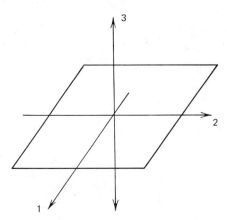

Fig. 6 Square has no interior in $V_3(R)$.

the function f from $V_n(R)$ to $V_m(R)$ is defined on U except possibly at X_0. Let L be a vector in $V_m(R)$. Then $\lim_{X \to X_0} f(X) = L$ if for each $\epsilon > 0$ there exists a corresponding $\delta > 0$, such that, whenever $0 < |X - X_0| < \delta$, it follows that $|f(X) - L| < \epsilon$.

Definition. *If f is in addition defined at X_0 and*

$$\lim_{X \to X_0} f(X) = f(X_0),$$

then f is said to be continuous at X_0. If f is continuous at each point of the open set U, then f is said to be continuous on U.

The set of vectors $Y \in V_m(R)$ satisfying $|Y - L| < \epsilon$ is just the ball $D_\epsilon(L)$. Similarly $\{X : |X - X_0| < \delta\}$ is just the ball $D_\delta(X_0)$. Therefore $\lim_{X \to X_0} f(X) = L$ if for each ball $D_\epsilon(L)$ there is a corresponding ball $D_\delta(X_0)$, such that $f(X) \in D_\epsilon(L)$ whenever $X \in D_\delta(X_0)$ and $X \neq X_0$. Thus f will be continuous at X_0 if for each ball B about $f(X_0)$, there is a corresponding ball C about X_0 such that f maps C into B, that is, $f(X) \in B$ whenever $X \in C$. See Figure 7.

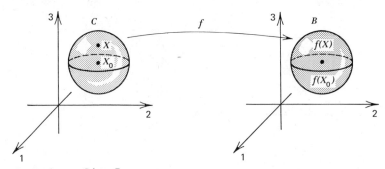

Fig. 7 f maps C into B.

If $f_1, \ldots, f_m$ are the coordinate functions of the vector function f, then $\lim_{X \to X_0} f(X) = L = (l_1, \ldots, l_n)$ if and only if $\lim_{X \to X_0} f_i(X) = l_i, i = 1, \ldots, n$. Thus we may write

$$\lim_{X \to X_0} f(X) = (\lim_{X \to X_0} f_1(X), \ldots, \lim_{X \to X_0} f_n(X)).$$

This result has an easy "ϵ, δ" proof.

LIMITS AND CONTINUITY

THEOREM 6.1 *Let f be a vector function from $V_n(R)$ to $V_m(R)$. Then*
$$\lim_{X \to X_0} f(X) = L = (l_1, \ldots, l_m)$$
if and only if
$$\lim_{X \to X_0} f_i(X) = l_i, \quad i = 1, \ldots, n$$
for each of the coordinate functions $f_1, \ldots, f_n$ for f.

Proof. Assume first that
$$\lim_{X \to X_0} f(X) = L = (l_1, \ldots, l_n).$$

We must show $\lim_{X \to X_0} f_i(X) = l_i$ for $i = 1, \ldots, m$. By definition of the notion of a limit this means that for a given $\epsilon > 0$ we must pick a corresponding $\delta > 0$ so that whenever $0 < |X - X_0| < \delta$ it follows that $|f_i(X) - l_i| < \epsilon$. By hypothesis for a given $\epsilon > 0$ we may pick $\delta > 0$ such that whenever $0 < |X - X_0| < \delta$ it follows that $|f(X) - L| < \epsilon$. Since for each $i = 1, \ldots, m$

$$|f_i(X) - l_i| \leq \left\{ \sum_{j=1}^{m} (f_j(X) - l_j)^2 \right\}^{1/2} = |f(X) - L|$$

we have $|f_i(X) - l_i| < \epsilon$. Therefore for each i $\lim_{X \to X_0} f_i(X) = l_i$.

Conversely if we assume that $\lim_{X \to X_0} f_i(X) = l_i$ for $i = 1, \ldots, n$, then for a given $\epsilon > 0$ we may pick $\delta > 0$ so that whenever $0 < |X - X_0| < \delta$ it follows that for each i

$$|f_i(X) - l_i| < \frac{\epsilon}{\sqrt{m}}.$$

Why we chose $\epsilon/\sqrt{m}$ rather than ϵ will be apparent momentarily. Indeed, if
$$|f_i(X) - l_i| < \frac{\epsilon}{\sqrt{m}},$$
then
$$|f_i(X) - l_i|^2 < \frac{\epsilon^2}{m}$$
and
$$|f(X) - L|^2 = \sum_{i=1}^{m} |f_i(X) - l_i|^2 < m \frac{\epsilon^2}{m} = \epsilon^2.$$

Hence for this choice of δ, $0 < |X - X_0| < \delta$ implies $|f(X) - L| < \epsilon$, and we have proved that $\lim_{X \to X_0} f(X) = L$.

COROLLARY. *Let f be a vector function from $V_n(R)$ to $V_m(R)$. Then f is continuous at X_0 if and only if each of the coordinate functions $f_1, \ldots, f_m$ is continuous at X_0.*

Just as for scalar functions we have the following fundamental limit theorem.

THEOREM 6.2 *Let f and g be vector functions from $V_n(R)$ to $V_m(R)$ which are defined on the open set U. Let $X_0 \in U$ and assume*
$$\lim_{X \to X_0} f(X) = L$$
and
$$\lim_{X \to X_0} g(X) = M.$$

Then
$$\lim_{X \to X_0} (af)(X) = aL$$
for each real number a, and
$$\lim_{X \to X_0} (f+g)(X) = L+M.$$

Proof. The arguments are exactly the same as for the similar theorems for scalar functions. We prove the second statement leaving the first as an exercise. We must show that for a given $\epsilon > 0$, we may pick a corresponding $\delta > 0$ so that whenever X belongs to an open ball of radius δ about X_0 and $X \neq X_0$, then $f(X) + g(X)$ belongs to an open ball of radius ϵ about $L+M$. Stated symbolically we must prove that whenever $X \in D_\delta(X_0)$ and $X \neq X_0$, it follows that $f(X) + g(X) \in D_\epsilon(L+M)$. Consider the ball of radius $\epsilon/2$ about L and M, respectively. Then if $Y \in D_{\epsilon/2}(L)$ and $Z \in D_{\epsilon/2}(M)$, it follows that $Y+Z \in D_\epsilon(L+M)$. This is clear since by the triangle inequality

$$|Y+Z-(L+M)| < |Y-L| + |Z-M| < \frac{\epsilon}{2} + \frac{\epsilon}{2} = \epsilon.$$

Since $\lim_{x \to X_0} f(X) = L$ and $\lim_{x \to X_0} g(X) = M$, we may pick a ball $D_\delta(X_0)$ such that if $X \in D_\delta(X_0)$ and $X \neq X_0$, then $f(X) \in D_{\epsilon/2}(L)$ and $g(X) \in D_{\epsilon/2}(M)$. (Why does one ball work for both f and g?) Hence by the above inequality $f(X) + g(X) \in D_\epsilon(L+M)$ and we have proved the result.

COROLLARY 1. *If f and g are continuous on the open set U, then $af + bg$ is continuous on U for each pair of scalars a and b.*

Proof. Exercise

COROLLARY 2. *If f is a linear function from $V_n(R)$ to $V_m(R)$, then f is continuous.*

Proof. Let $A = (a_{ij})$ be the matrix which represents the linear function f. Then

$$f(X) = AX = \begin{pmatrix} a_{11} & \cdots & a_{1n} \\ \vdots & & \vdots \\ a_{m1} & \cdots & a_{mn} \end{pmatrix} \begin{pmatrix} x_1 \\ \vdots \\ x_n \end{pmatrix}.$$

The ith coordinate function is just $f_i(X) = a_{i1}x_1 + \cdots + a_{in}x_n = (A_i, X)$ where $A_i = (a_{i1}, \ldots, a_{in})$. It follows from Theorem 6.2 that each function f_i is continuous. Hence by Theorem 6.1 the function f must be continuous. It is easy to give a direct "ϵ, δ" verification of the fact that each function f_i is continuous. To do this note that by the Schwartz inequality

$$|f_i(X-X_0)| = |(A_i, X-X_0)| \leq |A_i| \|X-X_0\|.$$

Hence if ϵ is given and we choose $\delta = \epsilon/|A_i|$, then $|X-X_0| < \delta$ implies

$$|f_i(X-X_0)| \leq |A_i| \|X-X_0\| < |A_i|\delta = \epsilon.$$

Hence the function f_i is continuous. This completes the proof.

If f and g are scalar-valued, then we have the full analogue of the limit theorems for scalar functions defined on intervals of the real line.

THEOREM 6.3 *Let f and g be functions from $V_n(R)$ to R which are defined on the open set U except possibly at the point X_0. If $\lim_{X \to X_0} f(X) = l$ and*

$\lim_{X \to X_0} g(X) = m$, then
$$\lim_{X \to X_0} f(X)g(X) = lm$$
and
$$\lim_{X \to X_0} \frac{f(X)}{g(X)} = \frac{l}{m} \quad \text{if } m \neq 0.$$

Thus if f and g are continuous at X_0, then $f \cdot g$ is continuous at X_0. If $g(X_0) \neq 0$, then the quotient f/g is continuous at X_0.

The proof of this theorem is exactly the same as for functions of one variable and is left as an exercise. We note also that if $\lim_{X \to X_0} g(X) = 0$, then, just as for functions defined on sets of real numbers, $\lim_{X \to X_0} (f(X)/g(X))$ may or may not exist. Whether or not this limit exists must be determined from further investigation of f and g. We can conclude nothing from Theorem 6.3 in this case.

The following corollary to Theorem 6.3 is often useful. The proof we leave as an exercise.

COROLLARY. *Under the same hypotheses as Theorem 6.3 if $\lim_{X \to X_0} (f(X)/g(X))$ exists and if $\lim_{X \to X_0} g(X) = 0$, then $\lim_{X \to X_0} f(X) = 0$.*

Theorems 6.2 and 6.3 provide us with a large class of continuous scalar-valued functions defined on sets in $V_n(R)$. Indeed any polynomial $p(x_1, \ldots, x_n)$ is always continuous. The quotient of two polynomials
$$\frac{p(x_1, \ldots, x_n)}{q(x_1, \ldots, x_n)}$$
is continuous at each point $X = (x_1, \ldots, x_n)$ at which q does not vanish. For example
$$p(x_1, x_2, x_3) = x_1^3 + (x_1 - x_2 + x_3)^3 - x_1 x_2 x_3$$
is continuous for all values of x_1, x_2, x_3. The function
$$r(x_1, x_2, x_3) = \frac{x_1 x_2 x_3}{x_1 + x_2 + x_3}$$
is continuous at all points $X = (x_1, x_2, x_3)$ lying off the plane $x_1 + x_2 + x_3 = 0$.

To construct more examples of continuous vector functions we need to know that the composition of two continuous vector functions is continuous. Just as for scalar-valued functions, this is an extremely important and useful fact.

THEOREM 6.4 *Let f be a vector function from $V_n(R)$ to $V_m(R)$ which is defined on an open set U, and let g be a vector function from $V_m(R)$ to $V_p(R)$ which is defined on an open set V. Assume f maps U into V. Then if f is continuous at X_0 and g is continuous at $f(X_0)$, it follows that $g \circ f$ is continuous at X_0.*

Proof. The proof is simple. Let $Y_0 = f(X_0)$, and let B be an open ball about $g(Y_0)$. The function $g \circ f$ will be continuous at X_0 if we can find an open ball D about X_0 with the property that $g \circ f$ maps D into B. By definition of continuity for g at Y_0, there is an open ball C about Y_0 such that g maps C into B. But by definition of continuity for f there is an open ball D about X_0 such that f maps D into C. Hence $g \circ f$ maps D into B, and consequently $g \circ f$ is continuous at X_0.

Example 1. Show that

$$f(X) = f(x_1, x_2) = \frac{e^{x_1+x_2}}{\log(x_1^2 + x_2^2)}$$

is continuous for all $X = (x_1, x_2)$ such that $|X| \neq 1, 0$. The functions $g(x_1, x_2) = x_1 + x_2$ and $h(x_1, x_2) = x_1^2 + x_2^2$ are continuous for all values of X by Theorems 6.2 and 6.3. Since it is known that e^x is continuous for all x and $\log x$ is continuous for all $x > 0$, we may conclude from Theorem 6.4 that $e^{(x_1+x_2)}$ is continuous for all X and $\log(x_1^2 + x_2^2)$ is continuous for all $X \neq 0$. That

$$\frac{e^{x_1+x_2}}{\log(x_1^2 + x_2^2)}$$

is continuous for all X such that $|X| \neq 1, 0$ now follows by Theorem 6.4.

Now if the scalar-valued function f is continuous on an open set U in $V_n(R)$, and we fix all of the variables x_i in $f(x_1, \ldots, x_n)$ except one, then the function f is continuous in that remaining variable and we say that f is *continuous in each variable separately*. To see this, fix all of the coordinates of the vector X except the first, say. Then if $X = (x_1, x_2, \ldots, x_n)$ and $X_0 = (x_1^0, x_2, \ldots, x_n)$, we have $X - X_0 = (x_1 - x_1^0, 0, \ldots, 0)$ and $|X - X_0| = |x_1 - x_1^0|$. It now follows immediately that the continuity of f implies that the function

$$g(x_1) = f(x_1, \ldots, x_n)$$

obtained by fixing the variables $x_2, \ldots, x_n$ is a continuous function of x_1. The argument proceeds similarly for the variables $x_2, \ldots, x_n$. The actual "ϵ, δ" verification we leave as an exercise.

It would be very convenient if the converse of the above statement were true. Namely, if $f(x_1, \ldots, x_n)$ is continuous in each variable or coordinate separately, then f is continuous. Unfortunately this result is false in general. The following example illustrates this. Let

$$f(x_1, x_2) = \frac{2x_1 x_2}{x_1^2 + x_2^2} \quad \text{and} \quad f(0, 0) = 0.$$

We claim f is not continuous at $X = 0$, yet f is continuous in each variable separately at each point X. To see the second statement observe that if $X \neq 0$, then f is continuous. This fact follows from Theorem 6.2. Hence for such X, f is continuous in each variable separately. However, if $X = 0$ then $x_1^2 + x_2^2 = 0$, and we can conclude nothing from Theorem 6.2. If we fix $x_2 = a$ and assume $a \neq 0$, then clearly

$$f(x_1, a) = \frac{2ax_1}{x_1^2 + a^2}$$

is continuous as a function of x_1. If $a = 0$, then $f(x_1, 0) = 0$ for all x_1, and f is certainly a continuous function of x_1. In a similar fashion we may conclude that $f(a, x_2)$ is a continuous function of x_2 for each fixed value of a. To see that f is not continuous at the origin we need only observe that if $x_1 = x_2$ and $x_1 \neq 0$, then

$$f(x_1, x_2) = \frac{2x_1^2}{x_1^2 + x_1^2} = 1.$$

Yet on either of the coordinate axes $x_1 = 0$, and $x_2 = 0$, f has the value 0. Hence if B is any disc containing the origin, there are vectors $Y, Z \in B$ Such

LIMITS AND CONTINUITY 201

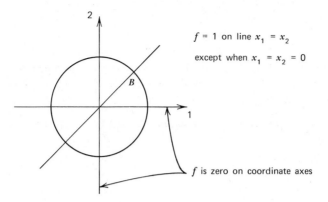

Fig. 8 $f(x_1 + x_2) = \dfrac{2x_1 x_2}{x_1^2 + x_2^2}$ is continuous in each variable separately but is not continuous at $X = 0$.

that $f(Y) = 1$ and $f(Z) = 0$. Hence $\lim_{X \to 0} f(X)$ can not possibly exist. Consequently f is not continuous at 0. See Figure 8.

To say that a function $f(X) = f(x_1, \ldots, x_n)$ is continuous in each variable separately means that f is continuous along each line parallel to the coordinate axes. The above example shows that this is not enough to guarantee the continuity of f. Indeed the functions

$$g(x_i) = f(x_1, \ldots, x_n)$$

obtained by fixing all variables except the ith may have derivatives of all orders and still f may not be continuous. We shall have more to say about these matters in the next section.

EXERCISES

Show on the basis of the theorems of this section that the following functions are continuous for all values of X.

1. $f(x_1, x_2) = (x_1 - x_2)^2$.
2. $f(x_1, x_2) = ((x_1 + x_2)^2, e^{x_1 + x_2})$.
3. $f(x_1, x_2) = \dfrac{1}{x_1^2 + 1} \log(1 + x_2^2)$.
4. $f(x_1, x_2, x_3) = \dfrac{\sin(x_1 + x_2)}{e^{x_1 + x_2 - x_3}}$.
5. $f(x_1, x_2, x_3) = \left(\cos(e^{x_1 - x_2}), \sin \dfrac{1}{1 + x_1^2 + x_3^2}\right)$.

Determine open sets on which the following functions are continuous.

6. $f(x_1, x_2) = \dfrac{x_1 + x_2}{x_1 - 2x_2}$.
7. $f(x_1, x_2) = \dfrac{x_1 - 2x_2}{x_1^2 - x_2^2}$.
8. $f(x_1, x_2) = \left(\dfrac{x_1}{x_1^2 - 4x_2^2}, \log(x_1^2 + x_2^2)\right)$.
9. $f(x_1, x_2) = \left(\dfrac{1}{1 - e^{x_1 + x_2}}, \dfrac{1 + x_1^2}{\log(x_1^2 + x_2^2)}\right)$.

Establish that the following functions are continuous directly from the definition.

10. $f(X) = |X|$.
11. $f(X) = (A, X)$ where A is a fixed vector.
12. Show that if $\lim_{X \to X_0} f(X) = L$, then $\lim_{X \to X_0} (af)(X) = aL$, where a is a real scalar.
13. Let f and g be real-valued functions defined on the open set U in $V_n(R)$. If $\lim_{X = X_0} f(X) = l$ and $\lim_{X = X_0} g(X) = m$ prove that $\lim_{X = X_0} (f \cdot g)(X) = lm$ and $\lim_{X = X_0} (f/g)(X) = l/m$ if $m \neq 0$.
14. Let f and g be real-valued functions defined on an open set U in $V_n(R)$. If $\lim_{X \to X_0} (f/g)(X)$ exists and $\lim_{X \to X_0} g(X) = 0$, show that $\lim_{X \to X_0} f(X) = 0$.

Draw a sketch of each of the following sets and determine which are open.

15. $\{X = (x_1, x_2) : 2x_1^2 + x_2^2 = 1\}$.
16. $\{X = (x_1, x_2) : x_1 > 0\}$.
17. $\{X = (x_1, x_2) : X \neq 0\}$.
18. $\{X = (x_1, x_2) : x_1 \geq 0, x_2 \geq 0\}$.
19. $\{X = (x_1, x_2) : x_1 \geq x_2\}$.
20. $\{X = (x_1, x_2, x_3) : |X| \neq 1\}$.
21. $\{X = (x_1, x_2) : |X| \neq 1, \frac{1}{2}, \frac{1}{3}, \ldots\}$.
22. $\{X = (x_1, x_2) : |X| \neq 1, \frac{1}{2}, \frac{1}{3}, \ldots, 0\}$.
23. Let U and V be open sets in $V_n(R)$. Show that $U \cup V$ and $U \cap V$ are open. (We define the empty set to be open.)
24. It follows from Exercise 23 and induction that the union and intersection of a finite number of open sets is open. Show that the union of an arbitrary collection of open sets is open. By examining the open intervals $I_n = (-1/n, 1/n)$ on the real line show that the intersection of an infinite number of open sets need not be open.
25. Show that the function defined by $f(x_1, x_2) = x_1 x_2 / (x_1^2 + 4x_2^2)$, $f(0, 0) = 0$ is continuous in each variable separately but is not continuous at the origin. (Examine the example in the text.)
26. Show that the set of all continuous functions mapping $V_n(R)$ into $V_m(R)$ is a vector space.

6.3 Partial Derivatives

Let f be a real-valued function defined on an open set U in $V_n(R)$, and let $I_1, \ldots, I_n$ be the unit coordinate vectors. Then

$$\lim_{h \to 0} \frac{f(X + hI_k) - f(X)}{h}$$

is called the kth *partial derivative of the function* f and is written $\partial f / \partial x_k (X)$. If we write $f(X) = f(x_1, \ldots, x_n)$, then we have

$$\frac{\partial f}{\partial x_k}(X) = \lim_{h \to 0} \frac{f(X + hI_k) - f(X)}{h}$$

$$= \lim_{h \to 0} \frac{f(x_1, \ldots, x_k + h, \ldots, x_n) - f(x_1, \ldots, x_n)}{h}.$$

Since the derivative measures the rate of change of a function f, the partial derivative $\partial f / \partial x_k$ may be thought of as the rate of change of the function f in the direction of the vector I_k.

To actually compute the kth partial derivative of f, one considers $f(x_1, \ldots, x_n)$ as a function only of the kth coordinate of the vector X. The resulting function of x_k is differentiated by the usual rules. The values of x_j for $j \neq k$ present in the expression for f are treated as constants.

Example 1. If
$$f(X) = f(x_1, x_2, x_3) = (2x_1 - x_2 + 4x_3)^3$$
then
$$\frac{\partial f}{\partial x_1}(X) = 3(2x_1 - x_2 + 4x_3)^2 \cdot 2$$

$$\frac{\partial f}{\partial x_2}(X) = 3(2x_1 - x_2 + 4x_3)^2 \cdot (-1)$$

$$\frac{\partial f}{\partial x_3}(X) = 3(2x_1 - x_2 + 4x_3)^2 \cdot 4.$$

Clearly
$$\frac{\partial f}{\partial x_1}(1, 1, 1) = 150$$

$$\frac{\partial f}{\partial x_2}(1, 1, 1) = -75$$
and
$$\frac{\partial f}{\partial x_3}(1, 1, 1) = 300.$$

Each partial derivative $\partial f/\partial x_k$ is a real-valued function defined on a subset of $V_n(R)$, namely, the set of those X in U for which

$$\lim_{h \to 0} \frac{f(X + hI_k) - f(X)}{h} = \lim_{h \to 0} \frac{f(x_1, \ldots, x_k + h, x_{k+1}, \ldots, x_n) - f(x_1, \ldots, x_n)}{h}$$

exists.

The geometric interpretation of the derivative of a function f of one variable is, of course, the slope of the tangent line to the graph of the function $y = f(x)$. See Figure 9.

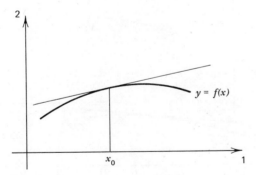

Fig. 9 Slope of tangent line $= f'(x_0)$.

For functions of two variables a similar interpretation may be given. If

(1) $$x_3 = f(x_1, x_2)$$

let S be the graph of this function, that is, the set of points $X = (x_1, x_2, x_3)$ satisfying (1). Then S is a surface in $V_3(R)$. If we set $x_2 = a$, then the set of points $\{X = (x_1, x_2, x_3) : x_2 = a\}$ is a plane π parallel to the 1, 3 plane. The intersection of this plane with the surface S defines a curve, and the slope of the tangent line to that curve is $\partial f/\partial x_1 (x_1, a)$. See Figure 10. Alternatively we may say that the partial derivative $\partial f/\partial x_1$ is the derivative of f in the direction of the vector $I_1 = (1, 0)$.

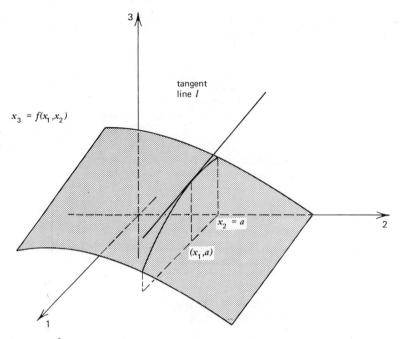

Fig. 10 $\dfrac{f}{x_1}(x_1, a)$ is the slope of the tangent line l at (x_1, a).

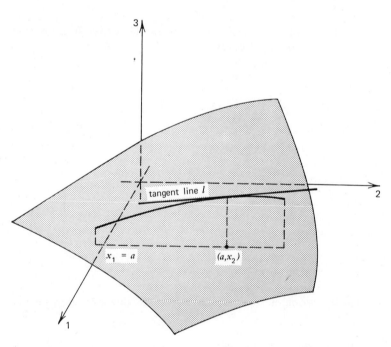

Fig. 11 $\dfrac{f}{x_2}(a, x_2)$ is slope of tangent line l at (a, x_2).

If we fix $x_1 = a$, we have a plane parallel to the 2, 3 plane, and we have a similar interpretation. See Figure 11.

Again we may think of $\partial f/\partial x_2(a, x_2)$ as the derivative of f in the direction of the vector $I_2 = (0, 1)$.

The notion of a partial derivative extends immediately to vector-valued functions. If f is a function from $V_n(R)$ to $V_m(R)$, then

$$\frac{\partial f}{\partial x_k}(X) = \lim_{h \to 0} \frac{f(x_1, \ldots, x_k+h, \ldots, x_n) - f(x_1, \ldots, x_n)}{h}$$

$$= \lim_{h \to 0} \frac{f(X+hI_k) - f(X)}{h}.$$

Just as for scalar functions, $\partial f/\partial x_k$ measures the rate of change of the vector function f in the direction of the unit coordinate vector I_k. It follows immediately from the definition of the partial derivative as a limit and Theorem 6.1 that if $f_1, \ldots, f_m$ are the coordinate functions of the vector function f, then

$$\frac{\partial f}{\partial x_k}(X) = \left(\frac{\partial f_1}{\partial x_k}(X), \ldots, \frac{\partial f_m}{\partial x_k}(X) \right).$$

Since all of the partial derivatives $\partial f/\partial x_1, \ldots, \partial f/\partial x_n$ of a function f are in turn functions defined on sets in $V_n(R)$, we may take the partial derivatives of these functions obtaining n^2 functions

$$\frac{\partial}{\partial x_1}\left(\frac{\partial f}{\partial x_1}\right), \frac{\partial}{\partial x_1}\left(\frac{\partial f}{\partial x_2}\right), \ldots, \frac{\partial}{\partial x_2}\left(\frac{\partial f}{\partial x_1}\right), \ldots, \frac{\partial}{\partial x_n}\left(\frac{\partial f}{\partial x_n}\right)$$

which are called the *second partial derivatives* of the function f. We abbreviate

$$\frac{\partial}{\partial x_k}\left(\frac{\partial f}{\partial x_j}\right) = \frac{\partial^2 f}{\partial x_k \partial x_j} \quad \text{if } k \neq j$$

$$= \frac{\partial^2 f}{\partial x_j^2} \quad \text{if } k = j.$$

It is in general not true that

$$\frac{\partial^2 f}{\partial x_k \partial x_j} = \frac{\partial^2 f}{\partial x_j \partial x_k}.$$

However, these "mixed" second partial derivatives are equal if they are known to be continuous. We state this result as our next theorem, but we omit the proof.

THEOREM 6.5 *Let f be a continuous real-valued function defined on an open set U in $V_n(R)$. If for each k and j, $\partial^2 f/\partial x_j \partial x_k$ is continuous on U, then*

$$\frac{\partial^2 f}{\partial x_j \partial x_k}(X) = \frac{\partial^2 f}{\partial x_k \partial x_j}(X).$$

Example 2. If $f(x_1, x_2) = 1/(2x_1 - x_2)$, compute the first and second partial derivatives of f. The first partial derivatives are

$$\frac{\partial f}{\partial x_1} = \frac{-2}{(2x_1 - x_2)^2}, \quad \frac{\partial f}{\partial x_2} = \frac{+1}{(2x_1 - x_2)^2}.$$

Hence

$$\frac{\partial^2 f}{\partial x_1^2} = \frac{\partial^2 f}{\partial x_1 \partial x_1} = \frac{8}{(2x_1 - x_2)^3}.$$

and

$$\frac{\partial^2 f}{\partial x_2^2} = \frac{\partial^2 f}{\partial x_2 \partial x_2} = \frac{2}{(2x_1 - x_2)^2}.$$

Moreover,

$$\frac{\partial^2 f}{\partial x_2 \partial x_1} = \frac{-4}{(2x_1 - x_2)^3}$$

and
$$\frac{\partial^2 f}{\partial x_1 \partial x_2} = \frac{-4}{(2x_1 - x_2)^3}.$$

Thus
$$\frac{\partial f}{\partial x_1}(1,1) = -2, \qquad \frac{\partial f}{\partial x_2}(1,1) = 1,$$

$$\frac{\partial^2 f}{\partial x_1^2}(1,1) = 8, \qquad \frac{\partial^2 f}{\partial x_2^2}(1,1) = 2, \quad \text{and} \quad \frac{\partial^2 f}{\partial x_2 \partial x_1} = -4.$$

Notice that the function f is clearly continuous for all points $X = (x_1, x_2)$ off the line $2x_1 - x_2 = 0$, and this is true for each of the first and second partial derivatives.

There are several notations for partial derivatives which are in current use in addition to the symbol $\partial f/\partial x_k$. For example, $D_{x_k}(f)$ or just f_{x_k}. If we write the coordinates of a vector X as (x, y, z) in place of (x_1, x_2, x_3), then we have

$$\frac{\partial f}{\partial x_1} = \frac{\partial f}{\partial x} = f_x,$$

$$\frac{\partial f}{\partial x_2} = \frac{\partial f}{\partial y} = f_y,$$

and
$$\frac{\partial f}{\partial x_3} = \frac{\partial f}{\partial z} = f_z.$$

The subscript notation is quite convenient for second and higher order partial derivatives. Writing $\partial f/\partial x_k = f_{x_k}$ we have

$$\frac{\partial^2}{\partial x_j \partial x_k} = \frac{\partial}{\partial x_j}\left(\frac{\partial f}{\partial x_k}\right) = \frac{\partial}{\partial x_j}(f_{x_k}) = f_{x_k x_j}.$$

The symbol $\partial/\partial x_j$ represents the operation of taking the partial derivative with respect to the variable x_j or, equivalently, in the direction of the vector I_j. Analogous to the dy/dx notation for ordinary derivatives, the following is quite common. If $y = f(t, x)$ then $\partial f/\partial t$ may be written $\partial y/\partial t$ or just y_t. Second-order partial derivatives then are written

$$\frac{\partial^2 f}{\partial x \partial t} = \frac{\partial}{\partial x}\left(\frac{\partial y}{\partial t}\right) = \frac{\partial^2 y}{\partial x \partial t} = y_{tx}.$$

EXERCISES

Compute $\partial f/\partial x_1$, $\partial f/\partial x_2$ for the following scalar functions.

1. $f(X) = (2x_1 + x_2)^2$.
2. $f(X) = \dfrac{x_1 + x_2}{x_1 - x_2}$.
3. $f(X) = \dfrac{x_1 \log x_2}{e^{x_1}}$.
4. $f(X) = \sin\left(\dfrac{x_1}{x_2}\right)$.
5. $f(X) = \arctan\left(\dfrac{1}{x_1 + x_2}\right)$.

Compute $f_x, f_y, f_{xx}, f_{yy},$ and f_{xy} for the following functions.

6. $f(X) = \dfrac{2x + y}{x - y}$.
7. $f(X) = x \cos y$.
8. $f(X) = e^{xy} \log(x - y)$.
9. $f(X) = \sin(ax + by)$.
10. $f(X) = \arcsin(x/y)$.

11. Consider the intersection of the ellipsoid $x_1^2 + 2x_2^2 + 4x_3^2 = 22$ with the plane

$x_1 = 2$. Compute the slope of the resulting ellipse at the points $(2, 1, 2)$. Sketch the surface and the curve. [Assume that $x_3 = f(x_2)$.]

12. Consider the intersection of the hyperboloid $4x_1^2 + x_2^2 - 3x_3^2 = 25$ with the plane $x_1 = 3$. Compute the slope of the resulting hyperbola at the points $(3, 1, \pm 2)$. [Assume that $x_3 = f(x_2)$.]

Show that $f_{xy} = f_{yx}$ for the following functions.

13. $f(X) = x\sqrt{x^2 + y^2}$.

14. $f(X) = \dfrac{y}{x + y}$.

15. $f(X) = x \sin y + y \cos x$.

16. If $f(x_1, x_2) = \dfrac{1}{x_1^2 + x_2^2} e^{x_1^2 + x_2^2}$, show that $x_2 f_{x_1}(X) - x_1 f_{x_2}(X) = 0$.

17. Let $f(x, t) = e^{x-at} \cos(x - at)$. Verify that $f_{tt} - a^2 f_{xx} = 0$.

18. Let $u = ye^x + xe^y$. Verify that $u_{xy} = u_{xxy} + u_{yyx}$.

19. If $f(x_1, x_2) = \log[(x_1 - a_1)^2 + (x_2 - a_2)^2]$ verify that for each $X = (x_1, x_2) \neq (a_1, a_2)$, $\partial^2 f/\partial x_1^2 + \partial^2 f/\partial x_2^2 = 0$. A function f which satisfies this partial differential equation on an open set U is said to be *harmonic* on U.

20. Let $f(x_1, x_2, x_3) = 1/|X - A|$, where A is a fixed vector. Verify that

$$\frac{\partial^2 f}{\partial x_1^2} + \frac{\partial^2 f}{\partial x_2^2} + \frac{\partial^2 f}{\partial x_3^2} = 0$$

for each $X \neq A$.

21. If $f(x, y) = 2xy/(x^2 + y^2)$, $f(0, 0) = 0$, convince yourself that f has partial derivatives of all orders at each point (x, y). [On the coordinate axes all partial derivatives vanish. Off the coordinate axes show that an nth partial derivative can be written $p(x, y)/(x^2 + y^2)^{2n}$ where $p(x, y)$ is a polynomial.] Conclude that all partial derivatives of f are continuous in each variable separately. Yet the function f is not continuous at the origin.

22. If $f(x_1, x_2)$ satisfies $f_{x_1 x_2} = 0$ and $g(x_1, x_2) = f(x_1, x_2) e^{x_1 + x_2}$ show that

$$g = g_{x_1} + g_{x_2} - g_{x_1 x_2}.$$

23. Let $u(r, t) = t^{-3/2} e^{r/4t}$. Show that

$$\frac{\partial u}{\partial t} = \frac{1}{r^2} \frac{\partial}{\partial r}\left(r^2 \frac{\partial u}{\partial r}\right).$$

24. Let $f(x_1, x_2) = [(x_1 + x_2)/(x_1 - x_2), e^{x_1 - x_2}]$. Compute $\partial f/\partial x_1$ and $\partial f/\partial x_2$ and evaluate these partial derivatives at $X = (2, 1)$.

25. Let $f(x_1, x_2, x_3) = (\sin(x_1/x_2), \cos x_1 x_2, \tan(x_3/x_2))$. Compute $\partial f/\partial x_1$, $\partial f/\partial x_2$, $\partial f/\partial x_3$ and evaluate these partial derivatives at $(\pi, 1, \pi)$.

6.4 Differentiable Functions

For a real-valued function f defined on an open set in $V_n(R)$ one might be tempted to call f differentiable at X_0 if all of the partial derivatives for f exist at X_0. Unfortunately, as was demonstrated by Exercise 21, p. 207, this is not enough to guarantee that f be continuous at X_0 unless, of course, $n = 1$. To guarantee the continuity of f at X_0 we need a stronger notion of differentiability. To see how to proceed for vector-valued functions let us examine the situation for $n = 1$.

Recall that a real-valued function f defined on an interval (a, b) is differentiable at the point x_0 in (a, b) if

$$\lim_{x \to x_0} \frac{f(x) - f(x_0)}{x - x_0} = l$$

exists. This may be rewritten

(1) $$\lim_{x \to x_0} \frac{|f(x) - f(x_0) - l(x - x_0)|}{|x - x_0|} = 0.$$

This formula is the one we wish to generalize. Note that if $h = x - x_0$ and l is a fixed constant then $k = lh$ defines a linear transformation from $V_1(R)$ to $V_1(R)$. Hence formula (1) expresses the fact that $f(x) - f(x_0)$ may be approximated by a linear function, namely, $l \cdot (x - x_0)$ with an error which can be made arbitrarily small even when divided by $|x - x_0|$. Now this statement makes perfectly good sense when f is a function from $V_n(R)$ to $V_m(R)$, and we take it as our definition of differentiability for vector functions.

Definition. *Let f be a function defined on an open set U in $V_n(R)$ with values in $V_m(R)$. Let $X_0 \in U$. Then f is said to be differentiable at X_0 if there exists a linear transformation L from $V_n(R)$ to $V_m(R)$ satisfying*

(2) $$\lim_{X \to X_0} \frac{|f(X) - f(X_0) - L(X - X_0)|}{|X - X_0|} = 0.$$

Thus for the vector function f to be differentiable at X_0 we require that $f(X) - f(X_0)$ can be approximated by a linear transformation L evaluated at $X - X_0$. If ϵ denotes the error in this approximation, that is,

$$\epsilon = |f(X) - f(X_0) - L(X - X_0)|,$$

then in addition we require that $\epsilon/|X - X_0|$ can be made arbitrarily small if X is chosen sufficiently close to X_0. The linear transformation L is called the *differential of f at X_0*. It, of course, depends on X_0 and f, and to indicate this dependence we write

$$L = d_{X_0} f.$$

It now may be immediately verified that differentiability at X_0 implies continuity at X_0.

THEOREM 6.6 *If the vector function f is differentiable at X_0, then f is continuous at X_0.*

Proof. We must show that

$$\lim_{X \to X_0} f(X) = f(X_0).$$

By the corollary to Theorem 6.3 and equation (2) we may infer that

$$\lim_{X \to X_0} |f(X) - f(X_0) - L(X - X_0)| = 0.$$

Since a linear transformation L is known to be continuous (corollary 2 to Theorem 6.2),

$$\lim_{X \to X_0} L(X - X_0) = 0.$$

Hence

$$\lim_{X \to X_0} f(X) = f(X_0).$$

Our next task is to determine the matrix which represents the linear transformation L with respect to the canonical basis in $V_n(R)$. Let $f_1, \ldots, f_m$ be the m coordinate functions for the vector function f.

THEOREM 6.7 *If the function f from $V_n(R)$ to $V_m(R)$ is differentiable* at X_0, then the matrix which represents the differential at X_0 is given by

(3)
$$\begin{pmatrix} \frac{\partial f_1}{\partial x_1}(X_0) & \cdots & \frac{\partial f_1}{\partial x_n}(X_0) \\ \vdots & & \vdots \\ \frac{\partial f_m}{\partial x_1}(X_0) & \cdots & \frac{\partial f_m}{\partial x_n}(X_0) \end{pmatrix}.$$

Proof. Let I_k be the kth coordinate vector in $V_n(R)$. Then

(4) $\lim_{h \to 0} \frac{f(X_0 + hI_k) - f(X_0)}{h} = \frac{\partial f}{\partial x_k}(X_0) = \left(\frac{\partial f_1}{\partial x_k}(X_0), \ldots, \frac{\partial f_m}{\partial x_k}(X_0) \right).$

Let $L = d_{X_0} f$ be the differential of f at X_0. To compute the matrix for L we must determine the scalars (a_{jk}) which satisfy

$$L(I_k) = \sum_j a_{jk} I_j.$$

By definition of the matrix for L, the kth column of this matrix is just

$$\begin{pmatrix} a_{1k} \\ \vdots \\ a_{nk} \end{pmatrix}.$$

We assert $a_{jk} = (\partial f_j / \partial x_k)(X_0)$. To prove this we evaluate (2) when $X = X_0 + hI_k$. Certainly as $h \to 0$, $X \to X_0$. Moreover

$$\frac{|f(X) - f(X_0) - L(X - X_0)|}{|X - X_0|} = \frac{|f(X_0 + hI_k) - f(X_0) - L(X_0 + hI_k - X_0)|}{|X_0 + hI - X_0|}$$

$$= \frac{|f(X_0 + hI_k) - f(X_0) - hL(I_k)|}{|h|} = \left| \frac{f(X_0 + hI_k) - f(X_0)}{h} - L(I_k) \right| \to 0$$

as $h \to 0$ since f is assumed to be differentiable. Therefore by (4)

$$L(I_k) = \frac{\partial f}{\partial x_k}(X_0) = \left(\frac{\partial f_1}{\partial x_k}(X_0), \ldots, \frac{\partial f_m}{\partial x_k}(X_0) \right) \quad \text{or} \quad a_{jk} = \frac{\partial f_j}{\partial x_k}(X_0)$$

which was to be shown.

There are several common notations for the matrix (3). We shall often use $(\partial f_i / \partial x_j)(X_0)$ or just $(\partial f_i / \partial x_j)$ if the point X_0 is understood. However, it is also convenient to write

$$f'(X_0) = \left(\frac{\partial f_i}{\partial x_j} \right)(X_0).$$

Just as in the one-variable case, we call $f'(X_0)$ the derivative of f at X_0. The matrix $(\partial f_i / \partial x_j)$ is called the *Jacobian matrix* of X_0. (Note that the row index of this matrix is i and the column index is j.) In terms of the matrix $f'(X_0)$, the condition for differentiability is just

(5) $\lim_{X \to X_0} \frac{|f(X) - f(X_0) - f'(X_0)(X - X_0)|}{|X - X_0|} = 0.$

Example 1. If $f(x_1, x_2) = (x_1 x_2, x_1^2, x_2^2)$ compute the Jacobian matrix for f.

Clearly f is a function from $V_2(R)$ to $V_3(R)$ and the matrix

$$\left(\frac{\partial f_i}{\partial x_j}\right) = \begin{pmatrix} \frac{\partial f_1}{\partial x_1} & \frac{\partial f_1}{\partial x_2} \\ \frac{\partial f_2}{\partial x_1} & \frac{\partial f_2}{\partial x_2} \\ \frac{\partial f_3}{\partial x_1} & \frac{\partial f_3}{\partial x_2} \end{pmatrix} = \begin{pmatrix} x_2 & x_1 \\ 2x_1 & 0 \\ 0 & 2x_2 \end{pmatrix}.$$

It is important to recognize that if f is a function from R to $V_n(R)$ then we may write the vector $f(x)$ as a row vector $(f_1(x), \ldots, f_n(x))$ or a column vector $\begin{pmatrix} f_1(x) \\ \vdots \\ f_n(x) \end{pmatrix}$ depending on the circumstances. However since $f'(x)$ is the matrix of a linear transformation from R to $V_n(R)$, we must write $f'(x)$ as a $n \times 1$ matrix. Thus

$$f'(x) = \begin{pmatrix} f_1'(x) \\ \vdots \\ f_n'(x) \end{pmatrix}.$$

Similarly if f is a function from $V_n(R)$ to R, then $f'(X)$ is the matrix of a linear transformation from $V_n(R)$ to R. Hence

$$f'(X) = \left(\frac{\partial f}{\partial x_1}, \ldots, \frac{\partial f}{\partial x_n}\right).$$

Example 2. If $f(t) = (t^3, t^2, t)$ then

$$f'(t) = \begin{pmatrix} 3t^2 \\ 2t \\ 1 \end{pmatrix}.$$

If we apply this linear function to a number h we obtain

$$f'(t)h = \begin{pmatrix} 3t^2 h \\ 2th \\ h \end{pmatrix}.$$

Hence

$$f'(2)h = \begin{pmatrix} 12h \\ 4h \\ h \end{pmatrix}.$$

Similarly if $f(x_1, x_2, x_3) = x_1 x_2 x_3$, then

$$f'(X) = (x_2 x_3, x_1 x_3, x_1 x_2).$$

Evaluating this linear function at a vector $Y = (y_1, y_2, y_3)$ we obtain

$$f'(X)Y = (x_2 x_3, x_1 x_3, x_1 x_2) \begin{pmatrix} y_1 \\ y_2 \\ y_3 \end{pmatrix}$$
$$= x_2 x_3 y_1 + x_1 x_3 y_2 + x_1 x_2 y_3.$$

Thus if $Y = (1, -1, 1)$ and $X = (2, 1, -1)$, we would have

$$f'(2, 1, -1)Y = -1 + 2 + 2 = 3.$$

For scalar-valued functions f, if $X = (x_1, \ldots, x_n)$ and $X_0 = (x_1^0, \ldots, x_n^0)$ then
$$f'(X_0)(X - X_0) = \frac{\partial f}{\partial x_1}(x_1 - x_1^0) + \cdots + \frac{\partial f}{\partial x_n}(x_n - x_n^0).$$

If we set $x_k - x_k^0 = dx_k$ then this expression becomes
$$f'(X_0)(X - X_0) = \frac{\partial f}{\partial x_1}dx_1 + \cdots + \frac{\partial f}{\partial x_n}dx_n.$$

The expression on the right is often called the *total differential of f*. If f is differentiable at X_0 then (5) implies that for X close to X_0, $f(X) - f(X_0)$ may be approximated by the total differential $f'(X_0)(X - X_0)$ with an error which is small even when divided by $|X - X_0|$. This is a very useful fact in many applied problems.

Example 3. A rectangular solid has length, height, width of 6, 5, 4 inches, respectively. Use the differential to estimate the error in the volume if each of these dimensions are subject to an error of .01 inches. If x, y, z denote the length, height, and width, respectively, the volume $V = xyz$. If $X = (x, y, z)$ and $X_0 = (6, 5, 4)$ then $V'(X) = (yz, xz, xy)$. Letting $x - x_0 = dx$, $y - y_0 = dy$, and $z - z_0 = dz$ the difference $V(X) - V(X_0)$ can be approximated by
$$V'(X_0)(X - X_0) = \frac{\partial V}{\partial x}dx + \frac{\partial V}{\partial y}dy + \frac{\partial V}{\partial z}dz = y_0 z_0 dx + x_0 z_0 dy + x_0 y_0 dz.$$

In our example $dx = dy = dz = 0.01$ and
$$V'(X_0)(X - X_0) = (20 + 24 + 30)(0.01) = 0.74.$$

Computing $V(X) - V(X_0)$ we obtain $V(X) - V(X_0) = 120.741501 - 120 = 0.741501$.

Lastly we note the following fact. The assertion that a vector function f is differentiable involves more than just the assertion that the Jacobian matrix exists. As we have already seen all the partial derivatives of a function may exist and the function still may not be continuous. In view of Theorem 6.6 such a function is not differentiable either. However, if all of the partial derivatives of a function f are continuous on an open set U, then f is differentiable at each point of the open set. We state this as our next theorem, but we omit the proof.

THEOREM 6.8 *Let f be a function from $V_n(R)$ to $V_m(R)$ with coordinate functions $f_1, \ldots, f_m$. If each of the partial derivatives $\partial f_i / \partial x_j$ is continuous on an open set U, then f is differentiable on U.*

A vector function f whose partial derivatives $\partial f / \partial x_i$ are all continuous on the open set U is said to be *continuously differentiable* on U. Thus Theorem 6.8 asserts that a continuously differentiable function on an open set U is differentiable on U.

EXERCISES

Compute the Jacobian matrices for the following functions.

1. $f(x_1, x_2) = (x_1 - x_2)^2$.
2. $f(x_1, x_2) = (x_1 + x_2, x_1 - x_2)$.
3. $f(x_1, x_2, x_3) = (x_1 x_3, x_2 - x_1)$.
4. $f(x_1, x_2) = \left(x_1 x_2, \frac{1}{x_1}, \frac{x_2 - x_1}{x_1 + x_2}\right)$.
5. $f(t) = (t, \sin t, \cos t)$.
6. $f(X) = |X|, X \in V_n(R)$.

7. $f(X) = (\partial g/\partial x_1(X), \ldots, \partial g/\partial x_n(X))$, $X \in V_n(R)$ where g is a scalar-valued function defined on an open set U in $V_n(R)$.

Compute the quantity

$$\frac{|f(X) - f(X_0) - f'(X_0)(X - X_0)|}{|X - X_0|}$$

for the following choices of f, X, X_0.

8. $f(x_1, x_2) = (2, x_1 - x_2)$, $X_0 = (1, 1)$, $X = (2, -1)$.
9. $f(x_1, x_2) = (x_1 x_2, x_1/x_2)$, $X_0 = (1, -1)$, $X = (2, 1)$.
10. $f(x_1, x_2) = (x_1 - x_2)^2$, $X_0 = (0, 1)$, $X = (1, 0)$.
11. $f(x_1, x_2) = (x_1 - 1, 2, (x_2 + 3)^2)$, $X_0 = (1, 1)$, $X = (0, -2)$.
12. $f(t) = (t, \sin t, \cos t)$, $t_0 = 0$, $t = \pi/2$.
13. Use the differential to estimate the error in the surface area of a rectangular solid if the length, height, and width are 5, 4, and 3 inches, respectively, and each linear dimension is subject to an error of 0.01 inches.
14. Platinum bricks are to be manufactured with linear dimensions 4, 2, and 1 inches. If the maximum allowable error in the volume is 10^{-4} cubic inches, use the differential to estimate the maximum permissible error in the linear dimensions.
15. A cylindrical soft drink can 2 inches in diameter and 4 inches high is manufactured of aluminum 0.01 inches thick. Use the differential to estimate the volume of material used in each can.
16. The range of a projectile is given by $R = g^{-1} v_0^2 \sin 2\varphi$ where φ is the angle of inclination to the horizontal, v_0 is the initial velocity and g is the gravitational constant. Assuming an initial velocity of 1000 feet per second, and an angle of 45°, estimate the error in range if there is an allowable error of 1% in the initial velocity and .01 radian in the determination of the angle of inclination. Take $g = 32$.
17. The area of the parallelogram determined by the vectors A and B is

$$|A \times B| = |A||B| \sin \theta$$

where θ is the angle between A and B. Estimate the error in the determination of the area when $|A| = 10$, $|B| = 25$, and $\theta = \pi/6$ if both $|A|$ and $|B|$ are subject to errors of 1% and θ is subject to an error of .01 radian.

18. The specific gravity S of an object can be determined from the formula

$$S = \frac{w_1}{w_1 - w_2}$$

where w_1 is the weight of the object in air and w_2 is the weight of the object when immersed in water. Assuming the errors of measurement of w_1 are 0.5% and of w_2 are 2%, estimate the maximum error in the determination of S if $w_1 = 100$ and $w_2 = 80$.

19. Let f be a constant vector function. Show that f is differentiable and that $f'(X) = 0$.
20. Let L be a linear function from V_n to V_m. Show that L is differentiable and compute the Jacobian matrix for L.
21. Suppose f, g are functions from V_n to V_m and assume f and g are differentiable at X_0. Let $h = af + bg$. Show that h is differentiable at X_0 and that $d_{X_0} h = a d_{X_0} f + b d_{X_0} g$.
21. Let

$$f(x_1, x_2) = \frac{x_1 x_2}{x_1^2 + x_2^2}$$

for $X \neq 0$ and $f(0, 0) = 0$. Show that the Jacobian matrix for f exists at $X = 0$ yet f is not differentiable at $X = 0$.

23. Show that there can be at most one linear transformation L satisfying

$$\lim_{X \to X_0} \frac{|f(X) - f(X_0) - L(X - X_0)|}{|X - X_0|} = 0.$$

6.5 Tangent Lines and Planes, Tangent Spaces

To interpret the notion of the differential geometrically let us consider two examples. First, if f is a continuous function from R to $V_2(R)$ or $V_3(R)$, then f is a curve. The requirement that f be differentiable at the point t_0 is just the statement that

$$\lim_{t \to t_0} \frac{|f(t) - f(t_0) - L(t - t_0)|}{|t - t_0|} = 0.$$

Thus near the point t_0, the function $f(t)$ may be approximated by the function $f(t_0) + L(t - t_0)$ with an error ϵ which has the property that

$$\lim_{t \to t_0} \frac{\epsilon}{|t - t_0|} = 0.$$

If $L \neq 0$, the range of the function $g(t) = f(t_0) + L(t - t_0)$ is, of course, a line in space passing through $f(t_0)$. Indeed $g(t)$ is just the tangent line to the curve $f(t)$ at $t = t_0$. Moreover the matrix for the linear transformation L is just

$$f'(t_0) = \begin{pmatrix} f_1'(t_0) \\ f_2'(t_0) \\ f_3'(t_0) \end{pmatrix}.$$

We may write

$$g(t) = f(t_0) + L(t - t_0)$$

in vector form as

$$g(t) = \begin{pmatrix} f_1(t_0) \\ f_2(t_0) \\ f_3(t_0) \end{pmatrix} + \begin{pmatrix} f_1'(t_0) \\ f_2'(t_0) \\ f_3'(t_0) \end{pmatrix} (t - t_0),$$

or in terms of row vectors,

$$g(t) = (f_1(t_0), f_2(t_0), f_3(t_0)) + (t - t_0)(f_1'(t_0), f_2'(t_0), f_3'(t_0)).$$

The vector $f'(t) = (f_1'(t_0), f_2'(t_0), f_3'(t_0))$ is called the *velocity vector* for f at the point t_0. The vector $f'(t_0)/|f'(t_0)|$ is the *unit tangent* to the curve at $f(t_0)$. Thus the tangent line to the curve at $t = t_0$ is the line passing through the point $f(t_0)$ which is parallel to the vector $f'(t_0)$. See Figure 12.

If f is a curve in the plane and $f(t) = (t, g(t))$ where g is a scalar-valued function of one variable, then the matrix for the differential L is $\begin{pmatrix} 1 \\ g'(t_0) \end{pmatrix}$. Hence

$$\frac{|f(t) - f(t_0) - L(t - t_0)|}{|t - t_0|} = \frac{\left| \begin{pmatrix} t \\ g(t) \end{pmatrix} - \begin{pmatrix} t_0 \\ g(t_0) \end{pmatrix} - \begin{pmatrix} 1 \\ g'(t_0) \end{pmatrix}(t - t_0) \right|}{|t - t_0|}$$

$$= \frac{\left| \begin{pmatrix} 0 \\ g(t) - g(t_0) - g'(t_0)(t - t_0) \end{pmatrix} \right|}{|t - t_0|}$$

$$= \frac{|g(t) - g(t_0) - g'(t_0)(t - t_0)|}{|t - t_0|}.$$

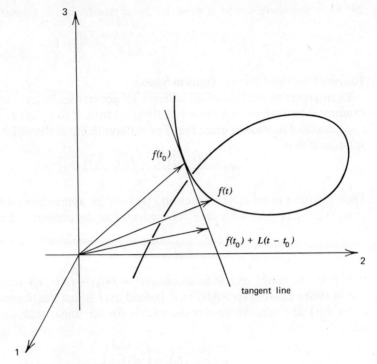

Fig. 12 Tangent line to a curve in space.

The numerator in the above expression is just the vertical distance $d(t)$ between the tangent line to the curve at t_0 and the function $f(t)$. See Figure 13. Hence f is differentiable at t_0 if

$$\lim_{t \to t_0} \frac{d(t)}{|t - t_0|} = 0.$$

This just means that g is a differentiable function of one variable.

Next let f be a continuous function defined on an open set U in $V_2(R)$ with values in $V_3(R)$. The requirement that f be differentiable at X_0 means that $f(X) - f(X_0)$ may be approximated by the linear function $L(X - X_0)$ with an error $\epsilon = |f(X) - f(X_0) - L(X - X_0)|$ such that $(\epsilon/|X - X_0|) \to 0$ as $X \to X_0$. If we set $g(X) = f(X_0) + L(X - X_0)$, then the set of vectors $Y = g(X)$, or the image of the function g, is called the *tangent space* to the function f at X_0. Such a space is called an *affine subspace* since it is obtained by

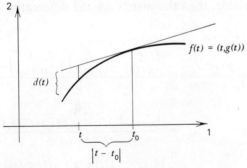

Fig. 13

TANGENT LINES AND PLANES, TANGENT SPACES

translating each of the vectors in the subspace spanned by the vectors $L(X-X_0)$ by the vector $f(X_0)$. The dimension of this tangent space is the dimension of the subspace spanned by the vectors $L(X-X_0)$. To compute the dimension of this subspace observe that the differential $L=d_{x_0}f$ has matrix

$$f'(X_0) = \begin{pmatrix} \dfrac{\partial f_1}{\partial x_1} & \dfrac{\partial f_1}{\partial x_2} \\ \dfrac{\partial f_2}{\partial x_1} & \dfrac{\partial f_2}{\partial x_2} \\ \dfrac{\partial f_3}{\partial x_1} & \dfrac{\partial f_3}{\partial x_2} \end{pmatrix}(X_0)$$

where f_1, f_2, f_3 are the component functions for f. Therefore

$$L(X-X_0) = \begin{pmatrix} \dfrac{\partial f_1}{\partial x_1} & \dfrac{\partial f_1}{\partial x_2} \\ \dfrac{\partial f_2}{\partial x_1} & \dfrac{\partial f_2}{\partial x_2} \\ \dfrac{\partial f_3}{\partial x_1} & \dfrac{\partial f_3}{\partial x_2} \end{pmatrix} \begin{pmatrix} x_1 - x_1^0 \\ x_2 - x_2^0 \end{pmatrix}$$

$$= (x_1 - x_1^0)\begin{pmatrix} \dfrac{\partial f_1}{\partial x_1} \\ \dfrac{\partial f_2}{\partial x_1} \\ \dfrac{\partial f_3}{\partial x_1} \end{pmatrix} + (x_2 - x_2^0)\begin{pmatrix} \dfrac{\partial f_1}{\partial x_2} \\ \dfrac{\partial f_2}{\partial x_2} \\ \dfrac{\partial f_3}{\partial x_2} \end{pmatrix}$$

$$= (x_1 - x_1^0)\dfrac{\partial f}{\partial x_1} + (x_2 - x_2^0)\dfrac{\partial f}{\partial x_2}.$$

Hence the tangent space is spanned by the vectors $\partial f/\partial x_1$ and $\partial f/\partial x_2$. If these vectors are linearly independent, this space is a plane. If the partial derivatives are dependent but not both zero, we have a line. In case $\partial f/\partial x_1$ and $\partial f/\partial x_2$ are linearly independent for each X_0 in U, we say that the function f is a *two-dimensional surface* defined on U. These partial derivatives will be linearly independent at X_0, of course, if and only if the rank of the matrix $f'(X_0)$ equals two.

One situation when this is always true is when

$$f(x_1, x_2) = (x_1, x_2, g(x_1, x_2)).$$

and g is a scalar-valued function. Then

$$f'(X) = \begin{pmatrix} 1 & 0 \\ 0 & 1 \\ \dfrac{\partial g}{\partial x_1} & \dfrac{\partial g}{\partial x_2} \end{pmatrix}$$

At the point $X_0 = (x_1^0, x_2^0)$ the tangent plane has vector equation

$$Y = f(X_0) + L(X - X_0)$$

or
$$\begin{pmatrix} y_1 \\ y_2 \\ y_3 \end{pmatrix} = \begin{pmatrix} x_1^0 \\ x_2^0 \\ g(x_1^0, x_2^0) \end{pmatrix} + \begin{pmatrix} 1 & 0 \\ 0 & 1 \\ \frac{\partial g}{\partial x_1} & \frac{\partial g}{\partial x_2} \end{pmatrix} \begin{pmatrix} x_1 - x_1^0 \\ x_2 - x_2^0 \end{pmatrix}$$
$$= \begin{pmatrix} x_1 \\ x_2 \\ g(x_1^0, x_2^0) + \frac{\partial g}{\partial x_1}(x_1 - x_1^0) + \frac{\partial g}{\partial x_2}(x_2 - x_2^0) \end{pmatrix}.$$

Thus $Y = f(X_0) + L(X - X_0)$ if $y_1 = x_1$, $y_2 = x_2$ and
$$y_3 = g(x_1^0, x_2^0) + \frac{\partial g}{\partial x_1}(x_1 - x_1^0) + \frac{\partial g}{\partial x_2}(x_2 - x_2^0).$$

Denoting the coordinates of points in $V_3(R)$ by (x_1, x_2, x_3) in place of (y_1, y_2, y_3) we see that the tangent plane to the surface at X_0 has equation

(2) $\quad -\frac{\partial g}{\partial x_1}(x_1 - x_1^0) - \frac{\partial g}{\partial x_2}(x_2 - x_2^0) + x_3 = g(x_1^0, x_2^0).$

The partial derivatives $\partial g/\partial x_1$ and $\partial g/\partial x_2$ are, of course, evaluated at the point (x_1^0, x_2^0).

Example 1. Determine the equation of the tangent plane to the hyperbolic paraboloid
$$x_3 = 2x_2^2 - x_1^2$$
at the point $(1, 1)$. If $g(x_1, x_2) = 2x_2^2 - x_1^2$, then $\partial g/\partial x_1 = -2x_1$ and $\partial g/\partial x_2 = 4x_2$ and $g(1, 1) = 1$. Therefore the scalar equation of the tangent plane is
$$x_3 + 2(x_1 - 1) - 4(x_2 - 1) = 1$$
or
$$2x_1 - 4x_2 + x_3 = -1.$$

Now a plane π in space may be described in two ways. It may be described as the affine space spanned by two linearly independent vectors T_1 and T_2. Alternatively we may compute a normal vector N to the plane π and describe π as the plane through Y_0 perpendicular to N. Moreover we know from Chapter 1 that the normal vector $N = T_1 \times T_2$. Thus we may describe the tangent plane to a function f at the point X_0 as the plane through $f(X_0)$ perpendicular to $\partial f/\partial x_1 \times \partial f/\partial x_2$. The vector $\partial f/\partial x_1 \times \partial f/\partial x_2$ is called the *normal vector* to the surface at the point $f(X_0)$. In the case that
$$f(x_1, x_2) = (x_1, x_2, g(x_1, x_2)),$$
$$\frac{\partial f}{\partial x_1} = \begin{pmatrix} 1 \\ 0 \\ \frac{\partial g}{\partial x_1} \end{pmatrix} \quad \text{and} \quad \frac{\partial f}{\partial x_2} = \begin{pmatrix} 0 \\ 1 \\ \frac{\partial g}{\partial x_2} \end{pmatrix}.$$

Hence
$$N = \frac{\partial f}{\partial x_1} \times \frac{\partial f}{\partial x_2} = D\begin{pmatrix} I_1 & I_2 & I_3 \\ 1 & 0 & \frac{\partial g}{\partial x_1} \\ 0 & 1 & \frac{\partial g}{\partial x_2} \end{pmatrix} = \left(-\frac{\partial g}{\partial x_1}, -\frac{\partial g}{\partial x_2}, 1\right).$$

Of course, this fact could have been obtained from formula (2) as well.

TANGENT LINES AND PLANES, TANGENT SPACES

When $f(x_1, x_2) = (x_1, x_2, g(x_1, x_2))$, the error of approximation by the differential may be given a simple geometric interpretation. If we let $\epsilon(X)$ be the error as a function of X, then

$$\epsilon(X) = |f(X) - f(X_0) - f'(X_0)(X - X_0)|$$
$$= \left| g(X) - g(X_0) - \frac{\partial g}{\partial x_1}(x_1 - x_1^0) - \frac{\partial g}{\partial x_2}(x_2 - x_2^0) \right|$$
$$= |g(X) - g(X_0) - g'(X_0)(X - X_0)|.$$

The $\epsilon(X)$ is the vertical distance at X from the surface defined by g to the tangent plane. See Figure 14. Thus f, or equivalently g, is differentiable if and only if

$$\lim_{X \to X_0} \frac{\epsilon(X)}{|X - X_0|} = \lim_{X \to X_0} \frac{|g(X) - g(X_0) - g'(X_0)(X - X_0)|}{|X - X_0|} = 0.$$

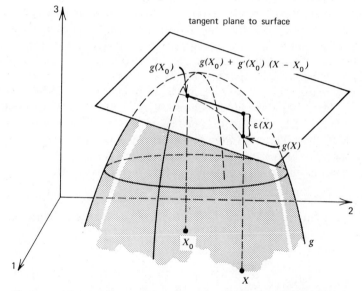

Fig. 14 $\epsilon(X) =$ vertical distance between tangent plane and surface = $|g(X) - g(X_0) - g'(X_0)(X - X_0)|$.

Example 2. If $f(x_1, x_2) = (x_1 + x_2, x_1 - x_2, x_1^2 x_2)$ determine the scalar equation of the tangent plane to the surface at the point $X_0 = (1, 1)$. First

$$f'(X) = \begin{pmatrix} 1 & 1 \\ 1 & -1 \\ 2x_1 x_2 & x_1^2 \end{pmatrix}.$$

Hence

$$f(X_0) = \begin{pmatrix} 2 \\ 0 \\ 1 \end{pmatrix} \quad \text{and} \quad f'(X_0) = \begin{pmatrix} 1 & 1 \\ 1 & -1 \\ 2 & 1 \end{pmatrix}.$$

Therefore

$$\begin{pmatrix} y_1 \\ y_2 \\ y_3 \end{pmatrix} = \begin{pmatrix} 2 \\ 0 \\ 1 \end{pmatrix} + \begin{pmatrix} 1 & 1 \\ 1 & -1 \\ 2 & 1 \end{pmatrix} \begin{pmatrix} x_1 - 1 \\ x_2 - 1 \end{pmatrix}$$

or

$$y_1 = x_1 + x_2$$
$$y_2 = x_1 - x_2$$
$$y_3 = 2x_1 + x_2 - 2.$$

We may eliminate the parameters x_1, x_2 in these three scalar equations to obtain the desired scalar equation for the tangent plane. A simpler technique is to observe that

$$N = \begin{pmatrix} 1 \\ 1 \\ 2 \end{pmatrix} \times \begin{pmatrix} 1 \\ -1 \\ 1 \end{pmatrix} = \begin{pmatrix} 3 \\ 1 \\ -2 \end{pmatrix}$$

is a normal vector to the plane. Hence the equation has the form

$$3y_1 + y_2 - 2y_3 = c.$$

Since $(2, 0, 1)$ lies on the plane, $c = 4$. Hence the desired equation is

$$3y_1 + y_2 - 2y_3 = 4.$$

We close this section with some remarks on the notion of tangency for a differentiable function mapping an open set U in $V_n(R)$ into $V_m(R)$. The set of vectors $Y = f(X_0) + f'(X - X_0)$ is called the *tangent space* to f at X_0 and the affine subspace $f'(X_0) + f'(X_0)(X - X_0)$ is spanned by the vectors $\partial f/\partial x_1, \ldots, \partial f/\partial x_n$. If the matrix $f'(X_0)$ has rank n at each point X_0 in U, then we say f is an n-dimensional surface defined on U. Just as in three-space, we may describe the tangent space to f in two ways. It is, of course, the affine space through $f(X_0)$ spanned by the vectors $\partial f/\partial x_1, \ldots, \partial f/\partial x_n$. Alternatively, if $m > n$ and $\partial f/\partial x_1, \ldots, \partial f/\partial x_n$ are linearly independent, then we may choose $m - n$ linearly independent vectors $N_1, \ldots, N_{m-n}$ each perpendicular to the subspace $f'(X_0)(X - X_0)$. The vectors $\partial f/\partial x_1, \ldots, \partial f/\partial x_n, N_1, \ldots, N_{m-n}$ span $V_m(R)$. Thus we may describe the tangent space as the affine space through $f(X_0)$ perpendicular to the vectors $N_1, \ldots, N_{m-n}$. In Section 6.11 we shall have more to say about these normal vectors N_k.

EXERCISES

For the following functions f from R to $V_n(R)$ determine the equation of the tangent line at the specified value of t.

1. $f(t) = (t^2 - 1, t^3 + 4)$, $\quad t = 2$.
2. $f(t) = (2 \sin t, 4 \cos t)$, $\quad t = \pi/2$.
3. $f(t) = (t^2, t - 1, t^3)$, $\quad t = 1$.
4. $f(t) = (t, \sin t, \cos t)$, $\quad t = \pi/4$.
5. $f(t) = (1, t, t^2, t^3)$, $\quad t = -1$.

Determine the scalar equation of the tangent planes to the surfaces defined by $f(x_1, x_2) = (x_1, x_2, g(x_1, x_2))$ at the designated points X_0 if:

6. $g(x_1, x_2) = 2x_1^2 - x_2^2$, $\quad X_0 = (1, -1)$.
7. $g(x_1, x_2) = 4x_1^2 - x_1 + x_2^3$, $\quad X_0 = (2, 1)$.
8. $g(x_1, x_2) = \dfrac{x_1 - x_2}{x_1 + x_2}$, $\quad X_0 = (1, 1)$.
9. If $x_1 x_2 + x_2 x_3 + x_3 x_1 = 1$ determine the equation of the tangent plane to the surface at the point $(1, 1, 0)$.

Determine the equations of the tangent planes to the surfaces defined by the following functions f at the indicated points X_0.

10. $f(x_1, x_2) = (x_1 x_2, x_1 + x_2, (x_1 + x_2)^2)$, $\quad X_0 = (1, -1)$.
11. $f(x_1, x_2) = \left(x_1 x_2^2, \dfrac{1}{x_1}, \dfrac{1}{x_2}\right)$, $\quad X_0 = (1, 1)$.
12. $f(x_1, x_2) = (x_1 \cos x_2, x_2 \sin x_1, x_1)$, $\quad X_0 = \left(\dfrac{\pi}{2}, \pi\right)$.
13. $f(x_1, x_2) = (\cos x_1 \sin x_2, \sin x_1 \cos x_2, \cos x_2)$, $\quad X_0 = \left(\dfrac{\pi}{4}, \dfrac{\pi}{2}\right)$.

14. Determine a normal vector to the surface defined by the function $f(x_1, x_2) = (x_1 x_2, x_1^2, x_2^2)$ at the point $(2, -3)$.
15. If f is a differentiable function from $V_2(R)$ to $V_3(R)$ show that the vector

$$\left(D\begin{pmatrix} \frac{\partial f_2}{\partial x_1} & \frac{\partial f_2}{\partial x_2} \\ \frac{\partial f_3}{\partial x_1} & \frac{\partial f_3}{\partial x_2} \end{pmatrix}, -D\begin{pmatrix} \frac{\partial f_1}{\partial x_1} & \frac{\partial f_1}{\partial x_2} \\ \frac{\partial f_3}{\partial x_1} & \frac{\partial f_3}{\partial x_2} \end{pmatrix}, D\begin{pmatrix} \frac{\partial f_1}{\partial x_1} & \frac{\partial f_1}{\partial x_2} \\ \frac{\partial f_2}{\partial x_1} & \frac{\partial f_2}{\partial x_2} \end{pmatrix} \right)$$

is normal to the surface at each point.
16. Determine the general form of the equation of the tangent plane to a surface defined by a function from $V_2(R)$ to $V_3(R)$.
17. A differentiable function f from $V_n(R)$ to $V_{n+1}(R)$ is called a *hypersurface* if $f'(X_0)$ always has rank n. The affine function

$$Y = f(X_0) + f'(X_0)(X - X_0)$$

is called the *tangent hyperplane* to the surface f at the point X_0. If $f(x_1, x_2, x_3) = (x_1 x_2, x_2 x_3, x_1 x_3, x_1 + x_2 + x_3)$ determine the vector equation of the tangent hyperplane to f at $X_0 = (1, 1, 1)$.
18. Determine a normal vector to the tangent hyperplane of Exercise 17 at $X_0 = (1, 1, 1)$. Hence determine a scalar equation for the tangent hyperplane to the surface at this point.
19. If $g(x_1, \ldots, x_n)$ is a differentiable scalar function defined on an open set U in $V_n(R)$, show that $f(x_1, \ldots, x_n) = (x_1, x_2, \ldots, x_n, g(x_1, \ldots, x_n))$ always defines a hypersurface in $V_{n+1}(R)$.
20. Determine a general expression for the normal vector to the hypersurface defined in Exercise 19.
21. Let f be a differentiable function from $V_n(R)$ to $V_m(R)$ such that $f'(X)$ always has rank n. Let $N_1, \ldots, N_{m-n}$ be linearly independent vectors each perpendicular to the tangent space to f at the point X. If $Y \in V_m(R)$ is perpendicular to the tangent space at the point X, show that there exist scalars $a_1, \ldots, a_{m-n}$ such that

$$Y = a_1 N_1 + \cdots + a_{m-n} N_{m-n}.$$

6.6 Directional Derivatives, the Mean Value Theorem

If f is a vector function, then we have seen that the kth partial derivative $\partial f / \partial x_k$ measures the rate of change of f in the direction of the kth coordinate vector I_k. A natural question to ask is the following. If a vector Y is specified, how does one measure the rate of change of the function f at the point X in the direction of Y? For this we need the notion of the derivative of a function f with respect to a vector Y.

Definition. *Let f be a function from $V_n(R)$ to $V_m(R)$ defined on the open set U. If $X, Y \subset V_n(R)$ and $X \in U$, then the derivative of f at X with respect to Y, which we write $\partial f / \partial Y(X)$, is defined by the following formula*

(1) $$\frac{\partial f}{\partial Y}(X) = \lim_{h \to 0} \frac{f(X + hY) - f(X)}{h}.$$

The function $\partial f / \partial Y$ has values in $V_m(R)$ and the domain of this function is just the set of points X for which

$$\lim_{h \to 0} \frac{f(X + hY) - f(X)}{h}$$

exists. The partial derivatives of f are just the special cases corresponding to

$Y = I_1, \ldots, I_n$. Indeed

$$\frac{\partial f}{\partial x_1}(X) = \lim_{h \to 0} \frac{f(x_1 + h, x_2, \ldots, x_n) - f(x_1, \ldots, x_n)}{h}$$

$$= \lim_{h \to 0} \frac{f(X + hI_1) - f(X)}{h} = \frac{\partial f}{\partial I_1}(X).$$

and

$$\frac{\partial f}{\partial x_k}(X) = \lim_{h \to 0} \frac{f(X + hI_k) - f(X)}{h} = \frac{\partial f}{\partial I_k}(X).$$

The actual computation of $\partial f/\partial Y(X)$ directly from the definition proceeds very much as in the one-variable case.

Example 1. Let $f(x_1, x_2, x_3) = x_1 x_2 x_3$.
Compute $\partial f/\partial Y(X)$ if $Y = (1, 2, 3)$. Now $X + hY = (x_1 + h, x_2 + 2h, x_3 + 3h)$. Hence $f(X + hY) = (x_1 + h)(x_2 + 2h)(x_3 + 3h)$. Therefore

$$\frac{f(X + hY) - f(X)}{h} = \frac{(x_1 + h)(x_2 + 2h)(x_3 + 3h) - x_1 x_2 x_3}{h}$$

$$= \frac{h(x_2 x_3 + 2x_1 x_3 + 3x_1 x_2) + h^2(2x_3 + 3x_2 + 6x_1) + 6h^3}{h}$$

$$= x_2 x_3 + 2x_1 x_3 + 3x_1 x_2 + h(2x_3 + 3x_2 + 6x_1) + 6h^2$$
$$\to x_2 x_3 + 2x_1 x_3 + 3x_1 x_2 \quad \text{as} \quad h \to 0.$$

Therefore if $Y = (1, 2, 3)$

$$\frac{\partial f}{\partial Y}(X) = x_2 x_3 + 2x_1 x_3 + 3x_1 x_2.$$

Next we observe that the derivative of a vector function f with respect to a scalar multiple of a vector Y is the product of the scalar with the derivative $\partial f/\partial Y$.

PROPOSITION 6.9 *If $\partial f/\partial Y(X)$ exists, then for each real number a, $\partial f/\partial (aY)(X)$ exists, and furthermore*

$$\frac{\partial f}{\partial (aY)}(X) = a \frac{\partial f}{\partial Y}(X).$$

Proof. If $a = 0$,

$$\frac{f(X + haY) - f(X)}{h} = \frac{f(X) - f(X)}{h} = 0.$$

Therefore

$$\lim_{h \to 0} \frac{f(X + haY) - f(X)}{h} = 0,$$

or $\partial f/\partial (aY) = 0$. If $a \neq 0$,

$$\frac{f(X + haY) - f(X)}{h} = a \cdot \frac{f(X + haY) - f(X)}{ha}.$$

Therefore

$$\frac{\partial f}{\partial (aY)}(X) = \lim_{h \to 0} \frac{f(X + haY) - f(X)}{h}$$

$$= a \cdot \lim_{ha \to 0} \frac{f(X + haY) - f(X)}{ha} = a \cdot \frac{\partial f}{\partial Y}(X).$$

If $Y \neq 0$ and we choose a unit vector J in the direction of Y then we define $\partial f/\partial J$ as the *derivative of f in the direction of Y*. We state this formally as a definition.

Definition. *Let f be defined on the open set U and let $X \in U$. Then we define the derivative of f at X in the direction of the nonzero vector Y to be*

$$\frac{\partial f}{\partial(Y/|Y|)}(X) = \lim_{h \to 0} \frac{f\left(X + h\frac{Y}{|Y|}\right) - f(X)}{h}.$$

In light of Proposition 6.9 we have

$$\frac{\partial f}{\partial(Y/|Y|)} = \frac{1}{|Y|} \frac{\partial f}{\partial Y}.$$

Example 2. If $f(X) = x_1 x_2 x_3$ and $Y = (1, 2, 3)$, we saw in the previous example that $\partial f/\partial Y(X) = x_2 x_3 + 2 x_1 x_3 + 3 x_1 x_2$. The derivative of f at X in the direction of Y would be

$$\frac{\partial f}{\partial(Y/|Y|)}(X) = \frac{1}{\sqrt{14}}(x_2 x_3 + 2 x_1 x_3 + 3 x_1 x_2).$$

A derivative $\partial f/\partial(Y/|Y|)(X)$ is called a *directional derivative*, and it measures the rate of change of f in the direction of the vector Y. If for example $x_3 = f(x_1, x_2)$ then the graph of this function is a surface in space. Intersect this surface with a plane π perpendicular to the 1, 2 coordinate plane and let Y be a vector along the line of intersection of π and the 1, 2 plane. Now the intersection of the plane π with the surface is a curve, and $\partial f/\partial(Y/|Y|)(X)$ is just the slope of this curve at the point X. See Figure 15.

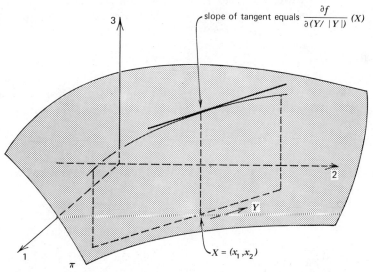

Fig. 15 A vector along the line of intersection of π and the 1, 2 plane.

If the partial derivatives $\partial f/\partial x_1, \ldots, \partial f/\partial x_n$ for f exist at the point X, then one would like to conclude that all the directional derivatives exist for f. Unfortunately this is not true. An example that illustrates this is the following. Let

$$f(x_1, x_2) = \frac{x_1 x_2}{x_1^2 + x_2^2} \quad \text{if} \quad X \neq 0, f(0, 0) = 0.$$

We have already seen that $\partial f/\partial x_1 = \partial f/\partial x_2 = 0$ at the origin. Yet if $Y = (1,1)$

$$\frac{\partial f}{\partial Y}(0) = \lim_{h \to 0} \frac{f(0+hY)-f(0)}{h} = \lim_{h \to 0} \frac{f(hY)}{h}$$

does not exist. To see this observe that

$$\frac{f(hY)}{h} = \frac{h \cdot h}{(h^2+h^2)h} = \frac{1}{2h}$$

which certainly has no limit as $h \to 0$.

However if f is differentiable at X, then $\partial f/\partial Y(X)$ does exist at X for each choice of the vector Y. Furthermore, for differentiable functions f we have a very convenient formula for computing $\partial f/\partial Y(X)$. Indeed we show in our next result that

(2) $$\frac{\partial f}{\partial Y}(X) = (d_X f)(Y) = f'(X)Y.$$

Thus $\partial f/\partial Y(X)$ may be computed by applying the Jacobian matrix of f at the point X to the vector Y. Thus in the previous example where $f(x_1, x_2, x_3) = x_1 x_2 x_3$ and $Y = (1, 2, 3)$ we have $f'(X) = (x_2 x_3, x_1 x_3, x_1 x_2)$. Hence

$$\frac{\partial f}{\partial Y}(X) = f'(X)Y = (x_2 x_3, x_1 x_3, x_1 x_2)\begin{pmatrix}1\\2\\3\end{pmatrix}$$

$$= x_2 x_3 + 2 x_1 x_3 + 3 x_1 x_2.$$

We now establish formula (2).

THEOREM 6.10 *Let f be a function from $V_n(R)$ to $V_m(R)$ which is defined on the open set U. If f is differentiable at the point $X_0 \in U$, then $\partial f/\partial Y(X_0)$ exists for each $Y \in V_n(R)$. Furthermore*

(2) $$\frac{\partial f}{\partial Y}(X_0) = f'(X_0)Y.$$

Proof. If $Y = 0$, then it follows immediately from the definition that $\partial f/\partial Y(X_0) = 0 = f'(X_0)Y$. In general, f differentiable at X_0 implies that

$$\lim_{X \to X_0} \frac{|f(X)-f(X_0)-f'(X_0)(X-X_0)|}{|X-X_0|} = 0.$$

But if $Y \neq 0$, set $X = X_0 + hY$. Then $X - X_0 = hY$ and as $h \to 0$, $X \to X_0$. Hence

$$\frac{|f(X)-f(X_0)-f'(X_0)(X-X_0)|}{|X-X_0|} = \frac{|f(X_0+hY)-f(X_0)-f'(X_0)(hY)|}{|h||Y|}$$

$$= \frac{1}{|Y|}\left|\frac{f(X_0+hY)-f(X_0)}{h} - \frac{1}{h}f'(X_0)(hY)\right|$$

$$= \frac{1}{|Y|}\left|\frac{f(X_0+hY)-f(X_0)}{h} - f'(X_0)Y\right| \to 0.$$

Hence $\partial f/\partial Y(X_0) = f'(X_0)Y$ and we are done.

Example 3. Let $f(x_1, x_2) = x_1 x_2 + x_1 - x_2$. Compute the directional derivatives of f at $X = (1, 2)$ in the direction of $Y = (2, -2)$. We must compute $\partial f/\partial Z(X)$ where

$$Z = \frac{Y}{|Y|} = \frac{1}{2\sqrt{2}}(2, -2) = \frac{1}{\sqrt{2}}(1, -1).$$

Hence by Theorem 6.10

$$\frac{\partial f}{\partial Z}(X) = f'(X)Z = \left(\frac{\partial f}{\partial x_1}, \frac{\partial f}{\partial x_2}\right)\binom{z_1}{z_2} = \frac{1}{\sqrt{2}}\frac{\partial f}{\partial x_1} - \frac{1}{\sqrt{2}}\frac{\partial f}{\partial x_2}.$$

Since $\partial f/\partial x_1 = x_2 + 1$ and $\partial f/\partial x_2 = x_1 - 1$,

$$\frac{\partial f}{\partial Z}(X) = \frac{1}{\sqrt{2}}(x_2 + 1) - \frac{1}{\sqrt{2}}(x_1 - 1) = \frac{1}{\sqrt{2}}(x_2 - x_1 + 2).$$

Hence if $X = (1, 2)$

$$\frac{\partial f}{\partial Z}(X) = \frac{3}{\sqrt{2}}.$$

Example 4. Let $f(x_1, x_2) = (x_1 + x_2, x_1 x_2)$. Compute the directional derivative of f at $(1, -1)$ toward the point $(2, 1)$. A vector in this direction is clearly

$$Y = (2, 1) - (1, -1) = (1, 2).$$

Letting $Z = Y/|Y| = 1/\sqrt{5}\,(1, 2)$ we must compute $\partial f/\partial Z\,(1, -1)$. But

$$f'(X) = \begin{pmatrix} 1 & 1 \\ x_2 & x_1 \end{pmatrix}.$$

Hence

$$f'(X)Z = \begin{pmatrix} 1 & 1 \\ x_2 & x_1 \end{pmatrix}\begin{pmatrix} \frac{1}{\sqrt{5}} \\ \frac{2}{\sqrt{5}} \end{pmatrix} = \frac{1}{\sqrt{5}}\begin{pmatrix} 3 \\ x_2 + 2x_1 \end{pmatrix}.$$

Therefore if $X = (1, -1)$ we have

$$\frac{\partial f}{\partial Z}(X) = f'(X)Z = \frac{1}{\sqrt{5}}\begin{pmatrix} 3 \\ 1 \end{pmatrix}.$$

We close this section with a mean value theorem for differentiable scalar-valued functions defined on open sets in $V_n(R)$. First let us recall the mean value theorem for functions of one variable. This states that if $f(x)$ is continuous on the closed interval $[a, b]$ and differentiable at each interior point, then there is a point x_0, $a < x_0 < b$, such that

(3) $$f(b) - f(a) = f'(x_0)(b - a).$$

For scalar-valued functions defined on open sets in $V_n(R)$ we have a similar formula. Namely, if f is differentiable along the line joining two points A and B, then there exists a point X_0 on the interval joining A and B such that

(4) $$f(B) - f(A) = f'(X_0)(B - A).$$

Formulas (3) and (4) appear formally the same. Now, however, $f'(X_0)$ is a $1 \times n$ matrix and $B - A$ is a vector in $V_n(R)$. Indeed, since

$$f'(X_0) = \left(\frac{\partial f}{\partial x_1}(X_0), \ldots, \frac{\partial f}{\partial x_n}(X_0)\right),$$

if $A = (a_1, \ldots, a_n)$ and $B = (b_1, \ldots, b_n)$ then

$$f'(X_0)(B - A) = \left(\frac{\partial f}{\partial x_1}(X_0), \ldots, \frac{\partial f}{\partial x_n}(X_0)\right)\begin{pmatrix} b_1 - a_1 \\ \vdots \\ b_n - a_n \end{pmatrix}.$$

The matrix $f'(X_0)$ is called the mean value of the derivative $f'(X)$ along the interval $[A, B]$.

224 THE CALCULUS OF VECTOR FUNCTIONS

THEOREM 6.11 *Let f be a real-valued function defined on an open set U in $V_n(R)$. Let A, B be two points in U such that the closed line segment $[A, B]$ joining A and B lies in U. Assume f is differentiable at each point X on this line segment. Then there exists an interior point X_0 of $[A, B]$ such that*

(3) $$f(B) - f(A) = f'(X_0)(B - A)$$

Proof. We establish (3) by defining an appropriate function $g(t)$ defined on the closed unit interval $[0, 1]$ to which the ordinary mean value theorem applies. Recall that a vector X lies on the line joining A and B if

$$X = A + t(B - A)$$

and this vector will lie on the closed line segment joining A and B if and only if $0 \leq t \leq 1$. See Figure 16.

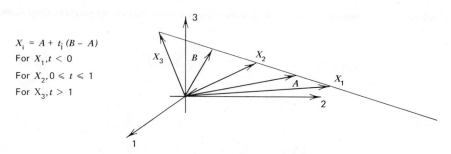

$X_i = A + t_i (B - A)$
For $X_1, t < 0$
For $X_2, 0 \leq t \leq 1$
For $X_3, t > 1$

Fig. 16

Define $g(t) = f(A + t(B - A))$, $0 \leq t \leq 1$. Clearly $g(0) = f(A)$ and $g(1) = f(B)$. We assert that g is continuous on the closed interval $[0, 1]$ and differentiable on the open interval $(0, 1)$ which are the hypotheses of the mean value theorem for functions of one variable. To prove this it is enough to show that

$$\lim_{h \to 0} \frac{g(t+h) - g(t)}{h}$$

exists for each t, $0 \leq t \leq 1$. But letting $X = A + t(B - A)$ we have

$$\frac{g(t+h) - g(t)}{h} = \frac{f(A + (t+h)(B-A)) - f(A + t(B-A))}{h}$$

$$= \frac{f(X + h(B-A)) - f(X)}{h}.$$

Therefore

$$\lim_{h \to 0} \frac{g(t+h) - g(T)}{h} = \lim_{h \to 0} \frac{f(X + h(B-A)) - f(X)}{h} = \frac{\partial f}{\partial (B-A)}(X).$$

But by Theorem 6.10

$$\frac{\partial f}{\partial (B-A)}(X) = f'(X)(B-A)$$

if f is differentiable at X. Hence for each X on the closed segment $[A, B]$ the above limit exists, and consequently g is differentiable for each t, $0 \leq t \leq 1$. Applying the mean value theorem to g, we infer that for some value of t_0, $0 < t_0 < 1$,

$$g(1) - g(0) = g'(t_0).$$

But
$$g(1) - g(0) = f(B) - f(A)$$
and
$$g'(t_0) = f'(X_0)(B-A)$$
where $X_0 = A + t_0(B-A)$. Hence
$$f(B) - f(A) = f'(X_0)(B-A)$$
as required.

Example 5. Let $A = (1, 1, 0)$, $B = (1, 0, -1)$, and $f(X) = x_1 x_2 x_3$. Compute the point X on the interval $[A, B]$ at which the mean value of the derivative of f along $[A, B]$ is taken on. This is the point X on the interval $[A, B]$ such that
$$f(B) - f(A) = f'(X)(B-A).$$
Now $B - A = (0, -1, -1)$ and $f(B) - f(A) = 0$. Hence
$$f'(X)(B-A) = -\frac{\partial f}{\partial x_2} - \frac{\partial f}{\partial x_3} = -x_1 x_3 - x_1 x_2 = 0.$$
Since X lies on the interval $[A, B]$,
$$X = A + t(B-A) = (1, 1, 0) + t(0, -1, -1)$$
and $0 \le t \le 1$. Therefore $x_1 = 1$, $x_2 = 1 - t$, and $x_3 = -t$. Substituting in the equation
$$x_1 x_3 + x_1 x_2 = 0$$
yields
$$-t + 1 - t = 0, \quad \text{or} \quad t = \frac{1}{2}.$$
Hence $X = (1, \frac{1}{2}, -\frac{1}{2})$ is the desired point.

For functions defined on intervals (a, b) of real numbers, the mean value theorem is used to prove the vanishing derivative theorem. Namely if $f'(x) = 0$ for each $x \in (a, b)$ then f is constant. A similar result for functions defined on sets in $V_n(R)$ follows immediately from Theorem 6.11.

COROLLARY. *Let f be a real-valued function which is differentiable on an open ball E in $V_n(R)$. If $f'(X) = 0$ for each $X \in E$, then f is constant on E.*

Proof. Let X_0 be the center of the ball E. If $X \in E$, then $f'(Y) = 0$ for each Y on the segment $[X_0, X]$. Hence the mean value formula
$$f(X) - f(X_0) = f'(Y)(X - X_0)$$
implies $f(X) = f(X_0)$.

This result may be easily extended to more general open sets. See for example Exercises 25 and 26 below.

EXERCISES

For the following functions f and vectors X, Y, compute $\partial f / \partial Y(X)$ directly from the definition.

1. $f(x_1, x_2) = x_1 - 3x_1 x_2$, $X = (2, 1)$, $Y = (-1, 1)$.
2. $f(x_1, x_2) = \dfrac{1}{x_1 - x_2}$, $X = (1, 2)$, $Y = (1, 1)$.
3. $f(x_1, x_2, x_3) = \dfrac{1}{x_1 - x_2 + 2x_3}$, $X = (1, 1, 1)$, $Y = (2, 1, 0)$.

Compute $\partial f / \partial Y(X)$ for the specified vectors Y by using Theorem 6.10.

4. $f(x_1, x_2) = \sin(x_1 - x_2)$, $Y = (2, 1)$.
5. $f(x_1, x_2) = \dfrac{x_1 - x_2}{x_1 + x_2}$, $Y = (-1, 2)$.
6. $f(x_1, x_2) = \dfrac{\log x_2}{e^{x_1}}$, $Y = (0, 1)$.
7. $f(x_1, x_2, x_3) = x_1 x_2 - 2x_2 x_3 + 3x_1 x_3$, $Y = (1, 0, 1)$.
8. $f(x_1, x_2, x_3) = (x_1 + x_2 - x_3)^2$, $Y = (2, 1, -1)$.
9. $f(x_1, x_2, x_3) = \dfrac{\sin(x_1 + x_2)}{\cos(x_2 - x_3)}$, $Y = (1, 0, -1)$.
10. If $f(x, y, z) = x(y^2 - z)$, $A = (1, -1, 1)$, and $B = (x, y, z)$, determine $\partial f / \partial B(A)$ and $\partial f / \partial A(B)$.
11. If $f(x, y, z) = x^2 + y^2 + z^2 = |X|^2$, $X = (x, y, z)$, and $A = (2, 1, -1)$, determine $\partial f / \partial A(X)$, $\partial f / \partial X(X)$, $\partial f / \partial (A + X)(A)$.

Determine $\partial f / \partial Y(X)$ for the following choices of f, X, Y.

12. $f(x_1, x_2) = (x_1 x_2, x_1/x_2)$, $X = (2, 1)$, $Y = (1, -1)$.
13. $f(x_1, x_2) = (\sin(2x_1 - x_2), \cos(x_1 + x_2))$, $X = (\pi/4, \pi/4)$, $Y = (2, 1)$.
14. $f(x_1, x_2, x_3) = (x_1 x_2, x_1 x_3, x_2 x_3)$, $X = (1, 0, -1)$, $Y = (2, 1, 0)$.
15. If $f(x_1, x_2) = 2x_1 + (4x_2/x_1)$, compute the directional derivative of f at $(1, -1)$ in the direction of the vector $(2, 3)$. For what direction J does $\partial f / \partial J = 0$ at the point $(1, -1)$?
16. Let $f(x_1, x_2, x_3) = x_1 x_2 + \sin x_3$. Compute the directional derivative of f at $(1, 2, \pi/4)$ along the line from $A = (4, -1, 2)$ to $B = (1, 1, -1)$.
17. Let $f(x, y, z) = (x - y)^2 \tan z$. Compute the directional derivative of f at $(-1, 1, \pi/3)$ along the line from $A = (4, -1, 2)$ to $B = (1, 1, -1)$.
18. If $f(x_1, x_2) = \sqrt{169 - x_1^2 - x_2^2}$, find the direction at the point $(3, 4)$ for which the directional derivative of f has the value 0.
19. If $f(x_1, x_2) = 4x_1^2 + 9x_2^2$ find the direction at the point $(2, 1)$ for which the directional derivative of f is zero.
20. If $f(x_1, x_2) = ((x_1 - x_2)^2, x_1 x_2)$ find the directional derivative of f at $(1, 2)$ toward $(2, -1)$.

For the following functions f and intervals $[A, B]$ find the point X_0 at which the mean value of $f'(X)$ along $[A, B]$ is taken on.

21. $f(x_1, x_2) = x_1^2 - x_2^2$, $A = (2, -1)$, $B = (1, 3)$.
22. $f(x_1, x_2) = x_1^3 + 2x_2^3$, $A = (1, 1)$, $B = (0, 1)$.
23. $f(x_1, x_2, x_3) = (2x_1 - x_3)x_2$, $A = (-1, 0, 2)$, $B = (1, -1, 0)$.
24. $f(x_1, x_2, x_3) = x_1^2 - 2(x_2 - x_3)^2$, $A = (1, -1, 1)$, $B = (2, 1, 1)$.
25. A set S in $V_n(R)$ is called convex if for each pair of vectors $X, Y \in S$ the closed line segment $[X, Y]$ also belongs to S. For example, in the plane, the set

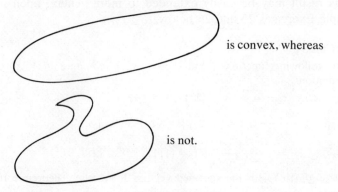

is convex, whereas

is not.

Extend the corollary of Theorem 6.11 to convex sets. That is, show that if f is differentiable at each X in the open convex set S and furthermore $f'(X) = 0$ for each $X \in S$, then f is constant on S.

26. A subset S of $V_n(R)$ is called *polygonally connected* if any two points A and B in S can be joined by a finite sequence of line segments $[A, X_1], [X_1, X_2], \ldots, [X_n, B]$ each of whic lies entirely in S. Extend the corollary of Theorem 6.11 to polygonally connected open sets.

27. Let $f(x_1, x_2) = 2x_1^2 x_2/(x_1^4 + x_2^2)$, $(x_1, x_2) \neq 0$, $f(0,0) = 0$. If $A = (a_1, a_2)$ verify that

$$\frac{\partial f}{\partial A}(0) = \frac{2a_1^2}{a_2} \quad \text{if } a_2 \neq 0$$

$$= 0 \quad \text{if } a_2 = 0.$$

Therefore $\partial f/\partial A\,(0)$ exists for each vector A. However, if $x_2 = x_1^2 \neq 0$ then $f(x_1, x_2) = 1$. Conclude from this that f is not continuous. This example shows that the directional derivative for a function f may exist at X for all possible directions yet the function f may not be differentiable.

6.7 The Gradient

We saw in the last section that if the scalar function f was differentiable, then the derivative of f with respect to a vector $Y = \sum y_k I_k$ could be written

(1) $$\frac{\partial f}{\partial Y}(X) = f'(X)Y = \sum \frac{\partial f}{\partial x_k}(X) y_k.$$

If we introduce the vector function

(2) $$\nabla f(X) = \left(\frac{\partial f}{\partial x_1}(X), \ldots, \frac{\partial f}{\partial x_n}(X) \right),$$

then this formula may be written

(3) $$\frac{\partial f}{\partial Y}(X) = (\nabla f(X), Y).$$

The vector function $\nabla f(X)$ defined in (2) is called the *gradient of the function f at the point X*. (The symbol, ∇f, is usually pronounced "grad f" or "del f.") In the study of scalar functions defined on open sets in $V_n(R)$ the gradient plays a fundamental role. In what follows we shall always assume that the function f is differentiable.

The geometrical significance of the gradient of a scalar function f can be readily determined from formula (3). If we compute the directional derivative $\partial f/\partial(Y/|Y|)(X)$ and apply the Schwartz inequality to (3), we obtain

$$\frac{\partial f}{\partial (Y/|Y|)}(X) = \left(\nabla f(X), \frac{Y}{|Y|} \right) \leq |\nabla f(X)| \frac{|Y|}{|Y|} = |\nabla f(X)|.$$

Therefore the length of the gradient at X dominates the value of each directional derivative of f at X. However, if $Y = \nabla f(X)$ then

$$\frac{\partial f}{\partial (Y/|Y|)}(X) = \left(\nabla f(X), \frac{\nabla f(X)}{|\nabla f(X)|} \right) = \frac{|\nabla f(X)|^2}{|\nabla f(X)|} = |\nabla f(X)|.$$

Hence we may say that the direction of the gradient at X is the direction which maximizes the directional derivative of f. Moreover, the length of the gradient is the maximum value of the directional derivative $\partial f/\partial (Y/|Y|)(X)$.

Example 1. If $f(x_1, x_2, x_3) = x_1 x_2 x_3$ then $\nabla f(X) = (x_2 x_3, x_1 x_3, x_1 x_2)$, and
$$|\nabla f(X)| = (x_2^2 x_3^2 + x_1^2 x_3^2 + x_1^2 x_2^2)^{1/2}.$$
The maximum value of the directional derivative of f at $X_0 = (1, 2, 3)$ is then
$$|\nabla f(X_0)| = (36 + 9 + 4)^{1/2} = 7,$$
and the direction at which this maximum is taken on is
$$\frac{\nabla f(X_0)}{|\nabla f(X_0)|} = \frac{1}{7}(6, 3, 2).$$

The mean value theorem of the previous section may be formulated in terms of the gradient. This theorem states that if f is a scalar valued function which is differentiable on the closed line segment $[A, B]$ then
$$f(B) - f(A) = f'(Z)(B - A)$$
where Z is a vector in the interval $[A, B]$. Using (3) this may be written
$$f(B) - f(A) = (\nabla f(Z), B - A).$$
Taking absolute values and applying the Schwartz inequality we have the following important estimate on the size of $|f(B) - f(A)|$. Namely,
$$(4) \qquad |f(B) - f(A)| = |(\nabla f(Z), B - A)| \leq |\nabla f(Z)| \cdot |B - A|.$$

The operation of computing the gradient may be viewed as a transformation. It may be easily verified that
$$\nabla(f + g) = \nabla f + \nabla g$$
and
$$\nabla(af) = a \nabla f.$$

Hence ∇ is a linear transformation. If V is the vector space of scalar functions with continuous first partial derivatives on the open set U, and W is the vector space of continuous functions defined on U with values in $V_n(R)$, then ∇ is a linear transformation from V to W. In addition ∇ has the differentiation property
$$\nabla(fg) = f \nabla g + g \nabla f.$$
We leave the verification of these facts as exercises.

EXERCISES

Compute the gradient of each of the following functions.

1. $f(x_1, x_2) = x_1^2 - x_2^2$.
2. $f(x_1, x_2) = \dfrac{x_1 + x_2}{x_1 - x_2}$.
3. $f(x_1, x_2) = x_1 \tan(x_1 + x_2)$.
4. $f(x_1, x_2, x_3) = \dfrac{x_1 + x_2 + x_3}{x_1 x_2 x_3}$.
5. $f(x_1, x_2, x_3) = \dfrac{x_3 \sin(x_1 - x_2)}{e^{x_1^2 + x_2^2}}$.
6. $f(x_1, x_2, x_3) = \dfrac{\log(x_1^2 + x_2^2)}{\tan(1 + x_3^2)}$.
7. If $f = \log(x^2 + y^2)$, determine the direction which minimizes the directional derivative of f.
8. Find the direction of maximum rate of change of the function $u = x^2 + 4y^2 + 9z^2$ at $(1, 2, -1)$.
9. Find the direction of minimum rate of change of the function $x_4 = 2x_1^2 - x_2^2 + x_3^2$ at $(2, -1, 1)$.
10. If V is a vector perpendicular to the gradient of a function f at X, show that the directional derivative of f in the direction of V vanishes.

11. If f and g are scalar functions, verify that
   (i) $\nabla(af + bg) = a\nabla f + b\nabla g$.
   (ii) $\nabla(fg) = f\nabla g + g\nabla f$.
   (iii) $\nabla(f/g) = (1/g^2)[g\nabla f - f\nabla g]$ if g does not vanish.

12. If $X \in V_n(R)$ verify that
$$\nabla(|X|^m) = m|X|^{m-2}X \qquad m = 1, 2, \ldots.$$

13. Let U be an open set containing X_0. If for each $X \in U, f(X) \leq f(X_0)$, verify that $\nabla f(X_0) = 0$. (Look at each variable of $f(X) = f(x_1, \ldots, x_n)$ separately.)

6.8 The Chain Rule

If f and g are differentiable functions of one variable and we form the composition $h = f \circ g$, then the chain rule asserts that $h'(x) = f'(g(x)) \cdot g'(x)$. In this section we shall determine the form of this theorem for vector valued functions f and g. Surprisingly enough it reads roughly the same. Indeed if M is the differential of g and X_0 and L is the differential of f at $Y_0 = g(X_0)$, then the differential of $h = f \circ g$ at X_0 is the composition or product of the linear transformations L and M. That is

(1) $$d_{X_0}h = L \cdot M = d_{Y_0}f \cdot d_{X_0}g.$$

Thus for functions of one variable the derivative of $f \circ g$ is the product of the derivatives $f'(g(x)) \cdot g'(x)$, whereas for vector functions the differential is the product of the linear transformation $d_{Y_0}f \cdot d_{X_0}g$. This product is again, of course, a linear transformation. To actually compute with these differentials we work with their respective matrix representations or Jacobian matrices.

We know from Chapter 3 that if A is the matrix of the transformation L and B is the matrix of the transformation M, then the matrix of the product $L \cdot M$ is just the matrix product $A \cdot B$. Applying this fact to the respective Jacobian matrices we have the fundamental chain rule formula for the derivatives of vector functions. If $Y = g(X)$ is a function from $V_n(R)$ to $V_m(R)$ and $Z = f(Y)$ is a function from $V_m(R)$ to $V_p(R)$, then $Z = h(X) = f(g(X))$ is a function from $V_n(R)$ to $V_p(R)$, and if both f and g are differentiable, the composition $h = f \circ g$ is differentiable. Furthermore at each point X

(2) $$\left(\frac{\partial h_i}{\partial x_j}\right)(X) = \left(\frac{\partial f_i}{\partial y_k}\right)(g(X)) \cdot \left(\frac{\partial g_k}{\partial x_j}\right)(X).$$

If we suppress the variable X this equation becomes

$$\left(\frac{\partial h_i}{\partial x_j}\right) = \left(\frac{\partial f_i}{\partial y_k}\right)\left(\frac{\partial g_k}{\partial x_j}\right).$$

In expanded form we have

$$\begin{pmatrix} \frac{\partial h_1}{\partial x_1} & \cdots & \frac{\partial h_1}{\partial x_n} \\ \vdots & & \vdots \\ \frac{\partial h_p}{\partial x_1} & \cdots & \frac{\partial h_p}{\partial x_n} \end{pmatrix} = \begin{pmatrix} \frac{\partial f_1}{\partial y_1} & \cdots & \frac{\partial f_1}{\partial y_m} \\ \vdots & & \vdots \\ \frac{\partial f_p}{\partial y_1} & \cdots & \frac{\partial f_p}{\partial y_m} \end{pmatrix} \begin{pmatrix} \frac{\partial g_1}{\partial x_1} & \cdots & \frac{\partial g_1}{\partial x_n} \\ \vdots & & \vdots \\ \frac{\partial g_m}{\partial x_1} & \cdots & \frac{\partial g_m}{\partial x_n} \end{pmatrix}.$$

Applying the definition of matrix product, equation (2) implies that for each i and j

$$\frac{\partial h_i}{\partial x_j} = \sum_{k=1}^{m} \frac{\partial f_i}{\partial y_k} \frac{\partial g_k}{\partial x_j}.$$

One should bear in mind that if $\partial g_k/\partial x_j$ is evaluated at the point X, then $\partial f_i/\partial y_k$ is evaluated at the point $g(X)$. For example if $z = f(y_1, \ldots, y_m)$ is one scalar function defined on an open set in $V_m(R)$, and for each k, $y_k = g_k(t)$, then $z = f(g_1(t), \ldots, g_m(t)) = h(t)$ is a real-valued function of one real variable. Formula (2) Asserts that

$$\frac{dz}{dt} = \frac{dh}{dt} = h'(t) = \sum_{k=1}^{m} \frac{\partial f}{\partial y_k} \frac{dg_k}{dt}.$$

It is common to write $\partial z/\partial y_k$ in place of $\partial f/\partial y_k$ and dy_k/dt for dg_k/dt. With this notation we have

$$\frac{dz}{dt} = \sum_{k=1}^{m} \frac{\partial z}{\partial y_k} \frac{dy_k}{dt}.$$

If instead $y_k = g_k(x_1, \ldots, x_n)$ then

$$z = f(g_1(x_1, \ldots, x_n), \ldots, g_m(x_1, \ldots, x_n)) = h(x_1, \ldots, x_n),$$

and formula (2) implies that for $j = 1, \ldots, n$

$$\frac{\partial z}{\partial x_j} = \frac{\partial h}{\partial x_j} = \sum_{k=1}^{m} \frac{\partial f}{\partial y_k} \frac{\partial g_k}{\partial x_j} = \sum_{n=1}^{m} \frac{\partial z}{\partial y_k} \frac{\partial y_k}{\partial x_j}.$$

To evaluate $\partial h/\partial x_j$ at X one must, of course, compute $\partial g_k/\partial x_j = \partial y_k/\partial x_j$ at the point X and $\partial f/\partial y_k = \partial z/\partial y_k$ at the point $Y = g(X)$. If we use the notation

$$h'(X) = \left(\frac{\partial h_i}{\partial x_j}\right)(X), \quad f'(g(X)) = \left(\frac{\partial f_i}{\partial y_k}\right)(g(X)) \quad \text{and} \quad g'(X) = \left(\frac{\partial g_k}{\partial x_j}\right)(X),$$

then (2) becomes

(3) $$h'(X) = f'(g(X)) \cdot g'(X).$$

Equation (3) is formally the same equation as that for the chain rule for function of one variable. Now, however, the quantities are *matrices* not numbers. Before proving (1) let us look at some examples.

Example 1. If $g(x_1, x_2) = (x_1 x_2, x_1 + x_2)$ and $f(y_1, y_2) = (2y_1^2, y_2^3)$ then

$$g'(X) = \left(\frac{\partial g_k}{\partial x_j}\right) = \begin{pmatrix} x_2 & x_1 \\ 1 & 1 \end{pmatrix}$$

and

$$f'(Y) = \left(\frac{\partial f_i}{\partial y_k}\right) = \begin{pmatrix} 4y_1 & 0 \\ 0 & 3y_2^2 \end{pmatrix}.$$

Evaluating $(\partial f_i/\partial y_k)$ at the point $Y = g(X)$ we obtain

$$f'(g(X)) = \left(\frac{\partial f_i}{\partial y_k}\right)(g(X)) = \begin{pmatrix} 4x_1 x_2 & 0 \\ 0 & 3(x_1 + x_2)^2 \end{pmatrix}.$$

Thus if $h = f \circ g$, we have

$$\left(\frac{\partial h_i}{\partial x_j}\right)(X) = \left(\frac{\partial f_i}{\partial y_k}\right)(g(X)) \cdot \left(\frac{\partial g_k}{\partial x_j}\right)(X)$$

$$= \begin{pmatrix} 4x_1 x_2 & 0 \\ 0 & 3(x_1 + x_2)^2 \end{pmatrix} \begin{pmatrix} x_2 & x_1 \\ 1 & 1 \end{pmatrix}$$

$$= \begin{pmatrix} 4x_1 x_2^2 & 4x_1^2 x_2 \\ 3(x_1 + x_2)^2 & 3(x_1 + x_2)^2 \end{pmatrix}.$$

If $p = g \circ f$, then
$$\left(\frac{\partial p_i}{\partial y_j}\right) = \left(\frac{\partial g_i}{\partial x_k}\right)\left(\frac{\partial f_k}{\partial y_j}\right).$$

Evaluating the matrix $(\partial g_i/\partial x_k)$ at the point $X = f(Y)$ we obtain
$$g'(f(Y)) = \left(\frac{\partial g_i}{\partial x_k}\right)(f(Y)) = \begin{pmatrix} y_2^3 & 2y_1^2 \\ 1 & 1 \end{pmatrix}.$$
Hence
$$\left(\frac{\partial p_i}{\partial y_j}\right) = \begin{pmatrix} y_2^3 & 2y_1^2 \\ 1 & 1 \end{pmatrix}\begin{pmatrix} 4y_1 & 0 \\ 0 & 3y_2^2 \end{pmatrix}$$
$$= \begin{pmatrix} 4y_1 y_2^3 & 6y_1^2 y_2^2 \\ 4y_1 & 3y_2^2 \end{pmatrix}.$$

When writing a Jacobian matrix $(\partial f_i/\partial x_j)$ one should bear in mind that i is the row index and j is the column index of the matrix. Hence when we write the matrix product
$$\left(\frac{\partial f_i}{\partial y_k}\right)\left(\frac{\partial g_k}{\partial x_j}\right),$$
we use the same symbol k to denote the column index of $(\partial f_i/\partial x_k)$ and the row index of $(\partial g_k/\partial x_j)$. We do this since each term of the product matrix $(\partial f_i/\partial g_k)(\partial g_k/\partial x_j)$ is summed on the index k. That is
$$\left(\frac{\partial f_i}{\partial y_k}\right)\left(\frac{\partial g_k}{\partial x_j}\right) = \left(\sum_{k=1}^{m} \frac{\partial f_i}{\partial y_k}\frac{\partial g_k}{\partial x_j}\right).$$

Example 2. If $f(t) = (1, t, t^2)$ and $g(x_1, x_2, x_3) = (x_1 x_2 x_3, x_1 + 2x_2 + 3x_3)$, compute the Jacobian matrix of the transformation $g \circ f$ and evaluate it at $t = 1$. Now
$$f'(t) = \left(\frac{\partial f_j}{\partial t}\right) = \begin{pmatrix} 0 \\ 1 \\ 2t \end{pmatrix}$$
and
$$g'(X) = \left(\frac{\partial g_i}{\partial x_j}\right) = \begin{pmatrix} x_2 x_3 & x_1 x_3 & x_1 x_2 \\ 1 & 2 & 3 \end{pmatrix}.$$

Evaluating $(\partial g_i/\partial x_j)$ at $X = f(t)$ we obtain
$$g'(f(t)) = \left(\frac{\partial g_i}{\partial x_j}\right)(f(t)) = \begin{pmatrix} t^3 & t^2 & t \\ 1 & 2 & 3 \end{pmatrix}.$$
Therefore if $h = g \circ f$,
$$\left(\frac{\partial h_i}{\partial t}\right) = \left(\frac{\partial g_i}{\partial x_j}\right)\left(\frac{\partial f_j}{\partial t}\right) = \begin{pmatrix} t^3 & t^2 & t \\ 1 & 2 & 3 \end{pmatrix}\begin{pmatrix} 0 \\ 1 \\ 2t \end{pmatrix} = \begin{pmatrix} 3t^2 \\ 2 + 6t \end{pmatrix}.$$
Hence at $t = 1$
$$\left(\frac{\partial h_i}{\partial t}\right) = \begin{pmatrix} 3 \\ 8 \end{pmatrix}.$$

The function h is a function from R to $V_2(R)$ and $(\partial h_i/\partial t)$ is the matrix of a linear transformation from R to $V_2(R)$.

Example 3. Let $f(x_1, x_2) = (x_1/x_2, x_1 x_2)$ and $g(x_1, x_2) = (x_1 - x_2)^2$. Then $h = g \circ f$ is a function from $V_2(R)$ to R, and
$$\left(\frac{\partial h}{\partial x_i}\right) = \left(\frac{\partial g}{\partial y_k}\right)\left(\frac{\partial f_k}{\partial x_i}\right).$$

The use of different symbols, i.e., $X = (x_1, x_2)$, $Y = (y_1, y_2)$, to denote points in the domains of f and g, respectively, is just a notational convenience. Thus

$g(y_1, y_2) = (y_1 - y_2)^2$ and
$$g'(Y) = \left(\frac{\partial g}{\partial y_k}\right) = 2(y_1 - y_2, y_2 - y_1).$$
Since
$$\left(\frac{\partial f_k}{\partial x_i}\right) = \begin{pmatrix} \frac{1}{x_2} & -\frac{x_1}{x_2^2} \\ x_2 & x_1 \end{pmatrix}, \quad \frac{\partial h}{\partial x_i} = \left(\frac{\partial g}{\partial y_k}\right)\left(\frac{\partial f_k}{\partial x_i}\right) = 2(y_1 - y_2, y_2 - y_1) \begin{pmatrix} \frac{1}{x_2} & -\frac{x_1}{x_2^2} \\ x_2 & x_1 \end{pmatrix}.$$

Evaluating $(\partial g/\partial y_k)$ at $Y = f(X)$ we have $(\partial g/\partial y_k) = 2((x_1/x_2) - x_1 x_2, x_1 x_2 - (x_1/x_2))$. Hence

$$\left(\frac{\partial h}{\partial x_i}\right) = 2\left(\frac{x_1}{x_2} - x_1 x_2, x_1 x_2 - \frac{x_1}{x_2}\right) \begin{pmatrix} \frac{1}{x_2} & -\frac{x_1}{x_2^2} \\ x_2 & x_1 \end{pmatrix}$$

$$= 2\left(\frac{x_1}{x_2^2} - x_1 + x_1 x_2^2 - x_1, -\frac{x_1^2}{x_2^3} + \frac{x_1^2}{x_2} + x_1^2 x_2 - \frac{x_1^2}{x_2}\right).$$

At $X = (1, 2)$
$$\left(\frac{\partial h}{\partial x_i}\right) = 2\left(-\frac{7}{4}, \frac{15}{8}\right) = \left(-\frac{7}{2}, \frac{15}{4}\right).$$

Since $(\partial h/\partial x_i)$ is the matrix of the linear transformation $L = d_X h$ from $V_2(R)$ to R, we may in turn compute $L(Y)$ for any vector $Y \in V_2(R)$. For example, letting $X = (1, 2)$

$$L(Y) = (d_{(1,2)} h)(Y) = \left(-\frac{7}{2}, \frac{15}{4}\right)\begin{pmatrix} y_1 \\ y_2 \end{pmatrix} = -\frac{7}{2} y_1 + \frac{15}{4} y_2.$$

We turn now to the proof of the chain rule.

THEOREM 6.12. *Let g be a function from $V_n(R)$ to $V_m(R)$ which is differentiable on an open set U, and let f be a function from $V_m(R)$ to $V_p(R)$ which is differentiable on an open set V. Let $W = \{W \in U : g(X) \in V\}$. Then W is an open subset of $V_n(R)$, and $h = f \circ g$ is differentiable at each point of W. Furthermore at each $X \in W$*

(3) $$d_X h = d_{g(X)} f \cdot d_X g.$$

Proof. Since g is differentiable, g is continuous by Theorem 6.6. From this it follows that W is open. To see this let $X \in W$ and set $Y = g(X)$. Since V is open there is a ball B about Y such that $B \subset V$. On the other hand since g is continuous there is a ball C about X such that g maps C into B. Thus $C \subset W$, and W is open.

To prove the statement about differentiability and establish formula (3) let $X_0 \in W$ and set $L = d_{g(X_0)} f$ and $M = d_{X_0} g$. We assert

$$\lim_{X \to X_0} \frac{|h(X) - h(X_0) - L \cdot M(X - X_0)|}{|X - X_0|} = 0.$$

Hence h is differentiable at X_0 and $d_{X_0} h = L \cdot M$. Now if $Y = g(X)$ and $Y_0 = g(X_0)$, then

$$\frac{h(X) - h(X_0) - LM(X - X_0)}{|X - X_0|}$$

$$= \frac{(f \circ g)(X) - (f \circ g)(X_0) - L(Y - Y_0) + L(Y - Y_0) - LM(X - X_0)}{|X - X_0|}$$

$$= \frac{f(Y) - f(Y_0) - L(Y - Y_0)}{|X - X_0|} + L\left(\frac{g(X) - g(X_0) - M(X - X_0)}{|X - X_0|}\right).$$

We assert that each of the above terms tends to zero as $X \to X_0$. Certainly

$$\lim_{X \to X_0} \frac{g(X) - g(X_0) - M(X - X_0)}{|X - X_0|} = 0,$$

since g is assumed to be differentiable at X_0. The linear transformation L is continuous (Corollary 2 to Theorem 6.2 p. 198) which implies

$$\lim_{X \to X_0} L\left(\frac{g(X) - g(X_0) - M(X - X_0)}{|X - X_0|}\right) = 0.$$

To complete the proof we need only show that

(4) $$\lim_{X \to X_0} \frac{f(Y) - f(Y_0) - L(Y - Y_0)}{|X - X_0|} = 0.$$

To prove (4) let us grant for the moment the fact that there are positive constants c and δ such that whenever $0 < |X - X_0| < \delta$ it follows that

$$\frac{|Y - Y_0|}{|X - X_0|} \leq c.$$

To estimate

(5) $$\frac{|f(Y) - f(Y_0) - L(Y - Y_0)|}{|X - X_0|}$$

one is tempted to multiply numerator and denominator of (5) by $|Y - Y_0|$ obtaining

$$\frac{|f(Y) - f(Y_0) - L(Y - Y_0)|}{|X - X_0|} = \frac{|f(Y) - f(Y_0) - L(Y - Y_0)|}{|Y - Y_0|} \cdot \frac{|Y - Y_0|}{|X - X_0|}$$

$$\leq c \frac{|f(Y) - f(Y_0) - L(Y - Y_0)|}{|Y - Y_0|} \quad \text{if } |X - X_0| < \delta.$$

As $X \to X_0$, it follows that $Y \to Y_0$ (why?). Applying the differentiability of f it follows that as $Y \to Y_0$

$$c \frac{|f(Y) - f(Y_0) - L(Y - Y_0)|}{|Y - Y_0|} \to 0.$$

However, we cannot guarantee that $|Y - Y_0| \neq 0$ without which the argument doesn't make sense. We get around this difficulty in the following way. Set

$$p(Y) = \frac{f(Y) - f(Y_0) - L(Y - Y_0)}{|Y - Y_0|} \quad \text{if } Y \neq Y_0$$

$$= 0 \quad \text{if } Y = Y_0.$$

Since f is differentiable at Y_0, we have $\lim_{Y \to Y_0} p(Y) = 0$. Moreover for all values of Y

(6) $$|Y - Y_0| p(Y) = f(Y) - f(Y_0) - L(Y - Y_0).$$

Dividing (6) by $|X - X_0|$ we obtain

$$\frac{|f(Y) - f(Y_0) - L(Y - Y_0)|}{|X - X_0|} = \frac{|Y - Y_0|}{|X - X_0|} |p(Y)| \leq c|p(Y)|$$

if $0 < |X - X_0| < \delta$. Now as $X \to X_0$, $Y \to Y_0$ and $|P(Y)| \to 0$. Hence

$$\lim_{X \to X_0} \frac{|f(Y) - f(Y_0) - L(Y - Y_0)|}{|X - X_0|} = 0,$$

and we are done.

The fact that $|Y - Y_0|/|X - X_0| \leq c$ if $0 < |X - X_0| < \delta$, now follows from the differentiability of g. We leave the verification of this as an exercise. (Exercises 16 and 17.)

Occasionally the fact that we are dealing with the composition of two functions is disguised.

Example 4. If $x_3 = 1/x_1 \, f(x_2/x_1)$ show that

$$x_1 \frac{\partial x_3}{\partial x_1} + x_2 \frac{\partial x_3}{\partial x_2} + x_3 = 0.$$

Set $g(x_1, x_2) = (1/x_1, x_2/x_1)$ and $h(y_1, y_2) = y_1 f(y_2)$. Then $x_3 = (h \circ g)(x_1, x_2)$, and

$$\left(\frac{\partial x_3}{\partial x_1}, \frac{\partial x_3}{\partial x_2} \right) = \left(\frac{\partial h}{\partial y_1}, \frac{\partial h}{\partial y_2} \right) \begin{pmatrix} \frac{\partial g_1}{\partial x_1} & \frac{\partial g_1}{\partial x_2} \\ \frac{\partial g_2}{\partial x_1} & \frac{\partial g_2}{\partial x_2} \end{pmatrix}$$

$$= (f(y_2), y_1 f'(y_2)) \begin{pmatrix} -\frac{1}{x_1^2} & 0 \\ -\frac{x_2}{x_1^2} & \frac{1}{x_1} \end{pmatrix}.$$

Therefore since

$$(y_1, y_2) = g(x_1, x_2) = \left(\frac{1}{x_1}, \frac{x_2}{x_1} \right)$$

$$\frac{\partial x_3}{\partial x_1} = -\frac{f(y_2)}{x_1^2} - y_1 \frac{x_2}{x_1^2} f'(y_2) = -\left(\frac{1}{x_1^2} \right) f\left(\frac{x_2}{x_1} \right) - \frac{x_2}{x_1^3} f'\left(\frac{x_2}{x_1} \right)$$

and

$$\frac{\partial x_3}{\partial x_2} = \frac{y_1}{x_1} f'(y_2) = \frac{1}{x_1^2} f'\left(\frac{x_2}{x_1} \right).$$

Hence

$$x_1 \frac{\partial x_3}{\partial x_1} + x_2 \frac{\partial x_3}{\partial x_2} + x_3 = -\frac{1}{x_1} f\left(\frac{x_2}{x_1} \right) - \frac{x_2}{x_1^2} f'\left(\frac{x_2}{x_1} \right) + \frac{x_2}{x_1^2} f'\left(\frac{x_2}{x_1} \right) + \frac{1}{x_1} f\left(\frac{x_2}{x_1} \right) = 0.$$

EXERCISES

1. If $f(x_1, x_2) = (x_1 - x_2)^2$ and $g(t) = (t^2, 1 + t^3)$, let $h = f \circ g$. Compute $h'(t)$ and $h'(2)$.
2. Let $u = \sin(x - y)$, $x = t\sqrt{1 + t}$, $y = t^2$. Compute $u'(t)$.
3. Let $f(y_1) = (y_1^2 - 3)^2$ and $g(x_1, x_2) = x_1 + x_2 - x_1 x_2$. If $h = f \circ g$, compute $\partial h/\partial x_1$ and $\partial h/\partial x_2$ and evaluate these partial derivatives at $X_0 = (2, 1)$.
4. Let $s = f(t) = t \sin t$ and $t = x^2 - y^2$, compute $\partial s/\partial x$, $\partial s/\partial y$.
5. If $u = z^2 \log z$ and $z = xy - y^2 x + t$, compute $\partial u/\partial t$ at $x = 2$, $y = 1$, $t = 1$.
6. If $f(x_1, x_2) = (x_1 x_2, 2x_1 - x_2)$ and $g(x_1, x_2) = (x_1 + x_2, x_1/x_2)$ let $h = f \circ g$ and $p = g \circ f$. Compute $h'(2, 1)$ and $p'(1, -2)$.
7. If $f(x, y) = (\sin(x + y), \cos(x - y))$ and $g(x, y) = (2x + y, x - 2y)$ compute $(f \circ g)'(x, y)$ and $(g \circ f)'(x, y)$.
8. If $f(x_1, x_2, x_3) = (x_1 x_2, x_2 x_3)$ and $g(x_1, x_2) = (x_1 x_2, x_1 + x_2, x_1 - x_2)$ compute $(f \circ g)'(2, -1)$.

THE CHAIN RULE

9. If $z = f(x, y)$ and $x = r \cos \theta$ and $y = r \sin \theta$ show that
$$\frac{\partial z}{\partial r} = \frac{\partial f}{\partial x} \cos \theta + \frac{\partial f}{\partial y} \sin \theta$$
and
$$\frac{\partial z}{\partial \theta} = -\frac{\partial f}{\partial x} r \sin \theta + \frac{\partial f}{\partial y} r \cos \theta.$$

10. If $x_3 = f(x_1 - x_2, x_2 - x_1) = g(x_1, x_2)$ show that $(\partial g/\partial x_1) + (\partial g/\partial x_2) = 0$.

11. If f is a differentiable scalar function and $g(x_1, x_2) = x_1 x_2 f[(x_1 + x_2)/(x_1 x_2)]$ show that
$$x_1^2 \frac{\partial g}{\partial x_1} - x_2^2 \frac{\partial g}{\partial x_2} = (x_1 - x_2) g(x_1, x_2).$$

12. A scalar function f defined on an open set in $V_n(R)$ is said to be *homogeneous of degree m* if for each real number t
$$f(tX) = t^m f(X).$$
Show that if f is differentiable and homogeneous, then
$$(\nabla f(X), X) = m f(X).$$
This result is known as Euler's theorem. (Fix X and differentiate the function $f(tX)$ with respect to t.)

13. Prove the converse to Euler's theorem. That is, if f is differentiable on an open set U and for each $X \in U$
$$(\nabla f(X), X) = m f(X),$$
then f is homogeneous of degree m. (Differentiate the function
$$g(t) = f(tX)/t^m$$
and show that the function g' is identically equal to zero. From this the result follows easily.)

14. If f and g are differentiable functions from $V_n(R)$ to $V_n(R)$ and for each X
$$(f \circ g)(X) = X$$
show that
$$f'(g(X)) = [g'(X)]^{-1}.$$

15. If f is differentiable and L is a linear function from $V_n(R)$ to $V_m(R)$ let $h = f \circ L$ and $p = L \circ f$. Show that
$$d_X h = d_{LX} f \cdot L$$
and
$$d_X p = L \cdot d_X f.$$

16. Let L be a linear transformation from $V_n(R)$ to $V_m(R)$. Show that there is a constant M, independent of X, such that
$$|LX| \leq M|X| \quad \text{for each } X \in V_n(R).$$
(Let A be the matrix for L and let A_i be the rows of A. Show that
$$|LX| \leq \{\Sigma |A_i|^2\}^{1/2} |X|.$$
Hence we may take $M = \{\Sigma |A_i|^2\}^{1/2}$.)

17. Let f be a differentiable vector function. Show that there exist constants N and δ such that if $0 < |X - X_0| < \delta$, then
$$\frac{|f(X) - f(X_0)|}{|X - X_0|} < N.$$

(Let L be the differential of f at X_0. Write

$$f(X) - f(X_0) = f(X) - f(X_0) - L(X - X_0) + L(X - X_0).$$

Now exploit the differentiability of f and Exercise 16 to establish the result.)

6.9 More on the Chain Rule: Coordinate Transformations

In the previous section we saw that if f was a differentiable function from $V_m(R)$ to R and g was a differentiable function from $V_n(R)$ to $V_m(R)$, then the partial derivatives $\partial h/\partial x_k$ of the composition $h = f \circ g$ could be computed using the formula

$$(1) \qquad \frac{\partial h}{\partial x_i} = \left(\sum_{k=1}^{m} \frac{\partial f}{\partial y_k} \frac{\partial g_k}{\partial x_i} \right).$$

Evaluating this expression at the point X we have

$$\frac{\partial h}{\partial x_i}(X) = \sum_{k=1}^{m} \frac{\partial f}{\partial y_k}(g(X)) \frac{\partial g_k}{\partial x_i}(X).$$

Suppose now we wish to compute

$$\frac{\partial}{\partial x_j}\left(\frac{\partial h}{\partial x_i}\right) = \frac{\partial^2 h}{\partial x_j \partial x_i}.$$

For the moment let us define $p_k(X) = \partial f / \partial y_k(g(X))$. Then

$$\frac{\partial h}{\partial x_i}(X) = \sum_{k=1}^{m} p_k(X) \frac{\partial g_k}{\partial x_i}(X).$$

If we differentiate this expression with respect to x_j and apply the product rule for differentiation, we obtain

$$(2) \qquad \frac{\partial^2 h}{\partial x_j \partial x_i}(X) = \sum_{k=1}^{m} \left[\frac{\partial p_k}{\partial x_j}(X) \frac{\partial g_k}{\partial x_i}(X) + p_k(X) \frac{\partial^2 g_k}{\partial x_j \partial x_i}(X) \right].$$

To compute $\partial p_k / \partial x_j(X)$ we must use the fact that p_k is the composition of the function $\partial f / \partial y_k$ and $g(X)$ and then apply the chain rule formula. Indeed replacing f by $\partial f / \partial y_k$ in (1) and changing the index of summation to l we obtain

$$\frac{\partial p_k}{\partial x_j} = \sum_{l=1}^{m} \frac{\partial}{\partial y_l}\left(\frac{\partial f}{\partial y_k}\right) \frac{\partial g_l}{\partial x_j}.$$

Evaluating this expression at the point X we have

$$(3) \qquad \frac{\partial p_k}{\partial x_j}(X) = \sum_{l=1}^{m} \frac{\partial}{\partial y_l}\left(\frac{\partial f}{\partial y_k}\right)(g(X)) \frac{\partial g_l}{\partial x_j}(X).$$

If we now substitute (3) in (2) we have the following formula for $\partial^2 h / \partial x_j \partial x_i$ evaluated at X

$$(4) \qquad \frac{\partial^2 h}{\partial x_j \partial x_i}(X) =$$

$$\sum_{k=1}^{m} \left\{ \left[\sum_{l=1}^{m} \frac{\partial}{\partial y_l}\left(\frac{\partial f}{\partial y_k}\right)(g(X)) \frac{\partial g_l}{\partial x_j}(X) \right] \frac{\partial g_k}{\partial x_i}(X) + \frac{\partial f}{\partial y_k}(g(X)) \frac{\partial^2 g}{\partial x_j \partial x_i}(X) \right\}.$$

No attempt should be made to memorize (4). The important thing to remember is that in expression (1), $\partial f/\partial y_k$ is really $\partial f/\partial y_k$ composed with g. Hence to compute $\partial/\partial x_j(\partial f/\partial y_k)$ we must in reality compute $\partial/\partial x_j(\partial f/\partial y_k \circ g)$ which will involve an application of formula (1).

We have stressed the fact that in (1) we have the partial derivatives of the function f and g. However, if we write $u = f(Y)$, and $Y = g(X)$, then formula (1) is often abbreviated

$$\frac{\partial u}{\partial x_i} = \sum_k \frac{\partial u}{\partial y_k} \frac{\partial y_k}{\partial x_i}.$$

This form of (1) obscures the roles played by the functions f and g. Nevertheless it is quite convenient and it is the usual way the chain rule is written, especially in applied work. We shall use both forms freely.

As an application of these ideas let us consider the coordinate change from rectangular to polar coordinates in the plane. As usual we define

(5)
$$x_1 = r \cos \theta$$
$$x_2 = r \sin \theta.$$

Then $r = \sqrt{x_1^2 + x_2^2}$ and $\tan \theta = x_2/x_1$. The geometric relation between the rectangular coordinates (x_1, x_2) and the polar coordinates (r, θ) of a point in the plane is illustrated in Figure 17. We shall consider (5) as defining a function $X = \varphi(r, \theta)$ from $V_2(R)$ onto $V_2(R)$. The function φ is of course not one-to-one, but if we restrict r, θ by requiring that $0 \leq r$ and $0 \leq \theta < 2\pi$, then φ is one-to-one except when $r = 0$. The function φ maps the strip $0 \leq r$, $0 \leq \theta < 2\pi$ onto the entire x_1, x_2 plane. (Figure 18.) Indeed φ maps

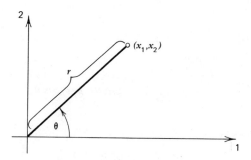

Fig. 17 Polar coordinates in the plane.

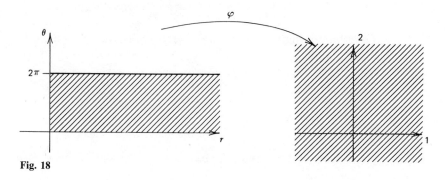

Fig. 18

the box $0 \leq r \leq a$, $0 \leq \theta \leq \theta_0$ onto a sector of a disc in the (x_1, x_2) plane as illustrated by Figure 19.

The transformation φ may be "pictured" as mapping the line $r = 0$ onto the origin in the (x_1, x_2) plane. Lines $\theta = $ constant are mapped onto spokes emanating from the origin as in Figure 20.

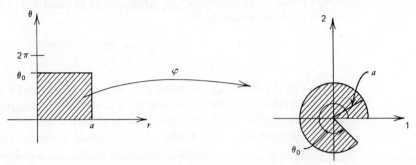

Fig. 19

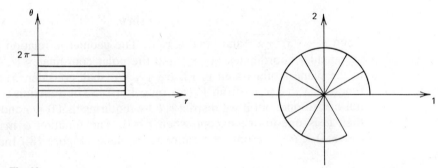

Fig. 20

The Jacobian matrix of the transformation φ is easily computed. Indeed

$$\phi'(r, \theta) = \begin{pmatrix} \dfrac{\partial x_1}{\partial r} & \dfrac{\partial x_1}{\partial \theta} \\ \dfrac{\partial x_2}{\partial r} & \dfrac{\partial x_2}{\partial \theta} \end{pmatrix} = \begin{pmatrix} \cos \theta & -r \sin \theta \\ \sin \theta & r \cos \theta \end{pmatrix}$$

and if $u = f(x_1, x_2) = f(\varphi(r, \theta)) = g(r, \theta)$ we may compute

$$\left(\frac{\partial g}{\partial r}, \frac{\partial g}{\partial \theta} \right) = \left(\frac{\partial u}{\partial r}, \frac{\partial u}{\partial \theta} \right) = \left(\frac{\partial u}{\partial x_1}, \frac{\partial u}{\partial x_2} \right) \begin{pmatrix} \cos \theta & -r \sin \theta \\ \sin \theta & r \cos \theta \end{pmatrix}$$

$$= \left(\cos \theta \frac{\partial u}{\partial x_1} + \sin \theta \frac{\partial u}{\partial x_2}, -r \sin \theta \frac{\partial u}{\partial x_1} + r \cos \theta \frac{\partial u}{\partial x_2} \right)$$

To compute higher order partial derivatives we apply the chain rule again. For example

$$\frac{\partial^2 u}{\partial \theta^2} = \frac{\partial}{\partial \theta} \left(\frac{\partial u}{\partial \theta} \right) = \frac{\partial}{\partial \theta} \left(-r \sin \theta \frac{\partial u}{\partial x_1} + r \cos \theta \frac{\partial u}{\partial x_2} \right)$$

$$= -r \cos \theta \frac{\partial u}{\partial x_1} - r \sin \theta \frac{\partial}{\partial \theta}\left(\frac{\partial u}{\partial x_1}\right) - r \sin \theta \frac{\partial u}{\partial x_2} + r \cos \theta \frac{\partial}{\partial \theta}\left(\frac{\partial u}{\partial x_2}\right).$$

Now each of the functions $\partial u / \partial x_1 = \partial f / \partial x_1$ and $\partial u / \partial x_2 = \partial f / \partial x_2$ is a composition of a function of x_1 and x_2 with $X = \varphi(r, \theta)$. Hence in reality

$$\frac{\partial}{\partial \theta}\left(\frac{\partial u}{\partial x_1}\right) = \frac{\partial}{\partial \theta}\left(\frac{\partial f}{\partial x_1} \circ \varphi\right)$$

and

$$\frac{\partial}{\partial \theta}\left(\frac{\partial f}{\partial x_2}\right) = \frac{\partial}{\partial \theta}\left(\frac{\partial f}{\partial x_2} \circ \varphi\right).$$

Therefore

$$\frac{\partial}{\partial \theta}\left(\frac{\partial u}{\partial x_1}\right) = \frac{\partial^2 f}{\partial x_1^2}\frac{\partial x_1}{\partial \theta} + \frac{\partial^2 f}{\partial x_2 \partial x_1}\frac{\partial x_2}{\partial \theta} = -r \sin \theta \frac{\partial^2 u}{\partial x_1^2} + r \cos \theta \frac{\partial^2 u}{\partial x_2 \partial x_1}$$

and

$$\frac{\partial}{\partial \theta}\left(\frac{\partial u}{\partial x_2}\right) = \frac{\partial^2 f}{\partial x_1 \partial x_2}\frac{\partial x_1}{\partial \theta} + \frac{\partial^2 f}{\partial x_2^2}\frac{\partial x_2}{\partial \theta} = -r \sin \theta \frac{\partial^2 u}{\partial x_1 \partial x_2} + r \cos \theta \frac{\partial^2 u}{\partial x_2^2}$$

Assuming $u = f(x_1, x_2)$ has continuous second partial derivatives so that

$$\frac{\partial^2 u}{\partial x_1 \partial x_2} = \frac{\partial^2 u}{\partial x_2 \partial x_1},$$

we have

$$\frac{\partial^2 u}{\partial \theta^2} = -r \cos \theta \frac{\partial u}{\partial x_1} + r^2 \sin^2 \theta \frac{\partial^2 u}{\partial x_1^2} - 2r^2 \sin \theta \cos \theta \frac{\partial^2 u}{\partial x_1 \partial x_2}$$

$$+ r^2 \cos^2 \theta \frac{\partial^2 u}{\partial x_2^2} - r \sin \theta \frac{\partial u}{\partial x_2}.$$

In space two useful nonlinear coordinate transformations are those specified by cylindrical coordinates and spherical coordinates. In the first instance the coordinate transformation $X = \varphi(r, \theta, x_3)$ is defined by

$$x_1 = r \cos \theta$$
$$x_2 = r \sin \theta$$
$$x_3 = x_3.$$

The relation between the rectangular coordinates (x_1, x_2, x_3) and the cylindrical coordinates (r, θ, x_3) of a point P in space is illustrated in Figure 21.

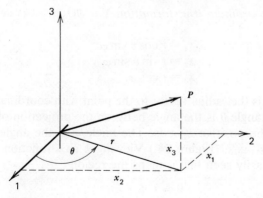

Fig. 21

Thus the infinite box $0 < r < \infty$, $0 \le \theta < 2\pi$, $-\infty < x_3 < \infty$, is mapped by the function φ onto all of $V_3(R)$. The transformation φ is one-to-one when $r > 0$. See Figures 22 and 23.

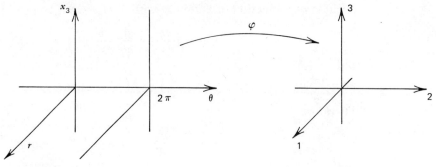

Fig. 22

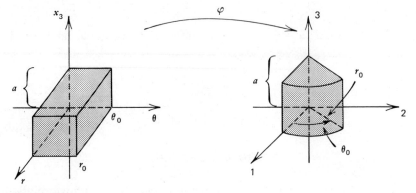

Fig. 23

The Jacobian matrix φ' is easily seen to be

$$\phi'(r, \theta, x_3) = \begin{pmatrix} \frac{\partial x_1}{\partial r} & \frac{\partial x_1}{\partial \theta} & \frac{\partial x_1}{\partial x_3} \\ \frac{\partial x_2}{\partial r} & \frac{\partial x_2}{\partial \theta} & \frac{\partial x_2}{\partial x_3} \\ \frac{\partial x_3}{\partial r} & \frac{\partial x_3}{\partial \theta} & \frac{\partial x_3}{\partial x_3} \end{pmatrix} = \begin{pmatrix} \cos\theta & -r\sin\theta & 0 \\ \sin\theta & r\cos\theta & 0 \\ 0 & 0 & 1 \end{pmatrix}.$$

The *spherical coordinate transformation* $X = \psi(r, \theta, \varphi)$ is defined by the three equations

$$x_1 = r\cos\theta \sin\varphi$$
$$x_2 = r\sin\theta \sin\varphi$$
$$x_3 = r\cos\varphi.$$

If $X = (x_1, x_2, x_3)$ is the radius vector to the point with coordinates x_1, x_2, x_3 then $r = |X|$. The angle θ is the angle between the projection of X onto the x_1, x_2 plane and the positive x_1 axis. The angle φ is the angle between X and the positive x_3 axis. (Figure 24.) Viewing ψ as a function from $V_3(R)$ onto $V_3(R)$ it is easily seen that ψ maps the region

$$0 \leq r < \infty$$
$$0 \leq \theta < 2\pi$$
$$0 \leq \varphi < \pi$$

onto all of $V_3(R)$. The function ψ is one-to-one when $r > 0$. (Figure 25.) Indeed the box $0 \leq r \leq r_0, 0 \leq \theta \leq \theta_0, 0 \leq \varphi \leq \varphi_0$ is mapped onto a spherical wedge in $V_3(R)$ as is illustrated in Figure 26.

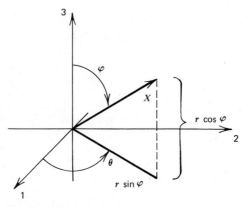

Fig. 24 Spherical coordinates.

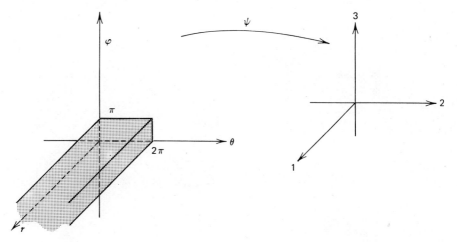

Fig. 25

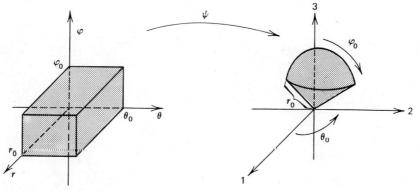

Fig. 26

To gain further insight into the geometry of ψ it is useful to examine the image of various planes in ρ, θ, φ space under the mapping ψ. The planes $r =$ constant are mapped onto concentric spherical shells in $V_3(R)$. The planes $\theta =$ constant go onto half planes perpendicular to the 1, 2 plane and at an angle θ from the 1, 3 plane. The planes $\varphi =$ constant are mapped onto conical shells. We illustrate this by Figure 27.

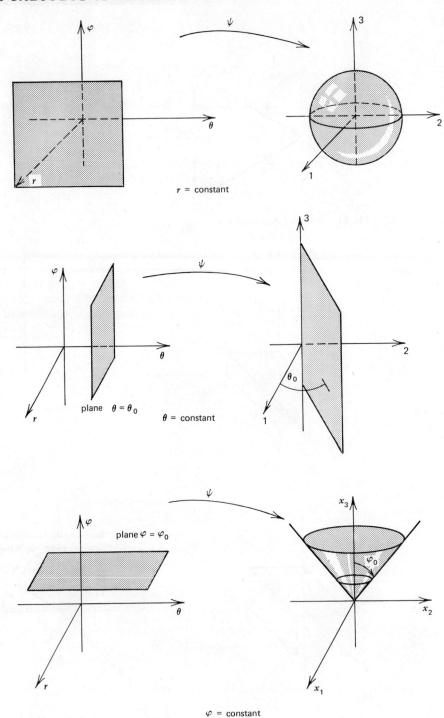

Fig. 27

The Jacobian matrix of ψ is given by
$$\psi'(r, \theta, \varphi) = \begin{pmatrix} \cos\theta\sin\varphi & -r\sin\theta\sin\varphi & r\cos\theta\cos\varphi \\ \sin\theta\sin\varphi & r\cos\theta\sin\varphi & r\sin\theta\cos\varphi \\ \cos\varphi & 0 & -r\sin\varphi \end{pmatrix}.$$

Thus if $u = f(x_1, x_2, x_3)$

$$\frac{\partial u}{\partial r} = \frac{\partial u}{\partial x_1}\frac{\partial x_1}{\partial r} + \frac{\partial u}{\partial x_2}\frac{\partial x_2}{\partial r} + \frac{\partial u}{\partial x_3}\frac{\partial x_3}{\partial r}$$

$$= \frac{\partial u}{\partial x_1}\cos\theta\sin\varphi + \frac{\partial u}{\partial x_2}\sin\theta\sin\varphi + \frac{\partial u}{\partial x_3}\cos\varphi.$$

The application of (1) yields similar formulas for $\partial u/\partial \theta$, $\partial u/\partial \varphi$. Formulas for the second partial derivatives of u with respect to r, θ, φ can also be computed. Such computations we leave for the exercises.

If in place of the angle φ we take $\gamma = \pi/2 - \varphi$ then the triple r, θ, γ are called the *geographical coordinates* of the point (x_1, x_2, x_3). The angles γ and θ are called angles of *latitude* and *longitude*, respectively. The latitude angle is measured from the plane of the equator (the x_1, x_2 plane). The angle γ is positive if x_3 is above the equatorial plane and negative if x_3 is below. See Figure 28. The x_1, x_3 plane is called the plane of the meridian. The equations for x_1, x_2, x_3 are easily seen to be

$$x_1 = r\cos\theta\cos\gamma$$
$$x_2 = r\sin\theta\cos\gamma$$
$$x_3 = r\sin\gamma.$$

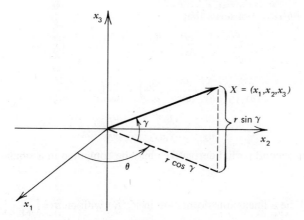

Fig. 28 Geographical coordinates.

The simplest change of coordinates in $V_n(R)$ is that accomplished by an invertible linear transformation. Indeed let L be a linear transformation from $V_n(R)$ onto $V_n(R)$ and set $Y = L(X)$. If $Q = (q_{ij})$ is the matrix of L with respect to the canonical basis in $V_n(R)$, then

$$\begin{pmatrix} y_1 \\ \vdots \\ y_n \end{pmatrix} = Q \begin{pmatrix} x_1 \\ \vdots \\ x_n \end{pmatrix}$$

If now f is a scalar-valued differentiable function, then

$$f(Y) = f(L(X)) = h(X)$$

and by the chain rule

$$\left(\frac{\partial h}{\partial x_i}\right)(X) = \left(\frac{\partial f}{\partial y_j}\right)(L(X))Q$$

and
$$\left(\frac{\partial f}{\partial y_j}\right)(L(X)) = \left(\frac{\partial h}{\partial x_i}\right)(X)Q^{-1}.$$

Writing out each entry $\partial h/\partial x_i$ we have

(6) $$\frac{\partial h}{\partial x_i} = \frac{\partial f}{\partial y_1} q_{1i} + \cdots + \frac{\partial f}{\partial y_n} q_{ni}.$$

Example 1. Let the linear transformation L in $V_2(R)$ have matrix $\begin{pmatrix} 3 & 2 \\ 1 & 1 \end{pmatrix}$. Let f be a differentiable scalar-valued function defined on $V_2(R)$. Let $Y = L(X)$ and set
$$h(X) = f(L(X)) = f(Y).$$
Then
$$h'(X) = f'(L(X)) \cdot L'(X)$$
or
$$\left(\frac{\partial h}{\partial x_1}, \frac{\partial h}{\partial x_2}\right) = \left(\frac{\partial f}{\partial y_1}, \frac{\partial f}{\partial y_2}\right)\begin{pmatrix} 3 & 2 \\ 1 & 1 \end{pmatrix}.$$

We may express $(\partial f/\partial y_1, \partial f/\partial y_2)$ in terms of $(\partial h/\partial x_1, \partial h/\partial x_2)$ by observing that
$$\left(\frac{\partial f}{\partial y_1}, \frac{\partial f}{\partial y_2}\right) = \left(\frac{\partial h}{\partial x_1}, \frac{\partial h}{\partial x_2}\right)\begin{pmatrix} 3 & 2 \\ 1 & 1 \end{pmatrix}^{-1} = \left(\frac{\partial h}{\partial x_1}, \frac{\partial h}{\partial x_2}\right)\begin{pmatrix} 1 & -2 \\ -1 & 3 \end{pmatrix}.$$

To compute $\partial^2 h/(\partial x_1)^2$ observe that
$$\frac{\partial h}{\partial x_1} = \frac{3\partial f}{\partial y_1} + \frac{\partial f}{\partial y_2}.$$
Hence
$$\frac{\partial^2 h}{(\partial x_1)^2} = \frac{3\partial}{\partial y_1}\left(\frac{3\partial f}{\partial y_1} + \frac{\partial f}{\partial y_2}\right) + \frac{\partial}{\partial y_2}\left(\frac{3\partial f}{\partial y_1} + \frac{\partial f}{\partial y_2}\right)$$
$$= \frac{9\partial^2 f}{\partial y_1^2} + \frac{3\partial^2 f}{\partial y_1 \partial y_2} + \frac{3\partial^2 f}{\partial y_2 \partial y_1} + \frac{\partial^2 f}{\partial y_2^2}.$$

The remaining second partial derivatives may be computed in a similar way.

EXERCISES

Let $L(X) = Y$ be a linear transformation in $V_2(R)$ with matrix $\begin{pmatrix} 2 & 1 \\ -1 & 1 \end{pmatrix}$. Let f be a scalar-valued function defined on $V_2(R)$ and define the function g by requiring that
$$g(X) = f(L(X)) = f(Y).$$

1. Express $\partial f/\partial y_1$, $\partial f/\partial y_2$ as linear combinations of $\partial g/\partial x_i$, $i = 1, 2$.
2. Express $\partial g/\partial x_1$, $\partial g/\partial x_2$ as linear combinations of $\partial f/\partial y_i$, $i = 1, 2$.
3. Express $\partial^2 f/\partial y_1^2$ as a linear combination of second partial derivatives of g with respect to x_1, x_2.
4. Express $\partial^2 f/\partial y_2^2$ and $\partial^2 f/\partial y_1 \partial y_2$ as linear combinations of second partial derivatives of g with respect to x_1, x_2.
5. Express $\partial^2 g/\partial x_1^2$ and $\partial^2 g/\partial x_2^2$ as linear combinations of second partial derivatives of f with respect to y_1, y_2.
6. Let r, θ be polar coordinates in the plane. If $u = f(x_1, x_2)$ is twice continuously differentiable, express $\partial^2 u/\partial r^2$ and $\partial^2 u/\partial r \partial \theta$ as linear combinations of first and second partial derivatives of f with respect to x_1, x_2.
7. Let f be a twice continuously differentiable function. Show that $u = f(x + at)$ is a solution of the partial differential equation
$$\frac{\partial^2 u}{\partial t^2} = a^2 \frac{\partial^2 u}{\partial x^2}.$$

Show also that $v = f(x - at)$ is a solution. Also if c, d are constants then
$$h(x, t) = cf(x + at) + df(x - at)$$
is a solution.

8. Let $X = (x_1, x_2)$ and $r = |X|$. If $u = f(r)$, show that
$$\frac{\partial^2 u}{\partial x_1^2} + \frac{\partial^2 u}{\partial x_2^2} = \frac{d^2 u}{dr^2} + \frac{1}{r}\frac{du}{dr}.$$

9. Let $X = (x_1, \ldots, x_n)$ and $r = |X|$. If $u = f(r)$, show that
$$\frac{\partial^2 u}{\partial x_1^2} + \cdots + \frac{\partial^2 u}{\partial x_n^2} = \frac{\partial^2 u}{\partial r^2} + \frac{n-1}{r}\frac{\partial u}{\partial r}.$$

10. If $x = r\cos\theta$ and $y = r\sin\theta$ are polar coordinates in the plane, show that if $u = f(x, y)$, then
$$\left(\frac{\partial u}{\partial x}\right)^2 + \left(\frac{\partial u}{\partial y}\right)^2 = \left(\frac{\partial u}{\partial r}\right)^2 + \frac{1}{r^2}\left(\frac{\partial u}{\partial \theta}\right)^2.$$

11. The second-order partial differential equation
$$\frac{\partial^2 u}{\partial x_1^2} + \cdots + \frac{\partial^2 u}{\partial x_n^2} = 0$$
is called the Laplace equation. Show that in polar coordinates in the plane Laplace's equation $\partial^2 u/\partial x_1^2 + \partial^2 u/\partial x_2^2 = 0$ becomes
$$\frac{\partial^2 u}{\partial r^2} + \frac{1}{r^2}\frac{\partial^2 u}{\partial \theta^2} + \frac{1}{r}\frac{\partial u}{\partial r} = 0.$$

12. Express $\partial^2 u/\partial r^2$, $\partial^2 u/\partial \theta^2$, $\partial^2 u/\partial \varphi^2$ in terms of the first and second partial derivatives of u with respect to x_1, x_2, x_3 if (r, θ, φ) are spherical coordinates in space.

13. If (r, θ, φ) are spherical coordinates in space, show that Laplace's equation $\partial^2 u/\partial x_1^2 + \partial^2 u/\partial x_2^2 + \partial^2 u/\partial x_3^2 = 0$ becomes
$$\frac{1}{r^2}\frac{\partial}{\partial r}\left(r^2 \frac{\partial u}{\partial r}\right) + \frac{1}{r^2 \sin\varphi}\frac{\partial}{\partial \varphi}\left(\sin\varphi \frac{\partial u}{\partial \varphi}\right) + \frac{1}{r^2 \sin^2\varphi}\frac{\partial^2 u}{\partial \theta^2} = 0.$$

14. Let $\psi(r, \theta, \gamma) = (x_1, x_2, x_3)$ be the geographical coordinate transformation defined on p. 243. Compute $\psi'(r, \theta, \gamma)$.

15. Let $X = f(r, \theta)$ be a transformation of $V_2(R)$ onto $V_2(R)$ defined by
$$x_1 = r\frac{e^\theta + e^{-\theta}}{2} \quad x_2 = r\frac{e^\theta - e^{-\theta}}{2}.$$
Determine the Jacobian matrix for f and sketch the curves $r =$ constant and $\theta =$ constant in the (x_1, x_2) plane.

16. If $X = f(r, \theta)$ is the coordinate transformation of Exercise 15 and $u = g(x, y)$, show that
$$\frac{\partial^2 u}{\partial x^2} - \frac{\partial^2 u}{\partial y^2} = \frac{\partial^2 u}{\partial r^2} - \frac{1}{r^2}\frac{\partial^2 u}{\partial \theta^2} + \frac{1}{r}\frac{\partial u}{\partial r}.$$

17. Let r, θ be polar coordinates in the plane. If f and g are two scalar-valued functions satisfying
$$\frac{\partial f}{\partial x_1} = \frac{\partial g}{\partial x_2}, \frac{\partial f}{\partial x_2} = \frac{\partial g}{\partial x_1}$$
show that
$$h(r, \theta) = f(r\cos\theta, r\sin\theta)$$
$$p(r, \theta) = g(r\cos\theta, r\sin\theta)$$
satisfy
$$\frac{\partial h}{\partial r} = \frac{1}{r}\frac{\partial p}{\partial \theta}; \quad \frac{\partial p}{\partial r} = -\frac{1}{r}\frac{\partial h}{\partial \theta}.$$

18. Let λ_1, λ_2 be the eigenvalues of the symmetric matrix $\begin{pmatrix} a & b \\ b & c \end{pmatrix} = A$. Let P be a rotation matrix which diagonalizes A. Then $P^t A P = \begin{pmatrix} \lambda_1 & 0 \\ 0 & \lambda_2 \end{pmatrix}$. For $X = \begin{pmatrix} x_1 \\ x_2 \end{pmatrix}$ define $Y = \begin{pmatrix} y_1 \\ y_2 \end{pmatrix} = P^t X$. If $u = f(X)$ is a scalar-valued function with continuous second partial derivatives, use the chain rule to show that

$$\lambda_1 \frac{\partial^2 u}{\partial y_1^2} + \lambda_2 \frac{\partial^2 u}{\partial y_2^2} = a \frac{\partial^2 u}{\partial x_1^2} + 2b \frac{\partial^2 u}{\partial x_1 \partial x_2} + c \frac{\partial^2 u}{\partial x_2^2}.$$

(Convince yourself that the right-hand side of the above equation behaves like a quadratic form.)

Let $u = f(X)$, where $X = (x_1, x_2)$, be a scalar function with continuous second partial derivatives. Determine rotation matrices P such that if $Y = P^t X$ then:

19. $5 \dfrac{\partial^2 u}{\partial x_1^2} + 8 \dfrac{\partial^2 u}{\partial x_1 \partial x_2} + 5 \dfrac{\partial^2 u}{\partial x_2^2} = 9 \dfrac{\partial^2 u}{\partial y_1^2} + \dfrac{\partial^2 u}{\partial y_2^2}$.

20. $2 \dfrac{\partial^2 u}{\partial x_1 \partial x_2} = \dfrac{\partial^2 u}{\partial y_2^2} - \dfrac{\partial^2 u}{\partial y_2^2}$.

21. $\dfrac{\partial^2 u}{\partial x_1^2} + 4 \dfrac{\partial^2 u}{\partial x_1 \partial x_2} + 4 \dfrac{\partial^2 u}{\partial x_2^2} = 5 \dfrac{\partial^2 u}{\partial y_1^2}$

(See Exercise 18.)

6.10 Inverse and Implicit Functions

Let f be a differentiable function from $V_n(R)$ to $V_n(R)$ defined on an open set U. If f is one-to-one, then for each Y in the range of f we may define a function g by requiring that $g(Y) = X$ if $f(X) = Y$. The function g reverses or inverts the action of f, and this function is called the *inverse* of f and we write $g = f^{-1}$. By definition of g, it is easy to check that

(1) $\qquad\qquad f^{-1} \circ f(X) = X \qquad$ for each X in U

and

(2) $\qquad\qquad f \circ f^{-1}(Y) = Y \qquad$ for each Y in the range of f.

If f^{-1} is known to be differentiable, then applying the chain rule to (1) we see that

(3) $\qquad\qquad d_{f(X)} f^{-1} \cdot d_X f = I.$

Formula (3) then asserts that the differential of f^{-1} is the inverse of the differential of f. That is,

$$d(f^{-1}) = (df)^{-1}.$$

Expressed in terms of the Jacobian matrices this equation reads

$$f^{-1\prime}(f(X)) \cdot f'(X) = I.$$

Thus the matrices $f'(X)$ and $f^{-1\prime}(f(X))$ are inverses of one another.

Now the $n \times n$ matrix $f'(X)$ will have an inverse if and only if its determinant $D(f'(X))$ is not zero. It is reasonable, therefore to ask if the fact that $D(f'(X)) \neq 0$ for each $X \in U$ is sufficient to guarantee that f has a differentiable inverse. If $n = 1$, this is certainly the case. Recall that if f' exists for each real number x in the interval (a, b) and f' never vanishes, then f is one-to-one. Furthermore if f^{-1} is the inverse of f, then f^{-1} is differentiable

and
$$f^{-1\prime}(f(x)) \cdot f'(x) = 1.$$

One may phrase this result in the following way. If the linear transformation $f'(x)$ has an inverse [namely, $1/f'(x)$] for each x in (a, b) then f has an inverse and
$$f^{-1\prime}(f(x)) = \frac{1}{f'(x)}.$$

For $n > 1$, however, this result is false. The trouble is that the assumption that $D(f'(X)) \neq 0$ for each X in the open set U is not enough to guarantee that f is one-to-one. When f is not one-to-one there clearly cannot exist a function g with the property that
$$(g \circ f)(X) = X \quad \text{for each } X \in U.$$

To see that f may not be one-to-one even though $D(f'(X)) \neq 0$, consider the following example.

Let $f(x_1, x_2) = (x_1 x_2, x_2{}^2 - x_1{}^2)$ and let U be the set of all vectors $X \neq 0$. Then
$$f'(X) = \begin{pmatrix} x_2 & x_1 \\ -2x_1 & 2x_2 \end{pmatrix},$$
and
$$D(f'(X)) = 2(x_1{}^2 + x_2{}^2)$$

whech never vanishes on U. However, $f(a, a) = (a^2, 0) = f(-a, -a)$ and f is not one-to-one.

If we are willing to add as a hypothesis that f is one-to-one on the open set U, then we have an analogue of the one-variable theorem.

THEOREM 6.13 *Let the function f from $V_n(R)$ to $V_n(R)$ be differentiable on the open set U. Assume that f is one-to-one on U and that $D(f'(X)) \neq 0$ for each $X \in U$. Then f maps U onto an open set V in $V_n(R)$. If $g = f^{-1}$ is the inverse of f, then g is differentiable and furthermore*

(2) $$d_{f(X)} g = (d_X f)^{-1}.$$

Thus if f is differentiable, one-to-one, and if $(df)^{-1}$ exists, then the inverse of f is also differentiable. The differential of f^{-1} at $f(X)$ is the inverse of the linear transformation $d_X f$. In terms of the Jacobian matrices (2) takes the familiar form
$$f^{-1\prime}(f(X)) \cdot f'(X) = I.$$

The proof of this theorem is in two parts. To show that f maps U onto an open set is quite difficult and is beyond the scope of this text. This fact is a consequence of the continuity of f and is a somewhat subtle theorem of topology. To actually show that f^{-1} is differentiable once it is known that V is open is straightforward and is virtually a repetition of the argument of Theorem 6.12. We leave it as an exercise.

The determinant of the Jacobian matrix $f'(X)$ is called the *Jacobian* of the transformation f. The customary notation, which is in terms of the component functions $f_1, \ldots, f_n$ of f, is

$$D(f'(X)) = \frac{\partial(f_1, \ldots, f_n)}{\partial(x_1, \ldots, x_n)}(X).$$

Suppressing X in this formula yields

$$D(f') = \frac{\partial(f_1,\ldots,f_n)}{\partial(x_1,\ldots,x_n)}$$

Although the fact that $D(f'(X)) \neq 0$ is not sufficient to guarantee that f is one-to-one, we can infer that f is one-to-one in some small ball about X if we strengthen the assumption on f. The next result is commonly called the *inverse function theorem*.

THEOREM 6.14 *Let f be continuously differentiable on the open set U and assume that at some point $X_0 \in U$,*

$$D(f'(X_0)) = \frac{\partial(f_1,\ldots,f_n)}{\partial(x_1,\ldots,x_n)}(X_0) \neq 0.$$

Then there is a ball B about X_0 on which f is one-to-one and on which

$$D(f') = \frac{\partial(f_1,\ldots,f_n)}{\partial(x_1,\ldots,x_n)} \neq 0.$$

Furthermore f maps B onto an open set C containing $f(X_0)$. The inverse function $g = f^{-1}$ is continuously differentiable on C and at each $Y \in B$

$$d_Y(f^{-1}) = (d_X f)^{-1} \quad \text{where} \quad Y = f(X).$$

The proof of this theorem is easier than 6.13. However, it involves an approximation argument which we prefer not to give. We do remark that since f is continuously differentiable, the Jacobian $D(f')$ is a continuous scalar-valued function. Hence $D(f') \neq 0$ on some small ball about X_0. The problem now is to show that f will be one-to-one on some smaller ball B about X_0, and further that f maps B onto an open set C.

Example 1. Let $f(x_1, x_2) = (x_1 x_2, x_2^2 - x_1^2)$ be the function considered previously. Then

$$f'(X) = \begin{pmatrix} x_2 & x_1 \\ -2x_1 & 2x_2 \end{pmatrix}$$

and

$$Df'(X) = 2(x_2^2 + x_1^2).$$

Certainly f is continuously differentiable everywhere, and if $X_0 \neq 0$, then $Df'(X_0) \neq 0$. Applying 6.15 we assert that in some *small* ball B about $X_0 \neq 0$ f is one-to-one. For each X in B, $g = f^{-1}$ exists at the point $f(X)$ and

$$g'(f(X)) = (f^{-1})'(f(X)) = \begin{pmatrix} x_2 & x_1 \\ -2x_1 & 2x_2 \end{pmatrix}^{-1} = \frac{1}{2(x_2^2 + x_1^2)} \begin{pmatrix} 2x_2 & -x_1 \\ 2x_1 & x_2 \end{pmatrix}.$$

For all values of $X \neq 0$ for which f has an inverse, this formula is correct. However, as we have seen, f does not have an inverse for all values of $X \neq 0$.

If $X \in S$ and S contains an open ball B about S, then S is called a *neighborhood* of X. In particular B itself is a neighborhood of X. Theorem 6.14 then asserts that if $D(f'(X)) \neq 0$, then f has an inverse in some neighborhood of $f(X)$. Whenever a phenomenon is asserted to hold in some neighborhood of a point X, then we say that the phenomenon holds *locally*. Thus f has an inverse *locally* at each point $f(X)$ if $D(f'(X)) \neq 0$. This does not imply that f has an inverse *globally*, i.e. on the entire domain of f, as the above example shows.

We shall apply Theorem 6.14 to the following problem of implicit functions. Suppose $x_3 = f(x_1, x_2)$ is a continuously differentiable scalar function defined on an open set U in the plane. Consider the set

$$S_c = \{X = (x_1, x_2) : f(x_1, x_2) = \text{constant } c\}.$$

Such a set S is called a *level set* or a *set of constancy* for the function f. We wish to specify conditions under which we can assert that there exists a function g from R to $V_2(R)$ such that for all values of t in some interval (a, b)

(4) $$f(g(t)) = f(g_1(t), g_2(t)) = c.$$

A function g, when it exists, and satisfies (4), is said to be *defined implicitly* by equation (4). Many functions may be defined implicitly by such an equation.

Example 2. If $f(x_1, x_2) = x_1^2 + x_2^2$, consider the equation

(5) $$f(x_1, x_2) = x_1^2 + x_2^2 = 4.$$

The functions $g(t) = (t, \sqrt{4 - t^2})$ $h(t) = (t, -\sqrt{4 - t^2})$ for $-2 \leq t \leq 2$ are defined implicitly by (4) since $t^2 + (\pm\sqrt{4 - t^2})^2 = 4$. Also $p(t) = (\sqrt{4 - t^2}, t)$ and $q(t) = (\sqrt{4 - t^2}, t)$ for $-2 \leq t \leq 2$ satisfy (4). Hence these functions also defined implicitly by (4).

In this example it was fairly obvious how to define functions $g(t)$ of the form $(t, g_2(t))$ or $(g_1(t), t)$ such that $f(g(t)) = 4$. However, if for example

$$f(x_1, x_2) = x_1 + \sin(x_1 x_2) + e^{x_1 + x_2} = 1$$

it is by no means obvious that there exists a function

$$g(t) = (g_1(t), g_2(t))$$

such that
$$g_1(t) + \sin(g_1(t) g_2(t)) + e^{g_1(t) + g_2(t)} = 1.$$

What we wish to do is to give conditions on f which guarantee that if $X_0 = (x_1^0, x_2^0)$ is a point on the level set $f(x_1, x_2) = c$, then there will exist a function g defined on an open interval $I = (a, b)$ such that for some $t_0 \in I$ $g(t_0) = X_0$ and $f(g(t)) = f(g_1(t), g_2(t)) = c$ for each $t \in I$. It will turn out that these functions g will be expressible in one of two ways. Either

$$g(t) = (t, g_2(t))$$

or
$$g(t) = (g_1(t), t).$$

If we think of $x_3 = f(x_1, x_2)$ as defining a surface or landscape over an open set in the plane, then the level sets are just the contour lines, or sets of constant altitude for the surface. See Figures 29 and 30.

Let us look naively at this "contour map" and assume that in some sense each level set has a smoothly turning tangent line. Let (a, b) be a point on the level set $f(x_1, x_2) = c$.

It is evident from Figure 31 that for t close to the point a there is a function $g(t)$ of the form $g(t) = (t, g_2(t))$ which satisfies $f(g(t)) = c$. This is clear because in a neighborhood of the point (a, b) each vertical line intersects the graph of $f(x_1, x_2) = c$ exactly once. This is precisely the condition we need to guarantee that we may write $x_2 = g_2(x)$ and have $f(x_1, g_2(x_1)) = c$ for x_1 near a. Similarly if t is close to b, then apparently there exists a function $g(t) = (g_1(t), t)$ satisfying $f(g(t)) = c$, since in a neighborhood of (a, b) each horizontal line intersects the graph of $f(x_1, x_2) = c$ exactly once.

250 THE CALCULUS OF VECTOR FUNCTIONS

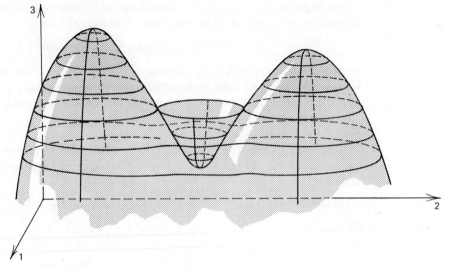

Fig. 29 Surface $x_3 = f(x_1, x_2)$ and its level sets.

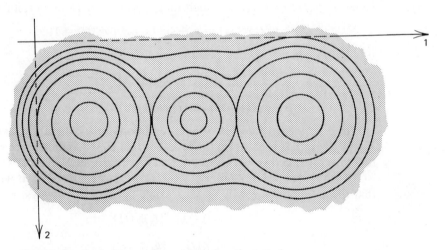

Fig. 30 Level sets for a surface projected on the 1, 2 plane.

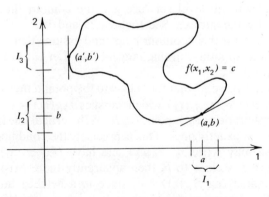

Fig. 31

INVERSE AND IMPLICIT FUNCTIONS 251

The situation in a neighborhood of the point (a', b') is somewhat different. In a neighborhood of this point, vertical lines intersect the graph of $f(x_1, x_2) = c$ either twice (for $x_1 > a'$) or not at all (for $x_1 < a'$). Indeed, there does not exist a function $g(t) = (t, g_2(t))$ defined on a neighborhood I of the point a' which satisfies $f(g(t)) = c$ for each $t \in I$. However, there does exist a function $g(t)$ of the form $g(t) = (g_1(t), t)$ defined in a neighborhood of b' and satisfying $f(g(t)) = c$.

The figure seems to indicate that if the tangent line is not vertical we can find the desired function in the form $g(t) = (t, g_2(t))$. If the tangent line is not horizontal then we may find a function of the form $g(t) = (g_1(t), t)$ satisfying $f(g(t)) = c$. The next result shows that if $\partial f/\partial x_2(a, b) \neq 0$ then a function $g(t) = (t, g_2(t))$ can be found satisfying $f(g(t)) = c$ for t in a neighborhood of a. The theorem only formalizes the discussion we have just given by providing a mathematical construction for the function $g_2(t)$.

THEOREM 6.15 *Let f be a continuously differentiable scalar function defined on an open set U in $V_2(R)$. Let $S_c = \{X \in U : f(x_1, x_2) = c\}$ be a level set for f and assume $(a, b) \in S_c$. If $\partial f/\partial x_2(a, b) \neq 0$, then there exists an open interval I containing the point a and a unique continuously differentiable function $x_2 = g_2(t)$ defined on I such that if $g(t) = (t, g_2(t))$ then $f(g(t)) = f(t, g_2(t)) = c$ for each $t \in I$.*

Proof. Our principal tool will be the inverse function theorem. Hence we must define a function F from $V_2(R)$ to $V_2(R)$ to which this theorem applies. To do this set
$$F(x_1, x_2) = (x_1, f(x_1, x_2)).$$
Then F maps the level set S_c into a portion of the line $y_2 = c$. Furthermore
$$F'(X) = \begin{pmatrix} 1 & 0 \\ \dfrac{\partial f}{\partial x_1} & \dfrac{\partial f}{\partial x_2} \end{pmatrix}.$$
Hence at the point $X_0 = (a, b)$ the Jacobian
$$D(F'(X_0)) = \frac{\partial f}{\partial x_2}(a, b) \neq 0 \qquad \text{by assumption.}$$

Applying Theorem 6.14 to F evaluated at X_0 we conclude the following. There exists an open disc B about $X_0 = (a, b)$ on which F is one-to-one. Furthermore F maps B onto an open set C containing
$$F(a, b) = (a, f(a, b)) = (a, c),$$
and if G is the inverse function for F on C, then G is continuously differentiable. The construction is illustrated in Figure 32.

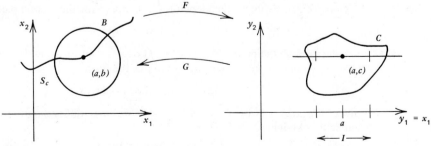

Fig. 32 $F(x_1, x_2) = (y_1, y_2) = (x_1, f(x_1, x_2))$, $F(a, b) = (a, c)$, $G = F^{-1}$.

Since C is open we may choose an interval I about the point a such that for each $t \in I$, $(t, c) \in C$. If we restrict the inverse function G to the points (t, c), $t \in I$ then $G(t, c) \in S_c$ since G is the inverse of F. Moreover this is the function $g(t)$ that we seek. To be precise, define

$$g(t) = G(t, c) \quad \text{for } t \in I.$$

Then

$$g(t) = (t, G_2(t, c))$$

where $G_2(t, c)$ is the second coordinate function for the inverse function G. If we set $g_2(t) = G_2(t, c)$ then $g(t) = (t, g_2(t))$. For $t \in I$, $g(t) \in S_c$. Hence $f(g(t)) = c$. The function $g(t)$ is continuously differentiable since G is. The uniqueness of g_2 we leave as an exercise. This completes the proof.

This result is clearly symmetric in x_1 and x_2. If $\partial f/\partial x_1(a, b) \neq 0$, then there exists a function $g(t)$ of the form $g(t) = (g_1(t), t)$ such that $f(g(t)) = c$. If we note that $(\nabla f)(X) = (\partial f/\partial x_1, \partial f/\partial x_2) \neq 0$ if and only if $\partial f/\partial x_1 \neq 0$ or $\partial f/\partial x_2 \neq 0$, then we may incorporate both results in the following theorem.

THEOREM 6.16 *Let f be a continuously differentiable scalar function defined on an open set U in $V_2(r)$. Let $S_c = \{X \in U : f(x_1, x_2) = c\}$ be a level set for f and assume $(a, b) \in S_c$. If $(\nabla f)(a, b) \neq 0$, then there exists an open interval I and a continuously differentiable function g mapping I into $V_2(R)$ satisfying $f(g(t)) = c$ and $g(t_0) = (a, b)$ for some point t_0 in I.*

The curves g defined by this theorem are called level curves for the function f. At a point (a, b) the tangent vector is in the direction of $g(t) = (g_1'(t), g_2'(t))$. This vector is never zero since $g'(t)$ can always be written in the form $(1, g_2'(t))$ or $(g_1'(t), 1)$. Note that since

$$f(g(t)) = c$$

we may apply the chain rule obtaining

(6) $$\frac{\partial f}{\partial x_1} g_1'(t) + \frac{\partial f}{\partial x_2} g_2'(t) = 0.$$

Thus the vector $\nabla f = (\partial f/\partial x_2, \partial f/\partial x_2)$ is normal to the curve g at the point (a, b). If $g(t) = (t, g_2(t))$, then $g'(t) = (1, g_2'(t))$, and we may deduce from (6) that

$$g_2'(t) = -\frac{\partial f/\partial x_1}{\partial f/\partial x_2}$$

which is a convenient formula for determining the slope of a level curve. This formula need not be memorized, however, since g_2' may be determined by differentiating $f(x_1, x_2) = c$ and applying the chain rule.

Example 3. Evaluate the slope of a level curve for the elliptic paraboloid

$$x_3 = 4x_1^2 + x_2^2$$

passing through the point $(1, 2, 8)$. Now $f(x_1, x_2) = 4x_1^2 + x_2^2$ and since $\partial f/\partial x_2 \neq 0$ at $(1, 2, 8)$ we may assume the level curve has the form

$$g(x_1) = (x_1, g_2(x_1)) = (x_1, x_2).$$

Since $f(g(x_1)) = 8$ or $4x_1^2 + g_2^2(x_1) = 8$ we may differentiate this expression implicitly obtaining

$$8x_1 + 2x_2 g_2'(x_1) = 0.$$

Hence $g_2'(x_1) = -4x_1/x_2$. At $(1, 2)$, $g_2'(1) = -2$.

EXERCISES

Determine the points X_0 at which $D(f'(X_0)) \neq 0$. For each such X_0 there exists by Theorem 6.14 an inverse g for f in a neighborhood of $f(X_0)$. Compute $g'(f(X))$ if $f(X)$ belongs to this neighborhood.

1. $f(x_1, x_2) = (2x_1 - x_2, x_1 + x_2)$.
2. $f(x_1, x_2) = (x_1 x_2, x_1^2 + x_2^2)$.
3. $f(x_1, x_2) = ((x_1 - x_2)^2, x_1^2/x_2)$.
4. $f(x_1, x_2) = (\cos(x_1 + x_2), \sin(x_1 - x_2))$.
5. $f(x_1, x_2, x_3) = (2x_1 - x_3, x_1 + x_2, x_2 - 2x_3)$.
6. $f(x_1, x_2, x_3) = (x_1 x_2, x_2 x_3, x_3 x_1)$.
7. Let $f(x_1, x_2) = (e^{x_1} \cos x_2, e^{x_1} \sin x_2)$. Show that the Jacobian of f does not vanish at any point $X \in V_2(R)$. Yet f is not one-to-one. Sketch the curves $f(c, x_2)$, $f(x_1, c)$ for a constant c.

Determine the points X on the following level sets $f(x_1, x_2) = c$ at which $\partial f/\partial x_2 \neq 0$. Hence by Theorem 6.15 in some interval containing the first coordinate of each such point X there exists a function $x_2 = g_2(x_1)$ satisfying $f(x_1, g_2(x_1)) = c$. Express g_2' as a function of x_1, x_2.

8. $2x_1^2 + 3x_2^2 = 1$.
9. $2x_1^2 + x_2^3 + x_2 = 1$.
10. $e^{x_1 - x_2} + x_2^2 = 1$.
11. $\cos(x_1 + x_2) - \log(x_1^2 + x_2^2) = 1$.
12. For the level sets of Exercises 8–11 determine the points X such that there exists level curves of the form $g(x_2) = (g_1(x_2), x_2)$. Express g_1' as a function of x_1 and x_2.

Determine by the use of Theorem 6.16 if the following level sets $S_c = \{f(x_1, x_2) = c\}$ implicitly define curves locally at the given points $A \in S_c$. If so determine a tangent vector to the curve at A. Use Theorem 6.16.

13. $x_1^2 + x_2^2 - x_1 + x_2 = 3$, $A = (2, 1)$.
14. $e^{x_1 + x_2} + x_1 = 1$, $A = (0, 0)$.
15. $\sin(x_1 + x_2) + x_2 = \pi/2$, $A = (\pi/2, \pi/2)$.
16. $(x_1 - 1)(x_2 - 1) = 0$, $A = (1, 1)$.
17. Let $x_3 = f(x_1, x_2)$. Show that the gradient of f at X is perpendicular to the tangent line to each level curve of $f(x_1, x_2) = c$ which passes through X.
18. Show that the directional derivative of f at X is a maximum in a direction perpendicular to each level curve of $f(x_1, x_2) = c$ which passes through X.
19. Let f satisfy the conditions of Theorem 6.16 and let g be a level curve for $f(X) = c$ passing through the point X_0. If N is a vector perpendicular to the tangent of g at X_0, show that N is a scalar multiple of ∇f at X_0.
20. Referring to Exercise 19, let $h(x_1, x_2)$ be another differentiable function and set $p(t) = h(g(t))$ where $g(t)$ is a level curve for f. Show that if $p'(t_0) = 0$, then at the point $g(t_0)$, ∇h is a scalar multiple of ∇f.
21. A continuously differentiable function $f(x_1, x_2)$ has a level set in the form of a figure 8. Explain why ∇f must vanish at the point where the level set crosses itself. (Assume the existence of two level curves passing through the point of intersection.)
22. Show that the function g_2 constructed in Theorem 6.15 is unique. That is show that if $\tilde{g}$ is any function satisfying $f(x_1, \tilde{g}(x_1)) = c$ and $(x_1, \tilde{g}(x_1)) \in B$, then $\tilde{g}(x_1) = \tilde{g}(x_1)$. (Use the fact that F is one-to-one on B.)

*6.11 Implicit Functions – Continued

We turn next to a vector analogue of Theorem 6.15. This result stated that if f is a function from $V_2(R)$ to R and $\partial f/\partial x_2 \neq 0$ at some point $A = (a_1, a_2)$

belonging to the level set $f(x_1, x_2) = c$ then there exists a function $x_2 = g_2(x_1)$ satisfying $a_2 = g_2(a_1)$ and $f(x_1, g_2(x_1)) = c$ for x_1 in a neighborhood of a_1.

This result may be generalized verbatim to functions f from $V_{n+m}(R)$ to $V_m(R)$. We shall consider such a function f as a function of two vector variables $X = (x_1, \ldots, x_n)$ and $Y = (y_1, \ldots, y_m)$. We then write

$$f(x_1, \ldots, x_n, y_1, \ldots, y_m) = f(X, Y)$$

and we shall abbreviate the $(n+m)$-tuple $(x_1, \ldots, x_n, y_1, \ldots, y_m)$ by $[X, Y]$. Let C be a constant vector in $V_m(R)$ and set

$$S_C = \{[X, Y] \in V_{n+m}(R) : f(X, Y) = C\}.$$

Such a set we call a *set of constancy* for the function f. Let $[A, B] \in S_C$. We seek conditions on f which guarantee that in a neighborhood of the point $A = (a_1, \ldots, a_n)$ there exist m functions $g_1, \ldots, g_m$ from V_n to R such that if

(1) $$y_k = g_k(x_1, \ldots, x_n)$$

then $(f(x_1, \ldots, x_n, g_1(x_1, \ldots, x_n), \ldots, g_m(x_1, \ldots, x_n))) = C$. Furthermore if $b_k = g_k(A)$ then $B = (b_1, \ldots, b_m)$. In other words we assert that in a neighborhood of the point A the equation

$$f(x_1, \ldots, x_n, y_1, \ldots, y_m) = (c_1, \ldots, c_m)$$

determines $y_1, \ldots, y_m$ as functions of $x_1, \ldots, x_n$. If we let $g_1, \ldots, g_m$ be the component functions of a vector function g, then we may phrase our problem in the following way. If $[A, B] \in S_C$, what is a condition on the vector function f which guarantees that in some neighborhood of the point A there exists a function g from V_n to V_m such that if $Y = g(X)$, then $f(X, g(X)) = C$ and $B = f(A)$? Such a vector function g is said to be *implicitly defined* by the equation $f(X, Y) = C$. The answer to this question is given by the following theorem.

THEOREM 6.17 Let $Z = f(X, Y)$ be a continuously differentiable function from $V_{n+m}(R)$ to $V_m(R)$ which is defined on the open set U. We assume $X \in V_n(R)$ and $Y \in V_m(R)$ and let

$$S_C = \{[X, Y] = (x_1, \ldots, x_n, y_1, \ldots, y_m) \in U : f(X, Y) = C\}$$

be a set of constancy for f. Let $[A, B] \in S_C$ and let $f_1, \ldots, f_m$ be the component functions of f. If the Jacobian

$$D\left(\left(\frac{\partial f_i}{\partial y_j}\right)(A, B)\right) = \frac{\partial(f_1, \ldots, f_m)}{\partial(y_1, \ldots, y_m)} \neq 0,$$

then there exists an open neighborhood V of A and a unique continuously differentiable function $Y = g(X)$ from $V_n(R)$ to $V_m(R)$ defined on V which satisfies

$$B = g(A)$$

and

$$f(X, g(X)) = C$$

for each $X \in V$.

Proof. Just as in the proof of 6.15 we must define a function F, this time from $V_{n+m}(R)$ to $V_{n+m}(R)$, to which the inverse function theorem applies.

IMPLICIT FUNCTIONS

To this end set
$$F(X, Y) = [X, f(X, Y)].$$

Then the Jacobian matrix
$$F'(X, Y) = \begin{pmatrix} I & 0 \\ \left(\dfrac{\partial f_i}{\partial x_k}\right) & \left(\dfrac{\partial f_i}{\partial y_j}\right) \end{pmatrix}.$$

This matrix is an $n+m$ square matrix, where I is the $n \times n$ identity matrix and $(\partial f_i/\partial x_k)$ and $(\partial f_i/\partial y_j)$ are $m \times n$ and $m \times m$ Jacobian matrices. Taking determinants we have immediately that

$$D\{F'(A, B)\} = D\left\{\left(\dfrac{\partial f_i}{\partial y_j}\right)(A, B)\right\}.$$

This quantity does not vanish by assumption, which means that we may apply Theorem 6.14 and conclude that there exists an open neighborhood D about $[A, B]$ on which F is one-to-one. Furthermore F maps D onto an open set E containing

$$F(A, B) = [A, f(A, B)] = [A, C].$$

If G is the inverse function for F on E, then G is continuously differentiable. With a little artistic license we may picture this exactly as we did in the one-variable case. See Figure 33.

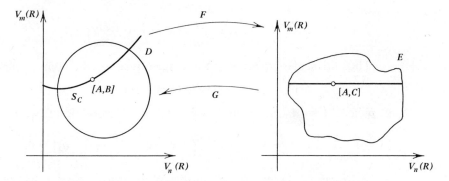

Now since E is open in $V_{n+m}(R)$, there exists an open ball V in $V_n(R)$ about A such that if $X \in V$, $[X, C] \in E$. (Why does this follow?) Moreover if $[X, Z] \in E$ and $G(X, Z) = [X, Y]$, then this defines a function $Y = G_2(X, Z)$. Just as in the one-variable case we set

$$g(X) = G_2(X, C)$$

If $X \in V$, and $Y = g(X)$, then $G(X, C) = [X, Y]$. Hence $F(X, Y) = [X, C]$ or
$$f(X, g(X)) = C.$$

The uniqueness follows just as before. We leave it as an exercise.

As our first illustration of this theorem we consider the problem of defining surfaces implicitly. To this end let f be a function from $V_3(R)$ to R and consider the set of constancy

$$S_c = \{X : f(x_1, x_2, x_3) = c\}.$$

If $A = (a_1, a_2, a_3) \in S_c$, and $\partial f/\partial x_3 \neq 0$ at A, then Theorem 6.17 asserts that there exists a function
$$x_3 = g(x_1, x_2)$$
such that $a_3 = g(a_1, a_2)$ and for each $X = (x_1, x_2)$ in a neighborhood of (a_1, a_2)
$$f(x_1, x_2, g(x_1, x_2)) = c.$$

The function $h(x_1, x_2) = (x_1, x_2, g(x_1, x_2))$ is a function from $V_2(R)$ to $V_3(R)$ and will define a two-dimensional surface if the tangent space is two-dimensional at each point. The tangent space is spanned by the two tangent vectors $\partial h/\partial x_1$ and $\partial h/\partial x_2$. Since
$$\frac{\partial h}{\partial x_1} = \left(1, 0, \frac{\partial g}{\partial x_1}\right) \quad \text{and} \quad \frac{\partial h}{\partial x_2} = \left(0, 1, \frac{\partial g}{\partial x_2}\right),$$
these vectors are clearly linearly independent. Hence the surface is two-dimensional at each point. A normal vector to the surface at a point X is given by $\nabla f(X)$ since $f(h(x_1, x_2)) = C$ implies $f'(h(x_1, x_2)) \cdot h'(x_1, x_2) = 0$. Writing out the matrices involved we have

$$\left(\frac{\partial f}{\partial x_1}, \frac{\partial f}{\partial x_2}, \frac{\partial f}{\partial x_3}\right) \begin{pmatrix} 1 & 0 \\ 0 & 1 \\ \frac{\partial g}{\partial x_1} & \frac{\partial g}{\partial x_2} \end{pmatrix} = 0$$

or
$$\left(\nabla f(X), \frac{\partial h}{\partial x_1}\right) = \left(\nabla f(X), \frac{\partial h}{\partial x_2}\right) = 0.$$

Thus the scalar equation of the tangent plane at the point A can be written

(2) $$(\nabla f(A), X - A) = 0$$

where $A = (a_1, a_2, a_3)$ and $a_3 = g(a_1, a_2)$. Expanding (2) we have

(3) $$\frac{\partial f}{\partial x_1}(x_1 - a_1) + \frac{\partial f}{\partial x_2}(x_2 - a_2) + \frac{\partial f}{\partial x_3}(x_3 - a_3) = 0.$$

Example 1. Determine the equation of the tangent plane to the surface defined implicitly by the equation

(4) $$x_1 x_2 + x_2 x_3 + x_3 x_1 = 3$$

at the point $(1, 1, 1)$. Here $f(x_1, x_2, x_3) = x_1 x_2 + x_2 x_3 + x_3 x_1$ and $\partial f/\partial x_3 = x_2 + x_1$. Hence at the point $A = (1, 1, 1)$ $\partial f/\partial x_3 = 2 \neq 0$ and a surface is defined implicitly by (4). Since $\nabla f(A) = (2, 2, 2)$ the equation of the tangent plane computed from (3) is

$$2(x_1 - 1) + 2(x_2 - 1) + 2(x_3 - 1) = 0$$

or
$$x_1 + x_2 + x_3 = 3.$$

It is easily seen (Section 5.7) that the surface is a one-sheeted hyperboloid.

These results are clearly symmetric in the variables x_1, x_2, x_3. Hence if $A = (a_1, a_2, a_3)$ is a point in the set of constancy

$$S_c = \{X : f(x_1, x_2, x_3) = c\},$$

and one of the partial derivatives $\partial f/\partial x_1, \partial f/\partial x_2, \partial f/\partial x_3$ fail to vanish, then this variable may be expressed locally as a function of the other two. From this we may deduce the following result.

THEOREM 6.18 *Let f be a continuously differentiable scalar function defined in an open set in $V_3(R)$. If A belongs to the set of constancy*

$$S_c = \{X : f(X) = c\}$$

and $\nabla f(A) \neq 0$, then there exists a continuously differentiable function h from $V_2(R) \to V_3(R)$ defined in a neighborhood V of a point B such that

$$h(B) = A.$$

Moreover $f(h(X)) = c$ for $X \in V$.

The proof of this result just consists in showing that G can be written

$$h(x_1, x_2) = (x_1, x_2, g_3(x_1, x_2)) \quad \text{if } \frac{\partial f}{\partial x_3} \neq 0$$

$$h(x_1, x_3) = (x_1, g_2(x_1, x_3), x_3) \quad \text{if } \frac{\partial f}{\partial x_2} \neq 0$$

$$h(x_2, x_3) = (g_1(x_2, x_3), x_2, x_3) \quad \text{if } \frac{\partial f}{\partial x_1} \neq 0.$$

We leave the details to the exercises. It is clear that in each case the two partial derivatives of h are linearly independent, hence h defines a two-dimensional surface at each point. If we write $h(s, t)$ for one of the three alternatives above, we see that the tangent plane to the surface h at the point $h(s_0, t_0)$ is the plane through $h(s_0, t_0)$ spanned by the vectors $\partial h/\partial s$ and $\partial h/\partial t$. This can also be characterized as the plane through $h(s_0, t_0)$ orthogonal to the vector ∇f. That ∇f is orthogonal to the partial derivatives $\partial h/\partial s$ and $\partial h/\partial t$ follows from the chain rule since

$$f(h(s, t)) = c.$$

Hence

$$f'(h(s, t)) \cdot h'(s, t) = 0$$

which implies

$$\left(\nabla f, \frac{\partial h}{\partial s}\right) = \left(\nabla f, \frac{\partial h}{\partial t}\right) = 0.$$

Hence if $A = h(s, t)$, then the tangent plane to the surface at the point A has equation

$$(\nabla f(A), X - A) = 0$$

or

$$\frac{\partial f}{\partial x_1}(x_1 - a_1) + \frac{\partial f}{\partial x_2}(x_2 - a_2) + \frac{\partial f}{\partial x_3}(x_3 - a_3) = 0.$$

Example 2. Let

$$f(x_1, x_2, x_3) = x_1 \sqrt{x_2 - e^{x_3}}.$$

Determine if a surface through the point $A = (1, 2, 0)$ is defined implicitly by the equation $f(X) = 1$. If so, determine the equation of the tangent plane to the surface at A. First

$$\nabla f(X) = \left(\sqrt{x_2 - e^{x_3}}, \frac{1}{2} \frac{x_1}{\sqrt{x_2 - e^{x_3}}}, \frac{1}{2} \frac{-x_1 e^{x_3}}{\sqrt{x_2 - e^{x_3}}}\right)$$

and

$$\nabla f(A) = \left(1, \frac{1}{2}, -\frac{1}{2}\right) \neq 0.$$

258   THE CALCULUS OF VECTOR FUNCTIONS

Hence there is a surface through A and its tangent plane has equation

$$(\nabla f(A), X - A) = 0.$$

Expanding we have

$$1(x_1 - 1) + \frac{1}{2}(x_2 - 2) - \frac{1}{2}(x_3) = 0$$

or

$$2x_1 + x_2 - x_3 = 4.$$

Example 3. Determine if the equations

(5)
$$\begin{aligned} 2x - y + u^3 - v^2 &= 1 \\ x + y + u^2 + v^3 &= 4 \end{aligned}$$

can be solved for functions $u = g(x, y)$, $v = h(x, y)$ in a neighborhood of the point $(1, 1)$ so that $g(1, 1) = h(1, 1) = 1$. If so, determine $\partial g/\partial x$ and $\partial h/\partial y$ at the point $(1, 1)$.

Define

$$f(x, y, u, v) = (2x - y + u^3 - v^2, x + y + u^2 + v^3).$$

Then if f_1, f_2 are the component functions of f we have

$$\frac{\partial(f_1, f_2)}{\partial(u, v)} = D\begin{pmatrix} 3u^2 & -2v \\ 2u & 3v^2 \end{pmatrix} = 9u^2v^2 + 4uv.$$

When $u = v = 1$, this Jacobian is different from zero. Applying Theorem 6.17 we conclude that the functions g and h exist. Differentiating equations (5) with respect to x by the chain rule remembering that both u and v are functions of x and y we obtain

$$2 + 3u^2 \frac{\partial u}{\partial x} - 2v \frac{\partial v}{\partial x} = 0$$

$$1 + 2u \frac{\partial u}{\partial x} + 3v^2 \frac{\partial v}{\partial x} = 0.$$

These linear equations may be solved in the usual way for the unknowns $\partial u/\partial x$ and $\partial v/\partial x$ obtaining

$$\frac{\partial u}{\partial x} = \frac{-6v^2 - 2v}{9u^2v^2 + 4uv} = -\frac{6v + 2}{9u^2v + 4u}.$$

When

$$u = v = 1, \frac{\partial u}{\partial x} = -\frac{8}{13}.$$

Differentiating (5) with respect to y we obtain

$$-1 + 3u^2 \frac{\partial u}{\partial y} - 2v \frac{\partial v}{\partial y} = 0$$

$$1 + 2u \frac{\partial u}{\partial y} + 3v^2 \frac{\partial v}{\partial y} = 0.$$

This implies that

$$\frac{\partial v}{\partial y} = -\frac{3u^2 + 2u}{9u^2v + 4u} = -\frac{3u + 2}{9uv + 4}.$$

When $u = v = 1$, $\partial v/\partial y = -5/13$.

The function $G(x, y) = (x, y, g(x, y), h(x, y))$ is a function from $V_2(R)$ to $V_4(R)$ and is the local surface in 4-space defined implicitly by equation (5). The vectors

$$T_1 = \left(1, 0, \frac{\partial g}{\partial x}, \frac{\partial h}{\partial x}\right)$$

and

$$T_2 = \left(0, 1, \frac{\partial g}{\partial y}, \frac{\partial h}{\partial y}\right)$$

IMPLICIT FUNCTIONS

are linearly independent tangent vectors. Two linearly independent normal vectors are ∇f_1 and ∇f_2. The reader should note that T_1, T_2, ∇f_1, ∇f_2 span $V_4(R)$.

EXERCISES

By using Theorem 6.18 determine if the following sets of constancy

$$S_c = \{f(x_1, x_2, x_3) = c\}$$

implicitly define surfaces locally at the given points $A \in S_c$. If so determine a normal vector to the tangent plane to the surface at the point A. Use Theorem 6.18.

1. $x_1 x_2^2 - x_2 x_3^2 + x_3 x_1^2 = 1$, $A = (1, 1, 1)$.
2. $\log(x_1^2 + x_2^2 + x_3^2) + x_1 x_2 x_3 = 0$, $A = (0, 0, 1)$.
3. $\sin(x_1 + x_2) + \sin(x_2 + x_3) + \sin(x_3 + x_1) = 3$, $A = (\pi/4, \pi/4, \pi/4)$.
4. $e^{x_1 + x_2 + x_3} + x_1 - x_2 + x_3 = 3$, $A = (1, 1, 2)$.
5. Let $f_1(x_1, x_2, x_3)$ and $f_2(x_1, x_2, x_3)$ be two continuously differentiable scalar functions. Let $A = (a_1, a_2, a_3)$ belong to the sets of constancy

$$\{X : f_1(X) = c_1\}$$

and

$$\{X : f_2(X) = c_2\}.$$

Show that there exist functions $x_2 = g(x_1)$, $x_3 = h(x_1)$ defined in a neighborhood I of $x_1 = a$, satisfying $a_2 = g(a_1)$, $a_3 = h(a_1)$, and

$$f_1(x_1, g(x_1), h(x_1)) = c_1$$
$$f_2(x_1, g(x_1), h(x_1)) = c_2$$

if

$$\frac{\partial(f_1, f_2)}{\partial(x_2, x_3)} \neq 0 \quad \text{at } A.$$

The function $G(x_1) = (x_1, g(x_1), h(x_1))$ is a curve defined implicitly by the equations $f_1(X) = c_1$, $f_2(X) = c_2$.

6. If in problem 5, $\partial(f_1, f_2)/\partial(x_2, x_3) \neq 0$ at A, show that

$$\begin{pmatrix} \dfrac{\partial f_1}{\partial x_2} & \dfrac{\partial f_1}{\partial x_3} \\ \dfrac{\partial f_2}{\partial x_2} & \dfrac{\partial f_2}{\partial x_3} \end{pmatrix} \begin{pmatrix} \dfrac{dg}{dx_1} \\ \dfrac{dh}{dx_1} \end{pmatrix} = - \begin{pmatrix} \dfrac{\partial f_1}{\partial x_1} \\ \dfrac{\partial f_2}{\partial x_1} \end{pmatrix}.$$

Conclude from this that $G'(x_1)$ is parallel to $\nabla f_1(X) \times \nabla f_2(X)$.

7. Determine the equation of the tangent line to the curve at $A = (1, 1, 1)$ defined implicitly by

$$x_1 - x_2 + x_3^2 = 1$$
$$2x_1 + x_2^3 - x_3 = 2.$$

8. Determine the equation of the normal plane to the curve defined implicitly by

$$x_1 x_2 + x_2 x_3 = 4$$
$$x_1^2 + 2x_2^2 + x_3^2 = 10$$

at the point $A = (1, 2, 1)$. (The normal plane through A is the plane through A perpendicular to the tangent vector to the curve at the point A.)

9. Exercise 5 may be generalized in the following way. Let $A = (a_1, a_2, a_3)$ be a point belonging to both of the following sets of constancy

$$\{X : f_1(x_1, x_2, x_3) = c_2\}$$
$$\{X : f_2(x_1, x_2, x_3) = c_2\}.$$

If there exists a function G from R to $V_3(R)$ defined on an interval I about a

point t_0 such that for $t \in I$
$$f_1(G(t)) = c_1$$
$$f_2(G(t)) = c_2$$
and $G(t_0) = A$, then the curve G is said to be defined implicitly by the equations $f_1(X) = c_1, f_2(X) = c_2$. Show that such a curve exists if the Jacobian matrix $(\partial f_i/\partial x_j)$ evaluated at the point A has rank equal to two. This will be the case if $\nabla f_1 \times \nabla f_2 \neq 0$. (The function G will have one of the following three forms $G(t) = (t, g_2(t), g_3(t))$ or $G(t) = (g_1(t), t, g_3(t))$ or $G(t) = (g_1(t), g_2(t), t)$.)

10. Determine an expression for the unit tangent vector T to the curve defined in Exercise 9. Hence determine the general form of the scalar equation of the tangent line to the curve at A.

11. In Exercise 9 show that $\nabla f_1(A)$, and $\nabla f_2(A)$ are linearly independent vectors each perpendicular to the tangent vector to the curve at the point A. Conclude from this that if N is perpendicular to the curve at A, then
$$N = \lambda_1 \nabla f_1(A) + \lambda_2 \nabla f_2(A)$$
for appropriate scalars λ_1, λ_2.

12. Do the equations
$$x_1^2 + \sin x_2 + e^{x_3} = e$$
$$e^{x_1} - x_2 + x_3 = 2$$
define implicitly a curve through $A = (0, 0, 1)$? If so, determine the equation of the tangent line.

13. Do the equations
$$x_1^2 - x_2^3 + x_3^2 = 1$$
$$4x_1 - 6x_2 + 4x_3 = 2$$
define implicitly a curve through the point $A = (1, 1, 1)$?

14. We say two surfaces intersect orthogonally at a point if their normal vectors at the given point are orthogonal. Determine the condition on constants a, r which will insure that the surfaces $x^2 + z^2 = 1$, $x^2 + (z - a)^2 = r^2$ intersect orthogonally if they intersect at all.

15. Determine the constant a so that the spheres $(x - a)^2 + y^2 + z^2 = 3$ and $x^2 + (y - 1)^2 + z^2 = 1$ intersect orthogonally if they intersect at all.

16. Find a Cartesian equation for the plane tangent to the surface $xyz = a^3$ at a point $X_0 = (x_0, y_0, z_0)$ on the surface.

17. Show that the sum of the intercepts of the tangent plane to the surface
$$x^{1/2} + y^{1/2} + z^{1/2} = a^{1/2}$$
with the coordinate axes is constant and equals a.

18. If functions $u = f(x, y)$ and $v = g(x, y)$ are defined implicitly by the equations
$$x + 2y - v + u = 1$$
$$2x - y + v - 2u = 0$$
determine $\partial v/\partial x$ and $\partial u/\partial y$.

19. If functions $u = f_1(x, y)$ and $v = f_2(x, y)$ are defined implicitly by the equations
$$g_1(x, y, u, v) = c_1$$
$$g_2(x, y, u, v) = c_2$$
show that
$$\begin{pmatrix} \dfrac{\partial g_1}{\partial u} & \dfrac{\partial g_1}{\partial v} \\ \dfrac{\partial g_2}{\partial u} & \dfrac{\partial g_2}{\partial v} \end{pmatrix} \begin{pmatrix} \dfrac{\partial f_1}{\partial x} & \dfrac{\partial f_1}{\partial y} \\ \dfrac{\partial f_2}{\partial x} & \dfrac{\partial f_2}{\partial y} \end{pmatrix} + \begin{pmatrix} \dfrac{\partial g_1}{\partial x} & \dfrac{\partial g_1}{\partial y} \\ \dfrac{\partial g_2}{\partial x} & \dfrac{\partial g_2}{\partial y} \end{pmatrix} = 0.$$

20. Show that the function g defined by Theorem 6.17 is unique in the sense that if $\tilde{g}$ is a function satisfying
$$f(X, \tilde{g}(X)) = C \qquad \text{for } X \in V$$
and
$$(X, \tilde{g}(X)) \in D, \qquad \text{then } g = \tilde{g}.$$

21. Show that in Theorem 6.18 if $f(X, g(X)) = C$ then the following matrix equation holds,
$$\left(\frac{\partial f_i}{\partial x_j}\right) + \left(\frac{\partial f_i}{\partial y_k}\right)\left(\frac{\partial g_k}{\partial x_j}\right) = 0.$$

The functions $f_1, \ldots, f_m$; $g_1, \ldots, g_m$ are the component functions of f and g, respectively.

CHAPTER SEVEN

TAYLOR POLYNOMIALS AND EXTREMAL PROBLEMS

7.1 Approximation by Polynomials

In this section we shall consider the approximation of scalar-valued functions by polynomials. To simplify the discussion we shall consider the case of a scalar-valued function defined on an open set in $V_2(R)$ which contains the origin. We shall seek an approximation to f by a polynomial $p(x, y)$ which is valid in some neighborhood of the origin.

Recall first that if f is a function of one variable, then the Taylor polynomial of degree n for f about zero is the unique polynomial $p(x)$ of degree n which satisfies

$$p(0) = f(0), p'(0) = f'(0), \ldots, p^{(n)}(0) = f^{(n)}(0).$$

The explicit formula for p is $p(x) = \Sigma_{k=0}^{n} (f^{(k)}(0)/k!) x^k$. Furthermore Taylor's theorem asserts that if $f^{(n+1)}(x)$ is continuous in an open interval (a, b) containing the origin, then

$$\lim_{x \to 0} \frac{f(x) - p(x)}{x^n} = 0.$$

Thus as x tends to zero the error of approximation tends to zero even when divided by x^n. We seek an analogue of this result for functions of two or more variables. To minimize the number of subscripts we shall write $f(x, y)$ in place of $f(x_1, x_2)$. Thus $X = (x, y)$ and $X_0 = (x_0, y_0)$, etc.

Consider the monomial $x^j y^k$. The *degree* of this term is defined to be the sum $j+k$, and the degree of a polynomial

$$p(x, y) = \sum a_{jk} x^j y^k$$

is the highest degree of the monomials present. There are, of course, $n+1$ monomials of degree n, namely $x^n, x^{n-1}y, \ldots, y^n$. Also if we compute the partial derivatives of a monomial $p_{jk}(x, y) = x^j y^k$, it is readily observed that

$$\left(\frac{\partial^{l+m}}{\partial x^l \partial y^m} p_{jk} \right)(0, 0) = 0 \quad \text{if } j \neq l \text{ or } k \neq m$$
$$= j! k! \quad \text{if } j = l \text{ and } k = m.$$

Furthermore a mixed partial derivative of p_{jk} is independent of the order in which the partial derivatives $\partial/\partial x$ and $\partial/\partial y$ are taken.

Now let f be an arbitrary function with continuous nth partial derivatives, and set

$$p(x, y) = \sum_{j+k=0}^{n} a_{jk} x^j y^k.$$

If we wish to assert that

$$\left(\frac{\partial^{j+k}}{\partial x^j \partial y^k}f\right)(0,0) = \left(\frac{\partial^{j+k}}{\partial x^j \partial y^k}p\right)(0,0)$$

for each j, k, then we must require that

$$\left(\frac{\partial^{j+k}}{\partial x^j \partial y^k}f\right)(0,0) = j!k!a_{jk}.$$

Thus the polynomial

$$p(x,y) = \sum_{j+k=0}^{n} \frac{1}{j!k!}\left(\frac{\partial^{j+k}}{\partial x^j \partial y^k}f\right)(0,0)x^j y^k$$

has the property that it and all its partial derivatives up to order n agree with the corresponding partial derivative of f at the origin. The polynomial p is called the *Taylor polynomial* of degree n for f about the origin. If in place of the origin we fix another point $(x_0, y_0) = X_0$, then the Taylor polynomial for f about X_0 is just

(1) $$p_{X_0}(x,y) = \sum_{j+k=0}^{n} \frac{1}{j!k!}\left(\frac{\partial^{j+k}}{\partial x^j \partial y^k}f\right)(X_0)(x-x_0)^j(y-y_0)^k.$$

As before p_{X_0} agrees with f at X_0 as do all of the partial derivatives up to order n. We shall always assume that the function f has continuous nth partial derivatives to insure that the partial derivatives $\partial^{j+k}/\partial x^j \partial y^k$ are independent of the order in which $\partial/\partial x$ and $\partial/\partial y$ are taken.

If we expand the sum (1) we obtain

$$p_{X_0}(x,y) = f(X_0) + \frac{\partial f}{\partial x}(X_0)(x-x_0) + \frac{\partial f}{\partial y}(X_0)(y-y_0) + \text{higher order terms}$$

$$= f(X_0) + (d_{X_0}f)(X-X_0) + \cdots$$

where $d_{X_0}f$ is the differential of f at X_0. We use the sum (1) as a means for defining the differentials of f of orders $m = 2, 3, \ldots, n$. Indeed we define the mth-order differential of f at the point X_0 to be the scalar-valued functions $d_{X_0}^{(m)}f$ defined by the formula

(2) $$(d_{X_0}^{(m)}f)(X-X_0) = \sum_{j+k=m} \frac{m!}{j!k!}\left(\frac{\partial^m}{\partial x^j \partial y^k}f\right)(X_0)(x-x_0)^j(y-y_0)^k.$$

We introduce the factor $m!$ so that in terms of the higher order differentials $d_{X_0}^{(m)}(f)$, formula (1) becomes

(1) $$p_{X_0}(x,y) = \sum_{m=0}^{n} \frac{1}{m!}(d_{X_0}^{(m)}f)(X-X_0)$$

where

$$(d_{X_0}^{(0)}f)(X-X_0) = f(X_0)$$

and

$$(d_{X_0}^{(1)}f)(X-X_0) = (d_{X_0}f)(X-X_0).$$

Formula (3) now closely resembles the formula for the Taylor polynomial of a function of one variable about the point x_0, namely,

$$p(x) = \sum_{m=0}^{n} \frac{f^{(m)}(x_0)}{m!}(x-x_0)^m.$$

Formula (2) can be formally simplified by the introduction of the differential operators $x(\partial/\partial x) + y(\partial/\partial y)$. Indeed we define

$$\left(x\frac{\partial}{\partial x} + y\frac{\partial}{\partial y}\right)_{X_0} f = x\frac{\partial f}{\partial x}(X_0) + y\frac{\partial f}{\partial y}(X_0)$$

$$\left(x\frac{\partial}{\partial x} + y\frac{\partial}{\partial y}\right)_{X_0}^2 f = \left(x^2\frac{\partial^2}{\partial x^2} + 2xy\frac{\partial^2}{\partial x \partial y} + y^2\frac{\partial^2}{\partial y^2}\right)_{X_0}(f)$$

$$= x^2\left(\frac{\partial^2 f}{\partial x^2}\right)(X_0) + 2xy\frac{\partial^2 f}{\partial x \partial y}(X_0) + y^2\frac{\partial^2 f}{\partial y^2}(X_0)$$

and

$$\left(x\frac{\partial}{\partial x} + y\frac{\partial}{\partial y}\right)_{X_0}^m (f) = \left(\sum_{k=0}^{m} \frac{m!}{k!(m-k)!} x^k y^{m-k} \frac{\partial^m}{\partial x^k \partial y^{m-k}}\right)_{X_0}(f)$$

$$= \left(\sum_{k=0}^{m} \frac{m!}{k!(m-k)!} x^k y^{m-k} \frac{\partial^m}{\partial x^k \partial y^{m-k}} f\right)(X_0).$$

The coefficients $m!/k!(m-k)!$ are called the binomial coefficients and we write $\binom{m}{k} = m!/k!(m-k)!$. Thus

$$\left(x\frac{\partial}{\partial x} + y\frac{\partial}{\partial y}\right)_{X_0}^m f$$

is computed by formally multiplying out

$$\left(x\frac{\partial}{\partial x} + y\frac{\partial}{\partial y}\right)^m$$

by the binomial formula obtaining

$$\left(x\frac{\partial}{\partial x} + y\frac{\partial}{\partial y}\right)^m = \sum_{k=0}^{m} \binom{m}{k} x^k y^{m-k} \frac{\partial^m}{\partial x^k \partial y^{m-k}}.$$

When this differential operator is then applied to f, the resulting partial derivatives are evaluated at the point X_0. In terms of these differential operators formula (2) becomes

(4) $$(d_{X_0}^{(m)} f)(X - X_0) = \left((x - x_0)\frac{\partial}{\partial x} + (y - y_0)\frac{\partial}{\partial y}\right)_{X_0}^m (f).$$

Hence

$$p_{X_0}(x, y) = \sum_{m=0}^{n} \frac{1}{m!}\left((x - x_0)\frac{\partial}{\partial x} + (y - y_0)\frac{\partial}{\partial y}\right)_{X_0}^m (f).$$

Example 1. Write the Taylor polynomial for $f(x, y) = xy^2$ about the point $X_0 = (1, 1)$. At this point

$$\frac{\partial f}{\partial x} = 1, \quad \frac{\partial f}{\partial y} = 2, \quad \frac{\partial^2 f}{\partial x^2} = 0, \quad \frac{\partial^2 f}{\partial x \partial y} = 2, \quad \frac{\partial^2 f}{\partial y^2} = 2, \quad \text{and} \quad \frac{\partial^3 f}{\partial x \partial y^2} = 2.$$

All remaining third and higher order partial derivatives clearly vanish, and hence

$$xy^2 = p_{(1,1)}(x, y) = 1 + (x-1) + 2(y-1) + \frac{1}{2}[4(x-1)(y-1) + 2(y-1)^2]$$

$$+ \frac{1}{6} \cdot 3 \cdot 2(x-1)(y-1)^2$$

$$= 1 + (x-1) + 2(y-1) + 2(x-1)(y-1) + (y-1)^2$$

$$+ (x-1)(y-1)^2.$$

Example 2. Determine the Taylor polynomial for $f(x, y) = \sin(x + y^2)$ up to the second-order terms about $X_0 = (0, 0)$. Now

$$\frac{\partial f}{\partial x} = \cos(x + y^2) \qquad \frac{\partial f}{\partial y} = 2y \cos(x + y^2),$$

$$\frac{\partial^2 f}{\partial x^2} = -\sin(x + y^2) \qquad \frac{\partial^2 f}{\partial x \partial y} = -2y \sin(x + y^2),$$

$$\frac{\partial^2 f}{\partial y^2} = 2 \cos(x + y^2) - 4y^2 \sin(x + y^2).$$

Therefore

$$p(x, y) = \sum_{k=0}^{2} \frac{1}{k!} (d_0^{(k)} f)(X) = x + \frac{1}{2} \cdot 2y^2 = x + y^2.$$

EXERCISES

Determine the Taylor polynomials for the following functions f about the specified points X_0.

1. $f(x, y) = 2xy + 1$, $X_0 = (1, 1)$.
2. $f(x, y) = x^2 y - 1$, $X_0 = (2, 1)$.
3. $f(x, y) = xy^3$, $X_0 = (1, -1)$.

Determine the Taylor polynomials about the origin through the terms of the second order for the following functions.

4. $f(x, y) = e^{x-y}$.
5. $f(x, y) = \sin x \cos y$.
6. $f(x, y) = \dfrac{1}{1 - x - y}$.
7. $f(x, y) = \log\left(\dfrac{1 + x}{1 + y}\right)$.
8. Let f be a function of one variable and define $g(x, y) = f(ax + by)$ where a and b are constants. Let $p(x, y)$ be the Taylor polynomial for g of degree n about 0. Show that

$$p(x, y) = \sum_{k=0}^{n} \frac{f^{(k)}(0)}{k!} (ax + by)^k$$

Compute the Taylor polynomials for the following functions $f(x, y)$ by the use of Exercise 8. In each case the Taylor polynomial for the appropriate function of one variable is specified.

9. $f(x, y) = \sin(x - 2y)$, $\sin x \sim \sum_{k=0}^{n} (-1)^k \dfrac{x^{2k+1}}{2k+1}!$
10. $f(x, y) = e^{(2x+y)}$, $e^x \sim \sum_{k=0}^{n} \dfrac{x^k}{k!}$
11. $f(x, y) = \arctan(x + y)$, $\arctan x \sim \sum_{k=0}^{n} (-1)^k \dfrac{x^{2k+1}}{2k+1}$
12. $f(x, y) = \dfrac{1}{1 + 3x + 2y}$, $\dfrac{1}{1 + x} \sim \sum_{k=0}^{n} (-1)^k x^k$.

Compute $(d_{X_0}^{(k)} f)(X - X_0)$ for the following choices of f, X_0, k.

13. $f(x, y) = xy^2$, $X_0 = (1, 0)$, $k = 2$.
14. $f(x, y) = x^3 y$, $X_0 = (1, -1)$, $k = 3$.
15. $f(x, y) = \dfrac{1}{1 + x + y}$, $X_0 = (0, 0)$, $k = 3$.
16. Let $p(x, y)$ be a polynomial of degree 2. Then

$$p(x, y) = \sum_{k=0}^{2} \frac{1}{k!} (d_0^{(k)} p)(X).$$

Show that if $X = (x, y) \neq 0$, then $p(x, y)$ always lies above the tangent plane τ to p at the origin if and only if
$$(d_0^{(2)}p)(X) > 0 \quad \text{for all } X \neq 0.$$
Similarly show that if $X \neq 0$, then $p(x, y)$ always lies below the tangent plane to p if and only if
$$(d_0^{(2)}p)(X) < 0 \quad \text{for all } X \neq 0.$$
Of course at $X = 0$, $p(x, y)$ coincides with its tangent plane.

17. Show that $(d_{X_0}^{(2)}f)(X - X_0)$ is a quadratic form by determining a matrix A such that
$$(d_{X_0}^{(2)}f)(X - X_0) = (X - X_0, A(X - X_0)).$$

18. (See Exercise 16.) Let $p(x, y)$ be a polynomial of degree 2 and let τ be the tangent plane to $p(x, y)$ at the origin. Show that $p(x, y)$ lies above the tangent plane τ, for $X \neq 0$ if and only if the matrix
$$\begin{pmatrix} \dfrac{\partial^2 p}{\partial x^2} & \dfrac{\partial^2 p}{\partial x \partial y} \\ \dfrac{\partial^2 p}{\partial x \partial y} & \dfrac{\partial^2 p}{\partial y^2} \end{pmatrix}$$
evaluated at zero has only positive eigenvalues. Formulate and prove the appropriate condition for p to lie below the tangent plane. [The differential $(d_0^{(2)}p)(x, y)$ is a quadratic form in the variables x and y.]

7.2 Taylor's Theorem

In this section we shall estimate the error of approximation by Taylor polynomials. The result, known as Taylor's theorem, states that if $p_{X_0}(X) = p_{X_0}(x, y)$ is the Taylor polynomial of degree n for f about X_0 and f has continuous $n + 1$ order partial derivatives in a neighborhood of X_0, then
$$\lim_{X \to X_0} \frac{|f(X) - p_{X_0}(X)|}{|X - X_0|^n} = 0.$$

To prove this result we reduce the question to a problem involving a function of one variable by considering the function φ defined by
$$\varphi(t) = f\left(X_0 + t \frac{(X - X_0)}{|X - X_0|}\right).$$

The function φ is just f restricted to the straight line running through X_0 and X. We must determine $\varphi^{(n)}(0)$ which we do in the following lemma.

LEMMA 7.1 *Let $f(x, y)$ have continuous partial derivatives of all orders along the line $Y = X_0 + tX$ where X_0 and X are fixed vectors. Set*
$$\varphi(t) = f\left(X_0 + t \frac{(X - X_0)}{|X - X_0|}\right).$$
Then

(1) $$\varphi^{(n)}(0) = \frac{1}{|X - X_0|^n} \sum_{k=0}^{n} \binom{n}{k} \frac{\partial^n f}{\partial x^k \partial y^{n-k}}(X_0)(x - x_0)^k (y - y_0)^{n-k}$$
$$= \frac{1}{|X - X_0|^n}(d_{X_0}^{(n)}f)(X - X_0).$$

Proof. We establish formula (1) by applying induction and the chain rule.

If
$$\varphi(t) = f\left(X_0 + t\frac{(X-X_0)}{|X-X_0|}\right)$$
$$= f\left(x_0 + t\frac{(x-x_0)}{|X-X_0|}, y_0 + t\frac{(y-y_0)}{|X-X_0|}\right),$$
then by the chain rule
$$\varphi'(0) = \frac{(x-x_0)}{|X-X_0|}\frac{\partial f}{\partial x}(X_0) + \frac{y-y_0}{|X-X_0|}\frac{\partial f}{\partial y}(X_0)$$
$$= \frac{1}{|X-X_0|}(d_{X_0}f)(X-X_0).$$

Differentiating again and applying the chain rule yields
$$\varphi''(0) = \frac{1}{|X-X_0|^2}\left((x-x_0)^2\frac{\partial^2 f}{\partial x^2}(X_0) + 2(x-x_0)(y-y_0)\frac{\partial^2 f}{\partial x \partial y}(X_0)\right.$$
$$\left. + (y-y_0)^2\frac{\partial^2 f}{\partial y^2}(X_0)\right)$$
$$= \frac{1}{|X-X_0|^2}(d_{X_0}^{(2)}f)(X-X_0).$$

We now assume (1) for $n = k$ and establish it for $n = k+1$. The computation is straightforward and we leave it as an exercise.

THEOREM 7.2 (*Taylor's Theorem*). *Let f have continuous $(n+1)$ order partial derivatives in an open set containing X_0. If $p_{X_0}(X)$ is the nth degree Taylor polynomial for f about X_0, then p_{X_0} is the unique polynomial of degree n satisfying*

(2) $$\lim_{X \to X_0} \frac{|f(X) - p_{X_0}(X)|}{|X-X_0|^n} = 0.$$

Proof. If
$$\varphi(t) = f\left(X_0 + t\frac{(X-X_0)}{|X-X_0|}\right).$$

Then $\varphi(0) = f(X_0)$ and $\varphi(|X-X_0|) = f(X)$. The Taylor polynomial for φ about 0 is, of course,

(3) $$q(t) = \sum_{k=0}^{n} \frac{\varphi^{(k)}(0)}{k!} t^k.$$

Since the $(n+1)$-order partial derivatives of f are all continuous, $\varphi^{(n+1)}(t)$ is continuous. As a result we may use the derivative estimate of the remainder $\varphi(t) - q(t)$. (See Theorem 10.15 of *Calculus, with an Introduction to Vectors*.) This states that if $t > 0$ then

(4) $$\varphi(t) - q(t) = \frac{\varphi^{(n+1)}(t_0) t^{n+1}}{(n+1)!}$$

where $0 < t_0 < t$.

Set $t = |X-X_0|$. Then combining (1) and (3) we have

(5) $$q(|X-X_0|) = \sum_{k=0}^{n} \frac{1}{k!}(d_{X_0}^{(k)}f)(X-X_0) = p_{X_0}(X).$$

By (4)

(6) $$\varphi(|X-X_0|) - q(|X-X_0|) = \frac{\varphi^{(n+1)}(t_0)}{(n+1)!}|X-X_0|^{n+1}$$

where $0 < t_0 < |X - X_0|$. Since
$$f(X) = \varphi(|X - X_0|) \text{ and } p_{X_0}(X) = q(|X - X_0|),$$
we have

(7) $$\frac{|f(X) - p_{X_0}(X)|}{|X - X_0|^n} = \frac{|\varphi(|X - X_0|) - q(|X - X_0|)|}{|X - X_0|^n} = \frac{|\varphi^{(n+1)}(t_0)||X - X_0|}{(n+1)!}.$$

We wish to conclude from (7) that
$$\lim_{X \to X_0} \frac{|f(x) - p_{X_0}(X)|}{|X - X_0|^n} = \lim_{X \to X_0} \frac{|\varphi^{(n+1)}(t_0)||X - X_0|}{(n+1)!} = 0$$

This will follow from (7) if we can show that $|\varphi^{(n+1)}(t_0)|$ remains bounded for small values of $|X - X_0|$. If for some $\delta > 0$, $|\varphi^{(n+1)}(t_0)| \leq M$ whenever $|X_0 - X| < \delta$, then it follows by the definition of a limit that

$$\lim_{X \to X_0} \frac{|\varphi^{(n+1)}(t_0)||X - X_0|}{(n+1)!} = 0.$$

Hence $\lim_{X = X_0} \frac{|f(X) - f(X_0)|}{|X - X_0|^n} = 0$ and we have proved (2). How to establish the necessary boundedness of $|\varphi^{(n+1)}(t_0)|$ for small values of $|X - X_0|$ is sketched in exercises 9–12 at the end of the section.

To show that the Taylor polynomial p_{X_0} is the unique polynomial satisfying (2), suppose that $\tilde{p}$ satisfies (2). Then

$$\lim_{X \to X_0} \frac{|f(X) - \tilde{p}(X)|}{|X - X_0|^n} = 0.$$

and consequently

(8) $$\lim_{X \to X_0} \frac{|p_{X_0}(X) - \tilde{p}(X)|}{|X - X_0|^n} = 0.$$

However, $\tilde{p}_{X_0} - p$ is a polynomial of degree n, and it is not hard to show that if (8) holds then $\tilde{p}_{X_0} - p = 0$. We leave the verification of this as an exercise.

COROLLARY. *If f is a polynomial of degree $\leq n$, and p_{X_0} is the nth degree Taylor polynomial for f about X_0, then for all values of X*

$$f(X) = p_{X_0}(X).$$

Proof. The proof which follows from the uniqueness part of Theorem 7.2 we leave as an exercise.

The facts of Sections 7.1 and 7.2 all have analogues for functions of more than two variables. If care is not taken, the notation and terminology can easily get out of hand. Perhaps the easiest way to proceed is to generalize the operators $x(\partial/\partial x) + y(\partial/\partial y)$. Indeed if $f(X) = f(x_1, \ldots, x_n)$ we define

$$(d_{X_0}^{(m)} f)(X) = \left(x_1 \frac{\partial}{\partial x_1} + \cdots + x_n \frac{\partial}{\partial x_n}\right)_{X_0}^m f$$

$$= \sum_{k_1 + k_2 + \cdots + k_n = m} \frac{m!}{k_1! \cdots k_n!} x_1^{k_1} \cdots x_n^{k_n} \cdot \frac{\partial^m f}{\partial x_1^{k_1} \cdots \partial x_n^{k_n}}(X_0).$$

The Taylor polynomial of degree n for f about X_0 is then defined to be

$$p_{X_0}(X) = \sum_{m=0}^{n} \frac{(d_{X_0}^{(m)} f)(X - X_0)}{m!}.$$

The functions f and p_{X_0} have the same partial derivatives up to order n, and the appropriate analogues of Theorems 7.1–7.2 can be established with much the same proof. We shall not go into these matters here.

EXERCISES
1. Complete the inductive proof of Lemma 7.1.
2. Show that if
$$\lim_{X \to 0} \sum_{k=0}^{n} \frac{a_k x^k y^{n-k}}{|X|^n} = 0$$
then $a_0 = a_1 = \cdots = a_n = 0$. [We argue indirectly. If not all the a_k's vanish, then there is an $X_0 = (x_0, y_0)$ such that
$$\sum_{k=0}^{n} \frac{a_k x_0^k y_0^{n-k}}{|X_0|^n} \neq 0.$$
Let $X = tX_0$, then as $t \to 0$, $X \to 0$. Show, however, that for this choice of X
$$\lim_{X \to 0} \sum_{k=0}^{n} \frac{a_k x^k y^{n-k}}{|X|^n} \neq 0$$
which is a contradiction.]
3. Show that
$$\lim_{X \to 0} \sum_{k=0}^{n} \frac{a_k x^k y^{n-k}}{|X|^l} = 0$$
for each integer $l = 0, 1, \ldots, n - 1$.
4. Using Exercises 2 and 3 show that if $p(x, y)$ is a polynomial of degree n and
$$\lim_{X \to 0} \frac{p(x, y)}{|X|^n} = 0,$$
then p is the zero polynomial.
5. Show on the basis of Exercise (4) that the Taylor polynomial p_{X_0} is the *unique* polynomial satisfying
$$\lim_{X \to X_0} \frac{|f(X) - p_{X_0}(X)|}{|X - X_0|^n} = 0.$$
6. Show that if q is a polynomial of degree $\leq n$ and p_{X_0} is the nth-degree Taylor polynomial for q, then $p_{X_0} = q$.
7. Let p be a polynomial of degree 2. Let τ_{X_0} be the tangent plane to the surface $Z = p(x, y)$ at the point X_0. Show that for $X \neq X_0$, $p(X)$ lies above the tangent plane τ_{X_0} if and only if $(d^{(2)}p)(X - X_0) > 0$ for $X \neq X_0$.
8. Formulate and prove the analogous condition for the surface in exercise 7 to lie below the tangent plane.

The following sequence of exercises shows that in the proof of Theorem 7.2 $|\varphi^{(n)}(t_0)|$ remains bounded for small values of $|X - X_0|$. To simplify the notation we consider $\varphi^{(n)}$ rather than $\varphi^{(n+1)}$.

*9. Using the notation of Lemma 7.1 show that for $n = 1, 2, \ldots, \varphi^{(n)}(t) = |X - X_0|^{-n}(d_Y^{(n)}f)(X - X_0)$ where $Y = X_0 + (t(X - X_0))/|X - X_0|$. (Apply the chain rule to the definition of φ.)

*10. Derive the following inequalities for real numbers a and b.
   (i) $2|a||b| \leq |a|^2 + |b|^2$
   (ii) $|a| + |b| \leq \sqrt{2}(|a|^2 + |b|^2)^{1/2}$.
(Square both sides of (ii). The resulting inquality now follows readily from (i).)

*11. If the nth partial derivatives of f are continuous at X_0 show that there exist constants M and δ_0 such that if $|X - X_0| < \delta$ then

$$\left| \frac{\partial^n f}{\partial x^k \partial y^{n-k}} (X) \right| \leq M, \quad k = 0, 1, \ldots, n.$$

*12. Show that if δ and M are the constants of exercise 11, then

$$|\varphi^{(n)}(t)| \leq 2^{n/2} M.$$

(By exercises (9) and (11))

$$|\varphi^{(n)}(t)| \leq M \frac{(|x - x_0| + |y - y_0|)^n}{|X - X_0|^n}.$$

Now apply exercise 10 (ii).

7.3 The Geometric Interpretation of the Second Differential

It is a familiar fact from the differential calculus for a function of a single variable that if $f''(x_0) > 0$, then the graph of the function lies above the tangent line for x in a neighborhood of the point x_0. If $f''(x_0) < 0$, then the graph lies below. See Figure 1.

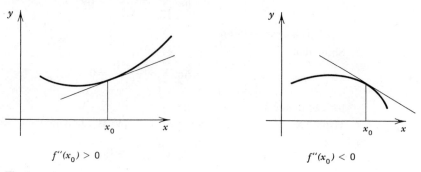

$f''(x_0) > 0$ $\qquad\qquad\qquad\qquad$ $f''(x_0) < 0$

Fig. 1

There is a similar interpretation for the second differential $d_{X_0}^{(2)} f$ of a function f of two variables. The graph of $z = f(x, y)$ is now a surface in $V_3(R)$. We seek a condition on $d_{X_0}^{(2)} f$ which tells when the surface defined by f lies either above the tangent plane at X_0 or below the tangent plane at that point. See Figure 2.

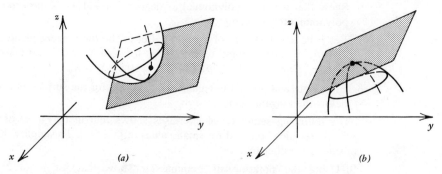

Fig. 2 (a) Graph of f above tangent plane for X in neighborhood of X_0, $X \neq X_0$. (b) Graph of f below tangent plane for X in neighborhood of X_0, $X \neq X_0$.

If $f(x, y)$ is a polynomial of the second degree, then

$$f(X) = f(X_0) + (d_{X_0} f)(X - X_0) + \frac{1}{2}(d_{X_0}^{(2)} f)(X - X_0).$$

INTERPRETATION OF THE SECOND DIFFERENTIAL

The first two terms of the right-hand side of this expression define the tangent plane τ_{X_0} to f at X_0. Clearly then f lies above τ_{X_0} when $X \neq X_0$ if and only if

$$(d_{X_0}^{(2)}f)(X-X_0) > 0, \qquad X \neq X_0.$$

Similarly f lies below τ_{X_0} if and only if $(d_{X_0}^{(2)}f)(X-X_0) < 0$, when $X \neq X_0$. For an arbitrary function with continuous partial derivatives of at least the third order, these two conditions are not enough to guarantee that f lies above (or below) the tangent plane for all $X \neq X_0$. They do suffice, however, to describe this behavior for X in a neighborhood of X_0. This is a consequence of Theorem 7.2 and a little analysis of the second differential $d_{X_0}^{(2)}(f)$.

Now,

$$d_{X_0}^{(2)}(f)(X-X_0) = \frac{\partial^2 f}{\partial x^2}(X_0)(x-x_0)^2 + 2\frac{\partial^2 f}{\partial x \partial y}(X_0)(x-x_0)(y-y_0)$$
$$+ \frac{\partial^2 f}{\partial y^2}(X_0)(y-y_0)^2$$

is a quadratic form in the variables $x-x_0$ and $y-y_0$. The matrix of this form is just

$$A_f = \begin{pmatrix} \frac{\partial^2 f}{\partial x^2} & \frac{\partial^2 f}{\partial x \partial y} \\ \frac{\partial^2 f}{\partial x \partial y} & \frac{\partial^2 f}{\partial y^2} \end{pmatrix}$$

and the condition that $d_{X_0}^2(f)(X-X_0) > 0$ for $X \neq X_0$ is just that the matrix A is *positive definite*, that is, the eigenvalues of A_f are positive. To see that this is the case recall that since A_f is symmetric there is a rotation matrix P such that

$$P^t A_f P = \Lambda_f$$

where $\Lambda_f = \begin{pmatrix} \lambda_1 & 0 \\ 0 & \lambda_2 \end{pmatrix}$ is the diagonal matrix of eigenvalues of A_f. Moreover if $U = \begin{pmatrix} u \\ v \end{pmatrix}$, $X - X_0 = \begin{pmatrix} x-x_0 \\ y-y_0 \end{pmatrix}$, and $\begin{pmatrix} u \\ v \end{pmatrix} = P^t \begin{pmatrix} x-x_0 \\ y-y_0 \end{pmatrix}$, then $PU = X - X_0$ and

$$(d_{X_0}^{(2)}f)(X-X_0) = (X-X_0, A_f(X-X_0))$$
$$= (PU, A_f PU)$$
$$= (U, P^t A_f PU) = (U, \Lambda_f U)$$
$$= \lambda_1 u^2 + \lambda_2 v^2.$$

This also implies that $(d_{X_0}^{(2)}f)(X-X_0) < 0$ for $X \neq X_0$ if and only if A_f has only negative eigenvalues. In this case we call the matrix, and equivalently the quadratic form, *negative definite*. To determine if A_f is positive or negative definite we can, of course, compute the eigenvalues for A_f. However, this fact may be ascertained by merely inspecting the matrix A_f as the next result shows.

PROPOSITION 7.3 *Let*

$$A = \begin{pmatrix} a & b \\ b & c \end{pmatrix}$$

be a symmetric 2×2 matrix. Then A is positive definite if and only if $D(A) > 0$ and $a > 0$. The matrix A is negative definite if and only if $D(A) > 0$ and $a < 0$.

Proof. Let λ_1, λ_2 be the eigenvalues of the matrix A. We assert first that

(1) $$\lambda_1 \lambda_2 = D(A) = ac - b^2$$
(2) $$\lambda_1 + \lambda_2 = a + c.$$

To establish (1) and (2) we use the fact that λ_1, λ_2 are the roots of the characteristic polynomial $D(A - xI)$. Hence

$$D(A - xI) = (x - \lambda_1)(x - \lambda_2) = x^2 - (\lambda_1 + \lambda_2) + \lambda_1 \lambda_2.$$

On the other hand

$$D(A - xI) = D\begin{pmatrix} a-x & b \\ b & c-x \end{pmatrix} = x^2 - (a+c) + ac - b^2.$$

Equating coefficients of these two polynomials we have (1) and (2). These two results do not depend on the fact that A is symmetric but hold for any 2×2 matrix A.

Proposition 7.3 now follows easily. Notice first that since $D(A) = ac - b^2$, if $D(A) > 0$, then a and c are both positive or both negative. Hence if λ_1, λ_2 are both positive, $D(A) > 0$ by (1) and by (2) $a > 0$. Similarly if λ_1, λ_2 are both negative, $D(A) > 0$ by (1) and by (2) $a < 0$. Conversely if $D(A) > 0$ and $a > 0$, then $c > 0$. Hence $\lambda_1 \lambda_2 > 0$ by (1) and $\lambda_1 + \lambda_2 > 0$ by (2). From this it follows that $\lambda_1 > 0$ and $\lambda_2 > 0$. In a similar fashion it follows that if $D(A) > 0$ and $a < 0$, then $\lambda_1 < 0$ and $\lambda_2 < 0$.

A symmetric matrix or quadratic form which is either positive definite or negative definite is called *definite*. To prove the desired result of this section we need the following lemma which estimates the size of (X, AX) for a symmetric matrix A. We shall use the result only for 2×2 matrices, but since it obviously holds in the $n \times n$ case as well, we present it in that form.

LEMMA 7.4 *Let A be a symmetric $n \times n$ matrix with eigenvalues $\lambda_1, \ldots, \lambda_n$. For each $X \in V_n(R)$*

$$\min(\lambda_1, \ldots, \lambda_n) |X|^2 \leq (X, AX) \leq \max(\lambda_1, \ldots, \lambda_n) |X|^2.$$

Proof. Let P be the rotation which diagonalizes A and set $Y = P^t X$. Then $X = PY$, and

$$(X, AX) = (PY, APY) = (Y, P^t APY) = \sum_{i=1}^n \lambda_i y_i^2.$$

Since P is orthogonal, $|X| = |PY| = |Y|$. Hence

$$\sum_{i=1}^n \lambda_i y_i^2 \leq \max(\lambda_1, \ldots, \lambda_n) \sum_{i=1}^n y_i^2$$
$$= \max(\lambda_1, \ldots, \lambda_n) |Y|^2$$
$$= \max(\lambda_1, \ldots, \lambda_n) |X|^2.$$

Similarly

$$\sum_{i=1}^m \lambda_i y_i^2 \geq \min(\lambda_1, \ldots, \lambda_n) |Y|^2 = \min(\lambda_1, \ldots, \lambda_n) |X|^2.$$

We now specialize to the 2×2 case for our desired theorem.

THEOREM 7.5 *Let $z = f(x, y)$ have continuous third partial derivatives and let τ_{X_0} be the tangent plane to the graph of f at X_0. If $(d_{X_0}^{(2)} f)(X - X_0)$ is positive definite, then for $X \neq X_0$ the graph of $f(X)$ lies above the tangent plane in some neighborhood of X_0. If $(d_{X_0}^{(2)} f)(X - X_0)$ is negative definite, the graph lies below the tangent plane for X in some neighborhood of X_0.*

INTERPRETATION OF THE SECOND DIFFERENTIAL

Proof. We prove the positive case and leave the negative case as an exercise. If the second differential is positive definite, we must show that

$$(3) \qquad f(X) > f(X_0) + (d_{X_0}f)(X - X_0)$$

for X in some neighborhood of X_0 and $X \neq X_0$. The graph of the right-hand side of (3) is, of course, the tangent plane for f.

We exploit Theorem 7.2. Let d be the minimum of the two eigenvalues of the matrix

$$A_f = \begin{pmatrix} \dfrac{\partial^2 f}{\partial x^2} & \dfrac{\partial^2 f}{\partial x \partial y} \\ \dfrac{\partial^2 f}{\partial x \partial y} & \dfrac{\partial^2 f}{\partial y^2} \end{pmatrix}$$

evaluated at X_0. It follows by Lemma 7.4 that for each $X \neq X_0$

$$(4) \qquad \frac{d_{X_0}^{(2)} f(X - X_0)}{|X - X_0|^2} \geq d.$$

Since by Theorem 7.2

$$\lim_{X \to X_0} \frac{f(X) - f(X_0) - (d_{X_0}f)(X - X_0) - \frac{1}{2}(d_{X_0}^{(2)}f)(X - X_0)}{|X - X_0|^2} = 0$$

it follows that for $\epsilon = d/4$ we may choose a number $\delta > 0$ such that if $0 < |X - X_0| < \delta$ then

$$\frac{d}{4} > \frac{f(X) - f(X_0) - (d_{X_0}f)(X - X_0) - \frac{1}{2}(d_{X_0}^{(2)}f)(X - X_0)}{|X - X_0|^2} > -\frac{d}{4}.$$

Using the second of these two inequalities and inequality (4) we see that if $0 < |X - X_0| < \delta$, then

$$\frac{f(X) - f(X_0) - d_{X_0}f(X - X_0)}{|X - X_0|^2} > \frac{1}{2} \frac{(d_{X_0}^{(2)}f)(X - X_0)}{|X - X_0|^2} - \frac{d}{4} \geq \frac{d}{2} - \frac{d}{4} = \frac{d}{4} > 0.$$

Hence if $0 < |X - X_0| > \delta$, it follows that

$$f(X) > f(X_0) + (d_{X_0}f)(X - X_0)$$

as required.

It is convenient to use the same terminology as in the one-variable case to describe the behavior of f at X_0. If the second differential of f is positive definite at X_0, then the graph of f is said to be *concave up* at X_0. If the second differential is negative definite then the graph of f is *concave down* at X_0.

Example 1. Consider the function $f(x, y) = x^3 + xy^2$. Determine the regions of concavity for the surface $z = f(x, y)$. Now

$$A_f = \begin{pmatrix} \dfrac{\partial^2 f}{\partial x^2} & \dfrac{\partial^2 f}{\partial x \partial y} \\ \dfrac{\partial^2 f}{\partial x \partial y} & \dfrac{\partial^2 f}{\partial y^2} \end{pmatrix} = \begin{pmatrix} 6x & 2y \\ 2y & 2x \end{pmatrix}$$

and $D(A_f) = 12x^2 - 4y^2$. Hence $D(A_f) > 0$ if $y^2 < 3x^2$. This is the shaded region in Figure 3. If in addition $x > 0$, the surface is concave up. If $x < 0$, the surface is concave down.

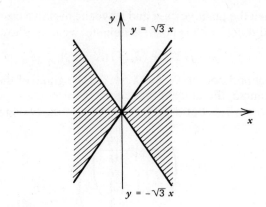

Fig. 3

If the second differential for f is definite at the point X_0, then the surface $z = f(x, y)$ is said to be *elliptic* at this point. The reason for this terminology is that if $z = f(x, y)$ is defined implicitly by the equation of an ellipsoid

$$a^2x^2 + b^2y^2 + c^2z^2 = 1,$$

then at each point (x, y, z) where the tangent plane is not vertical the surface $z = f(x, y)$ is either concave up or concave down. This is geometrically evident, and we leave it as an exercise to verify that if $z \neq 0$ then $D(A_f) > 0$. If $z > 0$, then the surface is concave down. If $z < 0$, the surface is concave up. See Figure 4.

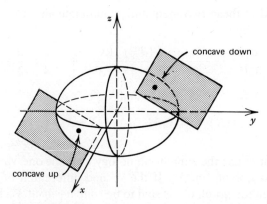

Fig. 4 Ellipsoid.

If now $D(A_f) < 0$, then since $D(A_f) = \lambda_1 \lambda_2$, the eigenvalues λ_1, λ_2 of A_f must have opposite sign. If Y_1, Y_2 are eigenvectors associated with the eigenvalues λ_1, λ_2, and we set $X_1 - X_0 = Y_1$ and $X_2 - X_0 = Y_2$, then

$$(d^{(2)}_{X_0} f)(X_1 - X_0) = (Y_1, A_f Y_1) = \lambda_1 (Y_1, Y_1)$$

and

$$(d^{(2)}_{X_0} f)(X_2 - X_0) = (Y_2, A_f Y_2) = \lambda_2 (Y_2, Y_2).$$

Assuming $\lambda_1 > 0$ and $\lambda_2 < 0$, then the argument of Theorem 7.5 shows that in a suitable neighborhood of X_0, the graph of $z = f(x, y)$ lies above the tangent plane in the direction of Y_1 and it lies below the tangent plane in the

INTERPRETATION OF THE SECOND DIFFERENTIAL 275

direction of Y_2. Thus at the point X_0 the surface $z = f(x, y)$ is saddle-shaped and the surface is said to be *hyperbolic* at the point X_0.

Example 2. Consider the hyperbolic paraboloid $z = y^2 - x^2$. The matrix $A_f = \begin{pmatrix} -2 & 0 \\ 0 & 2 \end{pmatrix}$ and $D(A_f) = -4$. Thus the surface is saddle-shaped, or hyperbolic, at each point. This is illustrated in Figure 5.

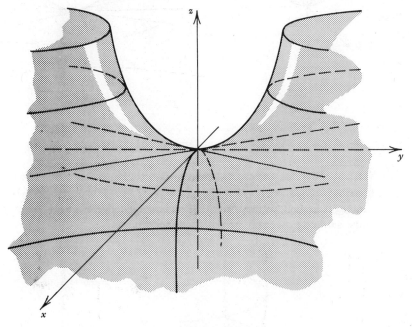

Fig. 5 The surface $z = x^2 - y^2$ is saddle shaped at each point.

The last possibility is that $D(A_f) = 0$. Then the second differential yields no information as to whether $f(X)$ lies above or below the tangent plane for X in a neighborhood of X_0. To determine this information, higher order differentials must be investigated. This situation is completely analogous to the one-variable case. Recall if $f''(x_0) = 0$, then the graph of f may be concave up at x_0, concave down, or neither. Whenever $D(A_f) = 0$ the surface is said to be *parabolic* at the point X_0.

For example,
$$f(x, y) = x^4 + y^4$$
is clearly concave up at the origin. Yet $A_f = 0$. Also
$$g(x, y) = x^4 - y^4$$
is saddle-shaped at the origin. The matrix A_f vanishes at the origin as well. See Figure 6.
Further investigation and elaboration of these ideas is the subject matter of differential geometry.

Example 3. Determine the shape of the surface
$$z = x^4 - y^3.$$
At each point X
$$A_f = \begin{pmatrix} 12x^2 & 0 \\ 0 & -6y \end{pmatrix}.$$

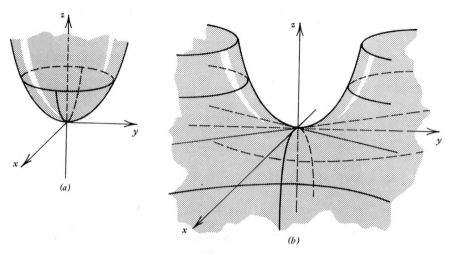

Fig. 6 (a) $f(x, y) = x^4 + y^4$. (b) $g(x, y) = x^4 - y^4$.

Hence $D(A_f) = -72x^2y$. Thus the surface is concave up when $x \neq 0$ and $y < 0$. It is saddle-shaped when $x \neq 0$ and $y > 0$. When $x = 0$ or $y = 0$, $D(A_f) = 0$ and the surface is parabolic at these points. From a knowledge of the curves $y^3 = x^4$ and $z = y^3$, the surface may easily be sketched. See Figure 7.

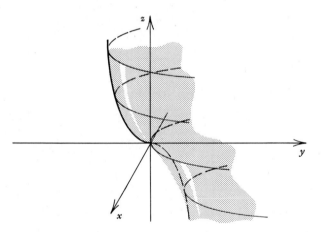

Fig. 7 $z = x^4 - y^3$.

It is evident from the sketch that on the coordinate axes $x = 0$ or $y = 0$ the surface is concave up when $x = 0$ and $y < 0$. At all other points it is saddle-shaped.

EXERCISES

For the following functions f determine the regions that the graph of f is elliptic, hyperbolic, and parabolic. When elliptic determine the sense of concavity. Sketch the surface and determine the shape of the surface at parabolic points by inspection.

1. $f(x, y) = x^2 - y$.
2. $f(x, y) = x + y^3$.
3. $f(x, y) = 2x^2 - y^3 + xy$.
4. $f(x, y) = 2x^3 + y^2$.

5. $f(x, y) = \dfrac{1}{x} + y + x^2 y$.

6. $f(x, y) = x - \dfrac{1}{y^2}$.

7. $f(x, y) = \dfrac{1}{x} + y^3$.

8. $f(x, y) = ax^2 + by^2$, all choices of a and b.

9. Let $z = f(x, y)$ be defined implicitly by the ellipsoid
$$ax^2 + by^2 + cz^2 = 1 \qquad a, b, c > 0.$$
Show that for $z \neq 0$ the graph of f is elliptic. Determine the sense of concavity when $z \neq 0$. (Compute the partial derivatives of f by implicit differentiation.)

10. Let $z = f(x, y)$ be defined implicitly by the one-sheeted hyperboloid
$$ax^2 + by^2 - cz^2 = 1 \qquad a, b, c > 0.$$
For $z \neq 0$, determine if the surface is elliptic or hyperbolic.

11. Let $z = f(x, y)$ be defined implicitly by the two-sheeted hyperboloid
$$ax^2 - by^2 - cz^2 = 1 \qquad a, b, c > 0.$$
For $z \neq 0$, determine if the surface is elliptic or hyperbolic.

12. Let $z = f(x, y)$ be a polynomial of degree 2 in x and y. Show that A_f is a constant matrix. What is the surface if $D(A_f) > 0$? If $D(A_f) < 0$?

13. Let $z = f(x, y)$ be a polynomial of degree 2 in x and y. If $D(A_f) = 0$ show that the graph of f and the tangent plane τ_{X_0} to f at X_0 must intersect in a line. If $A_f \neq 0$ what determines the direction of the line? What is the shape of the surface? What happens if $A_f = 0$?

14. Complete the proof of Theorem 7.5. That is, show that if $(d^2_{X_0} f)(X - X_0)$ is negative definite at X_0, then for $X \neq X_0$ and X in a suitably small neighborhood about X_0, $f(X)$ lies below the tangent plane to the surface at X_0.

15. Let X_0 be a point at which the function $z = f(x, y)$ is hyperbolic. Let λ_1, λ_2 be the eigenvalues of A_f at X_0 and assume $\lambda_1 > 0$ and $\lambda_2 < 0$. If Y_1, Y_2 are the respective eigenvectors show that for t sufficiently small $f(X_0 + tY_1)$ lies above the tangent plane whereas $f(X_0 + tY_2)$ lies below. (Adapt the argument of Theorem 7.5.)

7.4 Extreme Values

Let f be a real-valued function defined on an open set U in $V_n(R)$. The function f is said to have a *relative* or *local maximum* at the point X_0 if $f(X_0) \geq f(X)$ for each X in some neighborhood of the point X_0. Similarly f has a relative or local minimum at X_0 if $f(X_0) \leq f(X)$ for each X in some neighborhood of X_0. If f has either a local maximum or a local minimum at X_0, then f is said to have a local extreme at X_0 and the point X_0 is called a *local extreme point* for f.

If f is a function of one variable which is defined on the open interval (a, b) and has a local extreme at the point x_0, $a < x_0 < b$, then it is a well-known fact that $f'(x_0) = 0$. This result extends immediately to functions defined on open sets in $V_n(R)$.

THEOREM 7.6 *Let f be a real-valued function which is defined and continuous on an open set U in $V_n(R)$ and assume that f has a local extreme at the point X_0 in U. If f is differentiable at X_0, it follows that the differential $d_{X_0} f$ vanishes at the point X_0 or, equivalently, that*

$$\nabla f(X_0) = 0.$$

Proof. Since the gradient $(\nabla f)(X_0)$ is just the matrix of the linear transformation $d_{X_0}f$, it is clear that $d_{X_0}f = 0$ if and only if $(\nabla f)(X_0) = 0$. Now if f has a local extreme at the point $X_0 = (x_1^0, \ldots, x_n^0)$ and we fix all coordinates of the vector $X = (x_1, \ldots, x_n)$ except the ith, then f defines a function g of one variable by the formula
$$g(x_i) = f(X).$$
Moreover, g has a local extreme at the point $x_i = x_i^0$. Therefore $g'(x_i^0) = 0$.

A point X_0 at which $d_{X_0}f = 0$, or equivalently $\nabla f(X_0) = 0$, is called a *stationary point* for the function f. If f is defined on an open set in $V_2(R)$, then a stationary point for f is just a point where the tangent plane to the graph of f is horizontal. Theorem 7.6 just states that the local extremes for a continuous function f will occur either at stationary points or at points where f fails to be differentiable.

It is fairly evident that we can weaken the hypothesis of Theorem 7.6 from requiring that f is differentiable to just the requirement that $\nabla f(X_0)$ exists at X_0. We leave the verification of this as an exercise. It will not concern us in our applications.

Example 1. Let $f(x, y) = x^2 + y^3 - xy$. Determine the stationary points for f. Clearly $\partial f/\partial x = 2x - y$ and $\partial f/\partial y = 3y^2 - x$. The equations $2x - y = 0$ and $3y^2 - x = 0$ have the solutions $x = 0$, $y = 0$ and $x = 1/12$, $y = 1/6$. Thus $(0, 0)$ and $(1/12, 1/6)$ are the stationary points for f. To determine if these points are actually local extremes for f we must investigate further.

If f is defined on an interval (a, b) of real numbers and x_0 is a stationary point for f, then the second derivative provides a test for determining if f has a local extreme at x_0. Indeed if $f''(x_0) > 0$, then f has a local minimum at x_0. If $f''(x_0) < 0$, then f has a local maximum. If $f''(x_0) = 0$, then the second derivative yields no information. For functions defined on open sets in $V_n(R)$ we have a similar test involving the second differential $d_{X_0}^{(2)}f$. We state the result first for $n = 2$.

THEOREM 7.7 *Let f be three times continuously differentiable on the open set U in $V_2(R)$ and let $X_0 \in U$ be a stationary point for f.*

1. *If $d_{X_0}^{(2)}f$ is positive define at X_0, then f has a local minimum at X_0.*
2. *If $d_{X_0}^{(2)}f$ is negative definite at X_0, then f has a local maximum at X_0.*
3. *If f is hyperbolic at X_0, then f does not have a local extreme at X_0.*
4. *If f is parabolic at X_0, then the second differential yields no information.*

Proof. Since X_0 is a stationary point the tangent plane to the graph of f is horizontal. Hence Statements 1 and 2 are immediate consequences of Theorem 7.5. For Statement 3, if f is hyperbolic at X_0, then X_0 is a saddle point for f. The graph of f lies both above and below the tangent plane in any neighborhood of X_0. Hence X_0 is not a local extreme. The examples of the previous section show that if f is parabolic at X_0 then f may or may not have an extreme.

In terms of the matrix
$$A_f = \begin{pmatrix} \dfrac{\partial^2 f}{\partial x^2} & \dfrac{\partial^2 f}{\partial x \partial y} \\ \dfrac{\partial^2 f}{\partial x \partial y} & \dfrac{\partial^2 f}{\partial y^2} \end{pmatrix},$$

statements 1–4 about stationary points become the following:

1. If $D(A_f) > 0$ and $\partial^2 f/\partial x^2 > 0$ at X_0, then f has a local minimum at X_0.
2. If $D(A_f) > 0$ and $\partial^2 f/\partial x^2 < 0$ at X_0, then f has a local maximum at X_0.
3. If $D(A_f) < 0$ at X_0, then f does not have a local extreme at X_0.
4. If $D(A_f) = 0$ at X_0, then f may or may not have a local extreme at X_0.

If we apply this test to the example $f(x, y) = x^2 + y^3 - xy$ considered above, we have

$$A_f = \begin{pmatrix} 2 & -1 \\ -1 & 6y \end{pmatrix}.$$

Hence at the stationary point, $(1/12, 1/6)$, $D(A_f) = 1$, and since $\partial^2 f/\partial x^2 = 2 > 0$, the point is a local minimum. At the point $(0, 0)$, $D(A_f) = -1$, and this point is a saddle point.

For functions defined on open sets U in $V_n(R)$ where $n > 2$ the second differential may be used in the same way as above to classify the stationary points of f. Notice that the second differential

$$d_{X_0}^{(2)}(f)(X - X_0) = \left[(x_1 - x_1^0) \frac{\partial}{\partial x_1} + \cdots + (x_n - x_n^0) \frac{\partial}{\partial x_n} \right]_{X_0}^2 (f)$$

$$= \sum_{i=1}^{n} (x_i - x_i^0)^2 \frac{\partial^2 f}{\partial x_i^2}(X_0)$$

$$+ 2 \sum_{i<j}^{i,j \leq n} (x_i - x_i^0)(x_j - x_j^0) \frac{\partial^2 f}{\partial x_i \partial x_j}(X_0)$$

$$= (X - X_0, A_f(X - X_0))$$

where

$$A_f = \begin{pmatrix} \frac{\partial^2 f}{\partial x_1^2} & \cdots & \frac{\partial^2 f}{\partial x_1 \partial x_n} \\ & \ddots & \\ \frac{\partial^2 f}{\partial x_n \partial x_1} & \cdots & \frac{\partial^2 f}{\partial x_n^2} \end{pmatrix} = \left(\frac{\partial^2 f}{\partial x_i \partial x_j} \right).$$

Thus $d_{X_0}^{(2)}(f)(X - X_0)$ is a quadratic form in the n variables $x_1 - x_1^0, \ldots, x_n - x_n^0$. Hence for $X \neq X_0$, $d_{X_0}^{(2)}(f)(X - X_0) > 0$ if the matrix A_f has only positive eigenvalues. The quadratic form $d_{X_0}^{(2)}(f)(X - X_0) < 0$ when $X \neq X_0$ if A_f has only negative eigenvalues.

To get the desired test for local extreme values we need the analogue of Theorem 7.5. Indeed it can be shown that if f has continuous partial derivatives of orders up to $m + 1$, then

$$\lim_{X \to X_0} \frac{\left| f(X) - \sum_{k=0}^{m} \frac{1}{k!} d_{X_0}^{(k)}(f)(X - X_0) \right|}{|X - X_0|^m} = 0.$$

Letting $m = 2$ we have

$$\lim_{X \to X_0} \frac{\left| f(X) - f(X_0) - (d_{X_0} f)(X - X_0) - \frac{1}{2}(d_{X_0}^2 f)(X - X_0) \right|}{|X - X_0|^2} = 0.$$

At a stationary point X_0, $d_{X_0}f = 0$. Hence using exactly the same argument as in the case for $n = 2$ we conclude that $f(X) > f(X_0)$ for X in a neighborhood U of X_0 and $X \neq X_0$ if the second differential $(d^2_{X_0}f)(X - X_0)$ is positive definite. Similarly for $X \neq X_0$ and X in a suitable neighborhood of X_0, $f(X) < f(X_0)$ if the second differential is negative definite. If the matrix A_f has both positive and negative eigenvalues, then the stationary point X_0 is not a local extreme. This follows since in the direction of an eigenvector associated with a negative eigenvalue $f(X) < f(X_0)$ for X sufficiently close to X_0. Similarly in the direction of an eigenvector associated with a positive eigenvalue, $f(X) > f(X_0)$. By analogy with the case $n = 2$ we call such a point a saddle point. If one or more of the eigenvalues for A_f vanishes and A_f does not have both positive and negative eigenvalues, then just as in the two-variable case the second differential yields no information as to whether f has a local extreme at X_0.

Determining the eigenvalues for the matrix A_f can be a sizeable computational task if n is large. Indeed this is one area where large-scale computing machines have had great impact. However, for our purposes we may determine the algebraic signs of the eigenvalues by determining the changes in sign of the coefficients of the characteristic polynomial

$$D(A_f - xI).$$

(See p. 186.)

Example 2. If $f(x, y, z) = x^2 + y^2 - xz$, show that the origin is the only stationary point and show that this point is a saddle point for f.

First $(\nabla f)(X) = (2x - z, 2y, -x)$. Hence $(\nabla f)(X) = 0$ if and only if $X = 0$. But

$$A_f = \begin{pmatrix} 2 & 0 & -1 \\ 0 & 2 & 0 \\ -1 & 0 & 0 \end{pmatrix}$$

and

$$D(A_f - xI) = D \begin{pmatrix} 2 - x & 0 & -1 \\ 0 & 2 - x & 0 \\ -1 & 0 & -x \end{pmatrix}$$

$$= (2 - x)^2(-x) - (2 - x) = (2 - x)(x^2 - 2x - 1)$$
$$= -x^3 + 4x^2 - 3x - 2.$$

Since the coefficients of the polynomial $D(A_f - xI)$ have two sign changes, the matrix A_f has two positive and one negative eigenvalues. Hence the origin is a saddle point for f. In this case the eigenvalues for A_f are easily computed. They are $2, 1 \pm \sqrt{2}$.

To close this section we wish to emphasize the local character of the results we have discussed. If f is continuous on the open set U and has a local maximum at the point X_0, this does not imply that $f(X_0) \geq f(X)$ for each $X \in U$. Indeed, there may not be any points X_0 in U with this property.

If a function f is defined on a set S in $V_n(R)$ and there exists an $X_0 \in S$ such that

$$f(X_0) \geq f(X)$$

for each $X \in S$, then f is said to have an *absolute* or *global maximum* at X_0. Thus $f(X_0)$ is an absolute maximum for f if $f(X_0)$ is the maximum value of f over the entire domain of definition of f. The number $f(X_0)$ is a local maximum value for f if $f(X_0)$ is the maximum of f for X restricted to some neighborhood of X_0. Absolute minimum values are defined similarly.

EXTREME VALUES

For functions f defined on intervals I of real numbers, the extreme value theorem (p. 187, *Calculus; with an Introduction to Vectors*) states that f attains its maximum on I if (1) I is closed and (2) f is continuous on I.

This theorem has a precise counterpart for real-valued functions defined on sets S in $V_n(R)$. To state this we need to define first what is meant by a closed set in $V_n(R)$ and further we must define continuity for functions defined on sets which are not necessarily open.

Definition. *A point X is called a limit point for a set S in $V_n(R)$ if for each neighborhood N of X there is a point $Y \in S$ such that $Y \in N$ and $Y \neq X$. The set S is said to be closed if S contains all its limit points.*

Thus if $S = \{X \in V_2(R) : 0 < x_1 < 1 \text{ and } 0 < x_2 < 1\}$, then S is the open unit square in $V_2(R)$ and the limit points of S are the points $X = (x_1, x_2)$ satisfying $0 \leq x_1 \leq 1$ and $0 \leq x_2 \leq 1$. See Figure 8. Since the limit points on the edges of the square are not included in S, the set S is not closed. However, the set $\bar{S} = \{X \in V_2(R) : 0 \leq x_1 \leq 1 \text{ and } 0 \leq x_2 \leq 1\}$ is closed.

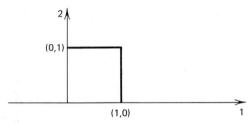

Fig. 8

Next we need the definition of continuity for a function defined on sets which are not necessarily open. This will be the same definition as before except we confine our attention just to the points $X \in S$.

Definition. *Let f be a real-valued function defined on a nonempty subset S of $V_n(R)$ and assume $X_0 \in S$. Then f is continuous at X_0 if for each $\epsilon > 0$ there exists a corresponding $\delta > 0$ such that whenever $|X - X_0| < \delta$ and $X \in S$ it follows that $|f(X) - f(X_0)| < \epsilon$.*

We need one further definition.

Definition. *A set $S \subset V_n(R)$ is bounded if there exists a constant M such that*

$$|X| \leq M \quad \text{for each } X \in S.$$

The extreme value theorem now reads as follows.

THEOREM 7.8 *Let S be a nonempty bounded closed set in $V_n(R)$. If f is a continuous real-valued function defined on S, then there exists a point $X_0 \in S$ such that*

$$f(X_0) \geq f(X) \quad \text{for each } X \in S.$$

Similarly there exists a point $X_1 \in S$ such that

$$f(X_1) \leq f(X) \quad \text{for each } X \in S.$$

We shall not prove this theorem although the argument is quite similar to that for the corresponding theorem for functions of one variable. In both cases the completeness of the real numbers must be exploited in an essential way.

282 TAYLOR POLYNOMIALS AND EXTREMAL PROBLEMS

If the global maximum for f on the set S occurs at an interior point X_0 of S and if f is differentiable there, then Theorem 7.6 applies and X_0 must be a stationary point for f. Thus if f is differentiable at each interior point X of S, the global extremes for f must either be stationary points for f or they must be boundary points of the set S. The second differential may then be used to test the stationary points to determine which, if any, are maximum or minimum points for f.

Example 3. Find the maximum product P of three nonnegative numbers x, y, z if $x + y + z = 1$. The set of nonnegative numbers x, y, z lying on the plane $x + y + z = 1$ is clearly a bounded closed set. Hence the product $P = xyz$ must have an absolute maximum by Theorem 7.8. Moreover this maximum must occur when x, y, and z are positive. Therefore if $z = 1 - x - y$.

$$P = x \cdot y \cdot z = x \cdot y(1 - x - y) = xy - x^2y - xy^2.$$

Furthermore the maximum of P must occur when $x > 0$, $y > 0$, and $x + y < 1$. Consequently this maximum must occur at a stationary point. Computing the partial derivatives we obtain

$$\frac{\partial P}{\partial x} = y - 2xy - y^2 = y(1 - 2x - y)$$

$$\frac{\partial P}{\partial y} = x - 2xy - x^2 = x(1 - 2y - x).$$

Therefore if $\partial P/\partial x = \partial P/\partial y = 0$, we may assume

$$1 - 2x - y = 0$$

and

$$1 - 2y - x = 0.$$

Consequently $x = y = 1/3$ and also $z = 1/3$. The maximum value of P is therefore $1/27$. Since there is only one stationary point we do not need to investigate the second differential.

EXERCISES

Determine the stationary points for the following functions. Classify them using the second differential.

1. $f(x, y) = x^2 + 2y^2 - 3xy$.
2. $f(x, y) = x^3 + y^3 + 12xy$.
3. $f(x, y) = x^3 - 2y^3 - 6xy + 4$.
4. $f(x, y) = -x^2 + xy - y^2 + 2x - y$.
5. $f(x, y) = x^4 + y^4 - 16x + 8y + 12$.
6. $f(x, y) = x^4 - y^4 + 8xy - 4y^2$.
7. $f(x, y) = x^3 + y^3 + 3xy^2 - 18(x + y)$.
8. $f(x, y) = \frac{xy}{8} + \frac{2}{x} + \frac{1}{y}$.
9. $f(x, y, z) = x^2 + y^2 - z^2 + xy$.
10. $f(x, y, z) = 2x^2 + y^2 + z^2 - xz + xy$.

Determine the points at which the following functions take on their absolute maximum and absolute minimum values over the specified sets S. Determine these extreme values.

11. $f(x, y) = x^2 + 2y^2 - xy$, $S = \{(x, y): 0 \leq x \leq 1, 0 \leq y \leq 2\}$.
12. $f(x, y) = x^3 + y^3 - xy$, $S = \{(x, y): |x| \leq 1, |y| \leq 2\}$.

13. $f(x, y) = x^4 + y^4$, $S = \{(x, y) : |x| \leq 1, |y| \leq 1\}$.
14. $f(x, y) = xy(1 - x - y)$, $S = \{(x, y) : x \geq 0, y \geq 0 \text{ and } x + y \leq 1\}$.

Solve the following extremal problems. Use Theorem 7.8 to guarantee the existence of the appropriate absolute maximum or minimum.

15. Find the maximum product of three nonnegative numbers having sum equal to 16.
16. Find the minimum sum of four nonnegative numbers if the product of these numbers is 81.
17. Find the maximum possible volume of a rectangular box without a top if the area of the sides and the bottom is 64 square inches.
18. Find the minimum possible surface area of a rectangular box without a top if the volume is 128 cubic inches.
19. Determine the volume of the largest rectangular box that can be inscribed in the ellipsoid
$$\frac{x^2}{a^2} + \frac{y^2}{b^2} + \frac{z^2}{c^2} = 1.$$

20. Let $X_i = (x_i, y_i)$ $i = 1, \ldots, n$ be n points in the plane. Determine the line $f(x) = ax + b$ which minimizes the quantity
$$\sum_{i=1}^{n} (f(x_i) - y_i)^2.$$

21. Let $x_1, \ldots, x_n$ be positive numbers with constant sum $c = x_1 + \cdots + x_n$. Show that the product $x_1 \cdots x_n$ is a maximum if and only if $x_1 = x_2 = \cdots = x_n = c/n$. Conclude that if $x_i > 0$ and $x_1 + \cdots + 1$, then $x_1 \cdots \cdots x_n \leq 1/n^n$.

22. Let $x_1, \ldots, x_n$ be positive numbers with constant product. Show that the sum is a minimum if and only if $x_1 = x_2 = \cdots = x_n$. Conclude that if $x_i > 0$ and $x_1 \cdots \cdots x_n = 1$, then $x_1 + \cdots + x_n \geq n$.

23. By using the result of 21 conclude that if $x_1, \ldots, x_n$ are nonnegative numbers then
$$(x_1 \cdots \cdots x_n)^{1/n} \leq \frac{x_1 + \cdots + x_n}{n}$$
with equality holding if and only if $x_1 = x_2 = \cdots = x_n$. (The left-hand side of this inequality is called the *geometric mean* of the nonnegative numbers $x_1, \ldots, x_n$. The right-hand side is the *arithmetic mean*. Thus the arithmetic mean of n nonnegative numbers always exceeds the geometric mean except when the numbers are all equal. Then the two means coincide.)

7.5 Constrained Extremal Problems

Let π be a plane in $V_3(R)$ with equation $ax + by + cz = 1$ and suppose we wish to find the distance from the plane π to the origin. To solve this problem we would minimize the function

(1) $$D^2 = x^2 + y^2 + z^2$$

subject to the condition that

(2) $$ax + by + cz = 1.$$

One procedure for solving this problem is the following. At least one of the coefficients a, b, c is not zero. Suppose for instance that $c \neq 0$. Then
$$z = \frac{1 - ax - by}{c}.$$

Substituting in (1) we obtain
$$D^2 = f(x, y) = x^2 + y^2 + \left(\frac{1 - ax - by}{c}\right)^2.$$

Then
$$\frac{\partial f}{\partial x} = 2x - \frac{2a}{c}\left(\frac{1-ax-by}{c}\right)$$
and
$$\frac{\partial f}{\partial y} = 2y - \frac{2b}{c}\left(\frac{1-ax-by}{c}\right).$$

Setting these partial derivatives equal to zero yields
$$c^2 x - a + a^2 x + aby = 0$$
and
$$c^2 y - b + abx + b^2 y = 0.$$

These two equations have solutions
$$x = \frac{ac^2}{(b^2+c^2)(a^2+c^2) - a^2b^2}$$
$$y = \frac{bc^2}{(b^2+c^2)(a^2+c^2) - a^2b^2}.$$

We may now compute z from (2) and then substitute these values of x, y, z in (1) obtaining after considerable computation
$$D^2 = \frac{1}{a^2+b^2+c^2}.$$

Suppose now we wish to find the distance from the origin to a surface defined implicitly by the equation

(3) $$h(x, y, z) = c.$$

The above technique is only valid when we can solve equation (3) explicitly for one variable in terms of the other two. As we have seen earlier this may not always be possible, and even when it is, the computation involved may be quite formidable. We wish instead to develop a technique for determining the extremes of a function $f(x, y, z)$ when x, y, z are constrained to satisfy an equation
$$h(x, y, z) = c$$
which does not rely on the explicit function $z = g(x, y)$ satisfying $h(x, y, z) = h(x, y, g(x, y)) = c$.

Suppose that a continuously differentiable function f is restricted to the set of points $S = \{h(x, y, z) = c\}$ and further assume that $\partial h/\partial z \neq 0$ at each point $X \in S$. Then we assert that the stationary points of f restricted to S are the points of S at which $\nabla f = \lambda \nabla h$ for some scalar λ. To see this we argue as follows. Since $\partial h/\partial z \neq 0$ at each point $X_0 \in S$, the implicit function theorem states that associated with each point X_0 there is a function $z = g(x, y)$ satisfying

(4) $$h(x, y, g(x, y)) = c$$

for all $X = (x, y)$ in some neighborhood of (x_0, y_0). The tangent plane to the surface defined by the function $z = g(x, y)$ at the point X_0 is spanned by the vectors $T_1 = (1, 0, \partial g/\partial x)$ and $T_2 = (0, 1, \partial g/\partial y)$ which are clearly linearly independent. Differentiating (4) with respect to x and y yields that
$$\frac{\partial h}{\partial x} + \frac{\partial h}{\partial z}\frac{\partial g}{\partial x} = 0$$

and
$$\frac{\partial h}{\partial y} + \frac{\partial h}{\partial z}\frac{\partial g}{\partial y} = 0,$$

or that $(\nabla h, T_1) = (\nabla h, T_2) = 0$. Therefore ∇h is a normal vector to the tangent plane to the surface at X. Furthermore any vector Y which is orthogonal to T_1, and T_2 must be a scalar multiple of ∇h.

Now if $u = f(x, y, z)$ is restricted to those points X satisfying $h(x, y, z) = c$, then
$$u = f(x, y, z) = f(x, y, g(x, y)).$$

The stationary points for f are the points at which $\partial u/\partial x = \partial u/\partial y = 0$. However, by the chain rule
$$\frac{\partial u}{\partial x} = \frac{\partial f}{\partial x} + \frac{\partial f}{\partial z}\frac{\partial g}{\partial x} = (\nabla f, T_1)$$
and
$$\frac{\partial u}{\partial y} = \frac{\partial f}{\partial y} + \frac{\partial f}{\partial z}\frac{\partial g}{\partial y} = (\nabla f, T_2).$$

Thus $\partial u/\partial x = \partial u/\partial y = 0$ if and only if $(\nabla f, T_1) = (\nabla f, T_2) = 0$. But we have observed above that this implies $\nabla f = \lambda \nabla h$. Hence the stationary points for f when f is restricted to the set $S = \{X : h(x, y, z) = c\}$ are just the points of S at which $\nabla f(x, y, z) = \lambda \nabla h(x, y, z)$ for some scalar λ.

Suppose now that the function f restricted to the set S has an extreme at point (x_0, y_0, z_0). Since $u = f(x, y, z) = f(x, y, g(x, y))$ for $X = (x, y)$ in a neighborhood of $X_0 = (x_0, y_0)$ it must follow that
$$\frac{\partial u}{\partial x} = (\nabla f(X_0), T_1) = 0 \quad \text{and} \quad \frac{\partial u}{\partial y} = (\nabla f(X_0), T_2) = 0.$$

Hence the extreme for f occurs at a stationary point. If there is only one such stationary point, this must be the extreme. If there is more than one, the values of f at these points must be inspected to determine which is the desired extreme. Let us return to the example discussed above. We have
$$f(x, y, z) = x^2 + y^2 + z^2$$
and the equation
$$h(x, y, z) = c$$
is the equation $ax + by + cz = 1$. Since
$$\nabla f = (2x, 2y, 2z)$$
and
$$\nabla h = (a, b, c).$$
we must solve the equations
$$(2x, 2y, 2z) = \lambda(a, b, c)$$
and
$$ax + by + cz = 1$$
for the unknown x, y, z and λ. The first of these equations implies that $x = \lambda a/2$, $y = \lambda b/2$, $z = \lambda c/2$. Substituting these values of x, y, z in the second equation yields
$$\frac{\lambda a^2}{2} + \frac{\lambda b^2}{2} + \frac{\lambda c^2}{2} = 1$$

or
$$\lambda = \frac{2}{a^2+b^2+c^2}.$$

Therefore the stationary point for $f(x, y, z) = x^2+y^2+z^2$ on the plane $ax+by+cz = 1$ is the point
$$\frac{1}{a^2+b^2+c^2}(a, b, c).$$

Since the minimum value of f clearly exists, it must be taken on at this stationary point. We conclude that the minimum value of f is
$$\frac{1}{(a^2+b^2+c^2)^2}(a^2+b^2+c^2) = \frac{1}{a^2+b^2+c^2}.$$

In the above discussion we assumed that $\partial h/\partial z$ never vanished. We may weaken this to just the assumption that $\nabla h \neq 0$ at each point $X_0 = (x_0, y_0, z_0)$ of S and still conclude that the stationary points for f on S are those points of S at which $\nabla f = \lambda \nabla h$. To see this note that if ∇h never vanishes on S, then we may conclude from the implicit function theorem that either

or
$$\begin{aligned} x &= g_1(y, z) \\ y &= g_2(x, z) \\ z &= g_3(x, y) \end{aligned}$$

for all vectors $X \in S$ belonging to a neighborhood N of X_0. Thus there must exist a function g from $V_2(R)$ into $V_3(R)$ such that if $X = g(u, v)$ then $h(g(u, v)) = c$. At a point $X_0 = g(u_0, v_0)$, the tangent plane to the surface defined by g is spanned by the linearly independent vectors $\partial g/\partial u$ and $\partial g/\partial v$. Furthermore $\nabla h(X)$ is a normal vector to this plane at this point. Letting
$$w = f(x, y, z) = f(g(u, v)),$$
we have by the chain rule that
$$\frac{\partial w}{\partial u} = \left(\nabla f, \frac{\partial g}{\partial u}\right)$$
and
$$\frac{\partial w}{\partial v} = \left(\nabla f, \frac{\partial g}{\partial v}\right).$$

Hence $\partial w/\partial u = 0 = \partial w/\partial v$ if and only if ∇f is orthogonal to the vectors $\partial g/\partial u$ and $\partial g/\partial v$. The only way this can happen is for ∇f to be a scalar multiple of ∇h. This is illustrated in Figure 9.

To generalize this situation suppose we wish to find the extremes of the function $f(x, y, z)$ when $X = (x, y, z)$ is constrained to satisfy the two equations

(5)
$$\begin{aligned} h_1(x, y, z) &= c_1 \\ h_2(x, y, z) &= c_2 \end{aligned}$$

where we assume that the rank of the Jacobian matrix

(6)
$$\begin{pmatrix} \frac{\partial h_1}{\partial x} & \frac{\partial h_1}{\partial y} & \frac{\partial h_1}{\partial z} \\ \frac{\partial h_2}{\partial x} & \frac{\partial h_2}{\partial y} & \frac{\partial h_2}{\partial z} \end{pmatrix}$$

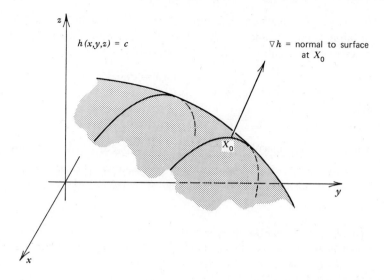

Fig. 9 If f, restricted to $h(x, y, z) = c$, has an extreme at X_0, then $\nabla f(x_0) = \lambda \nabla h(X_0)$.

equals 2 at each point X. We assert that these extremes occur at points X at which ∇f is a linear combination of ∇h_1 and ∇h_2. The argument is similar to those we have given before. Since the rank of (6) equals 2 at each point X satisfying (5), we conclude that at least one of the Jacobians

$$\frac{\partial(h_1, h_2)}{\partial(x, y)}, \frac{\partial(h_1, h_2)}{\partial(x, z)}, \frac{\partial(h_1, h_2)}{\partial(y, z)}$$

fails to vanish at each point X. Suppose for definiteness it is the last. Fixing the point $X = X_0$, the implicit function theorem asserts that there exist functions g_1 and g_2 such that if

$$G(x) = (x, g_1(x), g_2(x)) = (x, y, z)$$

for each x in a neighborhood of x_0, then

$$h_1(G(x)) = c_1$$

and

$$h_2(G(x)) = c_2.$$

Differentiating these two equations we see that

$$(\nabla h_1(X), G'(x)) = 0$$

and

$$(\nabla h_2(X), G'(x)) = 0.$$

Also $G'(x) = (1, g_1'(x), g_2'(x))$, and this vector is not zero. Since the rank of the matrix (6) equals 2, the vectors $\nabla h_1(X)$, $\nabla h_2(X)$ must be linearly independent and consequently they span the set of vectors Y which are orthogonal to $G'(x)$. Hence if $(Y, G'(x)) = 0$, then there must exist scalars λ_1, λ_2 such that

$$Y = \lambda_1 \nabla h_1 + \lambda_2 \nabla h_2.$$

Now if $u = f(x, y, z)$ is restricted to the set of points X satisfying (5), then $u = f(G(x))$, and at a stationary point

$$\frac{du}{\partial x} = (\nabla f, G'(x)) = 0.$$

As we have observed this can hold if and only if $\nabla f = \lambda_1 \nabla h_1 + \lambda_2 \nabla h_2$ for suitable scalars λ_1 and λ_2. Thus the stationary points for f restricted to (5) are precisely the points $X = (x, y, z)$ satisfying

$$h_1(X) = c_1$$
$$h_2(X) = c_2$$

and

$$\nabla f(X) = \lambda_1 \nabla h_1(X) + \lambda_2 \nabla h_2(X).$$

Moreover the desired extremes of f must occur at these stationary points.

Example 1. Find the distance from the point $(1, 1, 1)$ to the line which is the intersection of the planes π_1 and π_2 defined by the equations

(7) $\qquad\qquad 2x + y - z = 1 \qquad$ and $\qquad x - y + z = 2.$

It suffices to minimize the function $f(x, y, z) = (x - 1)^2 + (y - 1)^2 + (z - 1)^2$, which is the square of the distance subject to the constraining equations (7). If $h_1(x, y, z) = 2x + y - z$ and $h_2(x, y, z) = x - y + z$, then $\nabla h_1 = (2, 1, -1)$ and $\nabla h_2 = (1, -1, 1)$. See Figure 10.

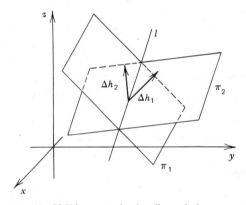

Fig. 10 If X is constrained to lie on l, then an extreme for $f(X)$ will occur at a point X where $\nabla f(X)$ is a linear combination of $\Delta h_1(X)$ and $\nabla h_2(X)$.

Hence $\nabla f = \lambda_1 \nabla h_1 + \lambda_2 \nabla h_2$ if and only if

$$2(x - 1) = 2\lambda_1 + \lambda_2$$
$$2(y - 1) = \lambda_1 - \lambda_2$$
$$2(z - 1) = -\lambda_1 + \lambda_2.$$

Hence

$$x = \frac{2\lambda_1 + \lambda_2}{2} + 1$$

$$y = \frac{\lambda_1 - \lambda_2}{2} + 1$$

$$z = \frac{-\lambda_1 + \lambda_2}{2} + 1$$

Substitution in (7) yields the two equations

$$2\lambda_1 + \lambda_2 + 2 + \frac{\lambda_1 - \lambda_2}{2} + 1 + \frac{\lambda_1 - \lambda_2}{2} - 1 = 1$$

and

$$\frac{2\lambda_1 + \lambda_2}{2} + 1 + \frac{-\lambda_1 + \lambda_2}{2} - 1 + \frac{-\lambda_1 + \lambda_2}{2} + 1 = 2.$$

These simplify to
$$3\lambda_1 = -1$$
and
$$\frac{3\lambda_2}{2} = 1.$$

Hence $\lambda_1 = -1/3$ and $\lambda_2 = 2/3$. Consequently $(x-1) = 0$, $y - 1 = -1/2$ and $z - 1 = 1/2$. Hence the minimum value of f is $1/2$ and the distance is $1/\sqrt{2}$.

We now state the general form of the rule for finding the stationary points for a function $f(x_1, \ldots, x_n)$ subject to k constraining conditions.

THEOREM 7.9 *Let $u = f(x_1, \ldots, x_n)$ be subject to the k constraining equations*

(8)
$$\begin{aligned} g_1(x_1, \ldots, x_n) &= c_1 \\ \vdots \quad\quad\quad &\quad \vdots \\ g_k(x_1, \ldots, x_n) &= c_k \end{aligned}$$

and suppose the Jacobian matrix $(\partial g_i/\partial x_j)$ has rank equal to k at each point X satisfying (8). Then the stationary points of f restricted to (8) are precisely those points $X = (x_1, \ldots, x_n)$ satisfying (8) and the equation

$$\nabla f(X) = \lambda_1 \nabla g_1(X) + \cdots + \lambda_k \nabla g_k(X)$$

for suitable scalars $\lambda_1, \ldots, \lambda_k$. Furthermore each local extreme for the function f constrained by (8) occurs at such a stationary point.

Proof. The argument proceeds exactly the same as in the special cases discussed above, and we only sketch it. Let G be a function from $V_{n-k}(R)$ to $V_n(R)$ defined implicitly by equations (8). The tangent space at $X = G(Z)$ has dimension $n - k$ and is spanned by the vectors $\partial G/\partial u_1, \ldots, \partial G/\partial u_{n-k}$. Furthermore the vectors $\nabla g_1(X), \ldots, \nabla g_k(X)$ are orthogonal to this tangent space. The condition that the rank of the matrix $(\partial g_i/\partial x_j)$ equals k guarantees that the vectors $\nabla g_1(X), \ldots, \nabla g_k(X)$ which are the rows of this matrix, are linearly independent. Now Z will be a stationary point for $f \circ G$ if and only if $\nabla f(X)$ is orthogonal to this tangent space. But then $\nabla f(X)$ must be in the span of the vectors $\nabla g_1(X), \ldots, \nabla g_k(X)$. Thus there exist scalars $\lambda_1, \ldots, \lambda_k$ so that

$$\nabla f(X) = \lambda_1 \nabla g_1(X) + \cdots + \lambda_k \nabla g_k(X).$$

Lastly if an extreme for f restricted to (8) occurs at $X_0 = G(Z_0)$, then letting $h(Z) = f(G(Z))$ we have by the chain rule

$$\frac{\partial h}{\partial u_i} = \left(\nabla f(X_0), \frac{\partial G}{\partial u_i} \right) = 0.$$

Thus the extreme occurs at a stationary point.

It is important to note that this theorem does not assert the existence of extremes for f. It only gives us a method for finding them if we know from other considerations that they exist. The constants $\lambda_1, \ldots, \lambda_n$ are called *Lagrange multipliers* in honor of the French mathematician who invented this technique for locating stationary points.

EXERCISES

Solve the following extremal problems by the method of Lagrange multipliers.
1. Determine the minimum distance from the point $(1, 1)$ to the line $2x - y = 3$.
2. Determine the minimum distance from the origin to the hyperbola $xy = 1$.

3. Determine the minimum and maximum distances from the point $(1, 0)$ to the ellipse $x^2 + 2y^2 = 4$.
4. Find the points on the hyperbola $x^2 - xy = 4$ closest to the origin.
5. What point on the sphere $x^2 + y^2 + z^2 = 4$ is furthest from $(1, -1, 1)$?
6. Find the points on the surface $z^2 - xy = 4$ closest to the origin.
7. If we attempt to maximize the function $f(x, y, z) = y(x + z)$ subject to the constraining equations $xy = 1$ and $x^2 + z^2 = 4$, the Lagrange multiplier method apparently fails. Why?
8. Find the rectangle of maximum perimeter which can be inscribed in the ellipse $(x^2/a^2) + (y^2/b^2) = 1$.
9. What is the maximum volume of a box if the sum of the lengths of the edges equals a?
10. Determine the radius and height of the cylinder of maximum surface area which can be inscribed in a sphere of radius a.
11. Determine the minimum distance from the line $y = x + 1$ to the parabola $x = y^2$.
12. Determine the distance from the intersection of the two hyperplanes
$$2x - y + z - w = 1$$
$$x + y - z + w = 1$$
to the origin in $V_4(R)$.
13. Derive the inequality
$$\sqrt[n]{x_1 \cdots x_n} \leq \frac{x_1 + \cdots + x_n}{n}$$
for nonnegative numbers x_i by solving an appropriate constrained extremal problem.
14. U.S. Postal regulations stipulate that a rectangular box is mailable if the length plus the girth does not exceed 100 inches. What is the maximum volume of such a box? (The girth is the distance around the box measured perpendicular to the longest side.)

CHAPTER EIGHT

MULTIPLE INTEGRALS

8.1 Introduction

We turn now to the notion of the integral for functions defined on sets in $V_n(R)$. Before giving the necessary definitions let us recall the situation for functions defined on intervals on the real line. We wish to make a direct generalization of these ideas, so it will be helpful to have the details clearly in mind.

The most elementary class of functions for which we define the integral is the set of step functions. Recall that a function f defined on the closed interval $[a, b]$ is called a *step function* if there is a partition π of the interval $[a, b]$ into subintervals $[x_{i-1}, x_i]$ where $a = x_0 < x_1 < \cdots < x_n = b$ so that f is constant on each of the open subintervals of this partition. Thus there exist constants c_k, $k = 1, \ldots, n$ such that

$$f(x) = c_k \quad \text{if} \quad x_{k-1} < x < x_k.$$

The graph of a step function is illustrated in Figure 1.

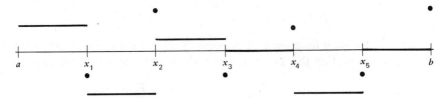

Fig. 1 Step function.

The dots indicate the values of the function at the division points x_i of the partition. We assume that the function f has been defined at these division points. The actual values are of no importance.

Next we define the integral of a step function f. Indeed if f is constant on the open subintervals of the partition π, then we define

$$\int_a^b f = \sum_{k=1}^n c_k (x_k - x_{k-1})$$

where

$$f(x) = c_k \quad \text{for} \quad x_{k-1} < x < x_k.$$

Letting $\Delta x_k = x_k - x_{k-1}$ this formula becomes

$$\int_a^b f = \sum_{k=1}^n c_k \Delta x_k.$$

To extend the notion of the integral to more general functions f we proceed as follows. Let f be a bounded function defined on the interval $[a, b]$. Let S

be the class of step functions g defined on $[a, b]$ satisfying

$$g(x) \leq f(x) \quad \text{for each} \quad x \in [a, b].$$

Let T be the class of step functions h defined on $[a, b]$ satisfying

$$f(x) \leq h(x) \quad \text{for each} \quad x \in [a, b].$$

Then if $g \in S$ and $h \in T$,

$$g(x) \leq h(x) \quad \text{for each} \quad x \in [a, b].$$

Moreover it follows immediately that

(1) $$\int_a^b g \leq \int_a^b h.$$

Inequality (1) implies that the sets of real numbers

$$\tilde{S} = \left\{ \int_a^b g : g \in S \right\}$$

and

$$\tilde{T} = \left\{ \int_a^b h : h \in T \right\}$$

are *separated* sets of real numbers. That is, if $s \in \tilde{S}$ and $t \in \tilde{T}$, then $s \leq t$. Now the completeness axiom for the real numbers states that if $\tilde{S}$ and $\tilde{T}$ are separated, then there exists at least one real number m with the property that

(1) $$s \leq m \leq t, \quad \text{for each } s \in \tilde{S} \text{ and } t \in \tilde{T}.$$

A real number m satisfying (1) is said to *separate* $\tilde{S}$ and $\tilde{T}$. Thus the completeness axiom guarantees that there exists at least one real number with the property that

$$\int_a^b g \leq m \leq \int_a^b h$$

for each function $g \in S$ and $h \in T$. Now if there exists precisely one value of m with this property, then f is said to be *integrable*, and we define

$$m = \int_a^b f.$$

In addition to just the definition of integrability we need a practical way of evaluating the integral $\int_a^b f$. The technique for doing this is contained in the fundamental theorem of the calculus. Let us state the facts for future reference.

THEOREM 8.1 *Assume f is continuous on $[a, b]$. Then f is integrable. If $F(x) = \int_a^x f$, then F is a primitive for f, that is, $F'(x) = dF/dx = f(x)$. Furthermore if G is any primitive for f, then*

$$\int_a^b f = G(b) - G(a).$$

If we wish to indicate the variable x, we write $\int_a^b f = \int_a^b f(x) \, dx$.

Our program will be first to generalize the notion of the integral to real-valued functions defined on sets in $V_2(R)$. This will indicate how to proceed for functions defined on sets in $V_n(R)$. The fundamental technique for computing such integrals is to reduce the computation to successive evaluations of integrals of functions of one variable. We shall examine several applications as we go along.

8.2 Step Functions and the Integral

Let S_1 and S_2 be sets of real numbers, then we define the *Cartesian product* of these two sets to be the set of points $X = (x_1, x_2)$ in the plane such that $x_1 \in S_1$ and $x_2 \in S_2$. We write this Cartesian product as $S_1 \times S_2$. If these two sets are closed intervals I_1 and I_2, then the Cartesian product $R = I_1 \times I_2$ is a closed rectangle. See Figure 2.

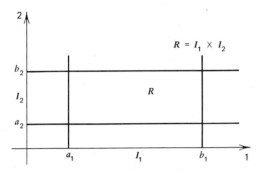

Fig. 2 Rectangle R is the cartesian product to two intervals I_1, I_2.

Let π_1 and π_2 be partitions of the intervals $I_1 = [a_1, b_1]$ and $I_2 = [a_2, b_2]$, respectively. That is,

$$\pi_1 = \{x_{10}, \ldots, x_{1m}\} \quad \text{and} \quad \pi_2 = \{x_{20}, \ldots, x_{2n}\}$$

where

$$a_1 = x_{10} < x_{11} < \cdots < x_{1m} = b_1 \quad \text{and} \quad a_2 = x_{20} < x_{21} < \cdots < x_{2n} = b_2.$$

Then we define a *partition* π of the rectangle R to be the Cartesian product $\pi_1 \times \pi_2$. If $I_{1j} = (x_{1j-1}, x_{1j})$ and $I_{2k} = (x_{2k-1}, x_{2k})$, then $R_{jk} = I_{1j} \times I_{2k}$ is called an open subrectangle of R defined by the partition π. See Figure 3.

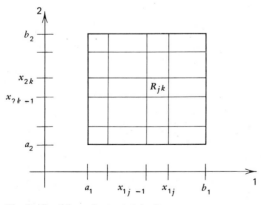

Fig. 3 Partition of a rectangle R.

A real-valued function f defined on the rectangle R is called a *step function* if the following two conditions are satisfied.

(1) There is a partition π of R such that f is constant on each of the open subrectangles R_{jk} defined by π.

(2) For each $X = (x_1, x_2) \in R$ the function f considered as a function of each variable separately is a step function defined on I_1 and I_2, respectively.

Condition (2) is only a technical convenience. If x_1 or x_2 belongs to one of the open subintervals of I_1 or I_2 then (2) is implied by (1). Condition (2) just insists that on each of the division lines of the partition π, f should be a step function considered as a function of one variable.

The generalization of the notion of the integral to step functions defined on R is now easy. Assume

$$f(X) = c_{jk} \quad \text{if} \quad X \in R_{jk}$$

Let Δ_{jk} be the area of the rectangle R_{jk}. Thus if $R_{jk} = I_{1j} \times I_{2k}$ where $I_{j1} = (x_{1j-1}, x_{1j})$ and $I_{k2} = (x_{2k-1}, x_{2k})$, then $\Delta_{jk} = (x_{1j} - x_{1j-1})(x_{2k} - x_{2k-1})$. We define

$$\int_R f = \sum_{j=1}^{m} \sum_{k=1}^{n} c_{jk} \Delta_{jk}.$$

Example 1. Suppose $I_1 = [-1, 3]$ and $I_2 = [2, 4]$. Let $f(x) = [x]$ be the greatest integer function. That is, for each integer k

$$f(x) = k \quad \text{if} \quad k \leq x < k + 1.$$

Now this function is a step function on any interval I. Moreover $f(x_1, x_2) = [x_1][x_2]$ is a step function on the Cartesian product of two intervals. If we consider $R = I_1 \times I_2$, then f is constant on the open subintervals defined by the partition

$$\pi = \{-1, 0, 1, 2, 3\} \times \{2, 3, 4\}.$$

The constant values of f on each of the rectangles R_{jk} are indicated in Figure 4. Now $\Delta_{jk} = 1$ for each subinterval R_{jk}. Hence

$$\int_R f = \sum c_{jk} \Delta_{jk} = 10.$$

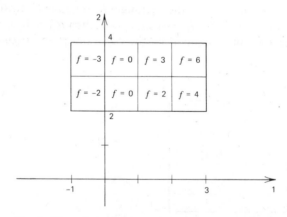

Fig. 4 The values of the step function $f(x_1, x_2) = [x_1][x_2]$ on $[-1, 3] \times [2, 4]$.

Let us denote by V_R the set of all step functions defined on the rectangle R. It follows almost from the definition that V_R is a vector space over the real numbers. To see this note that if $f \in V_R$ and a is a real number, then $af \in V_R$. The fact that $f + g \in V_R$ if $f, g \in V_R$ follows from the following observation. Let π_1 and π_2 be the partitions of R associated with f and g, respectively. Let π be the partition of R consisting of all of the lines belonging to either π_1 or π_2. We call this partition a *refinement* of π_1 and π_2. Indeed $\pi = \pi_1 \cup \pi_2$.

Clearly both f and g are constant on each open subrectangle defined by the new partition π. Thus $f+g$ must be constant as well. This proves that $f+g$ is a step function.

In general we say that the partition π_1 is a refinement of the partition π_2 if each of the division lines of π_2 belongs to π_1. See Figure 5. Equivalently, $\pi_2 \subset \pi_1$. Thus any function constant on the subrectangles defined by π_2 will also be constant on the subrectangles defined by π_1.

Next we have two important properties of the integral of step functions.

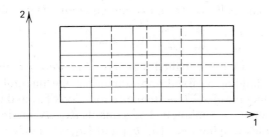

Fig. 5 π_1 = partition composed of solid lines. π_2 = partition composed of dotted lines. $\pi = \pi_1 \cup \pi_2$ = refinement of π_1 and π_2.

THEOREM 8.2 *Let V_R be the vector space of step functions defined on the rectangle R. If $f, g \in V_R$ and a and b are real numbers, then*

(1) $$\int_R (af+bg) = a\int_R f + b\int_R g$$

(2) $$\int_R f \geq \int_R g \quad \text{if} \quad f \geq g.$$

Equation (1) expresses the fact that $\int_R f$ is a linear function from the vector space V_R to the field of real numbers. Equation (2) is just the statement that the integral preserves order. Namely, if $f(X) \geq g(X)$ for each $X \in R$, then $\int_R f \geq \int_R g$.

To gain some insight as to how

$$\int_R f = \sum_{j,k} c_{jk} \Delta_{jk}$$

is computed, let us write out in full the formula for $\int_R f$. Since

$$\Delta_{jk} = (x_{1j-1} - x_{1j})(x_{2k-1} - x_{2k}) = \Delta x_{1j} \Delta x_{2k},$$

we have

(3) $$\int_R f = \sum_{j=1}^{m} \sum_{k=1}^{n} c_{jk} \Delta x_{1j} \Delta x_{2k}.$$

Let us fix x_1 momentarily. Then $f(x_1, x_2)$ considered as a function of x_2 is a step function on the interval $[a_2, b_2]$. Moreover if we compute the integral of this step function over this interval, we have by definition that

$$\int_{a_2}^{b_2} f(x_1, x_2) \, dx_2 = \sum_{k=1}^{n} c_{jk} \Delta x_{2k}.$$

Let us look now at the function of x_1 defined by this integral. Indeed, defining

(4) $$F(x_1) = \int_{a_2}^{b_2} f(x_1, x_2) \, dx_2$$

it is immediately apparent that F is a step function on the interval $[a_1, b_1]$. Let d_j be the values of F on the open subintervals of the partition of $[a_1, b_1]$. Clearly
$$d_j = \sum_{k=1}^{n} c_{jk} \Delta x_{2k}.$$

If we compute the integral of F over $[a_1, b_1]$ we obtain

(5) $$\int_{a_1}^{b_1} F = \int_{a_1}^{b_1} F(x_1) \, dx_1 = \sum_{j=1}^{m} d_j \Delta x_{1j} = \sum_{j=1}^{m} \left(\sum_{k=1}^{n} c_{jk} \Delta x_{2k} \right) \Delta x_{1j} = \int_R f.$$

Substituting the value of $F(x_1)$ from (4) into equation (5) we obtain

(6) $$\int_R f = \int_{a_1}^{b_1} \left[\int_{a_2}^{b_2} f(x_1, x_2) \, dx_2 \right] dx_1.$$

The right-hand side of (6) is called a *repeated or iterated integral*. It is evaluated by first holding x_1 fixed and computing the integral of $f(x_1, x_2)$ considered as a function of x_2. This defines a function of x_1 and the next the integral of this function is computed. The symbols dx_1, dx_2 are introduced to indicate the variables ranging over $[a_1, b_1]$ and $[a_2, b_2]$, respectively.

Since we may interchange the order of summation in (3), we also have
$$\int_R f = \sum_{k=1}^{n} \left(\sum_{j=1}^{m} c_{jk} \Delta x_{1j} \right) \Delta x_{2k} = \int_{a_2}^{b_2} \left[\int_{a_1}^{b_1} f(x_1, x_2) \, dx_1 \right] dx_2.$$

Thus we may integrate f first with respect to x_1 and then integrate the resulting function of x_2 to compute $\int_R f$.

Example 2. For $X \in R = [0, 2] \times [1, 3]$ define $f(x_1, x_2) = [2x_1][x_2]$. Compute
$$\int_0^2 \left[\int_1^3 f(x_1, x_2) \, dx_2 \right] dx_1,$$
and
$$\int_1^3 \left[\int_0^2 f(x_1, x_2) \, dx_1 \right] dx_2.$$

By definition of the greatest integer function on p. 294

$$\begin{aligned} [2x_1] &= 0 & 0 &\leq x_1 < \tfrac{1}{2} \\ &= 1 & \tfrac{1}{2} &\leq x_1 < 1 \\ &= 2 & 1 &\leq x_1 < \tfrac{3}{2} \\ &= 3 & \tfrac{3}{2} &\leq x_1 < 2 \end{aligned}$$

and
$$\begin{aligned}{} [x_2] &= 1 & 1 &\leq x_2 < 2 \\ &= 2 & 2 &\leq x_2 < 3. \end{aligned}$$

A sketch of the resulting partition of the rectangle R with the values of f on the open subrectangles is Figure 6. Hence if $F(x_1) = \int_1^3 f(x_1, x_2) \, dx_2$, we have

$$\begin{aligned} F(x_1) &= 0 & 0 &\leq x_1 < \tfrac{1}{2} \\ &= 3 & \tfrac{1}{2} &\leq x_1 < 1 \\ &= 6 & 1 &\leq x_1 < \tfrac{3}{2} \\ &= 9 & \tfrac{3}{2} &\leq x_1 < 2 \end{aligned}$$

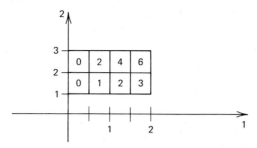

Fig. 6

and
$$\int_0^2 F(x_1)\, dx_1 = \frac{18}{2} = 9.$$

If
$$G(x_2) = \int_0^2 f(x_1, x_2)\, dx_1,$$

then
$$\begin{aligned} G(x_2) &= 3 & 1 \le x_2 < 2 \\ &= 6 & 2 \le x_2 < 3, \end{aligned}$$

and
$$\int_1^3 G(x_2)\, dx_2 = 9.$$

Thus
$$\int_R f = \int_0^2 F(x_1)\, dx_1 = \int_1^3 G(x_2)\, dx_2 = 9.$$

EXERCISES

Make a sketch of the partition of the following rectangles R and indicate the values of the following step functions f on the open subintervals.

1. $f(x_1, x_2) = [x_1 + 1][x_2]$, $R = [-1, 1] \times [0, 3]$.
2. $f(x_1, x_2) = [x_1/2][2x_2 + 1]$, $R = [0, 4] \times [0, 2]$.
3. $f(x_1, x_2) = [x_1^2][3x_2]$, $R = [0, 2] \times [0, 2]$.
4. $f(x_1, x_2) = [2x_1] + [x_2/3]$, $R = [0, 3] \times [0, 2]$.
5. Assuming $R = [a_1, b_1] \times [a_2, b_2]$ in Exercises 1–4, compute
$$F(x_1) = \int_{a_2}^{b_2} f(x_1, x_2)\, dx_2$$
for each of the functions f of Exercises 1–4.
6. Compute
$$G(x_2) = \int_{a_1}^{b_1} f(x_1, x_2)\, dx_1$$
for each of the functions in Exercises 1–4.
7. Compute $\int_R f$ for each of the functions f of Exercises 1–4.
8. Let $f(x_1, x_2) = [x_1 + x_2]$ be defined on the rectangle $R = [0, 2] \times [0, 2]$. Is f a step function?
9. In the following V_R denotes the vector space of step functions defined on the rectangle R. If $f, g \in V_R$, show that
$$\int_R f \le \int_R g$$
if $f(X) \le g(X)$ for each $X \in R$.
10. If $f, g \in V_R$ and a and b are real numbers, show that
$$\int_R (af + bg) = a \int_R f + b \int_R g.$$

11. If f and g are in V_R show that $f \cdot g \in V_R$.

8.3 The Integral of a Bounded Function

The extension of the idea of the integral from step functions to general bounded functions defined on a rectangle proceeds exactly the same as that for a bounded function defined on an interval of real numbers.

Let f be a bounded function defined on a rectangle R. Let

$$S = \{g \in V_R : g \leq f\}, T = \{h \in V_R : f \leq h\}.$$

The sets S and T will be called the sets of *lower and upper approximating step functions for f*. Let

$$\tilde{S} = \left\{ \int_R g : g \in S \right\}$$
$$\tilde{T} = \left\{ \int_R h : h \in T \right\}.$$

Since $g \in S$ and $h \in T$ implies $g \leq h$ and hence $\int_R g \leq \int_R h$, it follows that $\tilde{S}$ and $\tilde{T}$ are separated sets of real numbers. If there exists precisely one number m separating $\tilde{S}$ and $\tilde{T}$, then we say f is *integrable* and write $m = \int_R f$.

Thus a bounded function f is integrable if there exists precisely one number m with the property that

$$\int_R g \leq m \leq \int_R h$$

for each pair of step functions g, h satisfying $g \leq f \leq h$ on R.

It is an easy exercise to show that the integrable functions on the rectangle R form a vector space which we denote by I_R. Furthermore if $f, g \in I_R$ and a, b are real numbers, then

(1) $$\int_R (af + bg) = a \int_R f + b \int_R g$$

and

(2) $$\int_R f \leq \int_R g \quad \text{if} \quad f \leq g.$$

In order to calculate $\int_R f$ for an integrable function we need to be able to write $\int_R f$ as an iterated integral. That is, we must derive a formula

(3) $$\int_R f = \int_{a_2}^{b_2} \left[\int_{a_1}^{b_1} f(x_1, x_2) \, dx_1 \right] dx_2.$$

This enables us to evaluate $\int_R f$ by first computing

$$F(x_2) = \int_{a_1}^{b_1} f(x_1, x_2) \, dx_1$$

and then computing

$$\int_{a_2}^{b_2} F(x_2) \, dx_2.$$

Because of formula (3) $\int_R f$ is often written $\iint_R f$ to emphasize that two integrations must be performed. For the present we will not use this terminology, however.

Formula (3) is true for integrable functions but there are technical difficulties in establishing it. One problem is that

$$F(x_2) = \int_{a_1}^{b_1} f(x_1, x_2) \, dx_1$$

may not exist for *all* values of x_2. For example, $f(x_1, x_2)$ need not be integrable as a function of x_1 for x_2 one of the division points of a partition of $[a_2, b_2]$. To avoid such technical complications we restrict the class of functions which we consider.

Definition. *Let M_R be the class of bounded functions f defined on the interval R which can be approximated arbitrarily well by step functions. By this we mean that for each $\epsilon > 0$ there exists a step function g, such that*

$$|f(X) - g(X)| < \epsilon$$

for each $X \in R$.

Thus if $f \in M_R$, there is a step function g which approximates f as closely as desired all across the rectangle R. The class M_R includes most of the integrable functions one normally encounters. In particular, M_R includes the continuous functions. We will not prove that here, however. We shall prove that if $f \in M_R$, then f is integrable, and furthermore formula (3) holds.

THEOREM 8.3 *Let $f \in M_R$ where $R = I_1 \times I_2$. Then f is integrable. Furthermore for each $x_2 \in I_2$, $f(x_1, x_2)$ is integrable as a function defined on $I_1 = [a_1, b_1]$. If*

$$F(x_2) = \int_{a_1}^{b_1} f(x_1, x_2) \, dx_1,$$

then F is integrable on $[a_2, b_2] = I_2$ and

(3) $$\int_{a_2}^{b_2} F(x_2) \, dx_2 = \int_{a_2}^{b_2} \left[\int_{a_1}^{b_1} f(x_1, x_2) \, dx_1 \right] dx_2 = \int_R f.$$

Proof. Notice first that if g is a step function and $-C \leq g \leq C$ where C is a positive constant, then

$$-\int_R C \leq \int_R g \leq \int_R C.$$

But

$$\int_R C = C \int_R 1 = C(b_1 - a_1)(b_2 - a_2).$$

Hence

$$\left| \int_R g \right| \leq C(b_1 - a_1)(b_2 - a_2).$$

Now it follows immediately from the definition that f is integrable if and only if for each $\epsilon > 0$ there exists step functions g and h satisfying $g \leq f \leq h$ and

$$\int_R (h - g) < \epsilon.$$

For $f \in M_R$ we construct a pair of step functions g and h with this property in the following way. Let ϵ be a given positive number. Choose first a step function p such that

(4) $$|f(X) - p(X)| < \frac{\epsilon}{4(b_1 - a_1)(b_2 - a_2)}$$

for each $X \in R$. Then set

$$g(x) = p(x) - \frac{\epsilon}{4(b_1 - a_1)(b_2 - a_2)}$$

and
$$h(x) = p(x) + \frac{\epsilon}{4(b_1-a_1)(b_2-a_2)}.$$

The functions g and h are clearly step functions. Moreover (4) implies that $g(x) < f(x) < h(x)$ for each $X \in R$. To verify $\int_R (h-g) < \epsilon$, observe that

$$\int_R (h-g) = \int_R \left(p + \frac{\epsilon}{4(b_1-a_1)(b_2-a_2)} - \left(p - \frac{\epsilon}{4(b_1-a_1)(b_2-a_2)}\right)\right)$$

$$= \frac{\epsilon}{2(b_1-a_1)(b_2-a_2)} \int_R 1 = \frac{\epsilon}{2} < \epsilon.$$

Hence the function f is integrable.

This argument shows that if $f \in M_R$, then there exists approximating step functions g, h satisfying

(5) $\qquad \begin{cases} g \leq f \leq h & \text{and} \\ h(X) - g(X) < \epsilon, & \text{each } X \in R. \end{cases}$

To derive formula (3) observe that the same argument as above shows that $f(x_1, x_2)$ is integrable as a function of x_1 for a fixed value of x_2. Hence if g and h satisfy (5),

$$\int_{a_1}^{b_1} g(x_1, x_2)\, dx_1 \leq \int_{a_1}^{b_1} f(x_1, x_2)\, dx_1 \leq \int_{a_1}^{b_1} h(x_1, x_2)\, dx_1$$
$$\| \qquad\qquad\qquad \| \qquad\qquad\qquad \|$$
$$G(x_2) \qquad \leq \qquad F(x_2) \qquad \leq \qquad H(x_2).$$

Now G and H are approximating step functions for F. Since

$$\int_{a_2}^{b_2} [H(x_2) - G(x_2)]\, dx_2 = \int_R (h-g) = \epsilon,$$

the function F is integrable. But

$$\int_R g = \int_{a_2}^{b_2} G(x_2)\, dx_2 \leq \int_{a_2}^{b_2} F(x_2)\, dx_2 \leq \int_{a_2}^{b_2} H(x_2)\, dx_2 = \int_R h$$

and also

$$\int_R g \leq \int_R f \leq \int_R h.$$

Now there is only one number m with the property that

$$\int_R g \leq m \leq \int_R h$$

for each pair of step functions satisfying (5). Both $\int_R f$ and $\int_{a_2}^{b_2} F(x_2)\, dx_2$ have this property. Hence it must follow that

$$m = \int_R f = \int_{a_2}^{b_2} F(x_2)\, dx_2.$$

This is the desired formula (3).

This argument is clearly symmetric in x_1 and x_2. Hence

$$\int_R f = \int_{a_1}^{b_1} \left[\int_{a_2}^{b_2} f(x_1, x_2)\, dx_2\right] dx_1 = \int_{a_2}^{b_2} \left[\int_{a_1}^{b_1} f(x_1, x_2)\, dx_1\right] dx_2.$$

In what follows we shall freely use the fact that each continuous function on the rectangle R belongs to M_R.

Example 1. If $f(x_1, x_2) = (2x_1 + x_2)^2$, compute $\int_R f$ if $R = [0, 2] \times [1, 3]$. Now f is continuous. Hence $f \in M_R$, and

$$\int_R f = \int_0^2 \left[\int_1^3 (2x_1 + x_2)^2 \, dx_2 \right] dx_1.$$

We integrate first with respect to x_2, then with respect to x_1, obtaining

$$\int_R f = \int_0^2 \frac{(2x_1 + x_2)^3}{3} \bigg|_{x_2=1}^{x_2=3} dx_1 = \frac{1}{3} \int_0^2 [(2x_1 + 3)^3 - (2x_1 + 1)^3] \, dx_1$$

$$= \frac{1}{6} \left[\frac{(2x_1 + 3)^4}{4} - \frac{(2x_1 + 1)^4}{4} \right]_{x_1=0}^{x_1=2} = \frac{1}{24} [7^4 - 5^4 - 3^4 + 1^4] = \frac{212}{3}.$$

The next problem is to define the notion of integrability for functions defined on more general closed sets than just rectangles. This is easy, however. If f is defined on the bounded closed set E, enclose E in a rectangle R, and define a new function $\tilde{f}$ defined on all of R in the following way

$$\tilde{f}(X) = f(X) \quad \text{if} \quad X \in E$$
$$= 0 \quad \text{if} \quad X \notin E.$$

We say that f is integrable on E if the extended function $\tilde{f}$ is integrable on R. Moreover we define $\int_E f = \int_R \tilde{f}$. If $\tilde{f}$ is bounded and continuous on E, the extended function f does not necessarily belong to M_R. However, under relatively mild restrictions on f and on the boundary of the closed set E, it can be shown that the conclusions of Theorem 8.3 hold. Namely, $f(x_1, x_2)$ is integrable as a function of one variable for the other held fixed, and

(3) $$\int_E f = \int_R \tilde{f} = \int_{a_2}^{b_2} \left[\int_{a_1}^{b_1} \tilde{f}(x_1, x_2) \, dx_1 \right] dx_2$$

$$= \int_{a_1}^{b_1} \left[\int_{a_2}^{b_2} \tilde{f}(x_1, x_2) \, dx_2 \right] dx_1.$$

We shall assume the validity of (3) for all the functions f and sets E under consideration.

For certain types of sets E we may rewrite formula (3). Suppose E is the region between two lines $x_1 = a_1$ and $x_1 = b_1$ and the graphs of two continuous functions $x_2 = \varphi_1(x_1)$ and $x_2 = \varphi_2(x_1)$ where $\varphi_2(x_1) \geq \varphi_1(x_1)$ for $a \leq x_1 \leq b$. An example of such a region is illustrated in Figure 7.

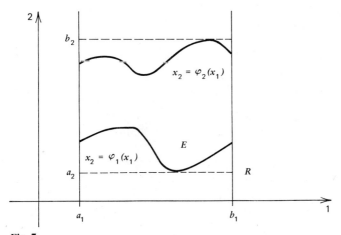

Fig. 7

If we enclose E in a rectangle $R = [a_1, b_1] \times [a_2, b_2]$ and define $\tilde{f} = f$ on E, $\tilde{f} = 0$ off E, then by definition

$$\int_E f = \int_R \tilde{f} = \int_{a_1}^{b_1} \left[\int_{a_2}^{b_2} \tilde{f}(x_1, x_2) \, dx_2 \right] dx_1.$$

Now, however, for a fixed value of x_1, $\tilde{f}(x_1, x_2) = 0$ if $x_2 > \varphi_2(x_1)$ or if $x_2 < \varphi_1(x_1)$. Hence

$$\int_{a_2}^{b_2} \tilde{f}(x_1, x_2) \, dx_2 = \int_{\varphi_1(x_1)}^{\varphi_2(x_1)} f(x_1, x_2) \, dx_2.$$

Thus

$$\int_E f = \int_{a_1}^{b_1} \left[\int_{\varphi_1(x_1)}^{\varphi_2(x_1)} f(x_1, x_2) \, dx_2 \right] dx_1.$$

Example 2. If $f(X) = x_1 + 1$, compute $\int_E f$ if E is the set of points in the first quadrant to the left of the line $x_1 = 2$, below the parabola $x_2 = 1 + x_1^2$ and above the line $x_2 = x_1$. A sketch of this region is Figure 8.

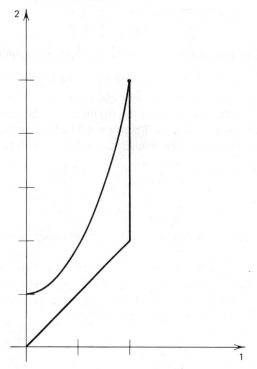

Fig. 8

Therefore

$$\int_E f = \int_0^2 \left[\int_{x_1}^{1+x_1^2} (x_1 + 1) \, dx_2 \right] dx_1$$

$$= \int_0^2 [x_1 x_2 + x_2] \Big|_{x_1}^{1+x_1^2} dx_1$$

$$= \int_0^2 [x_1(1 + x_1^2) + (1 + x_1^2) - x_1^2 - x_1] \, dx_1$$

$$= \int_0^2 (x_1^3 + 1) \, dx_1 = \frac{x_1^4}{4} + x_1 \Big|_0^2 = 6.$$

If E lies between two lines $x_2 = a_2$ and $x_2 = b_2$ and to the left of the curve $x_1 = \varphi_1(x_2)$ and to the right of the curve $x_1 = \varphi_2(x_2)$, then to evaluate $\int_E f$ we reverse the order of integration, integrating first with respect to x_1 and then with respect to x_2. Indeed

$$\int_E f = \int_{a_2}^{b_2} \left[\int_{\varphi_1(x_2)}^{\varphi_2(x_2)} f(x_1, x_2) \, dx_1 \right] dx_2.$$

A sketch of such a region is Figure 9.

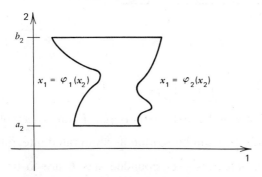

Fig. 9

Example 3. Integrate the function $f(x_1, x_2) = \cos(x_1 + x_2)$ over the trapezoid defined by connecting the points $(\pm\pi, \pi/2)$ and $(\pm\pi/2, 0)$ by straight lines. Sketching the region we obtain Figure 10.

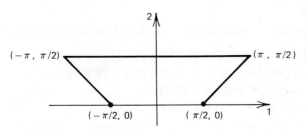

Fig. 10

The boundary curves are $x_2 = 0, \pi$ and $x_1 = x_2 - \pi/2$ and $x_1 = -x_2 - \pi/2$. Hence if E is the given trapezoid

$$\int_E f = \int_0^{\pi/2} \left[\int_{-x_2-\pi/2}^{x_2-\pi/2} \cos(x_1 + x_2) \, dx_1 \right] dx_2$$

$$= \int_0^{\pi/2} \sin(x_1 + x_2) \Big|_{-x_2-\pi/2}^{x_2-\pi/2} dx_2$$

$$= \int_0^{\pi/2} \left[\sin\left(2x_2 - \frac{\pi}{2}\right) + \sin\frac{\pi}{2} \right] dx_2 = \int_0^{\pi/2} \left[-\cos 2x_2 + 1 \right] dx_2$$

$$= \left[-\frac{1}{2} \sin 2x_2 + x_2 \right]_0^{\pi/2} = \frac{\pi}{2}.$$

For more complicated sets E we must partition the set into a union of sets of the above two types. Thus for the set E in Figure 11, $E = E_1 \cup \cdots \cup E_5$.

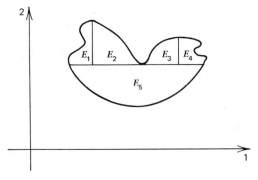

Fig. 11

The integral

$$\int_E f = \int_{E_1} f + \cdots + \int_{E_5} f.$$

The sets E_2, E_3, E_5 are of the first type, whereas E_1 and E_4 are of the second. Each of the integrals $\int_{E_i} f$ can be written as an iterated integral.

In practice when integrals over bounded sets E are to be computed a sketch of the set should first be made. The boundary curves can then be computed. At this point it can be determined if it is necessary to partition the set into smaller sets, each of one of the types discussed above. The appropriate order of integration for each of the necessary subsets can usually be determined by inspection.

EXERCISES

Compute $\int_R f$ for the following choices of functions f and rectangles R.

1. $f(x_1, x_2) = (x_1 + 2x_2)^2$, $R = [0, 2] \times [-1, 2]$.
2. $f(x_1, x_2) = \sin(2x_1 - x_2)$, $R = [-\pi/2, \pi/2] \times [0, \pi/4]$.
3. $f(x, y) = xe^{x^2+y}$, $R = [0, 1] \times [0, 1]$.
4. $f(X) = |X|^2$, $R = [-1, 1] \times [-1, 1]$.

Compute $\int_E f$ for the following choices of functions f and regions E.

5. $f(x_1, x_2) = 1$, E is the region in the first quadrant below the line $x_1 + x_2 = 4$.
6. $f(x_1, x_2) = 1$, E is the region in the first quadrant below the curve $x_2 = e^{x_1}$ and to the left of the line $x_1 = 4$.
7. $f(x_1, x_2) = x_1$, E is the region in the right half plane to the left of the parabola $x_2^2 = x_1$, and bounded by the lines $|x_2| = 2$.
8. $f(X) = |X|^2$, E is the region above $x_2 = |x_1|$ and below $x_2 = 3$.
9. $f(X) = 2x - y$, E is the region above $y = |x - 1|$ and below $y = 4 - |x|$.
10. $f(X) = x^2 + y^2$, E is the region in the first quadrant below $y = [2x]$ and to the left of $x = 4$.

Write the following integrals $\int_E f$ as integrated integrals in two ways, one in which the integration is taken first with respect to x, the other with respect to y. Be sure to indicate the limits of integration in both cases. Do not evaluate the integrals.

11. $f(X) = |X|$, E is the region in the upper half plane below the circle $x^2 + y^2 = 4$.
12. $f(X) = 1$, E is the region within the circle $x^2 + y^2 = 4$ and below the line $y = x$.
13. $f(X) = x$, E is the region below the curve $y = e^x$, above the curve $y - 2x + 5 = 0$ and within the strip $|x| \leq 2$.

14. $f(X) = x^2$, E is the region below $y = 4 - |2x|$, and above the curve $y = |1 - x|$.

15. Evaluate the integral $\int_E f$ where $f(x, y) = x$ and E is the region in the right half plane inside the circle $x^2 + y^2 = b^2$ and outside the circle $x^2 + y^2 = a^2$ where $a < b$.

16. Show that f is integrable on a rectangle R if and only if for each $\epsilon > 0$ there exist step functions g, h defined on R satisfying $g \leq f \leq h$ and $\int_R h - \int_R g < \epsilon$.

17. Show that the set of integrable functions I_R on the rectangle R is a vector space.

18. Show that the integral $\int_R f$ is a real-valued linear function defined on the vector space I_R.

19. Let E be a bounded open set contained in a rectangle R. Let π be a partition of R and let g be a step function taking on either the value zero or one on the open subintervals of the partition π. Define

$$\chi_E(X) = 1 \quad \text{if} \quad X \in U$$
$$ = 0 \quad \text{if} \quad X \notin U.$$

This function is called the characteristic function of the set E. Formulate a condition in terms of the step functions g which will guarantee that χ_E is integrable.

8.4 Applications

1. Area

Let S be a bounded set in the plane. The area of this set, if it exists, is defined to be $\int_S 1$. Thus S will have an area if the function χ_S defined by the condition

$$\chi_S(X) = 1 \quad X \in S$$
$$ = 0 \quad X \notin S,$$

is an integrable function. At first glance it seems strange to comtemplate the possibility that there should exist sets for which no area can be defined. However, examples of such sets are not hard to construct. One is given in Exercise 21 at the end of the section. Notice also that to say that the area of a set S does not exist or that S has no area is different from the assertion that the area of a set has value zero. For example, an interval $[a, b]$ on the real line has a perfectly well defined area. The value of this area is just the number zero.

That this definition of area is a reasonable one to take follows from the following considerations. Enclose S in a rectangle R, and consider a partition of R as in Figure 12.

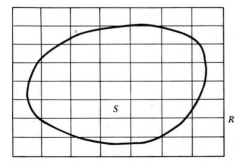

Fig. 12

Now if the number A is to be a candidate for the area of the set S, then it must satisfy the following two conditions.

(1) Let R_{ij} be an open subrectangle of the partition π lying totally within S. We call R_{ij} an *inner rectangle*. If A_{ij} is the area of this inner rectangle, then $A > A_{ij}$. Indeed, since these rectangles do not overlap,

$$A \geq \sum_{i,j} A_{ij}.$$

(2) Let R'_{ij} be an open subrectangle of π satisfying $R'_{ij} \cap S \neq \emptyset$. That is, some points of S belong to R'_{ij}. We call such a rectangle an *outer* rectangle. If we take the union of all the outer rectangles R'_{ij} we have $S \subset \cup R'_{ij}$. If A'_{ij} is the area of the outer rectangle R'_{ij}, we see that the area of the union of these nonoverlapping rectangles in $\sum A'_{ij}$. Hence

$$A \leq \sum A'_{ij}.$$

Figures 13 and 14 illustrate the definitions of these inner and outer rectangles R_{ij} and R'_{ij}.

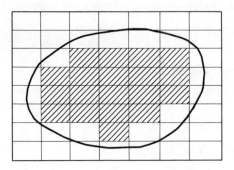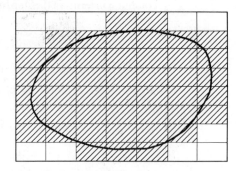

Fig. 13 Inner rectangles R_{ij} are shaded; $A \geq \Sigma A_{ij}$.   **Fig. 14** Outer ractangles R'_{ij} are shaded; $A \leq \Sigma A'_{ij}$.

If we take the position that these are the only conditions that can be required of area, then we have no choice but to say that S has an area A if and only if there is precisely one number A with the property that

$$\sum A_{ij} \leq A \leq \sum A'_{ij}$$

no matter how the partition π is chosen.

Now for X belonging to an open subrectangle of the partition π define

$$\begin{aligned} g(X) &= 1 &&\text{if} && X \in UR_{ij} \\ &= 0 &&\text{if} && X \notin UR_{ij} \end{aligned}$$

and

$$\begin{aligned} h(X) &= 1 &&\text{if} && X \in UR'_{ij} \\ &= 0 &&\text{if} && X \notin UR'_{ij}. \end{aligned}$$

If X belongs to one of the division lines of the partition π define

$$g(X) = 0 \quad \text{and} \quad h(X) = 1.$$

Then g and h are step functions defined on R. Furthermore

$$g \leq \chi_S \leq h,$$

$$\int_R g = \sum A_{ij},$$

and
$$\int_R h = \sum A'_{ij}.$$

Thus we are led to our definition that S has an area A if and only if χ_S is integrable. Furthermore if this is the case
$$\int_R \chi_S = \int_S 1 = A.$$

The function χ_S is called the *characteristic function* of the set S.

Now S will have an area if the boundary of S is the image of finitely many continuously differentiable curves. This will cover all of the situations that are met in practice. Before considering further applications, let us make some remarks about this area problem.

That there exist bounded open sets for which the area cannot be defined is to some extent contrary to one's intuition. Nevertheless such sets can be constructed. They may even have continuous boundary curves. If we wish to take the position that every bounded open set in the plane should have an area, then we must impose some condition stronger than (1) and (2). This problem was investigated around 1900 and was solved by the introduction of a more general integral due to the French mathematician, Henri Lebesgue. This integral, called the Lebesgue integral to distinguish it from the Riemann integral which we have been considering, applies to a wider class of functions. In particular, if E is any bounded open set in the plane then χ_E, the characteristic function for E, is Lebesgue integrable. Moreover, we may define the area of such an open set to be just $\int \chi_E$. We use the same symbol because it turns out that if a function is Riemann integrable, then it is Lebesgue integrable and the integrals have the same value.

We cannot go into any details of the Lebesgue integral here, except to say that in place of considering *finite* partitions of an interval $[a, b]$, and hence the rectangle R, one must consider *infinite* partitions of the interval into nonoverlapping subintervals.

2. Volume

Let f be a bounded nonnegative function defined on a bounded set S. The set of points $X = (x_1, x_2, x_3)$ satisfying $0 \leq x_3 \leq f(x_1, x_2)$ if $(x_1, x_2) \in S$ is called the *ordinate set* for the function f. This ordinate set O_S is just the set of points lying over the set S and beneath the graph of f, as in Figure 15.

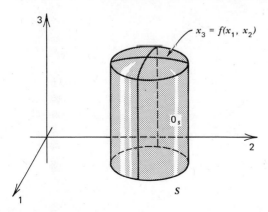

Fig. 15 Ordinate set for the function f.

308 MULTIPLE INTEGRALS

The volume of this set O_S if it exists, we define to be $\int_S f$. The justification of this definition proceeds in much the same way as for area. Since the volume of a rectangular box is the product of the length, width, and height, the volume of the ordinate set of a constant function f defined on a rectangle is $c \cdot A$ where c is the constant value of f and A is the area of the rectangle. Hence if f is a nonnegative step function defined on a rectangle R with constant values c_{ij} on each of the open subrectangles R_{ij}, the volume V of the ordinate set is given by

$$\sum c_{ij} A_{ij} = \int_R f$$

where A_{ij} is the area of the rectangle R_{ij}. For the general case let R be a rectangle containing S. Let $\tilde{f}$ be the extension of f to all of R obtained by defining $\tilde{f}(x) = f(x)$, $x \in S$; $\tilde{f}(x) = 0$, $x \notin S$. If g and h are nonnegative step functions defined on R and $g \leq \tilde{f} \leq h$, then if V is to be the volume of the ordinate set for f, we must have

$$\int_R g \leq V \leq \int_R h.$$

Hence if $\tilde{f}$ is integrable on R, we have no choice than to define

$$V = \int_R \tilde{f}.$$

Since $\tilde{f}$ vanishes off S, $\int_R \tilde{f} = \int_S f$.

Example 1. Determine the volume of the solid in the first octant bounded by the coordinate planes and the plane $2x + y + z = 6$. As in most problems of this type, we must determine the set S and the function f so that if V is the desired volume

$$V = \int_S f.$$

It is essential that a sketch be drawn of the region. Clearly if

$$z = 6 - 2x - y = f(x, y),$$

then we are seeking the volume of the ordinate set for f defined over the region in the first quadrant of the x, y plane bounded by the coordinate axes and the line $2x + y = 6$. See Figure 16.

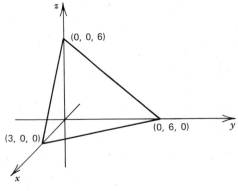

Fig. 16

Thus
$$V = \int_S f = \int_0^3 \left[\int_0^{6-2x} (6-2x-y)\, dy \right] dx$$
$$= \int_0^3 \left[(6-2x)y - \frac{y^2}{2} \right]_0^{6-2x} dx = \int_0^3 \left[(6-2x)^2 - \frac{1}{2}(6-2x)^2 \right] dx$$
$$= \frac{1}{2}\int_0^3 (6-2x)^2\, dx = -\frac{1}{12}(6-2x)^3 \Big|_0^3 = \frac{6^3}{12} = 18.$$

$\int_S f$ could have been computed by integrating first with respect to x then with respect to y. We would then obtain
$$V = \int_0^6 \left[\int_0^{3-y/2} (6-2x-y)\, dx \right] dy.$$

Example 2. Compute the volume of region below the paraboloid $z = 4 - x^2 - 2y^2$ and above the x, y plane. A sketch of the region is Figure 17.

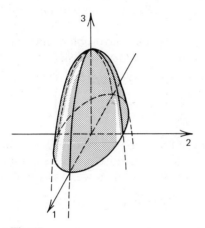

Fig. 17

The desired volume V is the volume of the ordinate set of $f(x, y) = 4 - x^2 - 2y^2$ defined above the set of points $X = (x, y)$ satisfying $x^2 + 2y^2 \leq 4$. Hence, integrating first with respect to y, we have
$$V = \int_{-2}^{2} \left[\int_{-(4-x^2)^{1/2}/\sqrt{2}}^{(4-x^2)^{1/2}/\sqrt{2}} (4 - x^2 - 2y^2)\, dy \right] dx$$
$$= 4 \int_0^2 \int_0^{(4-x^2)^{1/2}/\sqrt{2}} (4 - x^2 - 2y^2)\, dy\, dx$$
since f is symmetric about the origin. Therefore
$$V = 4 \int_0^2 (4-x^2)y - \frac{2y^3}{3} \Big|_0^{(4-x^2)^{1/2}/\sqrt{2}} dx = \frac{4}{3\sqrt{2}} \int_0^2 (4-x^2)^{3/2}\, dx.$$

The integral may be evaluated by a trigonometric substitution obtaining
$$V = \frac{4}{3\sqrt{2}} \left(\frac{x}{4}(10 - x^2)\sqrt{4 - x^2} + 6 \sin^{-1} \frac{x}{2} \right)_0^2$$
$$= \frac{8}{\sqrt{2}} \sin^{-1} 1 = \frac{4\pi}{\sqrt{2}} = 2\sqrt{2}\pi.$$

Before discussing the next application let us make some remarks about the approximation to the integral by finite sums. Let S be a bounded closed set such that the boundary is composed of the images of finitely many continuously differentiable curves. If f is continuous on S then f is integrable and we may make a strong statement as to how $\int_S f$ may be approximated by finite sums. Enclose S in a rectangle R. If π is a partition of R let $\bar{R}_{ij}$ be the closed rectangles of the partition. We assume as always that $f(X) = 0$ if $X \in R$ but $X \notin S$.

THEOREM 8.4 *Under the above assumptions, for each $\epsilon > 0$ there exists a partition π of R such that*

$$\left| \int_S f - \sum_{i,j} f(X_{ij}) A_{ij} \right| < \epsilon$$

where A_{ij} is the area of the rectangle R_{ij} and X_{ij} is any point in the closed rectangle $\bar{R}_{ij}$. Furthermore if f is a finite sum or product of continuous functions $f_1, \ldots, f_n$, then the points $X_{ij} \in \bar{R}_{ij}$ may vary with each of the functions $f_1, \ldots, f_n$.

As an application of this result consider the problem of defining the center of mass of a thin plate. If masses $m_1, \ldots, m_n$ are distributed at points in the plane with coordinates $X_1, \ldots, X_n$ then the center of mass of this system is the vector X satisfying

$$X = \frac{m_1 X_1 + \cdots + m_n X_n}{m_1 + \cdots + m_n}.$$

To generalize this we proceed as follows. Suppose a total mass M is spread over a bounded region S. If $f(X)$ denotes the density function, that is, the mass per unit area as a function of X, then we assert first that

$$M = \int_S f.$$

To see this, enclose S in a rectangle R and let R_{ij} be an open rectangle of a partition of R. We define f to be zero outside S as usual. Now if $c_{ij} \leq f(X) \leq d_{ij}$ for $X \in R_{ij}$, then the mass M_{ij} of the rectangle clearly must satisfy

$$c_{ij} A_{ij} \leq M_{ij} \leq d_{ij} A_{ij}$$

where A_{ij} is the area of the rectangle. Since the mass of disjoint sets (rectangles) is additive, the total mass of the system is the sum of the masses of the constituent rectangles. Therefore $M = \sum M_{ij}$, amd

$$\sum c_{ij} A_{ij} \leq M \leq \sum d_{ij} A_{ij}.$$

But we may define step functions g and h satisfying $g(X) = c_{ij}$ and $h(X) = d_{ij}$ for $X \in R_{ij}$ and $g \leq f \leq h$. Therefore

$$\int_S g = \sum c_{ij} A_{ij} \quad \text{and} \quad \int_S h = \sum d_{ij} A_{ij}.$$

Hence

$$\int_S g \leq M \leq \int_S h,$$

and if f is an integrable function, we have no choice but to define $M = \int_S f$.

To determine the center of mass we pose the following question. If R_{ij} are the subrectangles of a partition of the rectangle R and M_{ij}, M are the respective masses, can points $\bar{X}_{ij} \in R_{ij}$, $\bar{X} \in R$ be defined such that

$$M\bar{X} = \sum M_{ij}\bar{X}_{ij}?$$

If this identity is to hold for each partition of R we assert that if

$$\bar{X} = (\bar{x}, \bar{y}),$$

then

$$M\bar{x} = \int_R xf(x, y)\,dx\,dy$$

and

$$M\bar{y} = \int_R yf(x, y)\,dx\,dy.$$

To see this note that if

$$R_{ij} = (x_{i-1}, x_i) \times (y_{j-1}, y_j)$$

and

$$\bar{X}_{ij} = (\bar{x}_{ij}, \bar{y}_{ij}),$$

then since $\bar{X}_{ij} \in R_{ij}$, we have

$$x_{i-1} \leq \bar{x}_{ij} \leq x_i$$

and

$$y_{j-1} \leq \bar{y}_{ij} \leq y_j.$$

Let $f(X''_{ij})$ and $f(X'_{ij})$ be the maximum and minimum values of f over the closed rectangles $\bar{R}_{ij}$. Then

$$f_{ij}(X'_{ij})A_{ij} \leq M_{ij} \leq f(X''_{ij})A_{ij}$$

and

$$x_{i-1}f(X'_{ij})A_{ij} \leq M_{ij}\bar{x}_{ij} \leq x_i f(X''_{ij})A_{ij}.$$

Since $M\bar{x} = \sum M_{ij}\bar{x}_{ij}$ we have

$$\sum_{ij} x_{i-1}f(X'_{ij})A_{ij} \leq M\bar{x} \leq \sum_{ij} x_i f(X''_{ij})A_{ij}.$$

But by Theorem 8.4, $\int_R xf(x, y)$ can be approximated arbitrarily well by sums

$$\sum_{ij} x_i f(X''_{ij})A_{ij}.$$

Therefore we must have $M\bar{x} = \int_R xf(x, y)$. Similarly $M\bar{y} = \int_R yf(x, y)$. If the mass density function f has constant value c on our region S then we have

$$M = \int_S c = c\int_S 1$$

and

$$M\bar{x} = \int_S cx = c\int_S x.$$

Hence

$$\bar{x} = \frac{\int_S x}{\int_S 1}$$

and

$$\bar{y} = \frac{\int_S y}{\int_S 1}.$$

When the mass density function f is constant, the center of mass $(\bar{x}, \bar{y})$ is called the *centroid* of the region S.

Example 3. A thin plate S is in the shape of the region in the first quadrant bounded by the coordinate axes, the line $x = 1$ and the curve $y = e^x$. If the density of the plate is proportional to the distance from the x axis, determine the mass and center of mass of the plate.

A sketch of the region is Figure 18.

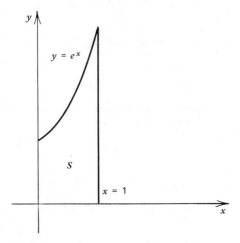

Fig. 18

Since the density function $f(x, y) = cy$, the mass

$$M = c \int_S y \, dx \, dy = c \int_0^1 \left[\int_0^{e^x} y \, dy \right] dx = \frac{c}{2} \int_0^1 e^{2x} \, dx = \frac{1}{4} e^{2x} \Big|_0^1 = \frac{c}{4}[e^2 - 1]$$

$$M\bar{x} = c \int_S xy \, dx \, dy = c \int_0^1 \left[\int_0^{e^x} xy \, dy \right] dx = \frac{c}{2} \int_0^1 x e^{2x} \, dx$$

$$= \frac{c}{4} \left[x e^{2x} - \frac{e^{2x}}{2} \right]_0^1 = \frac{c}{4} \left[\frac{e^2}{2} + \frac{1}{2} \right]$$

$$M\bar{y} = c \int_S y^2 \, dx \, dy = c \int_0^1 \left[\int_0^{e^x} y^2 \, dy \right] dx = \frac{c}{3} \int_0^1 e^{3x} \, dx = \frac{c}{9}[e^3 - 1].$$

Therefore

$$\bar{x} = \frac{1}{2} \frac{e^2 + 1}{e^2 - 1}$$

$$\bar{y} = \frac{9}{4} \frac{e + 1}{e^2 + e + 1}.$$

EXERCISES

1. Find the area of the region above the curve $y = x^2$ and below the line $y = 6 - 5x$.
2. Find the area of the region above the curve $y = x^2 - 1$ and below the lines $y = 1 + x$ and $y = 1$.
3. Find the area of the region enclosed between the two parabolas $y^2 = 4 - x$ and $y^2 = 4 - 2x$.

Sketch each of the regions for the following exercises.

4. Find the volume of the region in the first octant under the plane $2x + y + z = 6$ and above the triangle in the x, y plane bounded by the coordinate axes and the line $2y + x = 2$.
5. Find the volume of the region in the first octant under the plane $3x + y + 4z = 12$.
6. Find the volume of the region in the first octant under the surface $z = xy$ and above the region in the xy plane which lies within the circle $x^2 + y^2 = 1$ and to the right of the line $x + y = 1$.
7. Find the volume of the region in the first octant bounded by the cylinder $x^2 = 4 - z$ and the plane $4x + 3y = 12$.
8. Find the volume of the region above the xy plane, within the cylinder $x^2 + y^2 = 4$ and below the paraboloid $2z + x^2 + y^2 = 16$.
9. Find the volume of the intersection of the two cylinders $x^2 + y^2 = a^2$ and $x^2 + z^2 = a^2$.
10. Let S be a bounded set in the x, y plane with area A and let $X_0 = (x_0, y_0, z_0)$, $z_0 > 0$ be a fixed point. Form a conical region U by connecting X_0 to each point of S by a straight line segment. From the definition of the integral show that the area of each plane cross section of U parallel to the x, y plane is $A[(z_0 - z)/z_0]^2$ where z is the distance from the plane cross section to the x, y plane. Show also that the volume of the solid equals $Az_0/3$.
11. A thin plate is in the shape of the region in the first quadrant below the line $x + y = 1$. If the density of the plate is proportional to the square of the distance from the origin, find the mass of the plate and the center of mass.
12. Find the mass and center of mass of the plate in Exercise 11 if the density is proportional to the distance from the line $x = 1$.
13. Find the mass and center of mass of the plate in Exercise 11 if the density is proportional to the product of the distances from the coordinate axes.
14. Find the mass and center of mass of a plate bounded by $y = 1 - |x|$ and $y = |x| - 1$ if the density is proportional to the square of the distance from the line $x + y = 1$.
15. Find the centroid of the region in the first quadrant below the curve $\sqrt{x} + \sqrt{y} = 1$.
16. Let U_1, and U_2 be nonoverlapping regions with areas A_1, A_2 and centroids $\bar{x}_1$, $\bar{x}_2$. Let A and X be the area and centroid of the region $U_1 \cup U_2$. Prove that
$$A\bar{X} = A_1\bar{X}_1 + A_2\bar{X}_2.$$
This is the theorem of Pappas, a Greek geometer of the third century A.D.
17. If $R_1 = [1, 2] \times [3, 4]$ and $R_2 = [0, 2] \times [1, 3]$ and $R_3 = [-1, 1] \times [2, 3]$, use the theorem of Pappas to compute the centroid of $R_1 \cup R_2$ and $R_2 \cup R_3$.

If S is a set in the plane then
$$\int_S x f(x, y)\, dx\, dy, \int_S y f(x, y)\, dx\, dy$$
are called the *first moments* of the function f over the region S. The *second-order moments* of f are defined by the integrals
$$\int_S x^2 f(x, y)\, dx\, dy, \int_S y^2 f(x, y)\, dx\, dy.$$
If we interpret $f(x, y)$ as a mass density function for a thin plate, then these two integrals are called the *moments of inertia* about the y and x axis, respectively. The moment of inertia about the origin, or polar moment, is the sum of these integrals
$$\int_S (x^2 + y^2) f(x, y)\, dx\, dy.$$

This is usually written I_0 and

$$I_x = \int_S y^2 f(x, y)\, dx\, dy,$$

$$I_y = \int_S x^2 f(x, y)\, dx\, dy.$$

In general if l is a line in the plane and $d(x, y)$ is the distance from a point $X = (x, y)$ to the line l then the moment of inertia about the line l of the plate S with density function f is defined to be

$$I_l = \int_S d^2(x, y) f(x, y)\, dx\, dy.$$

18. Find the moment of inertia of the plate of Exercise 11 with respect to the coordinate axes.
19. Find the moment of inertia of the plate of Exercise 11 with respect to the line $x + y = 1$ if the density is constant.
20. Let l and l_0 be parallel lines in the plane. Assume l_0 passes through the center of mass of a thin plate S with total mass M. If I_l and I_{l_0} denote the respective moments of inertia show that

$$I_l = I_{l_0} + Md^2$$

where d is the distance from l to l_0. This is called the parallel axis theorem. (Assume the center of mass is at the origin and choose l parallel to one coordinate axis.)

*21. Let S_δ be a bounded closed set lying within the disc of radius δ centered at the point X. Assume that the area $A(S_\delta) > 0$. If f is an integrable function which is continuous at the point X, show that

$$\lim_{\delta \to 0} \frac{1}{A(S_\delta)} \int_{S_\delta} f = f(X).$$

[For a given ϵ, choose a δ so that if $|Y - X| < \delta$, $f(X) - \epsilon < f(Y) < f(X) + \epsilon$. Use this estimate to bound $\int_{S_\delta} f$.] This is the generalization of the first fundamental theorem of the calculus to functions defined on sets in $V_2(R)$.

*22. To define a set S in the plane for which no area can be defined we proceed as follows. Let f be the function defined on the interval $[0, 1]$ which satisfies

$$f(x) = 2 \quad \text{if } x \text{ is a rational number.}$$
$$f(x) = 1 \quad \text{if } x \text{ is an irrational number.}$$

Thus for example $f(1/2) = 2$ and $f(1/\sqrt{2}) = 1$. We set $S = \{(x, y) : 0 \leq x \leq 1 \text{ and } 0 \leq y \leq f(x)\}$. We cannot draw an accurate sketch of S. However, a rough indication is Figure 19.

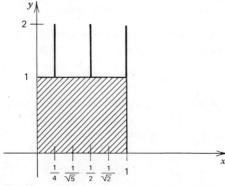

Fig. 19

One might call S the "square with the fringe on top." The fringe is that portion of S for which $1 < y \leq 2$. To show that S does not have an area we argue as follows. Let R be the rectangle $\{(x, y) : 0 \leq x \leq 1, 0 \leq y \leq 2\}$. Then S is contained in R. Partition R as in the discussion on p. 306. Let R_{ij} be an inner rectangle and R'_{ij} an outer rectangle of this partition. Also let $R_0 = \{(x, y) : 0 \leq x \leq 1, 0 \leq y \leq 1\}$. Then R_0 is completely contained in S. Show that the union of the inner rectangles is always contained in R_0, whereas the union of the outer rectangles is all of R. Conclude from this that S can have no area.

8.5 General Multiple Integrals

The extension of double integrals to n dimensions is straightforward. In place of functions defined on rectangles we consider functions defined on n-fold products of intervals. The set of points $X = (x_1, \ldots, x_n)$ where x_k belongs to an interval I_k of real numbers is called a *rectangular parallelepiped* or just a (n-dimensional) box. Thus B is a box if $B = I_1 \times \cdots \times I_n$ where I_k is an interval. If each interval I_k is partitioned into subintervals, the product of these subintervals determines a partition of the box B. A function f defined on the box B is a step function if there exists a partition of B such that f is constant on each of the sub-boxes of the partition. To avoid complications we also assume that $f(x_1, \ldots, x_n)$ is a step function of $n - 1$ variables if one variable is held fixed.

Now the volume of a box B is the product of the lengths of the edges. If we denote this by $v(B)$ we have $v(B) = l(I_1) \cdots l(I_n)$. If B_i are the sub-boxes of a partition of B, and f is a step function constant on B_i, then as in the two-dimensional case we define

$$\int_B f = \sum c_i v(B_i)$$

where c_i is the constant value of f on the box B_i. A general bounded function f defined on B is integrable if there is precisely one number m with the property that

$$\int_B g \leq m \leq \int_B h$$

for each pair of step functions g and h defined on B satisfying $g \leq f \leq h$.

Just as in the plane, a set S in $V_n(R)$ is bounded if it can be enclosed in a box. To define the integral of a bounded function f defined on the bounded set S we first enclose S in a box B. Then we define a new function $\tilde{f}$ by requiring that

$$\tilde{f}(X) = f(X) \quad X \in S$$
$$= 0 \quad X \notin S$$

Thus $\tilde{f}$ is the extension of f to the box B obtained by defining $\tilde{f}$ to be zero outside of S. The function f is then said to be integrable on S if the extended function $\tilde{f}$ is integrable on R. We then define

$$\int_S f = \int_R \tilde{f}.$$

The function f is called the *integrand*, and when f is written out explicitly, the integral $\int_S f$ is often written

$$\int_S f(x_1, \ldots, x_n)\, dx_1 \cdots dx_n, \qquad \int \cdots \int_S f(x_1, \ldots, x_n)\, dx_1 \cdots dx_n$$

$$\text{or} \quad \int_S f(X)\, dV$$

316 MULTIPLE INTEGRALS

where $dV = dx_1 \cdots dx_n$. For example if $S \subset V_3(R)$ and $f(x_1, x_2, x_3) = (x_1 - x_2 + 2x_3)^2$, we would write

$$\int_S (x_1 - x_2 + 2x_3)^2 \, dx_1 \, dx_2 \, dx_3$$

or

$$\int_S (x_1 - x_2 + 2x_3)^2 \, dV.$$

To actually calculate $\int_B f$, we must represent it as an iterated integral. In much the same way as in the case for $n = 2$ we may prove that

(1) $$\int_B f = \int_{a_n}^{b_n} \cdots \int_{a_2}^{b_2} \left[\int_{a_1}^{b_1} f(x_1, \ldots, x_n) \, dx_1 \right] dx_2, \ldots, dx_n$$

where $B = I_1 \times \cdots \times I_n$ and $I_k = [a_k, b_k]$. The actual integration may be performed in any order. To emphasize the particular order that has been chosen, formula (1) is often written

$$\int_B f = \int_{a_n}^{b_n} dx_n \cdots \int_{a_2}^{b_2} dx_2 \int_{a_1}^{b_1} f(x_1, \ldots, x_n) \, dx_1.$$

If f is defined on a bounded set S, then the actual calculation of (1) depends on the character of the boundary of S. We shall limit our examples to $n = 3$.

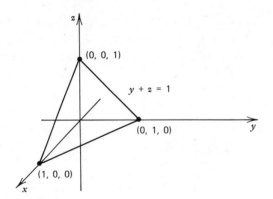

Fig. 20

Example 1. Let S be the set of points in the first octant lying below the plane $x + y + z = 1$. If $f(X) = yz$ for $X \in S$, compute $\int_S f$. Now S lies within the unit cube $B = [0, 1] \times [0, 1] \times [0, 1]$. (See Figure 20.) Defining $f = 0$ for $X \notin S$ we must compute

$$\int_0^1 \left[\int_0^1 \left[\int_0^1 f(x, y, z) \, dx \right] dy \right] dz.$$

But $f(X) = 0$ if $x \geq 1 - y - z$. Hence

$$\int_0^1 f(x, y, z) \, dx = \int_0^{1-y-z} yz \, dx = xyz \Big|_0^{1-y-z} = (1 - y - z)yz$$

Now

$$F(y, z) = \int_0^1 f(x, y, z) \, dx = (1 - y - z)yz$$

for those points (y, z) within the triangle formed by the coordinate axes and the line $y + z = 1$. Outside this region $F(y, z) = 0$. Hence $\int_0^1 F(y, z) \, dy = \int_0^{1-z} F(x, y) \, dy = \int_0^{1-z} (1 - y - z) \, dy$. Therefore

$$\int_0^1 \int_0^1 F(y,z)\,dy\,dz = \int_0^1 \left[\int_0^{1-z} (1-y-z)yz\,dy\right] dz$$

$$= \int_0^1 \int_0^{1-z} (yz - y^2 z - yz^2)\,dy\,dz$$

$$= \int_0^1 \left[\frac{y^2}{2}(z-z^2) - \frac{y^3}{3}z\right]_0^{1-z} dz$$
$$(1-z) \quad -$$

$$= \int_0^1 \left[\frac{z}{2}(1-z)^3 - \frac{z}{3}(1-z)^3\right] dz$$

$$= \frac{1}{6}\int_0^1 z(1-z)^3\,dz = \frac{1}{120}.$$

Thus
$$\int_S f = \int_0^1 dz \int_0^{1-z} dy \int_0^{1-y-z} yz\,dx.$$

The actual evaluation of 3-fold or n-fold iterated integrals can be quite tedious. We shall content ourselves in many instances to just setting up $\int_S f$ as an iterated integral. First let us consider some examples in $V_3(R)$, all of which have analogues in $V_n(R)$.

If S is a bounded closed set, then the characteristic function of S is the function χ_S defined by setting

$$\chi_S(X) = 1 \quad \text{if} \quad X \in S$$
$$ = 0 \quad \text{if} \quad X \notin S.$$

The set S has a volume V if χ_S is integrable. Furthermore

$$V = \int_S \chi_S.$$

If f is a mass density function defined on S, then the mass of this set is

$$M = \int_S f.$$

The center of mass is the vector $\bar{X} = (\bar{x}_1, \bar{x}_2, \bar{x}_3)$, the coordinates of which satisfy
$$M\bar{x}_i = \int_S x_i f(x_1, x_2, x_3)\,dx_1\,dx_2\,dx_3.$$

Second moments with respect to lines and planes may also be defined. If $d(x_1, x_2, x_3)$ is the distance from the point (x_1, x_2, x_3) to a plane π then the second moment with respect to π of the set S with mass density function f is

$$I_\pi = \int_S d^2(x_1, x_2, x_3) f(x_1, x_2, x_3)\,dx_1\,dx_2\,dx_3.$$

These second moments are called *moments of inertia* and can be defined with respect to lines l in $V_3(R)$, as well. The formula is the same as the above except $d(X)$ is now the distance from a point X to a line l.

Example 2. A region in $V_3(R)$ is that portion of the solid cylinder $x^2 + 2y^2 \leq 4$ lying above the xy plane and below the paraboloid $z = 8 - 2x^2 - y^2$. If the density is proportional to the distance from the z axis, write iterated integrals which represent the mass of the region, the center of mass and the moment of inertia with respect to the x axis. Now the density function $f(x, y, z) = c\sqrt{x^2 + y^2}$. Sketching the region we have Figure 21. The limits of integration are simplified if we integrate first with

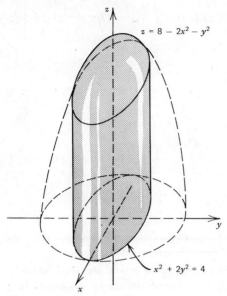

Fig. 21

respect to z. We thus obtain

$$M = c \int_{-\sqrt{2}}^{\sqrt{2}} \left[\int_{-(4-2y^2)^{1/2}}^{(4-2y^2)^{1/2}} \left[\int_0^{8-2x^2-y^2} \sqrt{x^2+y^2}\, dz \right] dx \right] dy.$$

The x coordinate of the center of mass is given by

$$\bar{x} = \frac{1}{M} \int_{-\sqrt{2}}^{\sqrt{2}} \left[\int_{-(4-2y^2)^{1/2}}^{(4-2y^2)^{1/2}} \left[\int_0^{8-2x^2-y^2} x\sqrt{x^2+y^2}\, dz \right] dx \right] dy.$$

Replacing the integrand $x\sqrt{x^2+y^2}$ by $y\sqrt{x^2+y^2}$ and $z\sqrt{x^2+y^2}$ yields the appropriate formula for the y and z coordinates of the center of mass. The moment of inertia I_x with respect to the x axis is the second moment of the function $f(x, y) = c\sqrt{x^2+y^2}$ with respect to the x axis. Since $d(X)$, the distance from X to the x axis, is $\sqrt{y^2+z^2}$, we have

$$I_x = c \int_{-\sqrt{2}}^{\sqrt{2}} dy \int_{-(4-2y^2)^{1/2}}^{(4-2y^2)^{1/2}} dx \int_0^{8-2x^2-y^2} (y^2+z^2)(\sqrt{x^2+y^2})\, dz.$$

EXERCISES

1. If $f(x, y, z) = xy$, compute $\int_S f$ where S is the region in the first octant lying below the plane $2x + y + z = 4$.
2. Compute the centroid of the tetrahedron formed by the coordinate planes and the plane $3x + 2y + z = 6$.
3. Compute the volume of the region bounded by the planes $z = 0$, $x = y$, $y = 2$ and the surface $z = xy$.
4. Compute the centroid of the tetrahedron formed by the coordinate planes and the plane $(x/a) + (y/b) + (z/c) = 1$ if $a, b, c > 0$. (Compute one coordinate and obtain the others by symmetry.)
5. Write an iterated integral that represents the volume of the region within the sphere $x^2 + y^2 + z^2 = 8$ and above the paraboloid $x^2 + y^2 = 7z$.
6. Write an iterated integral that represents the volume of the region within the ellipsoid $4x^2 + y^2 + 3z^2 = 6$ and below the paraboloid $4x^2 + y^2 = 3z$.
7. A solid is in the form of the region in the first octant within the cylinder

$x^2 + y^2 = a^2$ and below the plane $z = b$. If the density of the solid is equal to the product of the distances from the xz and xy planes, determine the mass.

8. A solid with unit density is in the form of a box with edges having lengths equal to a, b, c. Determine the moment of inertia of this box with respect to an edge. Determine the moment of inertia with respect to a line parallel to an edge if the line passes through the center of mass.

9. A solid is in the form of the region within the cylinder $y^2 + z^2 = 1$ and between the planes $x + y = 1$ and $2y + x = 4$. If the mass is proportional to the distance from the origin, write iterated integrals that represent the mass and the x coordinate of the center of mass of the solid.

10. A solid is in the form of the region in the first octant within the cylinder $z = 4 - y^2$ and above the paraboloid $z = 2x^2 + y^2$. If the density of the solid is proportional to the distance from the z axis, write an iterated integral that represents the moment of inertia of the solid with respect to the y axis.

11. Let S_δ be a bounded closed set in $V_n(R)$ within the n-dimensional ball of radius δ about the point $X = (x_1, \ldots, x_n)$. Assume that the volume $v(S_\delta) = \int_{S_\delta} 1 > 0$. Let f be an integrable function which is continuous at the point X. Show that

$$\lim_{\delta \to 0} \frac{1}{v(S_\delta)} \int_{S_\delta} f = f(X).$$

This is the generalization of the first fundamental theorem of the calculus to functions defined on sets in $V_n(R)$. (See Exercise 21, p. 314.)

12. Let π be the tangent plane to the surface defined by the equation $xyz = a^3$ at a point $X = (x_0, y_0, z_0)$. Show that the volume of the tetrahedron formed by this plane and the coordinate planes is $9a^3/2$. (See Exercise 16, p. 260.)

8.6 Change of Variable in Multiple Integrals

If f is an integrable function defined on the interval $[a, b]$ and φ is a differentiable one-to-one function mapping the interval $[c, d]$ onto $[a, b]$, then the the change of variable formula for definite integrals states that

$$(1) \qquad \int_a^b f(x) \, dx = \int_c^d f(\varphi(t)) \varphi'(t) \, dt$$

where $\varphi(c) = a$ and $\varphi(b) = d$. We seek the analogue of this formula for functions defined on sets in $V_n(R)$. Let us consider first the case for $n = 2$.

Suppose that Φ is a continuously differentiable one-to-one function defined on an open set U. If S is a bounded closed set in U, let T be the image of Φ on S (Figure 22). That is,

$$T = \{(x, y) \in V_2(R) : (x, y) = \Phi(u, v), (u, v) \in S\}.$$

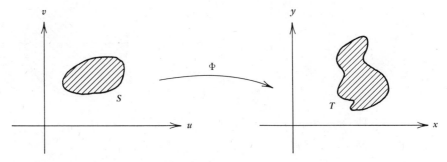

Fig. 22 Φ maps S onto T.

Thus Φ maps S in a one-to-one fashion onto the set T, and it is convenient to abuse the notation and write $T = \Phi(S)$. Denote the coordinate functions of Φ by φ_1 and φ_2. If f is an integrable function defined on T, we seek a formula that allows us to evaluate $\int_T f(x, y)\, dxdy$ by evaluating instead an integral over S.

The appropriate generalization of (1) is the following formula

(2) $$\int_T f(x, y)\, dxdy = \int_S f(\Phi(u, v)) \left| \frac{\partial(\varphi_1, \varphi_2)}{\partial(u, v)} \right| dudv$$

where

$$\left| \frac{\partial(\varphi_1, \varphi_2)}{\partial(u, v)} \right|$$

is the absolute value of the Jacobian of the transformation Φ. Recall

$$\frac{\partial(\varphi_1, \varphi_2)}{\partial(u, v)} = D(\Phi'(u, v)) = D\begin{pmatrix} \frac{\partial \varphi_1}{\partial u} & \frac{\partial \varphi_1}{\partial v} \\ \frac{\partial \varphi_2}{\partial u} & \frac{\partial \varphi_2}{\partial v} \end{pmatrix}.$$

The utility of this formula stems from the fact that we may choose the transformation Φ in such a way that the integral on the right of (2) is much easier to evaluate than the given one.

Example 1. Compute the area of a disc of radius a by evaluating an appropriate double integral. If A is the area of this disc, clearly

$$A = \int_{-a}^{a} \int_{-\sqrt{a^2-x^2}}^{\sqrt{a^2-x^2}} dydx = 4 \int_0^a \int_0^{\sqrt{a^2-x^2}} dydx = 4 \int_0^a \sqrt{a^2 - x^2}\, dx.$$

To evaluate this latter integral one needs to make a trigonometric substitution. However, the transformation

$$(x, y) = \Phi(r, \theta)$$

defined by

$$x = \varphi_1(r, \theta) = r \cos \theta$$
$$y = \varphi_2(r, \theta) = r \sin \theta$$

maps the rectangle $0 \leq r \leq a$, $0 \leq \theta \leq 2\pi$ onto the disc of radius a. See Figure 23.

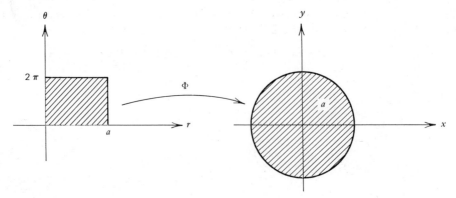

Fig. 23

Now

$$\frac{\partial(\varphi_1, \varphi_2)}{\partial(r, \theta)} = D\begin{pmatrix} \cos \theta & -r \sin \theta \\ \sin \theta & r \cos \theta \end{pmatrix} = r.$$

Therefore applying formula (2) we see that

$$A = \int_{-a}^{a} \int_{-\sqrt{a^2-x^2}}^{\sqrt{a^2-x^2}} dydx = \int_{0}^{2\pi} \int_{0}^{a} r\, drd\theta = \pi a^2.$$

The reader may object that the transformation Φ is not one-to-one on the closed rectangle $R = [0, a] \times [0, 2\pi]$ in the above example. This is indeed the case. However, formula (2) is valid as long as the set of points on which Φ is not one-to-one has zero area. The following theorem gives the precise result.

THEOREM 8.5 *Let Φ be a continuously differentiable function from $V_2(R)$ to $V_2(R)$ which is defined on a bounded open set U. Assume further that, except possibly for a set of zero area, the function Φ is one-to-one and*

$$\frac{\partial(\varphi_1, \varphi_2)}{\partial(u, v)} \neq 0.$$

Let S be a bounded closed subset of U such that the characteristic function of S is integrable, and set $T = \{\Phi(Y) : Y = (u, v) \in S\}$. Then if f is integrable on T, it follows that $f \circ \Phi$ is integrable on S. Moreover

(2) $$\int_T f(x, y)\, dxdy = \int_S f(\Phi(u, v)) \left|\frac{\partial(\varphi_1, \varphi_2)}{\partial(u, v)}\right| dudv.$$

In practice we are given the function f and the set T. The problem is to choose the function Φ in such a way that the right-hand side of (2) is tractable. The geometry of T is usually the clue to the appropriate choice of the function Φ.

Example 2. Compute the moment of inertia with respect to the x axis of the parallelogram formed by the vectors $X_1 = (3, 1)$ and $X_2 = (2, 3)$. We must compute

$$\int_T y^2\, dxdy$$

where T is the shaded region in Figure 24.

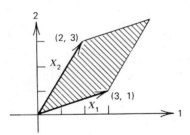

Fig. 24

Now, however, if Φ is the linear transformation taking $(1, 0)$ onto $(3, 1)$ and $(0, 1)$ onto $(2, 3)$, then Φ has matrix $\begin{pmatrix} 3 & 2 \\ 1 & 3 \end{pmatrix}$. Moreover Φ maps the unit square S onto T as in Figure 25. Since Φ is linear, the Jacobian matrix of Φ is just $\begin{pmatrix} 3 & 2 \\ 1 & 3 \end{pmatrix}$ and consequently $\partial(\varphi_1, \varphi_2)/\partial(u, v) = 7$. Moreover, since $\begin{pmatrix} x \\ y \end{pmatrix} = \begin{pmatrix} 3 & 2 \\ 1 & 3 \end{pmatrix}\begin{pmatrix} u \\ v \end{pmatrix}$, $y = u + 3v$, and hence

$$\int_S y^2\, dxdy = 7 \int_0^1 \int_0^1 (u + 3v)^2\, dudv = 7 \int_0^1 \left(\frac{1}{3} + 3v + 9v^2\right) dv = \frac{161}{6}.$$

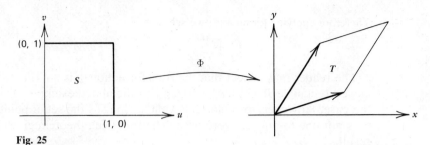

Fig. 25

The proof of Theorem 8.5 is rather long and involved and contains several approximation arguments. The fundamental idea underlying the proof, however, is a simple fact about linear transformations. First we recall that if a parallelogram is formed by the vectors $X = (x_1, x_2)$ and $Y = (y_1, y_2)$, then the area of this parallelogram is given by

$$\left| D \begin{pmatrix} x_1 & y_1 \\ x_2 & y_2 \end{pmatrix} \right|.$$

We have met this observation before on p. 41 in Chapter 1. The proof is very simple. Referring to Figure 26 we note that by elementary geometry the

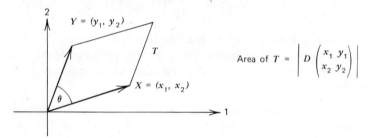

Fig. 26

area A of the parallelogram T is given by

$$A = |X||Y| \sin \theta.$$

Hence

$$A^2 = |X|^2|Y|^2 \sin^2 \theta$$
$$= |X|^2|Y|^2 (1 - \cos^2 \theta)$$
$$= |X|^2|Y|^2 - (X, Y)^2.$$

The last equality follows since

$$(X, Y) = |X||Y| \cos \theta.$$

We now easily check that

$$|X|^2|Y|^2 - (X, Y)^2 = (x_1 y_2 - x_2 y_1)^2.$$

Hence A is the absolute value of $(x_1 y_2 - x_2 y_1)$ or

$$A = \left| D \begin{pmatrix} x_1 & y_1 \\ x_2 & y_2 \end{pmatrix} \right|.$$

We leave the verification of this as an exercise.

Now if Φ is a linear transformation taking the vector U onto X and V onto Y, then Φ maps the parallelogram formed by U and V onto the parallelogram

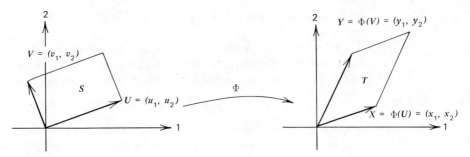

Fig. 27

formed by X and Y as in Figure 27. If we let $A(S)$ and $A(T)$ be the respective areas we have

(3) $$A(S) = \left| D\begin{pmatrix} u_1 & v_1 \\ u_2 & v_2 \end{pmatrix} \right|$$

and

(4) $$A(T) = \left| D\begin{pmatrix} x_1 & y_1 \\ x_2 & y_2 \end{pmatrix} \right|.$$

Since $X = \Phi(U)$ and $Y = \Phi(V)$, if $\begin{pmatrix} a_{11} & a_{12} \\ a_{21} & a_{22} \end{pmatrix}$ is the matrix of Φ, we see that

$$\begin{pmatrix} x_1 & y_1 \\ x_2 & y_2 \end{pmatrix} = \begin{pmatrix} a_{11} & a_{12} \\ a_{21} & a_{22} \end{pmatrix} \begin{pmatrix} u_1 & v_1 \\ u_2 & v_2 \end{pmatrix}.$$

Taking determinants and applying the multiplication theorem for determinants we have

$$D\begin{pmatrix} x_1 & y_1 \\ x_2 & y_2 \end{pmatrix} = D\begin{pmatrix} a_{11} & a_{12} \\ a_{21} & a_{22} \end{pmatrix} D\begin{pmatrix} u_1 & v_1 \\ u_2 & v_2 \end{pmatrix}$$

Taking absolute values and using (3) and (4) we obtain

(5) $$A(T) = \left| D\begin{pmatrix} a_{11} & a_{12} \\ a_{21} & a_{22} \end{pmatrix} \right| A(S).$$

But

$$\begin{pmatrix} a_{11} & a_{12} \\ a_{21} & a_{22} \end{pmatrix} = \Phi'$$

and

$$D\begin{pmatrix} a_{11} & a_{12} \\ a_{21} & a_{22} \end{pmatrix} = \frac{\partial(\varphi_1, \varphi_2)}{\partial(u, v)}.$$

Hence (5) may be rewritten

$$A(T) = \left| \frac{\partial(\varphi_1, \varphi_2)}{\partial(u, v)} \right| A(S).$$

Thus

(6) $$A(T) = \int_T 1 \, dx \, dy = \int_S 1 \left| \frac{\partial(\varphi_1, \varphi_2)}{\partial(u, v)} \right| du \, dv$$

which is a special case of (2).

If we assume now that T is a rectangle we may use the linearity of the integral together with (6) to verify (2) for step functions. The verification of (2) for arbitrary integrable functions f and linear transformations Φ can be accomplished by the usual approximation of f by step functions from above and below.

The passage to arbitrary one-to-one continuously differentiable transformations Φ depends on the fact that near a point $U_0 = (u_0, v_0)$, the function $\Phi(U)$ may be approximated by an affine function, namely, the function

$$\Phi(U_0) + (d_{U_0}\Phi)(U - U_0),$$

where $d_{U_0}\Phi$ is the differential of Φ at U_0. Indeed if R is a small rectangle, then its image under Φ does not differ too much from its image under the above affine transformation. This latter set is, of course, a parallelogram. See Figure 28.

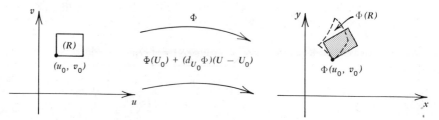

Fig. 28 Shaded region is the image under the affine transformation $\Phi(U_0) + (d_{U_0}\Phi)(U - U_0)$.

As an example, suppose

$$\Phi(u, v) = \left(u, \tfrac{1}{2}u^2 + v\right).$$

To determine the effect of Φ we examine the image of Φ on vertical lines $u = c$ and horizontal lines $v = c$. In the former case

$$\Phi(c, v) = \left(c, \tfrac{1}{2}c^2 + v\right)$$

and the trace of this curve is a vertical line through c. In the latter

$$\Phi(u, c) = \left(u, \tfrac{1}{2}u^2 + c\right)$$

which is a parabola through $(0, c)$. Sketching this for $c = 1, 2$ we obtain Figure 29.

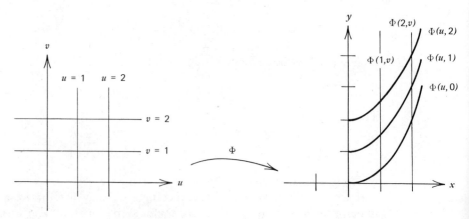

Fig. 29 $\Phi(u, v) = (u, \tfrac{1}{2}u^2 + v)$.

CHANGE OF VARIABLE IN MULTIPLE INTEGRALS

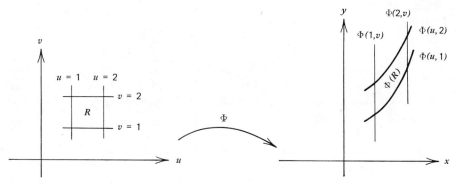

Fig. 30

Therefore Φ maps the unit square $R = [1, 2] \times [1, 2]$ onto the curved region in Figure 30.

Now the approximating affine function is just

$$\Phi(U_0) + d_{U_0}\Phi(U - U_0).$$

But the matrix for $d_{U_0}\Phi$ is just $\begin{pmatrix} 1 & 0 \\ u_0 & 1 \end{pmatrix}$. Hence if $U = \begin{pmatrix} u \\ v \end{pmatrix}$ and $U_0 = \begin{pmatrix} u_0 \\ v_0 \end{pmatrix}$

$$\Phi(U_0) + d_{U_0}\Phi(U - U_0) = \begin{pmatrix} u_0 \\ \frac{1}{2}u_0^2 + v_0 \end{pmatrix} + \begin{pmatrix} 1 & 0 \\ u_0 & 1 \end{pmatrix}\begin{pmatrix} u - u_0 \\ v - v_0 \end{pmatrix}.$$

The image of R can be computed by setting $(u_0, v_0) = (1, 1)$. Then

$$\Phi(U_0) + d_{U_0}\Phi(U - U_0) = \begin{pmatrix} 1 \\ \frac{3}{2} \end{pmatrix} + \begin{pmatrix} 1 & 0 \\ 1 & 1 \end{pmatrix}\begin{pmatrix} u - 1 \\ v - 1 \end{pmatrix}$$

and the square R is mapped onto the parallelogram in Figure 31, which is the desired approximation to $\Phi(R)$.

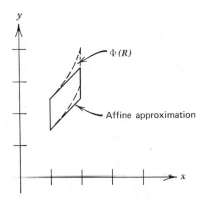

Fig. 31 Approximation to $\Phi(R)$ by affine transformation $\Phi(U_0) + d_{U_0}\Phi(U - U_0)$.

To prove Theorem 8.5 we must exploit the additivity of the integral to show that the total error involved by replacing $\Phi(U)$ by $\Phi(U_0) + d_{U_0}\Phi(U - U_0)$ in a neighborhood of U_0 can be made arbitrarily small. We shall not attempt to carry out the details.

For $n > 2$ we have an exact analogue of Theorem 8.5.

THEOREM 8.6 *Let Φ be a continuously differentiable transformation from $V_n(R)$ to $V_n(R)$ defined on the open set U. Assume that Φ is one-to-*

one on U and that

$$\frac{\partial(\varphi_1,\ldots,\varphi_n)}{\partial(y_1,\ldots,y_n)} \neq 0$$

on U except possibly for a set of zero n-dimensional volume. Let S be a bounded closed subset of U such that the characteristic function of S is integrable and set $\Phi(S) = T$. *If f is integrable on T, then* $f \circ \Phi$ *is integrable on S, and*

(7) $$\int_T f(X)\,dx_1,\ldots,dx_n = \int_S f(\Phi(u_1,\ldots,u_n))\left|\frac{\partial(\varphi_1,\ldots,\varphi_n)}{\partial(u_1,\ldots,u_n)}\right|du_1,\ldots,du_n.$$

Just as in the two-dimensional case, the proof of Theorem 8.6 rests on the fact that if T is the parallelopiped determined by the vectors $X_1,\ldots,X_n$ then the volume, $v(T)$, is the absolute value of the determinant of the matrix the columns of which are the vectors $X_1,\ldots,X_n$. Furthermore if Φ is a one-to-one linear transformation mapping another parallelopiped S onto T, then

$$v(T) = |D(\Phi')|v(S).$$

This is, of course, a special case of (7). If X_1, X_2, X_3 are vectors in $V_3(R)$, then the formula for the volume $v(T)$ of the parallelopiped determined by these vectors may be written

$$v(T) = |(X_1 \times X_2, X_3)|.$$

This we have already observed in Chapter 1, p. 43.

Example 3. Determine the centroid of the parallelopiped T defined by $X_1 = (1, 1, 0)$ $X_2 = (0, 1, 1)$, and $X_3 = (0, 0, 1)$. The region is sketched in Figure 32.

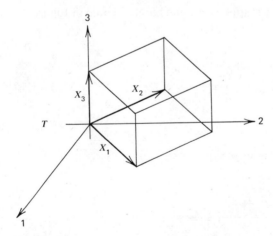

Fig. 32

Map the unit cube defined by the vectors $(1, 0, 0)$, $(0, 1, 0)$, $(0, 0, 1)$ onto T by the linear transformation Φ with matrix

$$\begin{pmatrix} 1 & 0 & 0 \\ 1 & 1 & 0 \\ 0 & 1 & 1 \end{pmatrix}.$$

Since the determinant of this matrix has value one, the volume of the parallelopiped is one. If $X = (x_1, x_2, x_3) = \Phi(y_1, y_2, y_3)$, then

CHANGE OF VARIABLE IN MULTIPLE INTEGRALS

$$x_1 = y_1$$
$$x_2 = y_1 + y_2$$
$$x_3 = y_2 + y_3.$$

Hence

$$\int_T x_1 \, dx_1 \, dx_2 \, dx_3 = \int_0^1 \int_0^1 \int_0^1 y_1 \, dy_1 \, dy_2 \, dy_3 = \frac{1}{2}$$

$$\int_T x_2 \, dx_1 \, dx_2 \, dx_3 = \int_0^1 \int_0^1 \int_0^1 (y_1 + y_2) \, dy_1 \, dy_2 \, dy_3 = 1$$

and

$$\int_T x_3 \, dx_1 \, dx_2 \, dx_3 = \int_0^1 \int_0^1 \int_0^1 (y_2 + y_3) \, dy_1 \, dy_2 \, dy_3 = 1$$

and the centroid has coordinates $(\frac{1}{2}, 1, 1)$.

Example 4. Determine the volume of the region T interior to the sphere $x^2 + y^2 + x^2 = a^2$ and above the cone $z^2 = x^2 + y^2$. This computation becomes very simple if we switch to spherical coordinates. Notice that the function $(x, y, z) = \Phi(r, \theta, \varphi)$ defined by

$$x = r \cos \theta \sin \varphi$$
$$y = r \sin \theta \sin \varphi$$
$$z = r \cos \varphi$$

maps the box $0 \leq r \leq a, 0 \leq \theta \leq 2\pi, 0 \leq \varphi \leq \pi/4$ onto the desired conical region T in $V_3(R)$. This is illustrated in Figure 33.

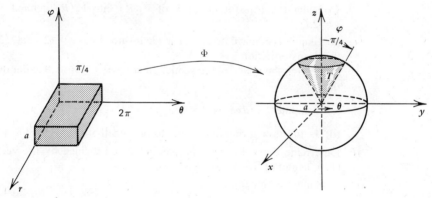

Fig. 33

Also

$$D(\Phi) = \frac{\partial(\varphi_1, \varphi_2, \varphi_3)}{\partial(r, \theta, \varphi)}$$

$$= D \begin{pmatrix} \cos \theta \sin \varphi & -r \sin \theta \sin \varphi & r \cos \theta \cos \varphi \\ \sin \theta \sin \varphi & r \cos \theta \sin \varphi & r \sin \theta \cos \varphi \\ \cos \varphi & 0 & -r \sin \varphi \end{pmatrix} = -r^2 \sin \varphi.$$

Hence the volume of the conical region T is given by

$$\int_T dx \, dy \, dz = \int_0^{\pi/4} d\varphi \int_0^{2\pi} d\theta \int_0^a r^2 \sin \varphi \, dr$$

$$= \frac{2\pi a^3}{3} \int_0^{\pi/4} \sin \varphi \, d\varphi = \frac{2\pi a^3}{3} \left(1 - \frac{\sqrt{2}}{2}\right).$$

EXERCISES

Solve the following problems by first setting up the appropriate integral and then evaluating it by making a change of variable.

1. Find the area of the parallelogram determined by the vectors (1, 2) and (3, 4).
2. Find the centroid of the parallelogram determined by the vectors $(-1, 3)$ and $(1, 4)$.
3. Find the centroid of a hemisphere of radius a.
4. Find the volume of a wedge of the sphere of radius a between the planes $z = 0$ and $z = ay$.
5. Find the volume of the region inside the sphere $x^2 + y^2 + z^2 = 4a^2$ and outside the cylinder $x^2 + y^2 = a^2$.
6. Find the moment of inertia of a disc of radius a about the center. Find the moment of inertia about a diameter.
7. Evaluate the integral
$$\int (2x + y) e^{x-y} \, dx \, dy$$
over the parallelogram determined by the vectors $(1, 1)$ and $(1, -2)$ by making an appropriate linear change of variable.
8. Let a transformation Φ from $V_2(R)$ to $V_2(R)$ be defined by
$$x = u$$
$$y = u(1 + v).$$
Sketch the image of the rectangle $0 \leq u \leq 2, 0 \leq v \leq 3$ under this transformation. Compute the centroid of this region.
9. Consider the transformation Φ of $V_2(R)$ into $V_2(R)$ defined by $x = u^2 - v^2$, $y = 2uv$.
   (a) Sketch the image of the rectangle formed by $(1, 0), (2, 0), (2, 2), (1, 2)$ under this transformation.
   (b) Sketch the image of the semicircle $u^2 + v^2 \leq 1, v \geq 0$ under this transformation.
   (c) Compute $\int_T xy \, dx \, dy$ if $T = \{(x, y): x^2 + y^2 \leq 1$.
   (d) Is the transformation Φ one-to-one on the disc $u^2 + v^2 \leq 1$?
10. Determine the volume of the parallelopiped defined by the vectors $(1, 1, 2)$, $(1, 0, 1)$ and $(2, 1, -1)$.
11. Evaluate the integral
$$\int_S \frac{1}{(5 + x - 3y - z)} \, dx \, dy \, dz$$
over the parallelepiped of Exercise 10.
12. Evaluate $\int xyz \, dx \, dy \, dz$ over the region in the first octant inside the sphere $x^2 + y^2 + z^2 = 1$.
13. Determine the volume of the ellipsoid $(x^2/a^2) + (y^2/b^2) + (z^2/c^2) < 1$. [Utilize the transformation $(x, y, z) = (au, bv, cw)$.]
14. Find the centroid of the region between two concentric hemispheres of radius r_1, r_2, where $r_1 > r_2$.
15. Find the moment of inertia of a spherical solid of radius a about a diameter if the density is constant.
16. Determine the center of mass of a cone with the height h if the density is proportional to the distance from the axis of the cone.

17. Let Φ be the transformation of $V_3(R)$ into $V_3(R)$ defined by cylindrical coordinates
$$x = r \cos \theta$$
$$y = r \sin \theta$$
$$z = z.$$
Determine the Jacobian $\partial(x, y, z)/\partial(r, \theta, z)$ of this transformation. Determine the moment of inertia with respect to the xz plane of the region within the cylinder $x^2 + y^2 = 1$ and between the planes $z = 0$, and $z = a$.

18. Find the volume of the 4-dimensional solid determined by the vectors $(1, 1, -1, 1)$, $(2, 1, -1, 0)$, $(1, 2, 1, 2)$, $(0, 1, 1, -1)$. What is the x coordinate of the centroid of this region?

CHAPTER NINE

LINE INTEGRALS

9.1 Line Integrals of Scalar Functions

We wish in this section to generalize the notion of the integral $\int_a^b f(t)\,dt$ of a function defined on an interval $[a, b]$ of real numbers to an integral of a function f defined along a curve. Such an integral is called the *line integral of f along α* and we shall write it $\int_\alpha f$. We shall formulate the definition for curves in two- or three-dimensional space. However, it is exactly the same for curves in $V_n(R)$.

Let us recall a few definitions. A *curve α* in $V_3(R)$ is a continuous function mapping a closed interval $[a, b]$ of real numbers into $V_3(R)$. If t stands for a point in the interval $[a, b]$ on which α is defined, then t is called the *parameter* of the curve α. The interval $[a, b]$ is called the *parameter interval* for the curve. The range or image of the function α, that is, the set of points X in $V_3(R)$ such that $X = \alpha(t)$ for some $t \in [a, b]$, is often called the *trace of the curve α*. The derivative $\alpha'(t) = (\alpha^{1'}(t), \alpha^{2'}(t), \alpha^{3'}(t))$ of the curve $\alpha(t) = (\alpha^1(t), \alpha^2(t), \alpha^3(t))$ is called the *velocity* of the curve. The vector $\alpha'(t)$ establishes a direction along the curve at the point $\alpha(t)$. Indeed, the tangent line to the curve at the point $\alpha(t_0)$ is just the line through $\alpha(t_0)$ parallel to the vector $\alpha'(t_0)$. This is illustrated in Figure 1.

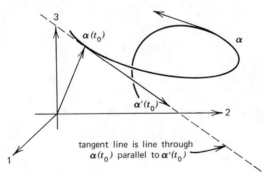

Fig. 1

If $\alpha'(t) \neq 0$, the vector

$$T(t) = \frac{\alpha'(t)}{|\alpha'(t)|}$$

is called the *unit tangent* to the curve at the point $\alpha(t)$.

If the velocity $\alpha'(t)$ is continuous for all values of t then α is called a *smooth curve*. If in addition α' never vanishes, so that α has a unit tangent at every point, we call α a *regular curve*. Such a curve α has a smoothly turning tangent vector at every point. A smooth curve with a nonvanishing velocity is called a regular curve. If the interval $[a, b]$ may be partitioned into subintervals $a = t_0 < t_1 < \cdots < t_n = b$ such that the curve α is regular on each of the subintervals, then α is said to be *piecewise regular*. Thus a piecewise regular curve has a smoothly turning tangent everywhere except at finitely many points. A piecewise regular curve is sketched in Figure 2.

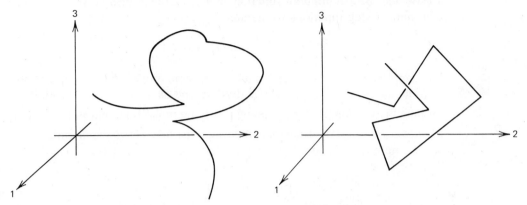

Fig. 2 Piecewise regular curve.   **Fig. 3** Polygonal curve.

An important class of piecewise regular curves are the *polygonal* curves. Such a curve is made up of finitely many straight line segments (Figure 3).

In this chapter we shall be concerned exclusively with piecewise regular curves. Hence from now on a curve will always mean a piecewise regular curve.

The *speed* $v(t)$ of a curve is the length of the velocity vector. Therefore $v(t) = |\alpha'(t)|$, and we recall from elementary calculus that if the curve α is defined on the interval $[a, b]$, then the length l of the curve is given by the formula

$$l = \int_a^b |\alpha'(t)|\, dt.$$

Thus the length of the curve is the integral of the speed over the parameter interval.

To motivate the definition of the line integral $\int_\alpha f$, suppose first that f is a step function defined on an interval $[a, b]$. Then $f(t) = c_i$ for $t_{i-1} < t < t_i$ where $a = t_0 < t_1 < \cdots < t_n = b$ is a partition of $[a, b]$, and

$$\int_a^b f = \sum c_i \Delta t_i$$

where $\Delta t_i = t_i - t_{i-1}$ is the length of each subinterval on which f is constant. If now f is a function defined on the trace of a curve α, then we call f a step function if there is a partition $a = t_0 < t_1 < \cdots < t_n = b$ of the parameter interval $[a, b]$ such that $f(\alpha(t))$ is constant on each of the open intervals (t_{i-1}, t_i). Indeed if $f(\alpha(t)) = c_i$ for $t_{i-1} < t < t_i$, we define the integral of f along the curve α to be

$$\int_\alpha f = \sum_{i=1}^n c_i l_i$$

where l_i is the length of the curve from $t = t_{i-1}$ to $t = t_i$. But

$$l_i = \int_{t_{i-1}}^{t_i} |\alpha'(t)|\, dt.$$

Hence

$$\int_\alpha f = \sum_{i=1}^n c_i \int_{t_{i-1}}^{t_i} |\alpha'(t)|\, dt = \sum_{i=1}^n \int_{t_{i-1}}^{t_i} c_i |\alpha'(t)|\, dt$$

$$= \sum_{i=1}^n \int_{t_{i-1}}^{t_i} f(\alpha(t))|\alpha'(t)|\, dt = \int_a^b f(\alpha(t))|\alpha'(t)|\, dt.$$

Now let f be a continuous function defined on the trace of α. In view of the fact that for step functions we define

(1) $$\int_\alpha f = \int_a^b f(\alpha(t))|\alpha'(t)|\, dt$$

it is tempting to use the same formula as the definition of the line integral of a continuous function f. We do exactly that, and as an added justification for formula (1) we note that if the integral is to be monotone, that is $\int_\alpha f \leq \int_\alpha g$ whenever $f \leq g$ on the trace of α then we must take (1) as the definition. To see this note that if g and h are step functions defined along α and $g \leq f \leq h$, then we require

(2) $$\int_\alpha g \leq \int_\alpha f \leq \int_\alpha h.$$

But

$$\int_\alpha g = \int_a^b g(\alpha(t))|\alpha'(t)|\, dt$$

$$\int_\alpha h = \int_a^b h(\alpha(t))|\alpha'(t)|\, dt.$$

Since we are assuming f to be continuous, it is not hard to show that there is exactly one number $\int_\alpha f$ satisfying (2). Furthermore this number $\int_\alpha f$ must be

$$\int_a^b f(\alpha(t))|\alpha'(t)|\, dt.$$

For technical reasons we usually assume in practice that the functions f are defined on open sets containing the trace of α.

Example 1. Compute $\int_\alpha f$ if $f(X) = (x^2 + y^2)$ and $\alpha(t) = (2t, t+1)$ for $0 \leq t \leq 1$. Clearly

$$f(\alpha(t)) = 4t^2 + (t+1)^2 = 5t^2 + 2t + 1.$$

Since $\alpha'(t) = (2, 1)$, $|\alpha'(t)| = \sqrt{5}$ and

$$\int_\alpha f = \int_0^1 (5t^2 + 2t + 1)\sqrt{5}\, dt = \frac{11}{3}\sqrt{5}.$$

Often the trace of a curve α is specified by an equation $g(x, y) =$ constant. Before computing the line integral $\int_\alpha f$, the curve α must be determined. For example, the circle $x^2 + y^2 = a^2$ is traced by the curve

$$\alpha(t) = (a \cos t, a \sin t), \qquad 0 \leq t \leq 2\pi.$$

The ellipse $\dfrac{x^2}{a^2} + \dfrac{y^2}{b^2} = 1$ is traced by the curve

$$\alpha(t) = (a \cos t, b \sin t), \qquad 0 \leq t \leq 2\pi.$$

The intersection of the sphere $x^2 + y^2 + z^2 = 4$ with the plane $z = 1$ is traced

by the curve
$$\alpha(t) = (\sqrt{3}\cos t, \sqrt{3}\sin t, 1), 0 \leq t \leq 2\pi.$$

Example 2. Let α be a curve running once around the unit circle $x^2 + y^2 = 1$ in a counterclockwise sense and let $f(x, y) = 1 + x$. Compute $\int_\alpha f$. Now
$$\alpha(t) = (\cos t, \sin t), \quad 0 \leq t \leq 2\pi,$$
traverses the unit circle once in a counterclockwise sense. Since $v(t) = |\alpha'(t)| = 1$,
$$\int_\alpha f = \int_0^{2\pi} (1 + \cos t)\,dt = 2\pi.$$

If α runs *twice* around the unit circle $x^2 + y^2 = 1$, then we may still take $\alpha(t) = (\cos t, \sin t)$. But now the parameter interval $[a, b] = [0, 4\pi]$ and
$$\int_\alpha f = \int_0^{4\pi} (1 + \cos t)\,dt = 4\pi.$$

When only the trace S of a curve α is specified, and the curve α is to be determined, it should be assumed that α runs through S only once.

When α is a piecewise smooth curve, then the computation of the line integral $\int_\alpha f$ is often simplified by considering the smooth portions of α separately. For example, suppose $a = t_0 < t_1 < \cdots < t_n = b$ is a partition of the parameter interval $[a, b]$ into subintervals $I_k = (t_{k-1}, t_k)$ such that α is smooth on each interval I_k. If we let α_k be the curve α restricted to the interval I_k, then
$$\int_\alpha f = \int_a^b f(\alpha(t))|\alpha'(t)|\,dt = \sum_{k=1}^n \int_{t_{k-1}}^{t_k} f(\alpha_k(t))|\alpha_k'(t)|\,dt = \sum_{k=1}^n \int_{\alpha_k} f.$$

Hence to compute $\int_\alpha f$ we compute the line integrals $\int_{\alpha_k} f$ separately and add the results.

Example 3. If $f(x, y) = x^2 y$ and α runs once in a counterclockwise direction around the square defined by the coordinate axes and the lines $x = 1, y = 1$, compute $\int_\alpha f$. Let $\alpha_1, \alpha_2, \alpha_3, \alpha_4$ be the four smooth portions of this curve as in Figure 4.

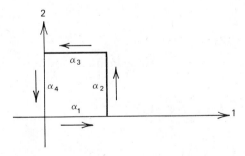

Fig. 4 α traverses the unit square in a counterclockwise direction.

We could define α_k in the following way
$$\begin{aligned}\alpha_1(t) &= (t, 0) & 0 \leq t \leq 1 \\ \alpha_2(t) &= (1, t - 1) & 1 \leq t \leq 2 \\ \alpha_3(t) &= (3 - t, 1) & 2 \leq t \leq 3 \\ \alpha_4(t) &= (0, 4 - t) & 3 \leq t \leq 4.\end{aligned}$$

It is easily checked that the curve α composed of these four smooth curves α_k traces out the unit square. Moreover,

$$\int_\alpha f = \sum_{k=1}^4 \int_{\alpha_k} f = \int_0^1 0\, dt + \int_1^2 (t-1)\, dt + \int_2^3 (3-t)^2\, dt + \int_3^4 0\, dt = 0 + \frac{1}{2} + \frac{1}{3} + 0$$
$$= \frac{5}{6}.$$

Often, however, it is easier to consider the curves α_k completely independently and not require that they be defined on consecutive intervals. Indeed it may be convenient to have them defined on the same interval. For example, we may take

$$\begin{aligned}\alpha_1(t) &= (t, 0) & 0 \le t \le 1 \\ \alpha_2(t) &= (1, t) & 0 \le t < 1 \\ \alpha_3(t) &= (1-t, 1) & 0 \le t \le 1 \\ \alpha_4(t) &= (0, 1-t) & 0 \le t \le 1.\end{aligned}$$

Then

$$\int_\alpha f = \sum_{i=1}^4 \int_{\alpha_k} f = \int_0^1 0\, dt + \int_0^1 t\, dt + \int_0^1 (1-t)^2\, dt + \int_0^1 0\, dt.$$

Hence adding these integrals we obtain

$$\int_\alpha f = \int_0^1 [t + (1-t)^2]\, dt = \int_0^1 (1 - t + t^2)\, dt = \frac{5}{6}.$$

This example show that line integrals over two curves having the same trace may have the same value. Whereas the previous example shows that they need not. We shall clarify this problem shortly.

The line integral $\int_\alpha f$ is often written $\int_\alpha f\, ds$ or $\int_\alpha f(x, y)\, ds$ where s stands for the arc length. The reason for this notation is the following. If $s = \int_a^t |\alpha'(u)|\, du$, then by the fundamental theorem of the calculus $ds/dt = |\alpha'(t)|$. Hence if we set $ds = |\alpha'(t)|\, dt$, it is reasonable to write $\int_\alpha f\, ds$ in place of $\int_\alpha f = \int_a^b f(\alpha(t))|\alpha'(t)|\, dt$.

Note that if the curve $\alpha(t)$ takes the form

$$\alpha(t) = (t, g(t)) \qquad a \le t \le b$$

or

$$\alpha(t) = (h(t), t) \qquad a \le t \le b,$$

then in the first case $|\alpha'(t)| = \sqrt{1 + g'^2(t)}$ and in the second $|\alpha'(t)| = \sqrt{1 + h'^2(t)}$. Hence the line integral

$$\int_\alpha f = \int_a^b f(\alpha(t))\sqrt{1 + g'^2(t)}\, dt$$

in the first case and

$$\int_\alpha f = \int_a^b f(\alpha(t))\sqrt{1 + h'^2(t)}\, dt$$

in the second.

EXERCISES

1. Determine curves α which traverse the following sets of points in $V_2(R)$. Specify the parameter interval.
    (a) $\{X : x^2 + y^2 = 4\}$.

(b) $\{X : x^2 + 4y^2 = 4\}$.
(c) $\{X : e^y = x\}$.
(d) $\{X : x^{2/3} + y^{2/3} = 1\}$.
(e) $\{X : 4x^2 + 2xy + 4y^2 = 1\}$. (Rotate the coordinate axes.)

2. Determine curves α which traverse the following sets in $V_3(R)$. Specify the parameter interval.
   (a) $\{X : 2x + y - z = 1, 3x + 2y + z = 4\}$.
   (b) $\{X : 3x + y + 2z = 4,$ and X lies on a coordinate plane$\}$.
   (c) $\{X : x^2 + y^2 + z^2 = 4, z = 1\}$.
   (d) $\{X : x^2 + 4y^2 + 2z^2 = 10, z = 1\}$.
   (e) $\{X : x^2 + y^2 + z^2 = 1, x + y + z = 1\}$. (Rotate the coordinate axes.)

Compute $\int_\alpha f$ for the following choices of functions f and curves α.

3. $f(x, y) = x$, $\quad \alpha(t) = (t, t^2)$, $\quad 0 \leq t \leq 1$.
4. $f(x, y) = x$, $\quad \alpha(t) = (\cos t, \sin t)$, $\quad 0 \leq t \leq \pi$.
5. $f(x, y) = y^3$, $\quad \alpha(t) = (t^3, t)$, $\quad 0 \leq t \leq 2$.
6. $f(x, y) = xy$, $\quad \alpha(t) = (4 \sin t, 4 \cos t)$, $\quad 0 \leq t \leq 2\pi$.
7. $f(x, y, z) = xyz$, $\quad \alpha(t) = (2 \cos t, 2 \sin t, t)$, $\quad 0 \leq t \leq \pi/4$.

Determine $\int_\alpha f$ if α is a curve running once in a counterclockwise sense around the given set of points S.

8. $f(x, y) = xy$, S is the triangle formed by the coordinate axes and the line $x + 2y = 1$.
9. $f(x, y) = x^2 y$, S is the rectangle formed by the coordinate axes and the lines $x = 2, y = 4$.
10. $f(x, y) = x^2 + y^2$, S is the semicircle formed by the x axis and the upper half of the circle $x^2 + y^2 = 4$.
11. $f(x, y) = xy - y^2$, $S = \{(x, y) : |x| + |y| = 1\}$.
12. $f(x, y) = (x - y)^2$, S is the quarter circle in the first quadrant formed by the circle $x^2 + y^2 = 4$ and the coordinate axes.

Determine $\int_\alpha f$ if the curve α has the following set of points S as a trace. Assume α traverses S once.

13. $f(x, y, z) = xy + z$, S is the straight line $2x + y - z = 1, x + y + z = 2$ between $(-1, 3, 0)$ and $(1, 0, 1)$.
14. $f(x, y, z) = xyz$, S is that portion of the line $x + y + z = 1, y - z = 0$ which lies in the first octant.
15. $f(x, y, z) = x^2 yz$, α traverses the intersection of the coordinate planes with the plane $2x + y + z = 1$.
16. $f(x, y, z) = x^2 + y^2 + z^2$, α runs once around the intersection of the sphere $x^2 + y^2 + z^2 = 4$ with the plane $z = 1$.
17. $f(x, y, z) = x$, α runs once around the intersection of the sphere $x^2 + y^2 + z^2 = 1$ and the plane $x + y + z = 1$. (Use problem 2e.)

9.2 Applications

Line integrals of scalar functions have many physical applications. For example we may represent the mass of a wire as a line integral. If α is a straight wire with constant density D, then we define the mass of the wire to be the product of D with the length of the wire. If α is a polygonal wire and D_i is the density of each straight line segment of length l_i, then the total mass M of the wire is

$$M = \sum_{i=1}^n D_i l_i.$$

336   LINE INTEGRALS

For a general mass density function f and a curve α we are led by the usual approximation arguments to define the *mass* of a wire traversed by the curve α to be

$$M = \int_\alpha f\,ds.$$

We may define the *center of mass* $(\bar{x}_1, \bar{x}_2, \bar{x}_3)$ to be the solution of the equation

$$M\bar{x}_i = \int_\alpha x_i f\,ds \qquad i = 1, 2, 3.$$

If $d(x_1, x_2, x_3)$ is the distance from a point $X = (x_1, x_2, x_3)$ to a line or plane then the corresponding moment of inertia of the curve α having mass density function f is defined to be

$$\int_\alpha d^2(x, y, z) f(x, y, z)\,ds.$$

Example 1. Determine the mass and the z coordinate of the center of mass of a helical wire described by the curve $\alpha(t) = (\cos t, \sin t, t)$ between $t = 0$ and $t = 2\pi$ if the density $f(x, y, z) = x^2 + y^2 + z^2$. First $M = \int_\alpha f\,ds$. If $s = l(t)$, denotes the length of the curve on the interval $[0, t]$, then

$$\frac{ds}{dt} = l'(t) = |\alpha'(t)| = \sqrt{2}.$$

Since $x^2 + y^2 + z^2 = f(\alpha(t)) = 1 + t^2$,

$$M = \int_\alpha f\,ds = \int_0^{2\pi} (x^2 + y^2 + z^2)\sqrt{2}\,dt = \sqrt{2}\int_0^{2\pi} (1 + t^2)\,dt$$

$$= \sqrt{2}\left(t + \frac{t^3}{3}\right)\Big|_0^{2\pi} = 2\sqrt{2}\left(\pi + \frac{4\pi^3}{3}\right).$$

The z coordinate of the center of mass satisfies $M\bar{z} = \int_\alpha z f\,ds$. At the point $\alpha(t)$, $z = t$, hence.

$$M\bar{z} = \sqrt{2}\int_0^{2\pi} t(1 + t^2)\,dt = \sqrt{2}\left(\frac{t^2}{2} + \frac{t^4}{4}\right)\Big|_0^{2\pi}.$$

$$= 2\sqrt{2}(\pi^2 + 2\pi^4)$$

and

$$\bar{z} = 3\,\frac{\pi + 4\pi^3}{3 + 4\pi^2}.$$

A helical wire is illustrated in Figure 5.

If the density function f has constant value c along the curve α, then the center of mass $\bar{x}$ is called the *centroid* of the curve. In this case the mass

$$M = \int_\alpha c\,ds = c\int_\alpha ds = cl$$

where l is the length of the curve. Hence the coordinates $\bar{x}_i$ of the centroid satisfy

$$\bar{x}_i = \frac{\int_\alpha cx_i\,ds}{cl} = \frac{\int_\alpha x_i\,ds}{l}, \quad i = 1, 2, 3.$$

EXERCISES

1. Find the mass and center of mass of a straight wire lying along the line $x + 3y = 1$ between the coordinate axes if the density $f(x, y) = x^2 + y^2$.

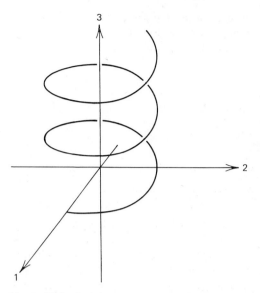

Fig. 5 Helical wire.

2. Find the mass and center of mass of a wire in the shape of the triangle formed by the line $2x + 3y = 6$ and the coordinate axes if the density $f(x, y) = x + y$.

3. Determine the moment of inertia of the triangular wire of Exercise 2 about the x axis, if the density $f(x, y) = 1$.

4. Determine the mass and center of mass of the helical wire traversed by the curve $\alpha(t) = (\cos t, \sin y, t), 0 \leq t \leq 2\pi$, if the density $f(x, y, z) = z$. Find the moment of inertia about the z axis.

5. Find the mass and center of mass of a semicircular wire of radius a having unit density.

6. Find the moment of inertia of a circular wire of unit density with respect to the origin if the wire has the shape of the intersection of the sphere $x^2 + y^2 + z^2 = 1$ with the plane $x + y + z = 1$.

7. Find the moment of inertia about the y axis of a semicircular wire having the shape of $x^2 + y^2 = 1$, $y \geq 0$ if the density $f(x, y) = |x| + |y|$. Determine the mass and center of mass of the wire.

9.3 Vector Fields and Line Integrals

A vector function F is a function defined on a set in $V_n(R)$ with values in $V_m(R)$. In applications the case that the dimension of the domain of F is the same as the dimension of the range occurs quite frequently. Such a vector function, that is, a function from $V_n(R)$ to $V_n(R)$ is called a *vector field*. If we consider the velocity vector α' of a curve α as a function of the position $X = \alpha(t)$, then $F(X) = \alpha'(t)$ defines a vector field. Of course, in order for a function to be defined by this technique the curve α should not cross itself. Another example is the velocity of a moving fluid. Consider a liquid or gas moving in space such that the motion is independent of time. Then the velocity of the fluid at a point X is a function only of the position. Hence this defines a vector field and we call this field a velocity field.

Vector fields F may be described visually in the following way. First, represent the vector $F(X)$ by an appropriate directed line segment. Then position the arrow as emanating from the point X. Thus the image of the function F is superimposed on the domain of F. This is especially graphic if F represents a velocity field as in Figure 6.

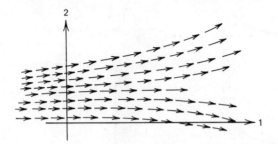

Fig. 6 Geometrical representation of a vector field in the plane.

If $F(X) = \alpha'(t)$ is the velocity vector along a curve α, then we have always represented $F(X)$ as an arrow having the appropriate length and direction which emanates from $X = \alpha(t)$. A velocity field along a curve is illustrated in Figure 7.

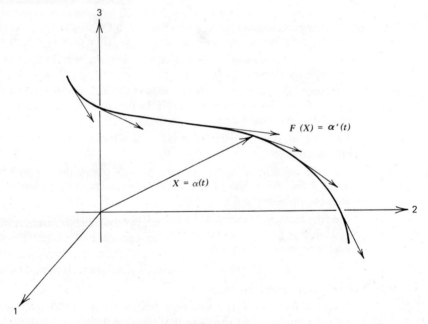

Fig. 7 A velocity field defined along a curve.

Next we wish to define the line integral of a vector field F along a curve α. To this end assume that the vector field F is defined on the trace of the curve α. Let $T(t)$ be the unit tangent to the curve at $X = \alpha(t)$. Then the inner product $(F(\alpha(t)), T(t))$ is a scalar-valued function defined on α. (Indeed this is the tangential component of the vector field F along α.) We define *the line integral of F along α*, which we write $\int_\alpha F$, by the formula

(1) $$\int_\alpha F = \int_\alpha (F(\alpha(t)), T(t)) \, ds$$

VECTOR FIELDS AND LINE INTEGRALS

Thus the line integral of the vector field F is just the line integral of the tangential component of F along α. If we observe that

$$\int_\alpha (F(\alpha(t)), T(t)) \, ds = \int_a^b (F(\alpha(t)), T(t)) |\alpha'(t)| \, dt$$

and

$$T(t) = \frac{\alpha'(t)}{|\alpha'(t)|},$$

then we have

(2) $$\int_\alpha F = \int_a^b \left(F(\alpha(t)), \frac{\alpha'(t)}{|\alpha'(t)|}\right) |\alpha'(t)| \, dt$$

$$= \int_a^b (F(\alpha(t)), \alpha'(t)) \, dt.$$

Thus if α is a curve with parameter interval $[a, b]$, the line integral of the vector field F along α is just the ordinary integral over $[a, b]$ of the inner product of $F(\alpha(t))$ and the velocity $\alpha'(t)$ of the curve.

Example 1. Compute the line integral of $F(x_1, x_2) = (x_1^2, x_2^2)$ along the parabola $\alpha(t) = (t, t^2)$ from $t = 0$ to $t = 2$.

Now $\alpha'(t) = (1, 2t)$ and $F(\alpha(t)) = (t^2, t^4)$. Hence

$$\int_0^2 (F(\alpha(t)), \alpha'(t)) \, dt = \int_0^2 (t^2 + 2t^5) \, dt = 24.$$

Example 2. Compute the line integral of $F(X) = (x_1, x_2, x_3)$ along the helix $\alpha(t) = (\cos t, \sin t, t)$ from $t = 0$ to $t = 2\pi$. Again

$$\alpha'(t) = (-\sin t, \cos t, 1)$$

and

$$F(\alpha(t)) = (\cos t, \sin t, t).$$

Hence $(F(\alpha(t)), \alpha'(t)) = t$ and the desired line integral is

$$\int_0^{2\pi} t \, dt = 2\pi^2.$$

There are many notations for line integrals. Assuming $n = 3$, if we write out F in component form we have

$$F(X) = (F_1(X), F_2(X), F_3(X)) = F_1(X) I_1 + F_2(X) I_2 + F_3(X) I_3.$$

The velocity $\alpha'(t) = \alpha^{1'}(t) I_1 + \alpha^{2'}(t) I_2 + \alpha^{3'}(t) I_3$. Hence the line integral

$$\int_a^b (F(\alpha)t), \alpha'(t)) \, dt = \int_a^b [F_1(\alpha(t)) \alpha^{1'}(t) + F_2(\alpha(t)) \alpha^{2'}(t)$$
$$+ F_3(\alpha(t)) \alpha^{3'}(t) \, dt.$$

This is often abbreviated

$$\int_\alpha [F_1 \, dx_1 + F_2 \, dx_2 + F_3 \, dx_3]$$

where α denotes the curve, and $dx_i = \alpha^{i'}(t) \, dt$.

We pause to make a comment about notation. We have defined a curve to be a continuous function from an interval $[a, b]$ into $V_n(R)$ and we shall use small Greek letters α, β, γ, to denote such functions. We shall write the component functions of the curve α with superscripts, such as $\alpha^1, \alpha^2, \ldots, \alpha^n$. We shall use subscripts $\alpha_1, \ldots, \alpha_m$ to denote *different* curves α_i. The line

340 LINE INTEGRALS

integral of the vector field F along α then may be written

$$\int_\alpha F, \quad \int_\alpha (F, dX), \quad \text{or} \quad \int_\alpha (F_1 dx_1 + F_2 dx_2 + \cdots + F_n dx_n).$$

To evaluate this integral one must compute

$$\int_a^b (F(\alpha(t)), \alpha'(t)) \, dt.$$

If α is a piecewise smooth curve, then there is a partition

$$a = t_0 < t_1 < \cdots < t_n = b$$

of the interval $[a, b]$ such that α is smooth on the interval $I_k = [t_{k-1}, t_k]$. If we denote by α_k the curve α defined on the interval I_k, then

$$\int_\alpha F = \int_a^b (F(\alpha(t)), \alpha'(t)) \, dt$$

$$= \sum_{k=1}^n \int_{t_{k-1}}^{t_k} (F(\alpha(t)), \alpha'(t)) \, dt$$

$$= \sum_{k=1}^n \int_{\alpha_k} F.$$

To evaluate $\int_{\alpha_k} F$ it is often convenient to consider the curves α_k as being defined on the same interval $[a, b]$.

Example 3. Compute the line integral $\int_\alpha F$ for $F(x, y) = (x, y^2)$ if α is the unit square traversed once in a counterclockwise direction. (Figure 8.)

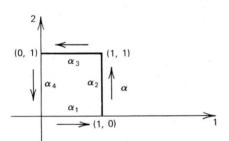

Fig. 8

Since

$$\int_\alpha F = \int_{\alpha_1} F + \cdots + \int_{\alpha_4} F,$$

it suffices to compute each of the integrals $\int_{\alpha_k} F$. Also it is more convenient to think of each curve α_k as defined on the unit interval $[0, 1]$ rather than to assume α_1 is defined on $[0, 1]$, α_2 on $[1, 2]$, etc. Hence for $0 \leq t \leq 1$ we may take, as in the example on p. 334,

$$\begin{aligned} \alpha_1(t) &= (t, 0) \\ \alpha_2(t) &= (1, t) \\ \alpha_3(t) &= (1-t, 1) \\ \alpha_4(t) &= (0, 1-t) \end{aligned} \quad \text{and} \quad \begin{aligned} \alpha_1'(t) &= (1, 0) \\ \alpha_2'(t) &= (0, 1) \\ \alpha_3'(t) &= (-1, 0) \\ \alpha_4'(t) &= (0, -1). \end{aligned}$$

Hence

$$\int_{\alpha_1} F = \int_0^1 ((t, 0), (1, 0)) \, dt = \int_0^1 t \, dt = \frac{1}{2}$$

$$\int_{\alpha_2} F = \int_0^1 ((1,t^2),(0,1))\,dt = \int_0^1 t^2\,dt = \frac{1}{3}$$

$$\int_{\alpha_3} F = \int_0^1 ((1-t,1),(-1,0))\,dt = \int_0^1 (t-1)\,dt = -\frac{1}{2}$$

$$\int_{\alpha_4} F = \int_0^1 ((0,(1-t)^2),(0,-1))\,dt = \int_0^1 -(1-t)^2\,dt = -\frac{1}{3}.$$

Therefore
$$\int_\alpha F = \frac{1}{2} + \frac{1}{3} - \frac{1}{2} - \frac{1}{3} = 0.$$

As an application of the line integral of a vector field let us consider the problem of work. If a constant force F is directed along the interval $[a,b]$, then the work done moving a particle from a to b is defined as the product $F \cdot (b-a)$. An elementary application of the definition of the definite integral shows that if $F(t)$ is a variable force function directed along the interval $[a,b]$, then the work done by the forcing function F is

$$\int_a^b F(t)\,dt.$$

In the plane or in space, however, force functions are vector fields. If a force field F is constant at each point on the line segment joining the points X_0 and X_1, then the component of the force along this line is the inner product of the vector F with a unit vector along the line from X_0 to X_1. Such a vector is
$$\frac{X_1 - X_0}{|X_1 - X_0|}.$$
Thus
$$\left(F, \frac{X_1 - X_0}{|X_1 - X_0|}\right)$$
measures the force along the line. The work done W is then the product of this force with the length of the line segment traversed, that is

$$W = \left(F, \frac{X_1 - X_0}{|X_1 - X_0|}\right)|X_1 - X_0| = (F, X_1 - X_0).$$

If α is a polygonal path joining points $X_0, \ldots, X_n$, and a force function F has a constant value F_k along each rectilinear portion of this curve, then the work W should be the sum of the work done on each portion. Hence we have

$$W = \sum_{k=1}^n (F_k, X_k - X_{k-1}).$$

If now α is a smooth curve and F is a variable force function defined at each point of an open set containing the trace of the curve, we wish to assert that the work done by the force moving along the curve should be defined to be

(3) $$W = \int_\alpha F = \int_a^b (F(\alpha(t)), \alpha'(t))\,dt.$$

To justify this we must make an assumption on how the work W can be approximated. Specifically let ϵ be given and let $a = t_0 < t_1 < \cdots < t_n = b$ be a partition of the interval $[a,b]$. Let $X_k = \alpha(t_k)$ and let $\tilde{\alpha}$ be a polygonal path joining the points X_k and satisfying $\tilde{\alpha}(t_k) = X_k$. This is illustrated in Figure 9.

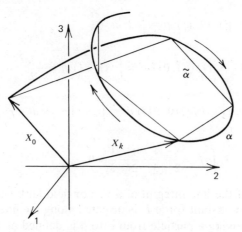

Fig. 9 Polygonal curve $\tilde{\alpha}$ approximating smooth curve α.

If we assume that for each $\epsilon > 0$ a partition of $[a, b]$ may be chosen so that W may be approximated to within ϵ by sums

(4) $$\sum_{k=1}^{n} (F(\tilde{\alpha}(t_k')), X_k - X_{k-1})$$

where t_k' is any point on the interval $[t_{k-1}, t_k]$, then we must have

$$W = \int_a^b (F(\alpha(t)), \alpha'(t)) dt$$

because it is not hard to show that the above integral can be approximated arbitrarily well by sums of the form (4). We omit the details of this verification, however.

Example 4. Determine the work done by a particle moving from $(0,0)$ to $(2,0)$ along a curve α which traces out the set $S = \{(x, y) : y = 1 - |1 - x|\}$ if the force field is given by $F(x, y) = (y^2, x)$. The trace of α is illustrated in Figure 10.

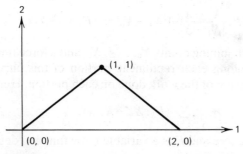

Fig. 10

Now if $0 \leq x \leq 1$, $y = 1 - |1 - x| = x$ and if $1 \leq x \leq 2$, $y = 1 - |1 - x| = 1 + (1 - x) = 2 - x$. Hence the curve α may be defined by

$$\alpha(t) = (t, t) \quad \text{if} \quad 0 \leq t \leq 1$$
$$\alpha(t) = (t, 2 - t) \quad \text{if} \quad 1 \leq t \leq 2.$$

The respective velocity vectors are given by

$$\alpha'(t) = (1, 1) \quad \text{if} \quad 0 \leq t \leq 1$$
$$\alpha'(t) = (1, -1) \quad \text{if} \quad 1 \leq t \leq 2.$$

Hence
$$(F(\alpha(t)), \alpha'(t)) = ((t^2, t), (1, 1)) = t^2 + t \quad \text{if} \quad 0 \leq t \leq 1$$
and
$$(F(\alpha(t)), \alpha'(t)) = ((2-t)^2, t), (1, -1)) = (2-t)^2 - t \quad \text{if} \quad 1 \leq t \leq 2.$$
Therefore
$$W = \int_\alpha F = \int_0^1 (F(\alpha(t)), \alpha'(t)) \, dt + \int_1^2 (F(\alpha(t)), \alpha'(t)) \, dt$$
$$= \int_0^1 (t^2 + t) \, dt + \int_1^2 ((2-t)^2 - t) \, dt = \frac{2}{3}.$$

EXERCISES

1. Determine the line integral of $F(x, y) = (x^2 - y^2, 2xy)$ along the parabola $y = x^2$ from $(-1, 1)$ to $(2, 4)$.
2. Determine the line integral of $F(x, y) = (x^2 - 2xy, y^2 - 2xy)$ from $(-1, 2)$ to $(2, 1)$ along the path defined by $y = |1 - x|$.
3. Compute $\int_\alpha [(x + y) \, dx + (x - y) \, dy]$ if α is the straight line from $(0, 0)$ to $(3, 4)$.
4. Compute $\int_\alpha [(x^2 - y) \, dx + xy^2 \, dy]$ if α is the straight line from $(2, 1)$ to $(3, -2)$.
5. Compute $\int_\alpha [y \, dx + (x^2 + y^2) \, dy]$ if α is the upper half of the circle $x^2 + y^2 = 16$ running from $(4, 0)$ to $(-4, 0)$.
6. Determine the line integral of $F(x, y) = (y, x)$ around the square $x = 0$, $y = 0$, $x = a$, $y = a$ in a clockwise direction.
7. Determine the line integral of $F(x, y, z) = (y, z, x)$ around the curve of intersection of the plane $x + y + z = 1$ with the three coordinate planes from the x axis to the y axis to the z axis to the x axis.
8. If the force field $F(x, y) = (x^2, y^2)$, determine the work done by a particle moving around the circle $x^2 + y^2 = a^2$ once in a counterclockwise direction.
9. Compute $\int_\alpha [y \, dx + z \, dy + x \, dz]$ along the straight line from $(0, 0, 0)$ to $(2, 4, -1)$.
10. Compute $\int_\alpha [z \, dx + (x - z) \, dy + y \, dz]$ along the straight line from $(2, 1, 0)$ to $(-1, 4, 2)$.
11. Compute $\int_\alpha [(y - z) \, dx + z \, dy + x \, dz]$ along the intersection of the planes $2x + y - z = 2$ and $x + y + 2z = 1$ from $(1, 0, 0)$ to $(7, -10, 2)$.
12. If $F(x, y, z) = (xy, yz, zx)$ determine the work done by this force moving a particle along the straight line from $(1, -1, 1)$ to $(2, 1, 3)$.
13. Determine the work done by the force field $F(x, y) = (2a - y, x)$ moving a particle along one arch of the cycloid $X(t) = (at - a \sin t, a - a \cos t)$, $0 \leq t \leq 2\pi$.
14. Compute
$$\int_\alpha \frac{(x + y) \, dx - (x - y) \, dy}{x^2 + y^2}$$
around the circle $x^2 + y^2 = 4$ in a counterclockwise direction.
15. Compute
$$\int_\alpha \frac{dx - dy}{|x| + |y|}$$
around the square $|y| = 1 - |x|$ in a counterclockwise direction.
16. Let $\alpha_n(t) = (t, t^n)$. Show that the work done by the force field $F(x, y) = (y, x)$ moving a particle from $(0, 0)$ to $(1, 1)$ along each of these paths is always equal to one.

17. Determine the work done by the force field $F(x, y, z) = (x^2, y^2, z^2)$ moving a particle around the curve of intersection of the sphere $x^2 + y^2 + z^2 = a^2$ and the cylinder $x^2 + y^2 = ay$ in a counterclockwise direction when viewed from above the sphere. [Determine the curve $\alpha(t)$.]

18. Compute
$$\int_\alpha (yz\,dx + xz\,dy + xy\,dz)$$
around the intersection of the sphere $x^2 + y^2 + z^2 = a^2$ with a plane $z = b$, $|b| \leq |a|$.

9.4 Properties of Line Integrals

We now discuss properties of line integrals. We confine our attention to line integrals of vector fields. The analogous results for line integrals of scalar functions we leave as exercises.

First we note that a line integral is a scalar-valued linear transformation. Hence if F and G are continuous vector fields defined on the trace of the curve α, then
$$\int_\alpha (aF + bG) = a\int_\alpha F + b\int_\alpha G.$$

This follows immediately from the linearity of the integral $\int_a^b f(t)\,dt$. We leave the verification as an exercise.

Next we discuss the dependence of the line integral $\int_\alpha F$ on the curve α. We have defined the trace or image of the curve α to be the set of points $\{\alpha(t) : a \leq t \leq b\}$, and have already noted that different curves can have the same trace. For example, a curve traversing the unit circle once in a counterclockwise direction is different from one traversing in a clockwise direction or one making two revolutions. An example of each of these three curves is

(1) $\begin{cases} \alpha_1 : \alpha_1(t) = (\cos t, \sin t) & 0 \leq t \leq 2\pi \\ \alpha_2 : \alpha_2(t) = (\cos t, -\sin t) & 0 \leq t \leq 2\pi \\ \alpha_3 : \alpha_3(t) = (\cos 2t, \sin 2t) & 0 \leq t \leq 2\pi. \end{cases}$

If F is a vector field defined on the unit circle, it is unreasonable to expect that
$$\int_{\alpha_1} F = \int_{\alpha_2} F = \int_{\alpha_3} F.$$

Indeed it will follow from our discussion that
$$\int_{\alpha_2} F = -\int_{\alpha_1} F \quad \text{and} \quad \int_{\alpha_3} F = 2\int_{\alpha_1} F.$$

For example, let $F(x, y) = (-y, x)$. Then for the three curves above $\alpha_1'(t) = (-\sin t, \cos t)$, $\alpha_2'(t) = (-\sin t, -\cos t)$, and $\alpha_3'(t) = (-2\sin 2t, 2\cos 2t)$. Hence
$$\int_{\alpha_1} F = \int_0^{2\pi} (\sin^2 t + \cos^2 t)\,dt = 2\pi$$
$$\int_{\alpha_2} F = -\int_0^{2\pi} (\sin^2 t + \cos^2 t)\,dt = -2\pi$$
and
$$\int_{\alpha_3} F = 2\int_0^{2\pi} (\sin^2 2t + \cos^2 2t)\,dt = 4\pi.$$

Our first problem will be to determine when two curves α_1, α_2 having the same trace have the property that $\int_{\alpha_1} F = \int_{\alpha_2} F$ for each continuous vector field defined on the common trace of α_1 and α_2. To this end we introduce the following definition.

Definition. *Two curves α_1 and α_2 with parameter intervals $[a, b]$ and $[c, d]$, respectively, where $a < b$ and $c < d$, are said to be* equivalent *if there exists a continuously differentiable function h mapping $[c, d]$ onto $[a, b]$ such that the following two conditions are satisfied.*

(i) *For each u in the interval $c \leq u \leq d$, $h'(u) > 0$*
(ii) *If $t = h(u)$, then $\alpha_1(t) = \alpha_1(h(u)) = \alpha_2(u)$.*

When α_1 and α_2 are equivalent we write $\alpha_1 \sim \alpha_2$.

The function h is called a *change of parameter transformation*. Thus α_1 and α_2 are equivalent if one curve may be obtained from the other by a change of parameter transformation with *positive* derivative. The next result shows that if α_1 and α_2 are equivalent, then $\int_{\alpha_1} F = \int_{\alpha_2} F$ for every continuous vector field. The proof is an immediate consequence of the change of variable formula for definite integrals.

THEOREM 9.1 *Let F be a continuous vector field defined on the trace of a curve α. If β is another curve which is equivalent to α then*

(2) $$\int_\alpha F = \int_\beta F.$$

Proof. Let $[a, b]$ be the parameter interval for α and $[c, d]$ be the parameter interval for β. By definition

$$\int_\alpha F = \int_a^b (F(\alpha(t)), \alpha'(t))\, dt$$

and

$$\int_\beta F = \int_c^d (F(\beta(u)), \beta'(u))\, du.$$

To verify (2) we must establish that

(3) $$\int_a^b F(\alpha(t), \alpha'(t))\, dt = \int_c^d (F(\beta(u)), \beta'(u))\, du.$$

Let h be the change of parameter transformation which maps $[c, d]$ onto $[a, b]$. Since $h'(u) > 0$, h is a strictly increasing function. Moreover if g is a scalar function defined on $[a, b]$, then the integration by substitution formula asserts that

(4) $$\int_a^b g(t)\, dt = \int_c^d g(h(u))h'(u)\, du.$$

If we define $g(t) = (F(\alpha(t)), \alpha'(t))$, then since $\beta(u) = \alpha(h(u))$, and consequently $\beta'(u) = \alpha'(h(u))h'(u)$, we have

$$g(h(u))h'(u) = (F(\alpha(h(u))), \alpha'(h(u)))h'(u)$$
$$= (F(\beta(u)), h'(u)\alpha'(h(u))) = (F(\beta(u)), \beta'(u)).$$

Hence

$$\int_c^d (F(\beta(s)), \beta'(s))\, ds = \int_c^d g(h(s))h'(s)\, ds$$
$$= \int_a^b g(t)\, dt \qquad \text{(by 4)}$$
$$= \int_a^b (F(\alpha(t)), \alpha'(t))\, dt \qquad \text{(by definition of } g\text{)}$$

which proves (3). This completes the proof.

If α is a given curve defined on $[a, b]$ and h is a change of parameter function with positive derivative which maps the interval $[c, d]$ onto $[a, b]$, then the curve β defined by $\beta(u) = \alpha(h(u))$ is often called a *reparametrization* of α. One very common reparametrization of α is obtained by replacing the parameter interval $[a, b]$ by the arc length $[0, l]$ along the curve. Indeed, for $a \leq t \leq b$ define

(5) $$s = l(t) = \int_a^t |\alpha'(u)|\, du.$$

Then s is the length of the curve α on the interval $a \leq u \leq t$. We define a new curve β on the interval $[0, l] = [l(a), l(b)]$ by the formula

$$\beta(s) = \alpha(t).$$

We claim that if the speed $|\alpha'(t)|$ never vanishes, then β and α are equivalent. To see this note that the change of parameter transformation h which maps $[0, l]$ onto $[a, b]$ is the inverse function for $l(t)$. Hence $h(l(t)) = t$, and differentiating with respect to t we obtain $h'(l(t))l'(t) = 1$. Setting $s = l(t)$ this yields

$$h'(s) = \frac{1}{l'(t)} = \frac{1}{|\alpha'(t)|} > 0,$$

and by definition

$$\beta(s) = \alpha(h(s)) = \alpha(t).$$

Hence α and β are equivalent.

Suppose now that h maps $[c, d]$ onto $[a, b]$, but now $h'(u) < 0$ for each u. Then h is a decreasing function and $h(c) = a, h(d) = b$. See Figure 11.

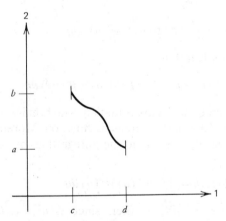

Fig. 11 h is a decreasing parameter transformation.

Furthermore we assert that if

$$\beta(u) = \alpha(h(u)) = \alpha(t),$$

then

(6) $$\int_\beta F = -\int_\alpha F$$

for each vector field F. Indeed when h is a parameter transformation with negative derivative the direction of motion along the curve α has been reversed. This is clear since if $\alpha(a) = X$ and $\alpha(b) = Y$, then $\beta(c) = \alpha(b) = Y$ and $\beta(d) = \alpha(a) = X$.

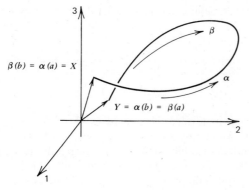

Fig. 12 The curves α and β have opposite directions.

As t moves from a to b, $\alpha(t)$ moves from X to Y. But as u moves from c to d, $\beta(u)$ moves from Y to X. This is illustrated in Figure 12. The verification of (6) is an immediate consequence of the change of variable formula for definite integrals.

It is convenient to have a standard way of reversing direction along a curve α. If $h(u) = a+b-u$ then $h(a) = b$ and $h(b) = a$. If we define the curve β by the formula $\beta(u) = \alpha(h(u))$, then using the chain rule it is easily checked that at each point $t = h(u)$, the unit tangent to β is the negative of the unit tangent to α. Moreover for each vector field F

(6) $$\int_\beta F = -\int_\alpha F.$$

It is convenient to denote this curve β by $-\alpha$. Hence (6) becomes

(7) $$\int_{-\alpha} F = -\int_\alpha F.$$

EXERCISES

1. Let $x^2 + y^2 = a^2$. Determine a curve α which traverses this circle once in a counterclockwise direction. Determine a curve which traverses the circle once in a clockwise direction. Determine a curve which traverses the circle twice in a counterclockwise direction.

2. Let $4x^2 + 2y^2 = 1$. Determine curves which traverse this ellipse once in a counterclockwise direction, three times in a counterclockwise direction.

3. Consider the triangle formed by the coordinate axes and the line $2x + y = 1$. Determine a curve which traverses this triangle once in a clockwise direction, twice in a counterclockwise direction, three times in a clockwise direction.

4. Determine a pair of equivalent curves traversing the following point sets in $V_2(R)$. Determine a pair of inequivalent curves which traverse the given sets.
   (a) $\{X : 4x^2 + y^2 = 1, x \geq 0, y \geq 0\}$.
   (b) $\{X : y = x^2, 0 \leq x \leq 4\}$.
   (c) The triangle formed by the coordinate axes and the line $x + y = 1$.
   (d) $\{X : |y| = 1 - |x|, |x| \leq 1, |y| \leq 1\}$.

5. Determine a curve α which traverses the triangle formed by the coordinate planes and the plane $x + 2y + z = 1$. Determine $-\alpha$.

6. Let $\alpha(t) = (a\cos t, a\sin t), \beta(t) = (a\cos nt, a\sin nt), 0 \leq t \leq 2\pi$. The curve α traverses the circle of radius a once, and β traverses it n times. If F is a vector field defined on $x^2 + y^2 = a^2$, show that

$$\int_\beta F = n \int_\alpha F.$$

7. Let F be a vector field defined on the trace of the curve α. Show that
$$\int_{-\alpha} F = -\int_\alpha F.$$

8. Let $\beta(u) = \alpha(h(t))$ where $h'(t) < 0$. If F is a vector field defined on the trace of α, show that
$$\int_\beta F = -\int_\alpha F$$

9. Let α traverse the square formed by the lines $x = 0, x = 1, y = 0, y = 1$ and let β traverse the square formed by $x = 0, x = 1, y = 1, y = 2$. Assume both curves run in the counterclockwise direction. Let γ traverse the rectangle formed by $x = 0, x = 1, y = 0, y = 2$. Assume $\gamma \sim \alpha$ on the bottom half of this rectangle and $\gamma \sim \beta$ on the top half. If F is a vector field defined on the rectangle $0 \leq x \leq 1, 0 \leq y \leq 2$ show that $\int_\gamma F = \int_\alpha F + \int_\beta F$. (Sketch the curves α, β, and γ.)

10. Let F, G be continuous vector fields defined on the curve α. Verify that
$$\int_\alpha (aF + bG) = a\int_\alpha F + b\int_\alpha G.$$
Verify the same equality if f and g are scalar functions.

11. If f is a continuous scalar function defined on the trace of α, and $\alpha \sim \beta$ show that
$$\int_\alpha f = \int_\beta f.$$

12. If f is a continuous *scalar* function defined on the trace of a curve α, show that $\int_\alpha f = \int_{-\alpha} f$. Why doesn't that contradict (7)?

13. If $\beta(u) = \alpha(h(u)) = \alpha(t)$ and $h'(u) < 0$ show that
$$\int_\alpha f = \int_\beta f$$
if f is a continuous *scalar* function defined on the trace of α.

9.5 Fundamental Theorems of the Calculus for Line Integrals

If g is a continuous scalar field, then the differential equation

(1) $$y' = g(x)$$

has the immediate solution

(2) $$y = f(x) = \int_0^x g(t)\,dt.$$

In this section we wish to discuss an analogue of this problem for vector fields. Namely, if the continuously differentiable vector field F is given, determine, if possible, a scalar function f satisfying the equation

(3) $$\nabla f = F.$$

If such a function f exists, then F is called a gradient field. The function f is called a *potential function* for the gradient field F. Although for $n = 1$ continuity of the function g was a sufficient condition to guarantee a solution to (1), this is no longer the case for $n > 1$. To see this let us examine equation (3). If $F_1, \ldots, F_n$ are the component functions for the vector field F, equation (3) asserts that for $i = 1, \ldots, n$

(4) $$\frac{\partial f}{\partial x_i} = F_i.$$

But if F is continuously differentiable, this implies that f must be twice continuously differentiable. Consequently for each i and j

THEOREMS OF THE CALCULUS FOR LINE INTEGRALS

$$\frac{\partial^2 f}{\partial x_i \partial x_j} = \frac{\partial^2 f}{\partial x_j \partial x_i}.$$

Differentiating (4) we infer that for each i and j

$$\frac{\partial^2 f}{\partial x_j \partial x_i} = \frac{\partial F_i}{\partial x_j} = \frac{\partial F_j}{\partial x_i} = \frac{\partial^2 f}{\partial x_i \partial x_j}.$$

Therefore the Jacobian matrices $(\partial F_i/\partial x_j)$ and $(\partial F_j/\partial x_i)$ must be equal. However, this is just the statement that the matrix $(\partial F_i/\partial x_j)$ is symmetric. Thus we see that if there exists a solution to (3), the Jacobian matrix $(\partial F_i/\partial x_j)$ for the vector field F must be symmetric. Thus for example if $F(x_1, x_2) = (x_1 + x_2, -x_1 + x_2)$, there can exist no solution to the equation $\nabla f = F$, since the Jacobian matrix $\partial F_i/\partial x_j = \begin{pmatrix} 1 & 1 \\ -1 & 1 \end{pmatrix}$ which is not symmetric.

The natural question to ask at this point is whether the symmetry of the Jacobian matrix $(\partial F_i/\partial x_j)$ is sufficient to guarantee that (3) has a solution. Again the answer in general is no. Whether or not the symmetry of $(\partial F_i/\partial x_j)$ is sufficient depends on geometric properties of the set U on which F is defined. Why this is so will become evident as we go along.

If we consider (3) as the generalization of (1) to higher dimensions, then since a one-dimensional line integral solves (1), it is perhaps reasonable to expect that an appropriate line integral might provide the solution to (3). To see that this is indeed the case let us assume that F is a continuously differentiable vector field defined on an open set U in $V_n(R)$. Further we shall assume that this open set U is *connected*. This means that any two points X, Y in U may be joined by a continuous curve whose trace lies entirely in U. Thus for the open set U to be connected we must be able to construct a continuous function α defined on an interval $[a, b]$ such that $\alpha(t) \in U$ for each t, $a \leq t \leq b$, and $\alpha(a) = X$ and $\alpha(b) = Y$. Naively speaking, an open set U is connected if it is all in one piece. Thus a disc or torus is connected, whereas the union of two disjoint open sets is not connected. These facts are illustrated in Figures 13 and 14.

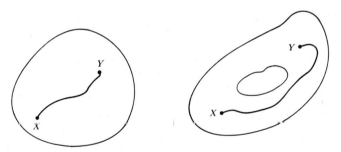

Fig. 13 Connected open sets.

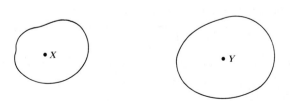

Fig. 14 The union of two disjoint open sets discs is not connected.

We first examine a line integral $\int_\alpha F$ assuming that it is known that $F = \nabla f$. In the one-dimensional situation if it is known that $f' = g$, then $\int_a^b g(t)\,dt = f(b) - f(a)$. This second fundamental theorem of the calculus has a direct generalization to line integrals. As always, all curves under consideration are assumed to be piecewise regular.

THEOREM 9.2 *Let f be a twice continuously differentiable scalar function defined on the connected open set U. Let $F = \nabla f$ and let X and Y be two points of U. If α is any piecewise smooth curve in U running from X to Y, then*

$$\int_\alpha \nabla f = \int_\alpha F = f(Y) - f(X).$$

Proof. This theorem states that the line integral between two points X and Y in the domain of a vector field F is independent of the path α joining these points if F is a gradient field. The result is somewhat surprising. However, the proof is immediate depending only on the chain rule for scalar functions.

Let us assume first that α is a smooth curve. Then by definition

$$\int_\alpha F = \int_\alpha \nabla f = \int_a^b (\nabla f(\alpha(t)), \alpha'(t))\,dt.$$

But if
$$g(t) = f(\alpha(t)),$$
then
$$g'(t) = (\nabla f(\alpha(t)), \alpha'(t))$$
and
$$\int_a^b (\nabla f(\alpha(t)), \alpha'(t))\,dt = \int_a^b g'(t)\,dt = g(b) - g(a) = f(Y) - f(X),$$

since $g(b) = f(Y)$ and $g(a) = f(X)$. For a piecewise smooth curve α, we need only partition the interval $[a, b]$ and apply the above result to each of the resulting subintervals.

As a corollary we have the analogue of the "vanishing derivative" theorem for functions of one variable.

COROLLARY. *If $\nabla f = 0$ on the connected open set U, then f is constant on U.*

Proof. Fix a point A in U and let α be a curve in U joining A and X. Then by Theorem 9.2, $f(X) - f(A) = \int_\alpha \nabla f$. But since $\nabla f = 0$; $\int_\alpha \nabla f = 0$, and hence $f(X) = f(A)$.

If α is a curve defined on $[a, b]$ then α is called a *closed curve* if $\alpha(a) = \alpha(b)$. If α is a closed curve and a and b are the only points t, t' for which $\alpha(t) = \alpha(t')$, then α is called a *simple closed* curve. Thus a simple closed curve intersects itself only at the end points of the interval $[a, b]$. An example of a simple closed curve is sketched in Figure 15. A closed curve which is not simple is sketched in Figure 16.

The next result is just the restatement of Theorem 9.2 in terms of closed curves.

COROLLARY *Let f be a twice continuously differentiable scalar function defined on the open set U. If α is a closed curve in U, and $F = \nabla f$, then*

$$\int_\alpha \nabla f = \int_\alpha F = 0.$$

THEOREMS OF THE CALCULUS FOR LINE INTEGRALS

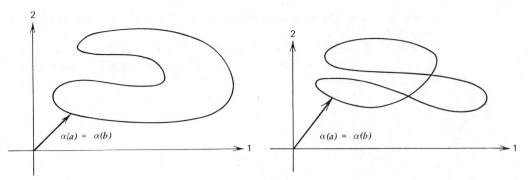

Fig. 15 A simple closed curve.   **Fig. 16** Closed curve which is not simple.

Proof. Since α is a closed curve $\alpha(a) = \alpha(b)$. But by Theorem 9.2
$$\int_\alpha \nabla f = f(\alpha(b)) - f(\alpha(a)) = 0.$$

Theorem 9.2 states the very important fact that if $\int_\alpha F$ is a line integral of the gradient field F between two points X and Y, then the value of the line integral depends only on the points X and Y. In particular the value of this line integral is independent of the curve α. It turns out that this "invariance of path" property characterizes those vector fields which are gradient fields. This is our next result. It is called the *first fundamental theorem of the calculus* for line integrals.

THEOREM 9.3 *Let F be a continuously differentiable vector field defined on the connected open set U. Assume that for each pair of points A and X in U, the line integral of F along a piecewise smooth curve α from A to X is independent of the curve α running from A to X. If we write $\int_A^X F$ in place of $\int_\alpha F$, then the function $f(X) = \int_A^X F$ satisfies $\nabla f(X) = F(X)$ for each $X \in U$.*

Proof. First we note the following fact. Let $X \in U$ and assume that E is a unit vector such that the straight line α from X to $X+tE$ lies in U. Then $\alpha'(u) = E$ for $0 \leq u \leq t$, and
$$\int_\alpha F = \int_0^t (F(X+uE), E) \, du.$$

Hence

(5) $$\lim_{t \to 0} \frac{1}{t} \int_0^t (F(X+uE), E) \, du = (F(X), E).$$

This follows from the first fundamental theorem of the calculus which asserts that if g is continuous at 0 and $h(t) = \int_0^t g(u) \, du$ then
$$h'(0) = \lim_{t \to 0} \frac{h(t)}{t} = \lim \frac{1}{t} \int_0^t g(u) \, du = g(0).$$

To prove the theorem we must show that if $f(X) = \int_A^X F$, then
$$\frac{\partial f}{\partial x_k}(X) = \lim_{t \to 0} \frac{f(X+tI_k) - f(X)}{t} = F_k(X)$$

where I_k is the kth unit coordinate vector and F_k is the kth coordinate function for F.

Since U is open, $X+tI_k \in U$ for all t sufficiently small. Choose a curve α running from A to X and then to $X+tI_k$ along the straight line joining X and $X+tI_k$ as in Figure 17.

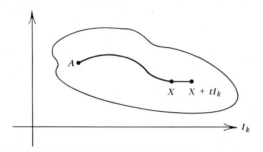

Fig. 17

Now
$$f(X+tI_k) - f(X) = \int_X^{X+tI_k} F = \int_0^t (F(X+uI_k), I_k)\, du.$$

Hence

(6) $$\frac{f(X+tI_k) - f(X)}{t} = \frac{1}{t} \int_0^t (F(X+uI_k), I_k)\, du,$$

and by (5),

$$\frac{\partial f}{\partial x_k}(X) = \lim_{t \to 0} \frac{1}{t} \int_0^t (F(X+uI_k), I_k)\, du = (F(X), I_k) = F_k(X)$$

This completes the proof.

COROLLARY. *Under the same assumptions as above if $\int_\alpha F = 0$ for each closed path α in U, then F is a gradient field.*

Proof. Let α_1 and α_2 be two curves running from the point A to the point X as in Figure 18. It is enough to show that $\int_{\alpha_1} F = \int_{\alpha_2} F$. If $-\alpha_2$ is the curve obtained by reversing the direction along α_2, we have already seen that $\int_{-\alpha_2} F = -\int_{\alpha_2} F$. Now let α be the curve running from A to X along α_1 and then back to A along α_2. By assumption $0 = \int_\alpha F$. But

$$\int_\alpha F = \int_{\alpha_1} F + \int_{-\alpha_2} F = \int_{\alpha_1} F - \int_{\alpha_2} F.$$

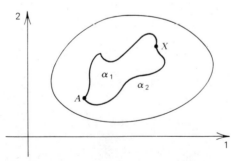

Fig. 18

THEOREMS OF THE CALCULUS FOR LINE INTEGRALS

Hence
$$\int_{\alpha_1} F = \int_{\alpha_2} F$$
and we are done.

It is convenient to note here that we may restrict the class of curves α considered in Theorem 9.3. We call a polygonal curve α a *step polygonal curve* or just a step polygon if each rectilinear portion of the trace of α is parallel to some coordinate axis. This is illustrated in Figure 19.

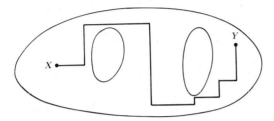

Fig. 19 Points X and Y connected by a step polygon.

An examination of the argument of Theorem 9.3 now shows that it is enough to assume that $\int_\alpha F$ is "path independent" just for step polygons α. Under this assumption formula (6) is valid, and we may pass to the limit in (6) obtaining
$$\frac{\partial f}{\partial x_k}(X) = F_k(X)$$
in exactly the same way as before.

If the gradient field F is interpreted as a force field, then we say this force field is *conservative*. This terminology results from the following considerations. First, if F is any force field, then the work done by F moving a particle along the curve α from $A = \alpha(a)$ to $B = \alpha(b)$, was defined in Section 9.3 to be $\int_\alpha F$. However, by Newton's second law if $X = \alpha(t)$ and m is the mass of the particle
$$F(X) = m\alpha''(t)$$
where α'' is the acceleration of the particle. Hence
$$\int_\alpha F = \int_a^b (F(\alpha(t)), \alpha'(t))\, dt = \int_a^b (m\alpha''(t), \alpha'(t))\, dt$$
$$= m \int_a^b (\alpha''(t), \alpha'(t))\, dt.$$

If we notice that
$$2(\alpha''(t), \alpha'(t)) = \frac{d}{dt}(\alpha'(t), \alpha'(t)) = \frac{d}{dt} v^2(t),$$
where $v(t) = |\alpha'(t)|$, then
$$m \int_a^b (\alpha''(t), \alpha'(t))\, dt = \frac{m}{2} \int_a^b \frac{d}{dt}(v^2(t))\, dt = \frac{m}{2}(v^2(b) - v^2(a)).$$

The quantity $\tfrac{1}{2}mv^2(t)$ is called the *kinetic energy* of the particle at the point $X = \alpha(t)$. Hence $\tfrac{1}{2}m(v^2(b) - v^2(a))$ represents the change in kinetic energy of the particle moving along the curve α from $\alpha(a)$ to $\alpha(b)$.

Now, however, if F is a gradient field and $F = -\nabla f$, then

$$\int_\alpha F = -\int_A^B \nabla f = f(A) - f(B).$$

The function f is called the *potential energy* of the particle. Note that this function is only determined up to an additive constant. If we equate the two expressions for $\int_\alpha F$, we have

$$f(A) - f(B) = \frac{1}{2} m(v^2(b) - v^2(a))$$

or

$$f(A) + \frac{1}{2} mv^2(a) = f(B) + \frac{1}{2} mv^2(b).$$

Hence the sum of the potential and kinetic energy is conserved. The force field is consequently called a conservative field.

Note that we chose f so that $-\nabla f = F$ in order to establish that the sum, rather than the difference, of the potential and kinetic energies is conserved. We shall not observe this convention in what follows. A *potential function* f for a gradient field F in our discussion will be a function f satisfying $\nabla f = F$.

Next we shall show that the symmetry of the Jacobian matrix $(\partial F_i / \partial x_j) = F'(X)$ is not in general sufficient to guarantee that $F = \nabla f$. Let

$$F(x, y) = \left(\frac{-y}{x^2 + y^2}, \frac{x}{x^2 + y^2} \right).$$

We shall show that $F'(X)$ is a symmetric matrix for all $(x, y) \neq 0$ yet there exists no potential function f for F which is defined for all $X \neq 0$. If F_1, F_2 are the coordinate functions for F, then

$$\frac{\partial F_1}{\partial y} = \frac{y^2 - x^2}{(x^2 + y^2)^2} = \frac{\partial F_2}{\partial x}.$$

Hence the Jacobian matrix $F'(X)$ is symmetric for $X \neq 0$. However, if we compute the line integral of F once around the unit circle we obtain the following. Let $\alpha(t) = (\cos t, \sin t)$, $0 \leq t \leq 2\pi$, then

$$\int_\alpha F = \int_0^{2\pi} (F(\alpha(t)), \alpha'(t)) \, dt$$

$$= \int_0^{2\pi} ((-\sin t, \cos t), (-\sin t, \cos t)) \, dt = \int_0^{2\pi} 1 \, dt = 2\pi.$$

Hence by the corollary to Theorem 9.2 there can exist no function f defined for all $X \neq 0$ which satisfies $\nabla f = F$.

Thus to show that the vector field F defined on a connected open set U is *not* a gradient field, we need only exhibit one closed curve α in U such that $\int_\alpha F \neq 0$. On the other hand, to show that F is a gradient field we must verify that $\int_\alpha F = 0$ for every closed path α in U. Even though we may confine our attention to smooth curves or even polygonal curves, this condition is very hard to check. However, for certain types of open sets the symmetry of $(\partial F_i / \partial x_j)$ is sufficient to guarantee that F is a gradient field. We verify this fact next.

Call a set S in $V_n(R)$ *star-shaped* if there is a point $A \in S$ such that for each $X \in S$ the closed line segment joining X and A lies entirely in S. More precisely S is called *star shaped about the point* A. See Figure 20.

THEOREMS OF THE CALCULUS FOR LINE INTEGRALS

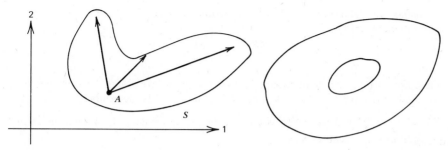

Fig. 20 S is star-shaped about A.   **Fig. 21** An annulus is not star-shaped about any point.

A disc, ellipse, or rectangle is an example of a set which is star shaped about each point of the set. A washer, or annulus, is not star-shaped about any point. (Figure 21.)

THEOREM 9.4 *Let U be an open set in V which is star-shaped about some point A. If F is a smooth vector field defined on U, and if the Jacobian matrix $F'(X) = (\partial F_j/\partial x_k)$ is symmetric for each X in U, then F is a gradient field on U. Indeed if*

$$f(X) = \int_0^1 (F(A + t(X-A)), X - A) \, dt, \tag{7}$$

then $\nabla f(X) = F(X)$. Furthermore if g satisfies $\nabla g = F$ then $g = f + C$ where C is a constant.

Proof. The integral (7) is just the line integral of the vector field F along the straight line joining A and X. If $A = 0$ this formula becomes

$$f(X) = \int_0^1 (F(tX), X) \, dt. \tag{8}$$

Assuming $A = 0$ let f be defined by (8). We verify that $\nabla f(X) = F(X)$. To do this we need first the following formula for the partial derivatives for f.

$$\frac{\partial f}{\partial x_k}(X) = \int_0^1 \frac{\partial}{\partial x_k}(F(tX), X) \, dt. \tag{9}$$

We shall not stop to prove (9) here. A sketch of how the argument goes is contained in Exercise 15, p. 357. Next, we need a formula for $(\partial/\partial x_k)(F(tX), X)$. Indeed it follows by the chain rule and the fact that the matrix $(\partial F_j/\partial x_k)$ is symmetric that

$$\frac{\partial}{\partial x_k}(F(tX), X) = \frac{d}{dt}[tF_k(tX)]. \tag{10}$$

Combining (9) and (10) we obtain

$$\frac{\partial f}{\partial x_k}(X) = \int_0^1 \frac{\partial}{\partial x_k}(F(tX), X) \, dt = \int_0^1 \frac{d}{dt}[tF_k(tX)] \, dt = tF_k(tX)\Big|_0^1 = F_k(X).$$

Therefore $\nabla f(X) = F(X)$. The proof for $A \neq 0$ is similar. We leave the verification of (10) as an exercise. To establish the last assertion observe that if $\nabla g = F$ then $\nabla(f-g) = F - F = 0$. Hence $f - g$ is constant by the corollary to Theorem 6.11, p. 225.

It follows from Theorem 9.4 that if $F'(X)$ is symmetric for all $X \in V_n(R)$, then a potential function f for the vector field F is given by

$$f(X) = \int_0^1 (F(tX), X) \, dt.$$

Example 1. Determine a potential function for $F(x, y) = (2y^2 + 1, 4xy)$ in the entire plane. Clearly

$$\frac{\partial F_1}{\partial y} = 4y = \frac{\partial F_2}{\partial x}.$$

Since the entire plane is star-shaped about the origin, we assert the existence of a potential function f as a consequence of Theorem 9.4. To determine f we evaluate $f(X) = \int_0^1 (F(tX), X) \, dt$ which is the line integral of F along the straight line from the origin to the point X. Now

$$F(tX) = (2t^2y^2 + 1, 4t^2xy)$$

and

$$(F(tX), X) = 2t^2xy^2 + x + 4t^2xy^2 = 6t^2xy^2 + x.$$

Consequently,

$$f(X) = \int_0^1 (6t^2xy^2 + x) \, dt = 2xy^2 + x$$

is a potential function for F.

EXERCISES

Determine which of the following vector fields F are gradient fields in the entire plane. If F is a gradient field, construct a potential function for F by using Theorem 9.4.

1. $F(x, y) = (x - y, -x)$.
2. $F(x, y) = (y \cos xy, x \sin xy)$.
3. $F(x, y) = (2x^2 + y^3, x^2 + 3xy^2)$.

Determine which of the following vector fields F are gradient fields in all of $V_3(R)$. If F is a gradient field, construct a potential function for F by using Theorem 9.4.

4. $F(x, y, z) = (2xyz, x^2z, x^2y)$.
5. $F(x, y, z) = (x^2y, xz^2, zy^2)$.
6. $F(x, y, z) = (2xy, x^2 - 2yz, -y^2)$.
7. $F(x, y, z) = (2xyz + y^2, x^2z + 2xy, x^2y + 2z)$.
8. A force field F is defined by $F(x, y, z) = (z, yz, y)$. Is this field conservative? Determine the work done by this force field if a particle is moved around the closed unit square in the yz plane in a counterclockwise direction.
9. Show that the force field $F(X) = f(|X|)X$ is a conservative force field if f is any continuously differentiable function of a real variable.
10. A fluid emanating from the origin in the x, y plane flows so that the velocity V at a point X is given by $V(X) = |X|^n X$ where n is a positive integer. Show that this velocity field is always a gradient field. Determine a potential function for V.
11. Let

$$F(x, y) = \frac{1}{|X|^2}(-y, x).$$

Show that F is not a gradient field in any annulus $r_1 < x^2 + y^2 < r_2^2$ but that F is a gradient field in any star-shaped open set not containing the origin.

12. Finish the proof of Theorem 9.4 by showing that if $(\partial F_j/\partial x_k)$ is symmetric, then for each k

(10) $$\frac{\partial}{\partial x_k}(F(tX), X) = \frac{d}{dt}(t(F_k(tX))).$$

[Verify that $(\partial/\partial x_k)(F(tX), X) = F_k(tX) + \Sigma_j (tx_j)(\partial F_k/\partial x_j)(tX)$. Now use the symmetry of $(\partial F_k/\partial x_j)$ to derive 10.]

13. Let U be the set of all points (x, y) in the plane except $(x, 0)$ where $x \leq 0$.

Consider the vector field

$$F(x, y) = \left(\frac{-y}{x^2 + y^2}, \frac{x}{x^2 + y^2}\right)$$

defined on U. Show that F is a gradient field on U by verifying that $f(x, y) = \theta$ where $-\pi < \theta < \pi$ and $\cos\theta = x/\sqrt{x^2 + y^2}$, $\sin\theta = y/\sqrt{x^2 + y^2}$ is a potential function for F. Express $f(x, y)$ as a line integral $\int_\alpha F$ where α is a suitable curve from $(1, 0)$ to (x, y) lying in U.

14. If $f(X) = -1/|X|$ for $X = (x, y, z)$, then f is the potential function for the Newtonian gravitational field. Determine this vector field and determine the work done by this field moving a particle from $(1, 1, 1)$ to $(-1, -1, -1)$ along a curve in the domain of definition of f.

*15. To establish (9) we need to verify the following formula for differentiation under the integral sign. Let $f(x, t)$ be a real-valued function with continuous first partial derivatives. Let

$$g(t) = \int_a^b f(x, t)\, dx.$$

Then we assert

(11) $$g'(t) = \int_a^b \frac{\partial f}{\partial t}(x, t)\, dx.$$

Now

$$g'(t) = \lim_{h \to 0} \frac{g(t + h) - g(t)}{h}.$$

Hence to prove (11) we must show

$$\lim_{h \to 0}\left\{\frac{g(t + h) - g(t)}{h} - \int_a^b \frac{\partial f}{\partial t}(x, t)\, dx\right\} = 0.$$

But

$$\frac{g(t + h) - g(t)}{h} = \int_a^b \left[\frac{f(x, t + h) - f(x, t)}{h}\right] dt.$$

Applying the mean value theorem of the differential calculus, we conclude that

$$\frac{f(x, t + h) - f(x, t)}{h} = \frac{\partial f}{\partial t}(x, t_h)$$

where $|t_h - t| < h$. Therefore

$$\frac{g(t + h) - g(t)}{h} - \int_a^b \frac{\partial f}{\partial t}(x, t)\, dx = \int_a^b \left[\frac{\partial f}{\partial t}(x, t_h) - \frac{\partial f}{\partial t}(x, t)\right] dx.$$

Since $\partial f/\partial t$ is assumed to be a continuous function from $V_2(R)$ to R, it can now be shown that

$$\left|\frac{\partial f}{\partial t}(x, t_h) - \frac{\partial f}{\partial t}(x, t)\right| < \epsilon$$

for all x, $a \leq x \leq b$ if h is small enough. From this (11) follows. The reader should satisfy himself that (9) now follows from (11).

9.6 Green's Theorem

We turn now to an important connection between line integrals and integrals over regions in the plane. Before stating the desired result we must make some preliminary observations about curves. Let α be a piecewise smooth simple closed curve in the plane. That is, α is a one-to-one function except at the end points of the parameter interval $[a, b]$. Then $\alpha(a) = \alpha(b)$.

Furthermore $\alpha'(t)$ is continuous except for finitely many points t_i. Now it is a fundamental theorem of topology that the simple closed curve α divides the plane into two regions, an interior region and an exterior region. This theorem is called the Jordan curve theorem, in honor of the French mathematician, Camille Jordan, who first pointed out that this seemingly obvious statement requires proof. Jordan himself attempted to supply a proof, which was later shown to be incomplete. The first complete proof was given in 1905 by the American mathematician, Oswald Veblen.

We shall accept the result and talk freely of the interior of a simple closed curve α. Now at each point $\alpha(t)$, if $\alpha'(t)$ exists and is different from zero, $\alpha'(t)$ establishes a direction along α. Let $T(t) = \alpha'(t)/|\alpha'(t)|$ be the unit tangent, and let $\tilde{N}(t)$ be the principal normal to the curve α at the point $\alpha(t)$. That is, $\tilde{N}(t)$ is a unit vector perpendicular to the tangent vector and, moreover, the vectors T, $\tilde{N}$, in that order, can be obtained by a rotation of the unit coordinate vectors I_1, I_2. Recall that if $E = (e_1, e_2)$ and $F = (f_1, f_2)$ are orthogonal unit vectors, then there exists a rotation R_θ such that $E = R_\theta(I_1)$ and $F = R_\theta(I_2)$ if and only if the determinant.

$$D\begin{pmatrix} e_1 & f_1 \\ e_2 & f_2 \end{pmatrix} = 1.$$

Hence if

$$T = \frac{1}{|\alpha'(t)|}(\alpha^{1'}(t), \alpha^{2'}(t))$$

is the unit tangent vector, and we define

$$\tilde{N} = \frac{1}{|\alpha'(t)|}(-\alpha^{2'}(t), \alpha^{1'}(t))$$

then T, $\tilde{N}$ are orthogonal unit vectors which can be obtained from I_1, I_2 by a rotation. When $\tilde{N}(t)$ always points toward the interior region of the simple closed curve α, we say that the curve α is running in a *counterclockwise* sense or that α is *positively oriented*. This is illustrated in Figure 22. If $\tilde{N}$ always points toward the exterior, then α runs in a clockwise direction and we say α is *negatively oriented*.

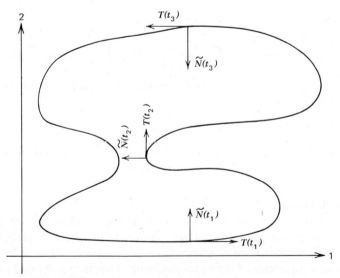

Fig. 22 Positively oriented simple closed curve.

GREEN'S THEOREM

When α is positively oriented, we shall indicate the line integral of the vector field F around α by $\oint_\alpha F$.

We are now in a position to state the result known as *Green's theorem*. This theorem expresses the line integral $\oint_\alpha F$ of a vector field F around a simple closed curve α as a suitable double integral over the interior region. To avoid confusion with line integrals we shall write the integral of a scalar function f over a region S in the plane as $\iint_S f(x, y) \, dx \, dy$.

THEOREM 9.5 *Let F be a continuously differentiable vector field defined on a connected open set U in the plane. Assume that the piecewise smooth simple closed curve α together with its interior region S lies entirely within U. Assume further that α traverses the boundary of S in a counterclockwise direction. Then, if F_1, F_2 are the coordinate functions for F,*

$$(1) \qquad \iint_S \left(\frac{\partial F_2}{\partial x} - \frac{\partial F_1}{\partial y} \right) dx \, dy = \oint_\alpha F.$$

Before examining the proof let us compute an example.

Example 1. If $F(x, y) = (x^2 y, x + y)$ compute $\oint_\alpha F$ if α traverses the unit square in a counterclockwise direction. Clearly

$$\frac{\partial F_2}{\partial x} = 1 \qquad \text{and} \qquad \frac{\partial F_1}{\partial y} = x^2.$$

Hence if we let $S = [0, 1] \times [0, 1]$, we have from (1)

$$\oint_\alpha F = \iint_S (1 - x^2) \, dx \, dy = \int_0^1 \int_0^1 (1 - x^2) \, dx \, dy = \frac{2}{3}.$$

If we use the notation

$$\oint_\alpha (F_1 \, dx + F_2 \, dy) \qquad \text{for} \qquad \oint_\alpha F,$$

equation (1) becomes

$$\iint_S \left(\frac{\partial F_2}{\partial x} - \frac{\partial F_1}{\partial y} \right) dx \, dy = \oint_\alpha (F_1 \, dx + F_2 \, dy).$$

A complete proof of Theorem 9.5 involves an approximation argument which we are not in a position to give. However, the idea of the proof is first to verify (1) for curves which traverse rectangles and triangles. Using this result one then proves (1) for closed polygonal curves α. The final step is to show that both integrals $\oint_\alpha F$ and

$$\int_S \left(\frac{\partial F_2}{\partial x} - \frac{\partial F_1}{\partial y} \right) dx \, dy$$

may be approximated arbitrarily well by integrals $\oint_{\alpha_n} F$ and

$$\iint_{S_n} \left(\frac{\partial F_2}{\partial x} - \frac{\partial F_1}{\partial y} \right) dx \, dy$$

where α_n is a polygonal curve and S_n is its interior region.

The proof of (1) for rectangles is immediate. Let $\alpha_1, \alpha_2, \alpha_3, \alpha_4$ be the four rectilinear portions of α as in Figure 23.

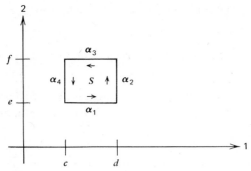

Fig. 23

Clearly
$$\iint_S \left(\frac{\partial F_2}{\partial x} - \frac{\partial F_1}{\partial y}\right) dx\, dy = \int_e^f \left[\int_c^d \left(\frac{\partial F_2}{\partial x} - \frac{\partial F_1}{\partial y}\right)\right] dx\, dy.$$

But if
$$\begin{aligned} \beta_1(x) &= (x, e) & c \leq x \leq d \\ \beta_2(y) &= (d, y) & e \leq y \leq f \\ \beta_3(x) &= (x, f) & c \leq x \leq d \\ \beta_4(y) &= (c, y) & e \leq y \leq f \end{aligned}$$

then $\alpha_1 \sim \beta_1$, $\alpha_2 \sim \beta_2$, $\alpha_3 \sim -\beta_3$ and $\alpha_4 \sim -\beta_4$. Hence

$$\oint_\alpha F = \sum_{i=1}^4 \int_{\alpha_i} F = \int_{\beta_1} F + \int_{\beta_2} F - \int_{\beta_3} F - \int_{\beta_4} F = \int_{\beta_1} F - \int_{\beta_3} F + \int_{\beta_2} F - \int_{\beta_4} F.$$

Furthermore
$$\int_{\beta_1} F = \int_c^d (F(\beta_1(x)), \beta_1'(x))\, dx = \int_c^d F_1(x, e)\, dx$$

and
$$\int_{\beta_3} F = \int_c^d (F(\beta_3(x)), \beta_3'(x))\, dx = \int_c^d F_1(x, f)\, dx$$

Therefore
$$\int_{\beta_1} F - \int_{\beta_3} F = \int_c^d [F_1(x, e) - F_1(x, f)]\, dx = -\int_c^d \int_e^f \frac{\partial F_1}{\partial y}\, dy\, dx.$$

The last equality holds by virtue of the fundamental theorem of the calculus. Similarly
$$\int_{\beta_2} F - \int_{\beta_4} F = \int_e^f [F_2(d, y) - F_2(c, y)]\, dy = \int_e^f \left[\int_c^d \frac{\partial F_2}{\partial x}\right] dx\, dy.$$

Hence adding these two results we have
$$\oint_\alpha F = \int_e^f \int_c^d \left[\frac{\partial F_2}{\partial x} - \frac{\partial F_1}{\partial y}\right] dx\, dy$$

which is the desired result. The verification of (1) for triangles proceeds similarly. We leave it as an exercise.

Next let α be a simple closed polygonal curve enclosing a region S. We may assume that S is the union of rectangles and triangles S_k having only edges in common as in Figure 24. Let γ_k be a curve traversing the boundary of S_k once in a counterclockwise direction.

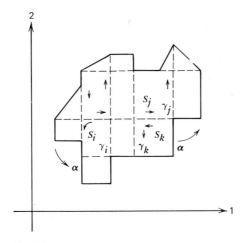

Fig. 24

Certainly $\iint_S \left(\frac{\partial F_2}{\partial x} - \frac{\partial F_1}{\partial y}\right) dx\, dy = \sum \iint_{S_k} \left(\frac{\partial F_2}{\partial x} - \frac{\partial F_1}{\partial y}\right) dx\, dy$ by the familiar additivity property of integrals over regions which overlap on sets of zero area. But also

$$\oint_\alpha F = \sum \oint_{\gamma_k} F.$$

This latter equality holds since all line integrals over interior line segments of the boundaries of the sets S_k cancel each other out in pairs. This follows since each is traversed once in one direction and once in the opposite. This leaves only the integral around the outside boundary which is $\oint_\alpha F$. The proof is completed, as we indicated earlier, by approximating the simple closed curve α by a polygonal curve α_n.

As a corollary of this result we may sharpen the sufficient condition for a two-dimensional vector field F to be a gradient field. Call a connected open set U in $V_2(R)$ *simply connected* if for each simple closed curve α in U the interior region of α also lies in U. Thus a disc is simply connected, but an annulus or washer-shaped region is not. See Figure 25.

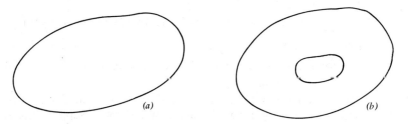

Fig. 25 (*a*) A simply connected open set. (*b*) A nonsimply connected open set.

A star-shaped open set is simply connected, but a simply connected open set is not necessarily star-shaped. Figure 26 is an example which illustrates that. Naively speaking, a connected open set is simply connected if it has no holes.

THEOREM 9.6 *Let F be a continuously differentiable vector field defined on a simply connected open set U in $V_2(R)$. If $(\partial F_i/\partial x_j)$ is symmetric on U, then F is a gradient field.*

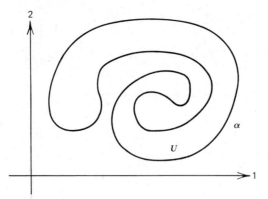

Fig. 26 A simply connected open set which is not star-shaped about any point.

Proof. Exploiting the remark following Theorem 9.3 it is enough to show that $\oint_\alpha F = 0$ for each closed step polygon α lying in U. If α is a simple closed step polygon, then since the interior region of α lies in U, $\oint_\alpha F = 0$. This follows from (1) since

$$\frac{\partial F_2}{\partial x} - \frac{\partial F_1}{\partial y} = 0$$

everywhere in U.

If α is not simple then we may break up α into finitely many pieces $\alpha_1, \ldots, \alpha_n$ each of which is either a simple closed polygon or a line segment traversed twice in opposite directions. For example, see Figure 27.

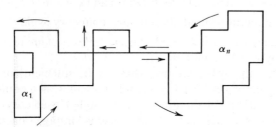

Fig. 27 Closed step polygon.

Now $\int_\alpha F = \sum_{i=1}^{n} \int_{\alpha_i} F$ and clearly $\int_{\alpha_i} F = 0$ for each curve α_i since α_i is either a simple closed polygon or a line segment traversed twice in opposite directions. Therefore $\int_\alpha F = 0$ for each closed stop polygon α. Consequently, F is a gradient field.

We may extend Green's theorem to more general sets than just simply connected ones by an extremely simple device. Suppose the set S has two simple closed curves α_1, α_2 as its boundary. Assume α_1, α_2 both are traversed in a counterclockwise sense relative to S. This means that the region S is always on the left as a particle moves around either α_1 or α_2. Connect α_1 and α_2 by two nonintersecting straight lines or polygonal paths as in Figure 28. The set S is now divided into two sets S_1, S_2 each having a simple closed boundary curve γ_1, and γ_2. By Theorem 9.5

$$\iint_{S_i} \left(\frac{\partial F_2}{\partial x} - \frac{\partial F_1}{\partial y} \right) dx\, dy = \oint_{\gamma_i} F \qquad i = 1, 2.$$

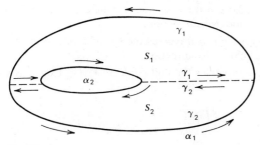

Fig. 28

Now

$$\iint_S \left(\frac{\partial F_2}{\partial x} - \frac{\partial F_1}{\partial y}\right) dx\,dy$$

is certainly the sum of the integrals over S_1 and S_2, respectively. Moreover

$$\int_{\alpha_1} F + \int_{\alpha_2} F = \int_{\gamma_1} F + \int_{\gamma_2} F$$

since the line integrals over the interior portions of γ_1 and γ_2 are taken twice in opposite directions. Hence they cancel. This argument extends immediately to sets S with finitely many holes. See Figure 29. We state the result which is the general form of Green's theorem.

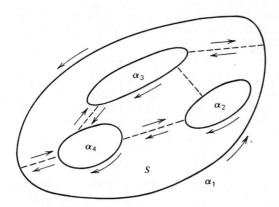

Fig. 29 Construction to prove Theorem 9.6 for a region with three holes.

THEOREM 9.7 *Let S be a bounded closed set in $V_2(R)$ such that the boundary is traversed by finitely many simple closed curves $\alpha_1, \ldots, \alpha_n$. Assume each curve α_k is oriented positively with respect to S. That is, S lies to the left of each curve α_k. If F is a smooth vector field defined in a neighborhood of S, then*

$$\iint_S \left(\frac{\partial F_2}{\partial x} - \frac{\partial F_1}{\partial y}\right) dx\,dy = \sum_{k=1}^n \oint_{\alpha_k} F.$$

EXERCISES

Evaluate the line integrals $\oint_\alpha F$ for the following choices of F and α by the use of Green's theorem.

1. $F(X) = (x^2 y, x^2 + y)$, α traverses the rectangle having vertices $(\pm 1, 0)$, $(\pm 1, 1)$ in a counterclockwise direction.

2. $F(X) = (x^3 - y^2 x, 2xy)$, α traverses the rectangle having vertices $(0, 0)$, $(2, 0)$, $(2, 3)$, $(0, 3)$ in a clockwise direction.
3. $F(X) = (2xy^2, (y + x)^2)$, α traverses the triangle formed by $(0, 0)$, $(1, 0)$ and $(0, 1)$ in a counterclockwise direction.
4. $F(X) = (y^2, x)$, α traverses the unit circle $x^2 + y^2 = 1$ once in a counterclockwise direction.

Show that
$$\iint_S \left(\frac{\partial F_2}{\partial x} - \frac{\partial F_1}{\partial y} \right) dx\, dy = \oint_\alpha F$$
by evaluating both integrals for the following choices of functions F and regions S with positively oriented boundary curve α.

5. $F(x, y) = (y, 2x)$, S is the rectangle with vertices $(0, 0)$, $(2, 0)$, $(2, 1)$, $(0, 1)$.
6. $F(x, y) = (x + y, xy)$, S is the triangle with vertices $(1, 1)$, $(4, 1)$, $(1, 3)$.
7. $F(x, y) = (x + y, y - x)$, S is the disc $x^2 + y^2 \le a^2$.
8. $F(x, y) = (-y, x)$, S is the annulus $a^2 \le x^2 + y^2 \le b^2$ where $0 < a < b$.
9. Prove formula (1) for an arbitrary smooth vector field F defined on a triangle with vertices (a, c), (b, c), (a, d).
10. Let S be a bounded closed set in the plane and assume that the boundary of S is traversed by a positively oriented simple closed curve α. Verify that the area $A(S)$ of the set S is given by
$$A(S) = \frac{1}{2} \oint_\alpha (x\, dy - y\, dx).$$
11. Let S be a bounded closed set in the plane and assume that the boundary of S is traversed by finitely many simple closed curves each oriented positively with respect to S. Show that the area
$$A(S) = \frac{1}{2} \sum_{k=1}^n \oint_{\alpha_k} (x\, dy - y\, dx).$$
12. Let S satisfy the conditions of Exercise 10. Let $\bar{X} = (\bar{x}, \bar{y})$ be the centroid of S. If $A(S)$ is the area of S, show that
$$\bar{x} = \frac{1}{4A(S)} \oint_\alpha (x^2\, dy - 2xy\, dx).$$
Write $\bar{y}$ as a suitable line integral around the boundary curve α.
13. Let $F(X) = (-y/(x^2 + y^2), x/(x^2 + y^2))$. If α is a positively oriented simple closed curve in the plane not passing through the origin, show that $\oint_\alpha F = 2\pi$ or 0 according as the origin does or does not lie in the interior region for α.
14. Let $F(X) = (F_1(X), F_2(X))$ be a vector field and assume $\partial F_2/\partial x = \partial F_1/\partial y$ at each point $X \ne 0$. If α_1 and α_2 are simple closed curves containing the origin in their interior show that $\oint_{\alpha_1} F = \oint_{\alpha_2} F$.
15. Let $f(x)$ be a continuously differentiable function defined on the interval $[0, a]$ and having a nonvanishing derivative. If $f(0) = b$ and $f(a) = 0$, let α be the curve enclosing the region S bounded by the coordinate axes and the graph of f as in Figure 30. If F is a continuously differentiable vector field defined on S, verify formula (1) directly assuming that α runs in a counterclockwise sense.
16. Let α_δ be a simple closed curve lying in a disc of radius δ about the point X. Let A_δ be the area of the region inside α_δ. Show that
$$\lim_{\delta \to 0} \frac{1}{A_\delta} \oint_{\alpha_\delta} F = \left(\frac{\partial F_2}{\partial x} - \frac{\partial F_1}{\partial y} \right)(X).$$

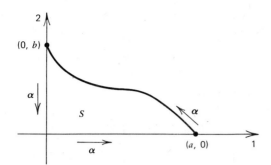

Fig. 30

(See Exercise 21, p. 314.)

17. Let f and g be continuously differentiable scalar functions defined on a connected open set U. If α is a closed curve in U, verify that

$$\int_\alpha f\nabla g = -\int_\alpha g\nabla f.$$

18. If the simple closed curve α traverses the boundary of a region S in a counter-clockwise direction, then we have seen that $\tilde{N}(t) = 1/|\alpha'(t)|(-\alpha^{2\prime}(t), \alpha^{1\prime}(t))$ is a unit normal vector to the curve pointing toward S. This vector $\tilde{N}$ is called the *inner* normal of α. The vector $N = -\tilde{N} = 1/|\alpha'(t)|(\alpha^{2\prime}(t), -\alpha^{1\prime}(t))$ is called the *outer* normal. The vector $T = 1/|\alpha'(t)|(\alpha^{1\prime}(t), \alpha^{2\prime}(t))$ is, of course, the unit tangent. If $F = (F_1, F_2)$ is a vector field defined on α, then by definition of the line integral

$$\int_\alpha (F_1\,dx + F_2\,dy) = \int_\alpha F = \int_\alpha (F, T)\,ds.$$

Verify the following formulas.

(i) $\int_\alpha (F_1\,dy - F_2\,dx) = \int_\alpha (F, N)\,ds.$

(ii) $\int_\alpha (F_2\,dx - F_1\,dy) = \int_\alpha (F, \tilde{N})\,ds.$

19. Use Exercise 18 together with Theorem 9.5 to prove Gauss' theorem. Namely, that

$$\iint_S \left(\frac{\partial F_1}{\partial x} + \frac{\partial F_2}{\partial y}\right) dx\,dy = \oint_\alpha (F, N)\,ds.$$

The quantities F, S, and α are as in Green's theorem.

20. Using the notation of Exercise 18 verify the following consequences of Green's theorem [Recall that for a scalar function f, $\nabla^2 f = [(\partial^2 f/\partial x^2) + (\partial^2 f/\partial y^2)]$ and also that for a vector X, $\partial f/\partial X = (\nabla f, X)$.]

(a) $\oint_\alpha \frac{\partial f}{\partial N} ds = \iint_S \nabla^2 f\,dx\,dy.$

(b) $\oint_\alpha g \frac{\partial f}{\partial N} ds = \iint_S [g\nabla^2 f + (\nabla f, \nabla g)]\,dx\,dy.$

(c) $\oint_\alpha \left(g\frac{\partial f}{\partial N} - f\frac{\partial g}{\partial N}\right) ds = \iint_S (g\nabla^2 f - f\nabla^2 g)\,dx\,dy.$

Formulas (b) and (c) are called *Green's identities* and occur very frequently in problems in mathematical physics.

21. A function f is called *harmonic* in the region U if $\nabla^2 f = 0$ everywhere in U. If f is harmonic in the region U and α is a simple closed curve in U, verify that

$$\oint_\alpha \frac{\partial f}{\partial N} ds = 0.$$

If f and g are both harmonic in U, verify that

$$\oint_\alpha g \frac{\partial f}{\partial N} ds = \oint_\alpha f \frac{\partial g}{\partial N} ds.$$

CHAPTER TEN

VECTOR FIELDS IN SPACE

10.1 Divergence and Curl of a Vector Field

In this chapter we shall study the extension of Green's theorem to functions from $V_3(R)$ to $V_3(R)$, that is, to three-dimensional vector fields. Before discussing the form this theorem takes, it is convenient to introduce two auxiliary functions which are useful in the study of three-dimensional vector fields. All vector fields considered in this chapter will be assumed to be continuously differentiable.

The first of these functions is a scalar function called the *divergence* of the vector field F. It is defined by the formula

(1) $$(\text{div } F)(X) = \frac{\partial F_1}{\partial x_1} + \frac{\partial F_2}{\partial x_2} + \frac{\partial F_3}{\partial x_3}.$$

If $F'(X) = (\partial F_i/\partial x_j)$ is the Jacobian matrix of the vector field F evaluated at the point X, then the divergence of F is the sum of the terms on the main diagonal of this matrix.

The defining formula for F is expressed in terms of the component functions F_1, F_2, F_3 of F. Hence, the divergence appears to depend on the canonical basis $I_1 = (1, 0, 0)$, $I_2 = (0, 1, 0)$, $I_3 = (0, 0, 1)$ of $V_3(R)$. In reality this is not the case. Indeed, if J_1, J_2, J_3 is any basis for $V_3(R)$, and for $X \in V_3(R)$ we write

$$X = \sum_{k=1}^{3} x_k I_k = \sum_{k=1}^{3} y_k J_k$$

and

$$F(X) = \sum_{k=1}^{3} F_k(x_1, x_2, x_3) I_k = \sum_{k=1}^{3} \tilde{F}_k(y_1, y_2, y_3) J_k,$$

then we leave it as an exercise to verify that

(?) $$(\text{div } F)(X) = \frac{\partial \tilde{F}_1}{\partial J_1} + \frac{\partial \tilde{F}_2}{\partial J_2} + \frac{\partial \tilde{F}_3}{\partial J_3}.$$

The verification of (2) is straightforward, involving only the chain rule and a few elementary calculations with matrices.

The second function associated with the vector field F is an associated vector field called the curl of F. It is defined by the formula

(3) $$(\text{curl } F)(X) = \left(\frac{\partial F_3}{\partial x_2} - \frac{\partial F_2}{\partial x_3}\right) I_1 - \left(\frac{\partial F_3}{\partial x_1} - \frac{\partial F_1}{\partial x_3}\right) I_2 + \left(\frac{\partial F_2}{\partial x_1} - \frac{\partial F_1}{\partial x_2}\right) I_3$$

$$= \left(\frac{\partial F_3}{\partial x_2} - \frac{\partial F_2}{\partial x_3}, \frac{\partial F_1}{\partial x_3} - \frac{\partial F_3}{\partial x_1}, \frac{\partial F_2}{\partial x_1} - \frac{\partial F_1}{\partial x_2}\right).$$

The curl of F may be written in terms of the Jacobian matrix $F'(X)$ in the following way. Let $(\partial F_i/\partial x_j)^t$ be the transpose of the Jacobian matrix for F, and set

$$A = \left(\frac{\partial F_i}{\partial x_j}\right) - \left(\frac{\partial F_i}{\partial x_j}\right)^t = \left(\frac{\partial F_i}{\partial x_j} - \frac{\partial F_j}{\partial x_i}\right).$$

Then A is a matrix of the form

$$A = \begin{pmatrix} 0 & -c & b \\ c & 0 & -a \\ -b & a & 0 \end{pmatrix}$$

where

$$a = \frac{\partial F_3}{\partial x_2} - \frac{\partial F_2}{\partial x_3}, \quad b = \frac{\partial F_1}{\partial x_3} - \frac{\partial F_3}{\partial x_1}, \quad c = \frac{\partial F_2}{\partial x_1} - \frac{\partial F_1}{\partial x_2}.$$

The vector (a, b, c) is thus the curl of F. Clearly curl $F = 0$ if and only if $F'(X) = (\partial F_i/\partial x_j)$ is symmetric.

The curl may be interpreted as a transformation which maps vector fields onto vector fields. For this the following matrix representation is useful.

$$(\text{curl } F)(X) = \begin{pmatrix} 0 & -\frac{\partial}{\partial x_3} & \frac{\partial}{\partial x_2} \\ \frac{\partial}{\partial x_3} & 0 & -\frac{\partial}{\partial x_1} \\ -\frac{\partial}{\partial x_2} & \frac{\partial}{\partial x_1} & 0 \end{pmatrix} \begin{pmatrix} F_1 \\ F_2 \\ F_3 \end{pmatrix}.$$

If we agree that $(\partial/\partial x_j) \cdot F_i = \partial F_i/\partial x_j$, then it is clear from the rules for matrix multiplication that the above matrix product is $(\text{curl } F)(X)$.

Just as for the divergence formula (3) appears to depend on the particular basis chosen for $V_3(R)$. It can be shown, however, that this is not true provided that the basis J_1, J_2, J_3 is orthonormal and can be obtained from the canonical basis I_1, I_2, I_3 by a rotation. We sketch the proof of this fact in the exercises.

Computations involving the divergence and curl are greatly facilitated by the introduction of the formal operator

$$\nabla = \left(\frac{\partial}{\partial x_1}, \frac{\partial}{\partial x_2}, \frac{\partial}{\partial x_3}\right).$$

We shall treat ∇, pronounced "del," just as if it were a vector in $V_3(R)$ and introduce formal rules for defining (∇, F) and $\nabla \times F$ when F is a vector field. Of course, we have already dealt with the gradient

$$\nabla f = \left(\frac{\partial f}{\partial x_1}, \frac{\partial f}{\partial x_2}, \frac{\partial f}{\partial x_3}\right)$$

which may be thought of as the operator ∇ multiplied by the scalar function f. For the vector field $F = (F_1, F_2, F_3) = \Sigma F_k I_k$ we define the inner product

(4) $$(\nabla, F) = \frac{\partial F_1}{\partial x_1} + \frac{\partial F_2}{\partial x_2} + \frac{\partial F_3}{\partial x_3} = \text{div } F$$

and the vector product

(5) $$\nabla \times F = D \begin{pmatrix} I_1 & I_2 & I_3 \\ \frac{\partial}{\partial x_1} & \frac{\partial}{\partial x_2} & \frac{\partial}{\partial x_3} \\ F_1 & F_2 & F_3 \end{pmatrix} = \text{curl } F.$$

The inner product (∇, F) and vector product $\nabla \times F$ are computed just as if the vector ∇ had scalar components. The product $(\partial/\partial x_j) \cdot F_i$ is, of course, defined to be $\partial F_i/\partial x_j$.

Since partial derivatives are linear operations, most of the algebra of scalar multiplication, inner product, and vector product extends verbatim to the use of ∇. Indeed, we have already observed that for scalar functions f and g

(6) $$\nabla(af+bg) = a\nabla f + b\nabla g$$

and

(7) $$\nabla(fg) = f\nabla g + g\nabla f.$$

If F and G are vector fields, then

(8) $$(\nabla, aF+bG) = a(\nabla, F) + b(\nabla, G),$$

(9) $$(\nabla, fF) = f(\nabla, F) + (\nabla f, F)$$

and

(10) $$(\nabla, F \times G) = (\nabla \times F, G) - (F, \nabla \times G).$$

Also

(11) $$\nabla \times (aF + bG) = a\nabla \times F + b\nabla \times G$$

and

(12) $$\nabla \times (fF) = f \cdot (\nabla \times F) + \nabla f \times F.$$

We leave the verification of these identities as exercises.

In each case the verification depends only on the linearity of partial differentiation and the product rule for differentiation of scalar functions.

It is sometimes convenient to define $F \times \nabla = -\nabla \times F$, which is in accord with the vector product of two vectors. However, the student should bear in mind that the vector product is *not* associative, so care should be taken with expressions like $F \times (\nabla \times G)$.

For vectors A, B, C we may form the scalar triple product

$$(A, B \times C) = D\begin{pmatrix} a_1 & a_2 & a_3 \\ b_1 & b_2 & b_3 \\ c_1 & c_2 & c_3 \end{pmatrix}.$$

Various identities such as $(A, B \times C) = -(B, A \times C)$ result from the definition of this scalar triple product as a determinant. Now expressions of the form $(\nabla, F \times G)$ and $(F, \nabla \times G)$ make sense but one should *not* expect that $(\nabla, F \times G) = -(F, \nabla \times G)$. The trouble is that for numbers a, b, c we have $a(bc) = (ab)c = (ba)c$. On the other hand, it is false that for functions f and g

$$f\left(\frac{\partial}{\partial x} \cdot g\right) = \frac{\partial}{\partial x}(fg).$$

However, the analogue of the fact that for vectors $(A, A \times B) = 0$ is indeed valid. If F is twice continuously differentiable, so that second partial derivatives may be taken in any order, then

(13) $$(\nabla, \nabla \times F) = 0.$$

Also
(14)
$$\nabla \times \nabla f = 0.$$

To verify (13) for the vector field $F = (F_1, F_2, F_3)$ write

$$(\nabla, \nabla \times F) = \begin{pmatrix} \frac{\partial}{\partial x_1} & \frac{\partial}{\partial x_2} & \frac{\partial}{x_3} \\ \frac{\partial}{\partial x_1} & \frac{\partial}{\partial x_2} & \frac{\partial}{\partial x_3} \\ F_1 & F_2 & F_3 \end{pmatrix}.$$

Since each of the functions F_i are twice continuously differentiable,

$$\frac{\partial^2 F_i}{\partial x_j \partial x_k} = \frac{\partial^2 F_i}{\partial x_k \partial x_j} \quad i,j,k = 1,2,3.$$

From this it follows readily that $(\nabla, \nabla \times F) = 0$. The proof of (14) follows for the same reason. The identities (13) and (14) are often written

$$\text{div}(\text{curl } F) = 0$$

and

$$\text{curl}(\text{grad } f) = 0.$$

Equations 13 and 14 have important converses. If for the vector field G, $\nabla \times G = 0$ on an open set U then it is clear from the definition of the curl that the Jacobian matrix $\partial G_i/\partial x_j$ is symmetric. Hence if U is star-shaped, we know from Theorem 9.4 that $G = \nabla f$ for some scalar function f. Indeed, assuming $X_0 = 0$ we may take

(15)
$$f(X) = \int_0^1 (G(tX), X) \, dt.$$

On the other hand, if $(\nabla, G) = 0$ over a star-shaped region U, it follows that there exists a vector field F such that $G = \nabla \times F$. This result is valid for more general regions than star-shaped ones, but some restriction on the open set U is necessary in order to conclude that $G = \nabla \times F$, if $(\nabla, G) = 0$. The formula for F is similar to (15) and is simplified if we introduce the notion of a *vector-valued integral*.

If α is a continuous curve and $\alpha(t) = (\alpha^1(t), \alpha^2(t), \alpha^3(t))$, $a \leq t \leq b$, then we define

$$\int_a^b \alpha(t) \, dt = \left(\int_a^b \alpha^1(t) \, dt, \int_a^b \alpha^2(t) \, dt, \int_a^b \alpha^3(t) \, dt \right).$$

The value of this integral is a vector, not a number, hence the term vector-valued integral.

Assume now that $(\nabla, G) = 0$ on a star shaped open set U containing the origin. For $X \in U$ let $\alpha(t) = tX$, $0 \leq t \leq 1$. Then α is the straight line joining 0 and X. If we define

$$\beta(t) = G(tX) \times (tX)$$

then β is a smooth curve joining $\beta(0) = 0$ and $\beta(1) = G(X) \times (X)$. It may be checked that

(16)
$$F(X) = \int_0^1 G(tX) \times (tX) \, dt$$

now has the property that
$$\nabla \times F(X) = G(X).$$

The verification of (16) is a straightforward but rather lengthy calculation. We leave it as an exercise.

Example 1. If $G(X) = (z - xy, y^2, x - yz)$, find a vector field F satisfying $\nabla \times F = G$. Clearly div $G = (\nabla, G) = -y + 2y - y = 0$. Moreover,

$$G(tX) \times (tX) = D\begin{pmatrix} I_1 & I_2 & I_3 \\ tz - t^2xy & t^2y^2 & tx - t^2yz \\ tx & ty & tz \end{pmatrix}$$

$$= (2t^3y^2z - t^2xy, -t^2(x^2 + z^2), t^2yz - 2t^3xy^2).$$

Hence
$$F(X) = \int_0^1 G(tX) \times (tX)\, dt = \left(\frac{y^2z}{2} - \frac{xy}{3}, -\frac{x^2 + z^2}{3}, \frac{yz}{3} - \frac{xy^2}{2} \right).$$

It is easily checked that $\nabla \times F = G$.

To summarize, it is always the case that $\nabla \times (\nabla f) = 0$ and $(\nabla, \nabla \times F) = 0$. Conversely, if $\nabla \times G = 0$ and $(\nabla, H) = 0$ on a star-shaped open set, then

$$G = \nabla f \quad \text{and} \quad H = \nabla \times F$$

where f and F are defined by equations (15) and (16), respectively. A vector field G such that $(\nabla, G) = 0$ is called *solenoidal*. If $\nabla \times H = 0$, then H is called *irrotational*. We shall discuss the physical reasons for this terminology later. These concepts are important both mathematically and physically because it can be shown that any vector field F can be written as the sum $F = G + H$ where G is solenoidal and H is irrotational.

Green's theorem has two important formulations in terms of the divergence and curl. If we assume the vector field $F(X)$ has a constant third component, then
$$F(X) = (F_1(x_1, x_2), F_2(x_1, x_2), c).$$

Consequently,

(17) $$\text{div } F = \frac{\partial F_1}{\partial x_1} + \frac{\partial F_2}{\partial x_2}$$

and

(18) $$\text{curl } F = \left(\frac{\partial F_2}{\partial x_1} - \frac{\partial F_1}{\partial x_2} \right) I_3.$$

Thus for a plane set S we may write

$$\iint_S \left(\frac{\partial F_2}{\partial x_1} - \frac{\partial F_1}{\partial x_2} \right) dx_1\, dx_2 = \iint_S (\text{curl } F, I_3)\, dx_1\, dx_2.$$

If α is the boundary curve of the set S, then

$$\oint_\alpha F = \oint_\alpha \left(F, \frac{\alpha'(t)}{|\alpha'(t)|} \right) |\alpha'(t)|\, dt = \oint_\alpha (F, T)\, ds,$$

and Green's theorem becomes

(19) $$\iint_S (\text{curl } F, I_3)\, dx_1\, dx_2 = \oint_\alpha (F, T)\, ds.$$

On the other hand, if N is the outer normal to the region S, then we may deduce from Green's theorem that

$$\iint_S \left(\frac{\partial F_1}{\partial x_1} + \frac{\partial F_2}{\partial x_2}\right) dx_1\, dx_2 = \oint_\alpha (F, N)\, ds.$$

(See Exercise 19, p. 365.) This may be rewritten

(20) $$\iint_S \operatorname{div} F\, dx_1\, dx_2 = \oint (F, N)\, ds.$$

In the next sections we shall extend the notion of an integral over a plane set S to a more general surface integral. The resulting extensions of formulas (19) and (20) are called Stokes' theorem and Gauss' theorem, respectively.

EXERCISES

1. If $G(X) = (2, 1, 3)$, verify that $(\nabla, G) = 0$ and determine a vector field F satisfying $\nabla \times F = G$.
2. If $G(X) = (x, y - z, -2z)$, determine a vector field F satisfying $\nabla \times F = G$.
3. If $G(X) = (y - x, z + x, y + z)$, determine a vector field F satisfying $\nabla \times F = G$.
4. If $G = F + \nabla f$, show that $\nabla \times G = \nabla \times F$. Conversely, if $\nabla \times G = \nabla \times F$ in a star-shaped region U, show that $G = F + \nabla f$ for some scalar function f.
5. Determine all vector fields F satisfying $\nabla \times F = (x^2 y, z - xy, zx - 2xyz)$.
6. Let f be a harmonic function defined on the star-shaped open set U. That is, $\nabla^2 f = 0$ everywhere on U. Show that $\nabla f = \operatorname{curl} F$ for some vector field F.
7. Let F and G be vector fields defined on the star-shaped open set U such that $\operatorname{curl} F = 0$ and $F = \operatorname{curl} G$. Show that F is the gradient of a harmonic function f.
8. Let F be a twice continuously differentiable vector field. Show that $\nabla \times (\nabla \times F) = \nabla(\nabla, F) - \nabla^2 F$ where $F = (F_1, F_2, F_3)$ and

$$\nabla^2 F = (\nabla^2 F_1, \nabla^2 F_2, \nabla^2 F_3).$$

9. Let F and G be vector fields defined in a star-shaped open set U. Suppose $(\nabla, F) = 0$ and $\nabla \times G = 0$. Then $F = \nabla \times H$ and $G = \nabla f$ for some vector field H and scalar function f. Show that $\nabla^2 f = \operatorname{div}(F + G)$ and $\nabla(\nabla, H) - \nabla^2 H = \nabla \times (F + G)$.
10. Let F and G be continuously differentiable vector fields. Verify that $(\nabla, F \times G) = (\nabla \times F, G) - (F, \nabla \times G)$.
11. If C is a constant vector and F is a continuously differentiable vector field, show that $\nabla \times (C \times F) = (\operatorname{div} F)C - F'(X)C$ where $F'(X)C$ is the Jacobian matrix $(\partial F_i / \partial x_j)$ applied to the vector C.
12. Let F be a vector field. If for the fixed vector $Y \neq 0$, $\nabla \times (Y \times F(X)) = 0$ at the point X, show that $(\operatorname{div} F)(X)$ is an eigenvalue for the matrix $F'(X)$ and that Y is an associated eigenvector. (See Exercise 11.)
13. If F and G are vector fields show that $\nabla \times (F \times G)(X) = (\operatorname{div} G)F - G'(X)F - (\operatorname{div} F)G + F'(X)G$.
14. To verify that if $(\nabla, G) = 0$ in a star-shaped open set U, then

$$F(X) = \int_0^1 G(tX) \times (tX)\, dt,$$

has the property that $\nabla \times F = G$ we proceed as follows. First we note that

$$(\nabla \times F)(X) = \int_0^1 \nabla \times (G(tX) \times (tX))\, dt.$$

The proof is similar to that of exercise 15 p. 357 and we omit it.

Next, using the chain rule and Exercise 13 show that

$$\nabla \times (G(tX) \times (tX)) = 2tG(tX) + t^2 G'(tX)X = \frac{d}{dt}(t^2 G(tX)).$$

From this conclude that

$$\int_0^1 \nabla \times (G(tX) \times (tX))\, dt = G(X).$$

The following sequence of exercises indicates the independence of div F and curl F from the particular basis chosen in $V_3(R)$. Let I_1, I_2, I_3 be the canonical basis in $V_3(R)$, and let J_1, J_2, J_3 be another basis. For $X \in V_3(R)$ write

$$X = \sum_{k=1}^{3} x_k I_k = \sum_{k=1}^{3} y_k J_k$$

and

$$F(X) = \sum_{j=1}^{3} F_j(x_1, x_2, x_3) I_j = \sum_{j=1}^{3} \tilde{F}_j(y_1, y_2, y_3) J_j.$$

Let C be the change of basis matrix from the basis I_1, I_2, I_3 to the basis J_1, J_2, J_3.

15. Show that at the point X

$$\left(\frac{\partial \tilde{F}_i}{\partial J_j}\right) = C^{-1}\left(\frac{\partial F_i}{\partial x_j}\right)C.$$

[Remember that $\partial F_i/\partial x_j = \partial F_i/\partial I_j$.]

16. For an $n \times n$ matrix $A = (a_{ij})$ define the *trace* of A by the formula tr $(A) = a_{11} + a_{22} + \cdots + a_{nn}$. Verify that if A and B are $n \times n$ matrices, then

$$\text{tr}(AB) = \text{tr}(BA).$$

Hence conclude that if A and B are similar, then tr $(A) = $ tr (B).

17. Using Exercise 16 conclude that

$$(\text{div } F)(X) = \frac{\partial \tilde{F}_1}{\partial J_1} + \frac{\partial \tilde{F}_2}{\partial J_2} + \frac{\partial \tilde{F}_3}{\partial J_3}.$$

Assume now that the basis J_1, J_2, J_3 is a right-handed orthonormal basis. As a result, if C is the orthogonal matrix having J_1, J_2, J_3 as columns, then the determinant $D(C) = 1$, and C is a rotation matrix. Assume now that C is a rotation matrix, that is C is orthogonal and $D(C) = 1$.

18. If $C = (c_{ij})$ is a rotation matrix, show that $c_{ij} = (-1)^{i+j}D(C_{ij})$ where $D(C_{ij})$ is the i,jth minor of C. (Use the fact that $C^{-1} = C^t$. See Chapter 4, p. 145.)

19. Let

$$A = \left(\frac{\partial F_i}{\partial x_j}\right) - \left(\frac{\partial F_i}{\partial x_j}\right)^t = \begin{pmatrix} 0 & -c & b \\ c & 0 & -a \\ -b & a & 0 \end{pmatrix}$$

and

$$B = \left(\frac{\partial \tilde{F}_i}{\partial J_j}\right) - \left(\frac{\partial \tilde{F}_i}{\partial J_j}\right)^t = \begin{pmatrix} 0 & -\tilde{c} & \tilde{b} \\ \tilde{c} & 0 & -\tilde{a} \\ -\tilde{b} & \tilde{a} & 0 \end{pmatrix}.$$

Verify that $B = C^tAC$.

20. Using the notation of Exercise 19 show that if $(\text{curl } F)(X) = aI_1 + bI_2 + cI_3$ then $(\text{curl } F)(X) = \tilde{a}J_1 + \tilde{b}J_2 + \tilde{c}J_3$. [It suffices to prove that

$$\begin{pmatrix} \tilde{a} \\ \tilde{b} \\ \tilde{c} \end{pmatrix} = C^t \begin{pmatrix} a \\ b \\ c \end{pmatrix}.$$

To establish this you may assume that $(a, b, c) = (1, 0, 0), (0, 1, 0), (0, 0, 1)$ in turn.]

10.2 Surfaces

We defined a smooth curve in space to be a continuously differentiable mapping of an interval $[a, b]$ into $V_3(R)$. A surface, like a curve, is also a continuously differentiable function φ. Indeed φ is defined on a connected open set U in $V_2(R)$ and has values in $V_3(R)$. However, to avoid complications we must further restrict the function φ. Let E be a bounded closed subset of U such that the boundary of E is traversed by finitely many piecewise smooth curves α_i which we assume are oriented positively. This means that the curves traverse the boundary of E in the counterclockwise sense. Then we say that φ defines a *simple smooth surface* with parameter set E if the following two conditions are satisfied.

(1) φ is one-to-one on E.

(2) At each point (u_0, v_0) of E the vectors $\partial \varphi / \partial u$ and $\partial \varphi / \partial v$ are linearly independent.

If either of these conditions fail on a subset of E having zero plane area, then we say that φ defines a surface but not a simple surface. Points where (1) or (2) fail are called *singular points* for the surface. A simple surface is illustrated in Figure 1. A cylinder is an example of a surface which is not simple.

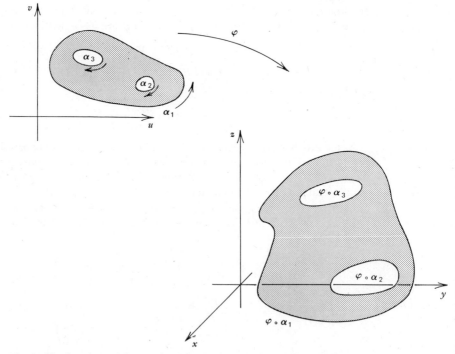

Fig. 1 The function φ defines a simple surface.

For if φ denotes the mapping as in Figure 2, then φ may be thought of as taking a rectangle in the u, v plane and then gluing it together along the edges $u = 0$ and $u = 2\pi$ to form the cylinder. Clearly φ is not one-to-one along the edges $u = 0$ and $u = 2\pi$.

If φ is a simple surface then φ maps the boundary curves α_i of the set E onto curves $\beta_i(t) = \varphi(\alpha_i(t))$ in space. These curves β_i are called the *border* of the simple surface φ.

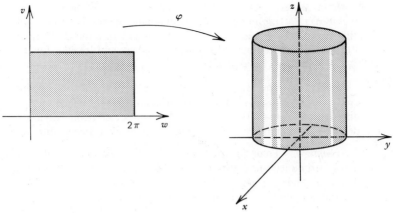

Fig. 2 Cylinder is not a simple surface.

We have already seen that the linear independence of $\partial\varphi/\partial u$ and $\partial\varphi/\partial v$ means that a tangent plane can be defined for the surface at the point $\varphi(u, v)$. This plane is spanned by the vectors $\partial\varphi/\partial u$ and $\partial\varphi/\partial v$. A normal vector to the surface φ at the point $\varphi(u, v)$ is given by the *fundamental vector product* $(\partial\varphi/\partial u) \times (\partial\varphi/\partial v)$. It is well known that $\partial\varphi/\partial u$ and $\partial\varphi/\partial v$ are linearly independent if and only if $(\partial\varphi/\partial u) \times (\partial\varphi/\partial v) \neq 0$. The variable $U = (u, v)$ standing for a point in the domain E of the function φ is often called the *parameter* of the surface φ. The set E is then called the *parameter set* for φ and the set of points $\{\varphi(U) : U \in E\}$ is called the *trace* of φ.

Example 1. If f is a continuously differentiable scalar function defined on a connected open set U, then a simple surface φ may be defined on a bounded closed subset E by setting

$$\varphi(x, y) = (x, y, f(x, y)) \quad \text{for each} \quad X = (x, y) \in E.$$

The paraboloid defined by $z = 1 - x^2 - y^2$ is an example of such a surface. Indeed $\varphi(x, y) = (x, y, 1 - x^2 - y^2)$. The trace of φ is illustrated in Figure 3.

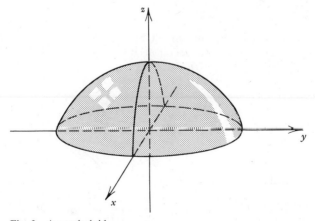

Fig. 3 A paraboloid.

If $\varphi(x, y) = (x, y, f(x, y))$, then $\partial\varphi/\partial x = (1, 0, \partial f/\partial x)$ and $\partial\varphi/\partial y = (0, 1, \partial f/\partial y)$. It is clear that in this case $(\partial\varphi/\partial x) \times (\partial\varphi/\partial y)$ never vanishes.

Example 2. The equation $f(x, y, z) = c$ for a scalar function f may or may not determine a simple surface. If, however, $\nabla f(X_0) \neq 0$ then the implicit function theorem asserts that there is a simple surface φ such that $\varphi(u_0, v_0) = X_0$ and $f(\varphi(u, v)) = c$ for all points (u, v) in a neighborhood of (u_0, v_0). Furthermore if we differentiate the equation $f(\varphi(u, v)) = c$ applying the chain rule we conclude that $\nabla f(X_0)$ is perpendicular to $(\partial \varphi / \partial u)(u_0, v_0)$ and $(\partial \varphi / \partial v)(u_0, v_0)$. Hence ∇f and $(\partial \varphi / \partial u) \times (\partial \varphi / \partial v)$ are parallel.

Example 3. The function $\varphi(x, y) = (x, y, \sqrt{1 - x^2 - y^2})$ defines a spherical surface over any closed parameter set in the open unit disc $x^2 + y^2 < 1$. Since φ is not differentiable on the set $x^2 + y^2 = 1$, φ does not define a surface on this set. To describe the hemisphere or the entire sphere we proceed some what differently.

Example 4. To describe a spherical surface we shall use the geographical coordinates introduced in Section 6-9. In place of the symbols θ, ψ for the angles we shall use u, v. Then, referring to Figure 4, u measures the angle of longitude counterclockwise from the x, z plane and runs between 0 and 2π.

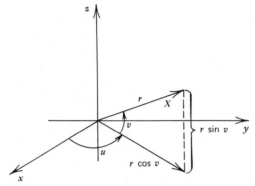

Fig. 4 Geographical coordinates.

The angle of latitude from the equatorial plane (the x, y plane) is measured by v, which runs between $-\pi/2$ and $\pi/2$. If $X = (x, y, z)$ then $r = |X| = \sqrt{x^2 + y^2 + z^2}$. Expressing (x, y, z) in terms of r, u, v we have

$$x = r \cos u \cos v$$
$$y = r \sin u \cos v$$
$$z = r \sin v.$$

Therefore, if we define the function φ by the formula $\varphi(u, v) = (r \cos u \cos v, r \sin u \cos v, r \sin v)$, φ maps the rectangle $0 \leq u \leq 2\pi$, $-\pi/2 \leq v \leq \pi/2$ onto the sphere $x^2 + y^2 + z^2 = r^2$ as in Figure 5.

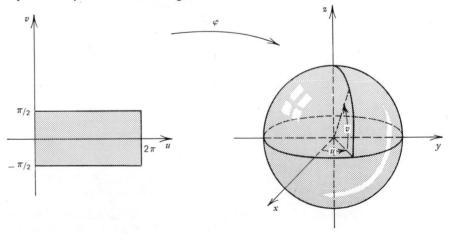

Fig. 5 φ defines a spherical surface.

$x^2 + y^2 + z^2 = r^2$

If we think of the parameter rectangle $0 \leq u \leq 2\pi$, $-\pi/2 \leq v \leq \pi/2$ as a sheet of rubber we may construct our spherical surface φ by gluing together the edges $u = 0$ and $u = 2\pi$ to form the meridian of the sphere. The line $v = \pi/2$ is pinched together to form one point, the north pole. Similarly the points $v = -\pi/2$ are all identified to form the south pole. The function φ maps the line $v = 0$ onto the equator. The upper half of the parameter rectangle is mapped onto the upper hemisphere, the lower onto the lower hemisphere.

Computing $\partial \varphi / \partial u$ and $\partial \varphi / \partial v$ we obtain

$$\frac{\partial \varphi}{\partial u} = (-r \sin u \cos v, r \cos u \cos v, 0)$$

and

$$\frac{\partial \varphi}{\partial v} = (-r \cos u \sin v, -r \sin u \sin v, r \cos v).$$

A short calculation shows that

$$\frac{\partial \varphi}{\partial u} \times \frac{\partial \varphi}{\partial v} = r \cos v (r \cos u \cos v, r \sin u \cos v, r \sin v)$$
$$= r \cos v \, \varphi(u, v).$$

The spherical surface φ is not a simple surface since φ is not one-to-one on the edges of the parameter rectangle. Furthermore, $(\partial \varphi / \partial u) \times (\partial \varphi / \partial v)$ vanishes when $v = \pi/2$ or $v = -\pi/2$. Notice also that φ has no border.

Example 5. For a fixed real number θ, $0 < \theta < \pi/2$, let $\varphi(u, v) = (v \sin \theta \cos u, v \sin \theta \sin u, v \cos \theta)$ be defined on the rectangle $0 \leq u \leq 2\pi$, $0 \leq v \leq h$. It is easily verified that at interior points of the rectangle $(\partial \varphi / \partial u) \times (\partial \varphi / \partial v) \neq 0$. Furthermore, if $\varphi(u, v) = (x, y, z)$ then

$$x^2 + y^2 = v^2 \sin^2 \theta$$

and

$$z^2 = v^2 \cos^2 \theta.$$

Hence $x^2 + y^2 = z^2 \tan^2 \theta$, and the trace of φ is a cone with central angle θ. See Figure 6. The mapping φ can be thought of as gluing together the two vertical edges of the rectangle and then identifying all the points $(u, 0)$ on the bottom.

Example 6. We generalize the preceding example in the following way. Let $\alpha(v) = (\alpha^1(v), \alpha^2(v))$, $a \leq v \leq b$ be a curve in the x, z plane. If this curve is rotated about the z axis, then α determines a surface as in Figure 7. Indeed the function φ which defines this surface is just

$$\varphi(u, v) = (\alpha_1(v) \cos u, \alpha_1(v) \sin u, \alpha_2(v)),$$

where $0 \leq u \leq 2\pi$. The surface φ is not a simple surface since for $a \leq v \leq b$ $\varphi(0, v) = \varphi(2\pi, v)$.

Often in applications a set of points S satisfying an equation $f(x, y, z) = c$ will be specified. The problem then will be to determine a surface having that set S as a trace. This is completely analagous to specifying a set of points in the plane satisfying $f(x, y) = c$ and then asking for a curve α having this set as trace. The curve α is called a parametrization of the given set, and similarly if the trace of φ is S, we say that φ is a *parametrization* of the set S. The set S may be parametrized in many ways, and just as for curves, we must define when two parametrizations are equivalent. We postpone that discussion momentarily.

378 VECTOR FIELDS IN SPACE

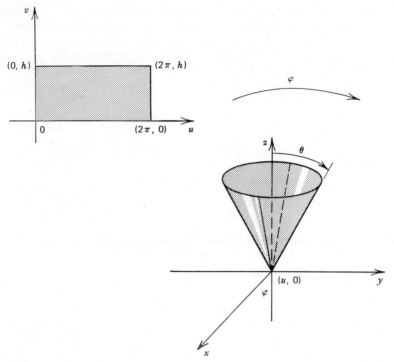

Fig. 6 A conical surface.

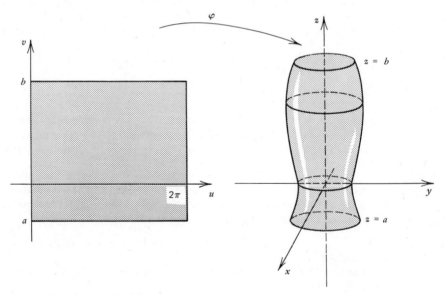

Fig. 7 A surface of revolution.

EXERCISES

1. Consider the spherical surface $\varphi(u, v) = (r \cos u \cos v, r \sin u \cos v, r \sin v)$ defined on the rectangle $0 \leq u \leq 2\pi$, $-\pi/2 \leq v \leq \pi/2$. Why are the boundary points of the rectangle singular points?

2. The conical surface $\varphi(u, v) = (v \sin \theta \cos u, v \sin \theta \sin u, v \cos \theta)$ is defined on the rectangle $0 \leq u \leq 2\pi$, $0 \leq v \leq h$. Determine the singular points for this surface.

3. The function $\varphi(u,v) = (av \cos u, bv \sin u, v^2)$ defines an elliptic paraboloid. De-

termine an equation $f(x, y, z) = k$ satisfied by the trace of φ. Show that $(\partial\varphi/\partial u) \times (\partial\varphi/\partial v)$ never vanishes. Define an appropriate parameter set for φ and determine the singular points.

4. The function $\varphi(u, v) = (v, a \sin u, b \cos u)$ defines an elliptical cylinder. Determine an equation $f(x, y, z) = k$ satisfied by the trace of φ. Show that $(\partial\varphi/\partial u) \times (\partial\varphi/\partial v)$ never vanishes. Define an appropriate parameter set for φ and determine the singular points.

5. Let $\varphi(u, v) = (x_0 + a_1 u + b_1 v,\ y_0 + a_2 u + b_2 v,\ z_0 + a_3 u + b_3 v)$. Determine necessary and sufficient conditions that φ defines a plane through (x_0, y_0, z_0). If φ defines a plane, determine a normal vector to this plane.

6. The function $\varphi(u, v) = (f(v) \cos u, f(v) \sin u, v)$, $f(v) \geq 0$, defines a surface of revolution. Compute $(\partial\varphi/\partial u) \times (\partial\varphi/\partial v)$ and determine when this vector product vanishes.

7. The function $\varphi(u, v) = ((R + r \cos v) \cos u,\ (R + r \cos v) \sin u,\ r \sin v)$ where $r < R$ defines a torus. The parameter set is the rectangle $0 \leq u \leq 2\pi$, $0 \leq v \leq 2\pi$. Sketch the trace of φ and compute $(\partial\varphi/\partial u) \times (\partial\varphi/\partial v)$. If the parameter set is thought of as a sheet of rubber, describe how the surface is formed.

(In Exercises 8-11, determine both the function φ and its parameter set E).

8. Determine a surface φ the trace of which is the ellipsoid $2x^2 + 4y^2 + z^2 = 1$.

9. Determine a surface φ the trace of which is the elliptic cone $2x^2 + y^2 = z^2$.

10. Determine a surface φ the trace of which is obtained by rotating the parabola $z = x^2$ around the z axis.

11. Determine a surface φ the trace of which is obtained by rotating the circle $(x - 2)^2 + z^2 = 1$ around the z axis.

12. If $\varphi(u, v) = (u + v, u - v, uv)$ determine $(\partial\varphi/\partial u) \times (\partial\varphi/\partial v)$ and determine those points (u, v) for which $(\partial\varphi/\partial u)$ and $(\partial\varphi/\partial v)$ are linearly independent.

13. If $\varphi(u, v) = (u + v, u^2 + v^2, u^3 + v^3)$ determine $(\partial\varphi/\partial u) \times (\partial\varphi/\partial v)$ and determine those points (u, v) for which $(\partial\varphi/\partial u)$ and $(\partial\varphi/\partial v)$ are linearly independent.

14. If $\varphi(u, v) = (x_1 + a_1 u + b_1 v,\ x_2 + a_2 u + b_2 v,\ x_3 + a_3 u + b_3 v,\ x_4 + a_4 u + b_4 v)$, determine necessary and sufficient conditions on the coefficients a_i, b_i which insure that φ defines a plane in $V_4(R)$ through the point $X = (x_1, x_2, x_3, x_4)$. If φ does define a plane, what is the maximum number of linearly independent normal vectors to this plane?

10.3 Surface Area

If A and B are vectors in space, then we verified on p. 39 that

(1) $$|A \times B|^2 = |A|^2 |B|^2 - (A, B)^2.$$

If θ is the angle between the vectors A and B, we may rewrite (1) in the following form

$$|A \times B|^2 = |A|^2 |B|^2 (1 - \cos^2 \theta) = |A|^2 |B|^2 \sin^2 \theta.$$

Hence $|A \times B| = |A| |B| \sin \theta$, and this is the area of the parallelogram formed by A and B. See Figure 8.

If now φ is a linear transformation from $V_2(R)$ to $V_3(R)$ with matrix

$$\begin{pmatrix} a_{11} & a_{12} \\ a_{21} & a_{22} \\ a_{31} & a_{33} \end{pmatrix},$$

then φ maps the vector $(1, 0)$ onto $A_1 = (a_{11}, a_{21}, a_{31})$ and $(0, 1)$ onto the vector $A_2 = (a_{12}, a_{22}, a_{32})$. Moreover, the rectangle formed by the unit coordinate vectors I_1 and I_2 is mapped onto the parallelogram formed by A_1 and A_2.

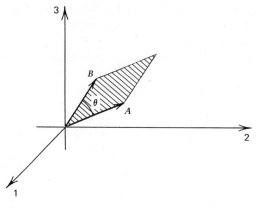

Fig. 8

If we consider this linear transformation φ to define a surface, then the columns of the matrix for φ are just the partial derivatives $\partial \varphi/\partial u$ and $\partial \varphi/\partial v$. Furthermore, if φ is defined on the unit rectangle, then the area A of the resulting surface is given by

$$A = \left|\frac{\partial \varphi}{\partial u} \times \frac{\partial \varphi}{\partial u}\right| = \int_0^1 \int_0^1 \left|\frac{\partial \varphi}{\partial u} \times \frac{\partial \varphi}{\partial v}\right| du\, dv.$$

Indeed if the linear function (surface) is defined on any rectangle R, then the area A of the resulting surface is

$$A = \iint_R \left|\frac{\partial \varphi}{\partial u} \times \frac{\partial \varphi}{\partial v}\right| du\, dv.$$

Now let φ be an arbitrary continuously differentiable function defined on a parameter set E (Figure 9). To define the area of this surface φ we proceed

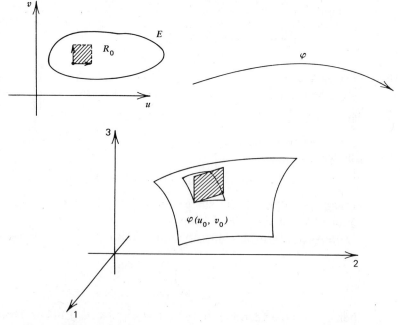

Fig. 9 The affine approximation to φ at (u_0, v_0) maps R_0 onto the shaded parallelogram.

as follows. First we observe that φ may be approximated locally by the affine function $\varphi(u_0, v_0) + d_{U_0}\varphi(U - U_0)$ where $U = (u, v)$ and $U_0 = (u_0, v_0)$. Hence it is reasonable to assert that the area of the surface φ confined to a small rectangle R_0 should be a number which can be approximated by

$$\left| \frac{\partial \varphi}{\partial u} \times \frac{\partial \varphi}{\partial v} \right| (u_0, v_0) \cdot A(R_0)$$

where $A(R_0)$ is the area of the rectangle R_0. Since area is additive, the total area of the surface φ should be a number A which can be approximated arbitrarily well by sums of the form.

(1) $$\sum_i \left| \frac{\partial \varphi}{\partial u} \times \frac{\partial \varphi}{\partial v} \right| (u_i, v_i) A(R_i).$$

where the rectangles R_i form a grid enclosing the parameter set E and each (u_i, v_i) is a point in R_i. Replacing the sums (1) by an appropriate integral we are led to define the area A of the surface φ defined on the parameter set E to be

$$A = \iint_E \left| \frac{\partial \varphi}{\partial u} \times \frac{\partial \varphi}{\partial v} \right| du\, dv.$$

If the function φ takes the form

$$\varphi(u, v) = (u, v, f(u, v))$$

then

$$\left| \frac{\partial \varphi}{\partial u} \times \frac{\partial \varphi}{\partial v} \right| = \left| D \begin{pmatrix} I_1 & I_2 & I_3 \\ 1 & 0 & \frac{\partial f}{\partial u} \\ 0 & 1 & \frac{\partial f}{\partial v} \end{pmatrix} \right| = \sqrt{\left(\frac{\partial f}{\partial u}\right)^2 + \left(\frac{\partial f}{\partial v}\right)^2 + 1}.$$

Hence letting $(u, v) = (x, y)$ we have that the area A of such a surface has the formula

$$A = \iint_E \left(1 + \left(\frac{\partial f}{\partial x}\right)^2 + \left(\frac{\partial f}{\partial y}\right)^2\right)^{1/2} dx\, dy$$

We may represent such a surface as in Figure 10.

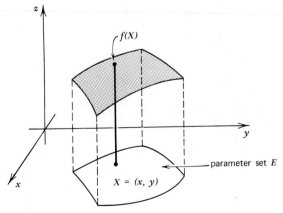

Fig. 10

In the next section we shall develop a notion of equivalence for surfaces similar to the idea of equivalence of curves discussed in Section 9.4. It will follow from this discussion that if two surfaces φ and ψ are equivalent, then they have the same surface area. Before giving some examples, note that if the trace of a surface is defined by an equation $f(x, y, z) = c$, one must first determine a parametrization φ of this set before the surface area can be computed.

Example 1. Determine the area of that portion of the hyperbolic paraboloid $z = y^2 - x^2$ lying within the cylinder $x^2 + y^2 = 1$. In this case the surface φ is defined by

$$\varphi(x, y) = (x, y, y^2 - x^2)$$

and the parameter set E is the unit disc $x^2 + y^2 \leq 1$. The fundamental vector product

$$\left| \frac{\partial \varphi}{\partial x} \times \frac{\partial \varphi}{\partial y} \right| = \sqrt{1 + 4(x^2 + y^2)}$$

and

$$A = \iint_E \sqrt{1 + 4(x^2 + y^2)} \, dx \, dy.$$

This integral may be evaluated directly but it is more convenient to transform it to polar coordinates. Since $x = r \cos \theta$ and $y = r \sin \theta$, we have by the change of variable formula for double integrals,

$$A = \int_0^1 \int_0^{2\pi} \sqrt{1 + 4r^2} \, r \, dr \, d\theta = 2\pi \int_0^1 \sqrt{1 + 4r^2} \, r \, dr$$

$$= \frac{\pi}{6} (1 + 4r^2)^{3/2} \Big|_0^1 = \frac{\pi}{6} (5\sqrt{5} - 1).$$

This surface is illustrated in Figure 11.

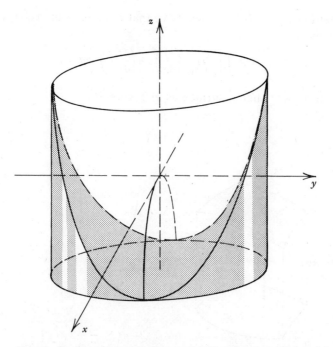

Fig. 11 Portion of a hyperbolic paraboloid within the cylinder $x^2 + y^2 = 1$.

SURFACE AREA

Example 2. Determine the surface of a sphere of radius a. We have already seen that the set of points $x^2 + y^2 + z^2 = a^2$ is parametrized by the surface

$$\varphi(u, v) = (a \cos u \cos v, a \sin u \cos v, a \sin v).$$

where $0 \le u \le 2\pi$ and $-\pi/2 \le v \le \pi/2$ (see Figure 5, p. 376). Moreover, we noted on p. 377 that

$$\frac{\partial \varphi}{\partial u} \times \frac{\partial \varphi}{\partial v} = a \cos v \, \varphi(u, v).$$

Since $|\varphi(u, v)| = a$, $|(\partial \varphi/\partial u) \times (\partial \varphi/\partial v)| = a^2 |\cos v|$. Hence the surface area A is given by

$$A = \int_{-\pi/2}^{\pi/2} \int_0^{2\pi} a^2 |\cos v| \, du \, dv = 2\pi a^2 \int_{-\pi/2}^{\pi/2} |\cos v| \, dv = 4\pi a^2.$$

EXERCISES

Some of the integrals in the following exercises may be simplified by a change to polar coordinates.

1. Determine the area of that portion of the plane $x + 2y - z = 1$ lying within the cylinder defined by $-1 \le x, y \le 1$.

2. Determine the area of that portion of the plane $x + y - z = 0$ which lies within the right circular cylinder $x^2 + y^2 + ax = 0$.

3. Determine the area of a right circular cone of radius r and central angle θ. (The central angle is the angle between the axis of the cone and a line in the surface which passes through the vertex.)

4. Determine the area of that portion of the elliptic paraboloid $z = x^2 + y^2$ lying below the plane $z = 2$.

5. Compute the area of that portion of the cone $z^2 = x^2 + y^2$ lying above the xy plane and within the sphere $x^2 + y^2 + z^2 - 4y = 0$.

6. Compute the area of that portion of the cylinder $x^2 + y^2 = 8y$ which lies within the sphere $x^2 + y^2 + z^2 = 64$.

7. Let $z = f(x)$ be a continuously differentiable function defined on the interval $[a, b]$, $0 < a < b$. Define a curve α in the x, z plane by the formula $\alpha(x) = (x, f(x))$ $a \le x \le b$. Let L be the length of this curve and let $\bar{x}$ be the x coordinate of the centroid of this curve. A surface is formed by rotating this curve about the z axis. Determine a function φ which describes this surface. Show that the area of the resulting surface is given by $2\pi L \bar{x}$. This is a special case of a theorem of Pappus which states that if a plane curve is rotated about a line in the plane of the curve which does not intersect the curve, then the area of the resulting surface is given by $2\pi L d$ when L is the length of the curve and d is the distance from the centroid of the curve to the axis of revolution.

8. Compute the area of the torus

$$\varphi(u, v) = ((R + r \cos v) \cos u, (R + r \cos v) \sin u, r \sin v)$$

where $0 \le u, v \le 2\pi$ and $0 < r < R$. Check your answer by applying the theorem of Pappus (Exercise 7).

9. One arch of the cycloid $\alpha(t) = (a(t - \sin t), a(1 - \cos t)) = (x, z), 0 \le t \le 2\pi$, is revolved about the z axis. Write a function φ which defines this surface. Determine an integral which respesents the area of the resulting surface. Do not attempt to evaluate the integral.

10. Let $z = f(x) > 0$ for $a < x < b$. If $\alpha(x) = (x, z) = (x, f(x))$ for $a \le x \le b$, form a surface by rotating this curve about the x axis. Determine a function φ which describes this surface and verify the theorem of Pappus (Exercise 7).

11. A sphere of radius a is cut by two parallel planes of distance d apart. Determine

the area of that portion of the surface of the sphere lying between the two planes. Show that this area does not depend on the position of the planes but depends only on the distance d between them. (Utilize the theorem of Pappus to write down the formula for the area.)

12. (The plank problem.) Planks of width $w_1, \ldots, w_n$ are laid down on a disc of radius a in random order as in Figure 12.

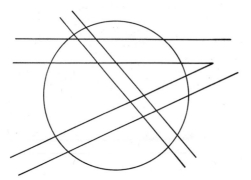

Fig. 12

Show that if the planks cover the disc, then $w_1 + \cdots + w_n \geq 2a$. (This result seems obvious. However, once a proof is thought to be necessary, it is by no means clear how to proceed since there is so little information to go on. However, the result follows readily from Exercise 11.)

This "plank problem" was posed by Alfred Tarski in 1932 for convex sets. Define the width w of a convex set S to be the width of the smallest plank containing S. If S is covered by n planks (strips) of width $w_1, \ldots, w_n$ then must $w_1 + \cdots + w_n \geq w$? This is much harder than the argument necessary to prove Exercise 12. The question was answered affirmatively by the Danish mathematician, Thøger Bang. His proof appears in the *Proceedings of the American Mathematical Society*, volume 2, 1951.

10.4 Surface Integrals

If α is a curve in space, and F is a vector field defined on the trace of α, then we have defined the line integral $\int_\alpha F$ of the vector field F by the formula

(1) $$\int_\alpha F = \int_a^b (F(\alpha(t)), \alpha'(t))\, dt$$

where $[a, b]$ is the parameter interval of the curve α. If f is a scalar function, then the line integral $\int_\alpha f$ is defined to be $\int_a^b f(\alpha(t))|\alpha'(t)|\, dt$. In addition, we have verified that both line integrals have the same value if α is replaced by an equivalent curve β. Furthermore, if we reverse the direction along the curve α, then the sign of the integral in (1) is reversed. Our next project is to extend this notion of an integral along a curve to an integral over a surface.

Before discussing the notion of a surface integral we must define the notion of *orientation* for the border of a simple surface φ. We say that the border of φ is *positively oriented* if the boundary curves α_i of the parameter set E for φ are positively oriented. (See p. 358). This means that the curves α_i traverse the boundary of E in a counterclockwise direction relative to E. If the boundary curves α_i are negatively oriented then we say the border of φ is *negatively oriented*.

We may interpret this notion of orientation for a simple surface geometrically in another way. At each point of the border of φ we construct a right-handed coordinate system consisting of the vectors N, T and $N \times T$ where

$$N = \frac{\dfrac{\partial \varphi}{\partial u} \times \dfrac{\partial \varphi}{\partial v}}{\left|\dfrac{\partial \varphi}{\partial u} \times \dfrac{\partial \varphi}{\partial v}\right|}$$

and T is the unit tangent to the border curve. It follows by the continuity of φ that at each point of the border $N \times T$ always points *toward* the surface or always points *away* from the surface. If the border of φ is positively oriented, then $N \times T$ points towards the surface. If it is negatively oriented, $N \times T$ points away. Thus at a point on the border of φ if N is considered to be the upward direction and T the forward direction then the surface will lie to the left if the border of φ is positively oriented and to the right if it is negatively oriented. See Figures 13 and 14. The proof of these assertions are not hard but we omit them principally because we wish to avoid making precise what we mean by the statement "$N \times T$ points in the direction of the surface."

If φ is a given surface, then it is useful to have a standard way of constructing a surface $\bar\varphi$ having the same trace as φ but the border has opposite

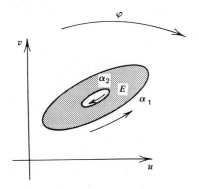

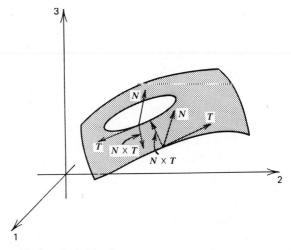

Fig. 13 A surface with positively oriented border.

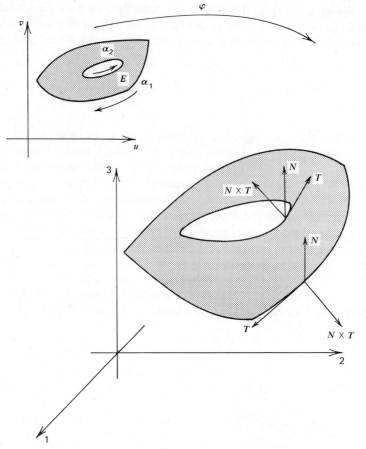

Fig. 14 A surface with negatively oriented border.

orientation. If E is the parameter set for φ, let $E_- = \{(u, -v) : (u, v) \in E\}$. For $(u, v) \in E_-$ define $\varphi_-(u, v) = \varphi(u, -v)$. It is easily checked that φ and φ_- have the same trace. The border curves run in the same direction but the normals N and N_- have opposite direction. See Figure 15. The verification of these facts we leave as an exercise.

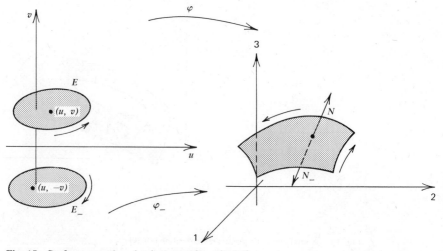

Fig. 15 Surfaces φ and φ_- having opposite orientation.

We now define surface integrals. First, if f is a continuous scalar function defined on the trace of φ we define

$$\text{(1)} \quad \iint_\varphi f = \iint_E f(\varphi(u,v)) \left| \frac{\partial \varphi}{\partial u} \times \frac{\partial \varphi}{\partial v} \right| du\, dv$$

where E is the parameter set for φ. If we write

$$dS = \left| \frac{\partial \varphi}{\partial u} \times \frac{\partial \varphi}{\partial v} \right| du\, dv.$$

then we have

$$\iint_\varphi f = \iint_E f\, dS.$$

For a vector field F defined on the trace of φ we define

$$\iint_\varphi F = \iint_E (F, N)\, dS.$$

Since

$$N = \frac{\dfrac{\partial \varphi}{\partial u} \times \dfrac{\partial \varphi}{\partial v}}{\left| \dfrac{\partial \varphi}{\partial u} \times \dfrac{\partial \varphi}{\partial v} \right|},$$

$$\iint_E (F, N)\, dS = \iint_E \left(F(\varphi(u,v)), \frac{\dfrac{\partial \varphi}{\partial u} \times \dfrac{\partial \varphi}{\partial v}}{\left| \dfrac{\partial \varphi}{\partial u} \times \dfrac{\partial \varphi}{\partial v} \right|} \right) \left| \frac{\partial \varphi}{\partial u} \times \frac{\partial \varphi}{\partial v} \right| du\, dv$$

$$= \iint_E \left(F(\varphi(u,v)), \frac{\partial \varphi}{\partial u} \times \frac{\partial \varphi}{\partial v} \right) du\, dv.$$

Hence

$$\text{(2)} \quad \iint_\varphi F = \iint_E \left(F(\varphi(u,v)), \frac{\partial \varphi}{\partial u} \times \frac{\partial \varphi}{\partial v} \right) du\, dv.$$

To actually conclude that the integrals in (1) and (2) exist we must know that the functions $(F, (\partial \varphi/\partial u) \times (\partial \varphi/\partial v))$ and $f \cdot |(\partial \varphi/\partial u) \times (\partial \varphi/\partial v)|$ are integrable functions on the parameter set E. This will be guaranteed if F, f, and $(\partial \varphi/\partial u) \times (\partial \varphi/\partial v)$ are continuous on E. This continuity condition may of course be relaxed somewhat. Our position throughout this discussion will be to assume that the functions $(F, (\partial \varphi/\partial u) \times (\partial \varphi/\partial v))$ and $f \cdot |(\partial \varphi/\partial u) \times (\partial \varphi/\partial v)|$ are always continuous over the parameter sets in question.

Surface integrals may be interpreted in much the same way as line integrals. If the scalar function f is a mass density function for the surface φ, then $\iint_\varphi f$ is the total mass of the surface. The center of mass of φ is defined to be that point $\bar{X} = (\bar{x}_1, \bar{x}_2, \bar{x}_3)$ satisfying the equation

$$\bar{x}_i \iint_\varphi f = \iint_\varphi x_i f.$$

The right-hand side of this expression is the integral of the function

$$g_i(X) = x_i f(x_1, x_2, x_3) \qquad i = 1, 2, 3$$

over the surface φ. If the density f is constant, then $\bar{X}$ is called the *centroid* of the surface φ. The moment of inertia of the surface φ with respect to a line or plane is the surface integral $\iint_\varphi f \cdot d^2$ where f is the mass density function for φ and $d(X)$ is the distance from a point X on the surface to the specified line or plane.

If the vector field F is interpreted as the velocity field of a fluid, then at a point $\varphi(u, v)$ on the surface, the inner product

$$(F, N) = \left(F, \frac{\frac{\partial \varphi}{\partial u} \times \frac{\partial \varphi}{\partial v}}{\left| \frac{\partial \varphi}{\partial u} \times \frac{\partial \varphi}{\partial v} \right|} \right)$$

is that component of the velocity F in the direction of the normal N to the surface at the point $\varphi(u, v)$. The integral

$$\iint_\varphi F = \iint_E \left(F, \frac{\partial \varphi}{\partial u} \times \frac{\partial \varphi}{\partial v} \right) du\, dv$$

$$= \iint_E \left(F, \frac{\frac{\partial \varphi}{\partial u} \times \frac{\partial \varphi}{\partial v}}{\left| \frac{\partial \varphi}{\partial u} \times \frac{\partial \varphi}{\partial v} \right|} \right) \left| \frac{\partial \varphi}{\partial u} \times \frac{\partial \varphi}{\partial v} \right| du\, dv$$

measures the total rate of flow of the fluid through the surface φ. This quantity is called the *flux* of the vector field F over the surface φ.

If we reverse the orientation of a surface φ we leave it as an exercise (number 11) to show that this reverses the sign of a surface integral $\iint_\varphi F$ of a vector field. The sign of the surface integral of a scalar function, however, is left unchanged.

Next we introduce a notion of equivalence for surfaces in much the same way as we did for curves. Let φ and ψ be surfaces with closed parameter sets E_1 and E_2, respectively. The surfaces φ and ψ are said to *be equivalent* if the following condition is satisfied. There exists a continuously differentiable function $G(u, v) = (G_1(u, v), G_2(u, v))$ mapping E_1 onto E_2 in a one-to-one fashion such that at each point X of E_1 the Jacobian $\partial(G_1, G_2)/\partial(u, v) > 0$. Furthermore for each $U = (u, v) \in E$,

$$\varphi(U) = \psi(G(U)).$$

If φ and ψ are equivalent, written $\varphi \sim \psi$, then the function G is called a change of parameter for the surfaces. Hence two surfaces are equivalent, if one can be obtained from the other by a change of parameter having a *positive* Jacobian. The next result shows that if φ and ψ are equivalent surfaces and F is a vector field defined on the common trace then $\int_\varphi F = \int_\psi F$. The result we obtain follows readily from the change of variable formula for double integrals.

THEOREM 10.1 *Let φ and ψ be surfaces defined on bounded closed parameter sets E_1 and E_2, respectively. Assume that there exists a continuously differentiable function G mapping E_1 onto E_2 and let G_1, G_2 be the coordinate functions for G. Assume further that, except possibly for a set of zero plane area, G is one-to-one and*

$$\frac{\partial(G_1, G_2)}{\partial(u, v)} > 0.$$

If for each point $U = (u, v)$ in G, $\varphi(U) = \psi(G(U))$, then

$$\iint_\varphi F = \iint_\psi F \quad \text{and} \quad \iint_\varphi f = \iint_\psi f$$

for each continuously differentiable vector field F or scalar function f defined on the common trace of φ and ψ.

Proof. We verify the equality of the integrals for a vector field F leaving the analogous verification for a scalar function f as an exercise. By definition

$$\iint_\varphi F = \iint_{E_1} \left(F, \frac{\partial \varphi}{\partial u} \times \frac{\partial \varphi}{\partial v} \right) du\, dv$$

and

$$\iint_\psi F = \iint_{E_2} \left(F, \frac{\partial \psi}{\partial x} \times \frac{\partial \psi}{\partial y} \right) dx\, dy$$

where $U = (u, v)$ and $X = (x, y)$ are points of E_1 and E_2, respectively. Applying the change of variable formula for double integrals, Theorem 8.5, we see that

$$\iint_\psi F = \iint_{E_2} \left(F, \frac{\partial \psi}{\partial x} \times \frac{\partial \psi}{\partial y} \right) dx\, dy$$

$$= \iint_{E_1} \left(F, \frac{\partial \psi}{\partial x} \times \frac{\partial \psi}{\partial y} \right) \frac{\partial (G_1, G_2)}{\partial (u, v)} du\, dv.$$

This latter integral will equal

$$\iint_{E_1} \left(F, \frac{\partial \varphi}{\partial u} \times \frac{\partial \varphi}{\partial v} \right) du\, dv = \iint_\varphi F$$

if we can show that at each point $X = G(U)$

(3) $$\left(F, \frac{\partial \varphi}{\partial u} \times \frac{\partial \varphi}{\partial v} \right) = \left(F, \frac{\partial \psi}{\partial x} \times \frac{\partial \psi}{\partial y} \right) \frac{\partial (G_1, G_2)}{\partial (u, v)}.$$

However,

(4) $$\frac{\partial \varphi}{\partial u} \times \frac{\partial \varphi}{\partial v} = \frac{\partial (\varphi_2, \varphi_3)}{\partial (u, v)} I_1 - \frac{\partial (\varphi_1, \varphi_3)}{\partial (u, v)} I_2 + \frac{\partial (\varphi_1, \varphi_2)}{\partial (u, v)} I_3.$$

For each pair of indices i and j it follows by the chain rule (p. 229) that

$$\begin{pmatrix} \frac{\partial \varphi_i}{\partial u} & \frac{\partial \varphi_i}{\partial v} \\ \frac{\partial \varphi_j}{\partial u} & \frac{\partial \varphi_j}{\partial v} \end{pmatrix} = \begin{pmatrix} \frac{\partial \psi_i}{\partial x} & \frac{\partial \psi_i}{\partial y} \\ \frac{\partial \psi_j}{\partial x} & \frac{\partial \psi_j}{\partial y} \end{pmatrix} \begin{pmatrix} \frac{\partial G_1}{\partial u} & \frac{\partial G_1}{\partial v} \\ \frac{\partial G_2}{\partial u} & \frac{\partial G_2}{\partial v} \end{pmatrix}.$$

Taking the determinants of these matrices and using the multiplication theorem for determinants we have

(5) $$\frac{\partial (\varphi_i, \varphi_j)}{\partial (u, v)} = \frac{\partial (\psi_i, \psi_j)}{\partial (x, y)} \frac{\partial (G_1, G_2)}{\partial (u, v)}.$$

Substituting (5) for approprimate pair of indices i and j in (4) yields that

$$\frac{\partial \varphi}{\partial u} \times \frac{\partial \varphi}{\partial v} = \frac{\partial (G_1, G_2)}{\partial (u, v)} \frac{\partial \psi}{\partial x} \times \frac{\partial \psi}{\partial y}.$$

Taking the inner product of (4) with F we have (3).

Example 1. Determine the z coordinate of the centroid of the hemisphere $x^2 + y^2 + z^2 = a^2$, $z \geq 0$. If $\varphi(u,v) = (a \cos u \cos v,\ a \sin u \cos v,\ a \sin v)$ and $0 \leq u \leq 2\pi$, $0 \leq v \leq \pi/2$, then φ has the hemisphere as trace. (See p. 376.) We have already seen that the area of the hemisphere is $2\pi a^2$. Hence to compute $\bar{z}$ we must compute

$$2\pi a^2 \bar{z} = \iint_\varphi z.$$

On the surface of the hemisphere $z = a \sin v$. Since $|(\partial \varphi/\partial u) \times (\partial \varphi/\partial v)| = a^2 \cos v$

$$\iint_\varphi z = \int_0^{\pi/2} \int_0^{2\pi} a \sin v \left| \frac{\partial \varphi}{\partial u} \times \frac{\partial \varphi}{\partial v} \right| du\, dv$$

$$= \int_0^{\pi/2} \int_0^{2\pi} a \sin v\, a^2 |\cos v|\, du\, dv$$

$$= 2\pi a^3 \int_0^{\pi/2} \sin v \cos v\, dv = \pi a^3.$$

Therefore

$$\bar{z} = \frac{1}{2\pi a^2} \iint_\varphi z = \frac{\pi a^3}{2\pi a^2} = \frac{a}{2}.$$

Example 2. If φ is the hemispherical surface of the previous example, and $F(X) = X$; compute $\iint_\varphi F$. Since $F(X) = X$,

$$\left(F, \frac{\partial \varphi}{\partial u} \times \frac{\partial \varphi}{\partial v} \right) = \left(\varphi, \frac{\partial \varphi}{\partial u} \times \frac{\partial \varphi}{\partial v} \right)$$

$$= D \begin{pmatrix} a \cos u \cos v & a \sin u \cos v & a \sin v \\ -a \sin u \cos v & a \cos u \cos v & 0 \\ -a \cos u \sin v & -a \sin u \sin v & a \cos v \end{pmatrix}$$

$$= a^3 \cos^2 u \cos^3 v + a^3 \sin^2 u \cos v + a^3 \cos^2 u \sin^2 v \cos v$$
$$= a^3 \cos v.$$

Therefore

$$\iint_\varphi F = a^3 \int_0^{\pi/2} \int_0^{2\pi} \cos v\, du\, dv = 2\pi a^3 \int_0^{\pi/2} \cos v\, dv = 2\pi a^3.$$

If we interpret $F(X) = X$ as the velocity of a fluid flowing through the sphere, then the total flux of the flow through the hemispherical surface is $\iint_\varphi F = 2\pi a^3$.

Often a set of points in space is made up of the traces of several (simple) surfaces $\varphi_1, \ldots, \varphi_n$ which overlap only on their respective borders. A cube, a box, or a cylindrical can are three familiar examples. We say that the surfaces $\varphi_1, \ldots, \varphi_n$ define a *piecewise smooth* surface φ if the following two conditions are satisfied.

(1) The borders of each of the surfaces $\varphi_1, \ldots, \varphi_n$ have the same orientation.

(2) The border curves α_i of the surfaces φ_i run in opposite directions along the points of intersection of the surfaces.

We shall call the surfaces $\varphi_1, \ldots, \varphi_n$ making up the piecewise smooth surface φ, the *faces* of the surface φ. See Figure 16.

If F is a vector field defined on the trace of a piecewise smooth surface φ, then we define the integral of F over φ to be the sum of the integrals of F over the faces φ_i. Thus

$$\iint_\varphi F = \sum_{i=1}^n \iint_{\varphi_i} F.$$

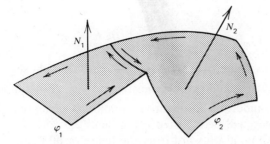

Fig. 16 A piecewise smooth surface φ.

We adopt the same convention for scalar functions as well. Hence

$$\iint_\varphi f = \sum_{i=1}^n \iint_{\varphi_i} f.$$

Example 3. Consider a cylindrical can with lateral surface $x^2 + y^2 = 1$, $0 \leq z \leq 4$, no top, and bottom in the x, y plane. Let φ be a piecewise smooth surface with outward pointing normal which parametrizes this set. Compute $\iint_\varphi F$ if $F(X) = X$.

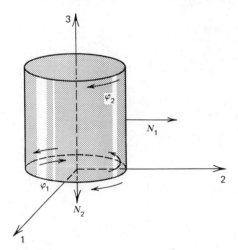

Fig. 17

Let φ_1 parametrize the side of the can and φ_2 the bottom as in Figure 17. Then we may take

$$\varphi_1(u, v) = (\cos u, \sin u, v) \qquad 0 \leq u \leq 2\pi, 0 \leq v \leq 4$$

and $\varphi_2(u, v) = (u, -v, 0)$. Consequently

$$\frac{\partial \varphi_1}{\partial u} \times \frac{\partial \varphi_1}{\partial v} = (\cos u, \sin u, 0)$$

and

$$\frac{\partial \varphi_2}{\partial u} \times \frac{\partial \varphi_2}{\partial v} = (0, 0, -1).$$

Clearly both normals point outward from the can. Now

$$\left(F, \frac{\partial \varphi_1}{\partial u} \times \frac{\partial \varphi_1}{\partial v}\right) = \cos^2 u + \sin^2 v = 1$$

and

$$\left(F, \frac{\partial \varphi_2}{\partial u} \times \frac{\partial \varphi_2}{\partial v}\right) = 0.$$

Hence
$$\iint_\varphi F = \iint_{\varphi_1} F = \int_0^4 \int_0^{2\pi} 1 \cdot du\,dv = 8\pi.$$

If in the above example the surfaces φ_1, φ_2 had been chosen with inward pointing normals, this would reverse the sign of $\iint_\varphi F$. For simple examples such as a cube, cylinder, or tetrahedron, conditions (1) and (2) for a piecewise smooth surface will be satisfied if the faces $\varphi_1, \ldots, \varphi_n$ are chosen so that the normals all point outward from the figure or all point inward.

There are several common notations for the surface integral
$$\iint_\varphi F = \iint_E (F, N)\,dS.$$

One is
$$\iint_\varphi (F, N)\,dS.$$

Another is
$$\iint_\varphi (F_1\,dx_2\,dx_3 + F_2\,dx_3\,dx_1 + F_3\,dx_1\,dx_2).$$

To justify this notation write
$$\frac{\partial \varphi}{\partial u} \times \frac{\partial \varphi}{\partial v} = \left(\frac{\partial(\varphi_2, \varphi_3)}{\partial(u, v)}, \frac{\partial(\varphi_3, \varphi_1)}{\partial(u, v)}, \frac{\partial(\varphi_1, \varphi_2)}{\partial(u, v)} \right)$$
$$= \left(\frac{\partial(x_2, x_3)}{\partial(u, v)}, \frac{\partial(x_3, x_1)}{\partial(u, v)}, \frac{\partial(x_1, x_2)}{\partial(u, v)} \right)$$

and set
$$dx_i\,dx_j = \frac{\partial(x_i, x_j)}{\partial(u, v)}\,du\,dv.$$

Then
$$\iint_\varphi F = \iint_E \left(F, \frac{\partial \varphi}{\partial u} \times \frac{\partial \varphi}{\partial v} \right) du\,dv$$
$$= \iint_E \left(F_1 \frac{\partial(x_2, x_3)}{\partial(u, v)} + F_2 \frac{\partial(x_3, x_1)}{\partial(u, v)} + F_3 \frac{\partial(x_1, x_2)}{\partial(u, v)} \right) du\,dv$$
$$= \iint_\varphi (F_1\,dx_2\,dx_3 + F_2\,dx_3\,dx_1 + F_3\,dx_1\,dx_2).$$

In this notation
$$\iint_\varphi \operatorname{curl} F = \sum_{i \neq j} \iint_\varphi \frac{\partial F_i}{\partial x_j}\,dx_j\,dx_i.$$

We leave the verification of this formula as an exercise.

EXERCISES

1. Determine the centroid of that portion of the sphere $x^2 + y^2 + z^2 = a^2$ lying in the first octant.
2. Determine the centoid of the conical surface $x^2 + y^2 = a^2 z^2$, if $0 \leq z \leq h$.
3. If $F(X) = (x, y, z)$ compute $\iint_\varphi F$ over that portion of the plane $3x + 2y + z = 1$ lying in the first octant, assuming the normal is pointing up.
4. If $F(X) = X$ denotes the velocity of a fluid flow at the point X, compute the total

flux of the flow through the surface of the unit cube $[0, 1] \times [0, 1] \times [0, 1]$. Assume the normal is pointing out.

5. Find the moment of inertia about the x axis of that portion of the plane $x + y + z = 1$ lying in the first octant.

6. Determine the moment of inertia with respect to the yz plane of that portion of the plane $z - x = 0$ lying within the cylinder $x^2 + y^2 = 1$.

7. A tetrahedron is formed by the coordinate planes and the plane $x + y + z = 1$. Find the centroid of this surface and the moment of inertia of the surface with respect to a coordinate axis.

8. Determine the total flux of the vector field $F(X) = X$ over the tetrahedron of Exercise 7, assuming an outward pointing normal.

9. If the surfaces φ and ψ satisfy the hypothesis of Theorem 10.1, show that $\iint_\varphi f = \iint_\psi f$ for each continuous scalar function defined on the common trace of φ and ψ.

10. Show that $\iint_\varphi \operatorname{curl} F = \sum_{i \neq j} \iint_\varphi \frac{\partial F_i}{\partial x_j} dx_j dx_i$.

11. Let F and f be a vector field and a scalar function defined on the trace of a surface φ. Show that $\iint_\varphi F = -\iint_{\varphi_-} F$, and $\iint_\varphi f = \iint_{\varphi_-} f$.

10.5 Stokes' Theorem

Let E be a connected bounded closed set in the plane and assume that the boundary of E is traversed in the counterclockwise direction by finitely many simple closed curves α_i. Denoting the boundary of E by α, we showed in Section 10.1 that Green's theorem could be written

$$\iint_E (\operatorname{curl} F, I_3) \, dx \, dy = \oint_\alpha F.$$

This extension of this result to surface integrals is called Stokes' theorem. We shall restrict ourselves at first to the consideration of simple surfaces. Indeed, if we denote the *border* of the simple surface φ by $\partial \varphi$ and if F is a smooth vector field defined on a neighborhood of the trace of φ, then Stokes' theorem asserts that

$$\iint_\varphi \operatorname{curl} F = \oint_{\partial \varphi} F.$$

The left-hand side of this equation is the *surface* integral of curl F over φ, whereas the right-hand side is the *line* integral of F over the border curves which we denote by $\partial \varphi$. Thus Stokes' theorem is the assertion that the line integral of a vector field F over the border of a surface is the same as the integral over the surface of the curl of F. In order to simplify the proof of this result we shall assume that the simple surface φ has continuous partial derivatives of order two.

THEOREM 10.2 *Let φ be a twice continuously differentiable simple surface with positively oriented border $\partial \varphi$. If F is a continuously differentiable vector field defined on a neighborhood of the trace of φ, then*

(1) $$\iint_\varphi \operatorname{curl} F = \oint_{\partial \varphi} F.$$

Proof. We establish this result by reducing it to Green's theorem in the plane. Let E be the parameter set for φ. Let $\alpha_1, \ldots, \alpha_n$ be the positively oriented curves which make up the boundary of E. Then $\beta_k = \varphi \circ \alpha_k$ are the curves comprising the border $\partial\varphi$ of the surface φ. If we abbreviate

$$\frac{\partial(\varphi_i, \varphi_j)}{\partial(u, v)} du\, dv \quad \text{by} \quad dx_i dx_j$$

then we have noted that (Exercise 10, p. 393)

$$\iint_\varphi \operatorname{curl} F = \sum_{i \neq j} \iint_\varphi \frac{\partial F_i}{\partial x_j} dx_j dx_i.$$

where $F = (F_1, F_2, F_3)$. Moreover

$$\oint_{\partial\varphi} F = \sum_i \oint_{\partial\varphi} F_i dx_i$$

where

$$\oint_{\partial\varphi} F_i dx_i = \sum_{k=1}^n \int_{a_k}^{b_k} F_i(\beta_k(t)) \beta_k^{i\prime}(t)\, dt$$

and $\beta_k(t) = \varphi(\alpha_k(t))$. We prove that for each $i = 1, 2, 3$ and $j \neq i, l \neq i$

(2) $$\iint_\varphi \frac{\partial F_i}{\partial x_j} dx_j dx_i + \iint_\varphi \frac{\partial F_i}{\partial x_l} dx_l dx_i = \oint_{\partial\varphi} F_i dx_i.$$

Setting $i = 1$ for example, this means that we must show that

$$\oint_{\partial\varphi} F_1 dx_1 = \iint_\varphi \frac{\partial F_1}{\partial x_2} dx_2 dx_1 + \iint_\varphi \frac{\partial F_1}{\partial x_3} dx_3 dx_1$$

$$= -\iint_\varphi \frac{\partial F_1}{\partial x_2} dx_1 dx_2 + \iint_\varphi \frac{\partial F_1}{\partial x_3} dx_3 dx_1$$

$$= \iint_E \left(\frac{\partial F_1}{\partial x_3} \cdot \frac{\partial(\varphi_3, \varphi_1)}{\partial(u, v)} - \frac{\partial F_1}{\partial x_2} \cdot \frac{\partial(\varphi_1, \varphi_2)}{\partial(u, v)} \right) du\, dv.$$

Set $G_1 = F_1 \circ \varphi$. Since φ is assumed to be twice continuously differentiable, it follows by the chain rule that

$$\frac{\partial}{\partial u}\left(G_1 \frac{\partial \varphi_1}{\partial v}\right) - \frac{\partial}{\partial v}\left(G_1 \frac{\partial \varphi_1}{\partial u}\right) = \frac{\partial F_1}{\partial x_3} \frac{\partial(\varphi_3, \varphi_1)}{\partial(u, v)} - \frac{\partial F_1}{\partial x_2} \frac{\partial(\varphi_1, \varphi_2)}{\partial(u, v)}.$$

Now α_k are the counterclockwise boundary curves of the parameter set E of φ. Hence we may infer by Green's theorem that

$$\iint_E \left[\frac{\partial F_1}{\partial x_3} \frac{\partial(\varphi_3, \varphi_1)}{\partial(u, v)} - \frac{\partial F_1}{\partial x_2} \frac{\partial(\varphi_1, \varphi_2)}{\partial(u, v)} \right] du\, dv$$

$$= \iint_E \left[\frac{\partial}{\partial u}\left(G_1 \frac{\partial \varphi_1}{\partial v}\right) - \frac{\partial}{\partial v}\left(G_1 \frac{\partial \varphi_1}{\partial u}\right) \right] du\, dv$$

$$= \sum_{k=1}^n \int_{a_k}^{b_k} (G(\alpha_k(t)), \alpha_k'(t))\, dt$$

where the vector field G is defined by

$$G(u, v) = \left(G_1 \frac{\partial \varphi_1}{\partial u}, G_1 \frac{\partial \varphi_1}{\partial v} \right).$$

It now may be easily checked that for each k

(3) $$(G(\alpha_k(t)), \alpha'_k(t)) = F_1(\beta_k(t))\beta_k^{1'}(t)$$

where, of course, $\beta_k = \varphi \circ \alpha_k$. The verification of (3), which we leave as an exercise, establishes (2) for $i = 1$. An analogous argument for $i = 2$ and 3 completes the proof.

This theorem holds for nonsimple surfaces and for surfaces where the differentiability conditions on φ are relaxed. However, the validity of this result does depend on the ability to establish an orientation for the border of the surface φ. If φ is one-to-one, then

$$N(u,v) = \frac{\frac{\partial \varphi}{\partial u} \times \frac{\partial \varphi}{\partial v}}{\left|\frac{\partial \varphi}{\partial u} \times \frac{\partial \varphi}{\partial v}\right|}$$

is uniquely defined for each point (u, v) in the parameter set. However, if φ is not one-to-one, then at points (u, v) and $(\bar{u}, \bar{v})$ which are mapped onto the same point by φ the vectors $N(u, v)$, $N(\bar{u}, \bar{v})$ may be different. This is certainly the case if the surface is not smooth at a point $X_0 = \varphi(u, v) = \varphi(\bar{u}, \bar{v})$. For example the cylinder in Figure 18 exhibits this behavior.

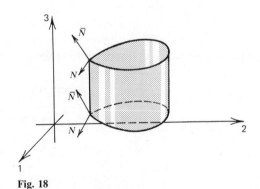

Fig. 18

However, even if the surface φ is smooth at such points the normals N, and $\bar{N}$ may be different. In particular, they may be pointing in opposite directions. The Möbius band in Figure 19 is a familiar example of such a surface.

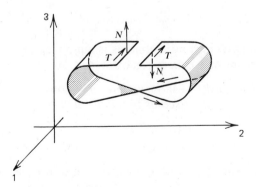

Fig. 19 Möbius band.

If this strip is joined together smoothly, the normals are pointing in opposite direction. Such a surface is called a *nonorientable* surface. Integrals can be defined over such surfaces, but Stokes' theorem is no longer valid. An intuitive reason for this, in the case of the Möbius band, is that the tangent vectors along the edges which are joined together are pointing in the same direction. Hence the line integrals along this common edge do not cancel.

Stokes' theorem can be extended to nonsimple surfaces so long as the border curves which are joined together are running in opposite directions. Such a surface is also orientable. For the following nonsimple but orientable surfaces the validity of Stokes' theorem and formula (1) can be assumed.

As an application of the theorem of Stokes let us return to the problem of determining when a vector field F is the gradient of a scalar function f provided that curl $F = 0$. If F is defined on an open set D in the plane, this will be the case provided that D is simply connected. This result we wish to generalize to three-space.

Assume D is a connected open set in $V_3(R)$. Now $F = \nabla f$ in D if $\oint_\alpha F = 0$ for every closed curve α in D. We have already noted that it is enough to verify that $\oint_\alpha F = 0$ for every closed polygonal curve α. Moreover, the argument on p. 362 shows that it is enough to prove that $\oint_\alpha F = 0$ for every simple closed polygonal curve α lying in D.

We call an open set D in $V_3(R)$ *simply connected* if each simple closed polygonal path in D is the border of a simple surface φ the trace of which lies entirely in D. Intuitively, an open ball or convex set is simply connected, whereas that part of a sphere outside a cylinder running through it is not. See Figure 20.

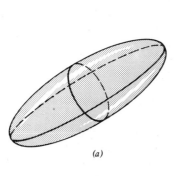
(a)

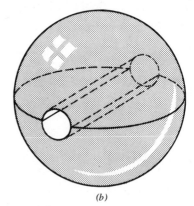
(b)

Fig. 20 (a) Simply connected. (b) Not simply connected.

The region between two concentric spheres is simply connected as is all of three-space with a point or a sphere removed.

THEOREM 10.3 *If curl $F = 0$ throughout a simply connected open set D in $V_3(R)$, then there exists a scalar function f defined on D such that*

$$F(X) = \nabla f(X)$$

for each X in D.

Proof. Let α be a simple closed polygonal path α in D. By the observations made earlier it is enough to show that $\oint_\alpha F = 0$. Let φ be the simple surface with trace in D having α as its border. By reversing the orientation along α if necessary, we may assume φ is positively oriented. Then by Stokes' theorem

$$0 = \iint_\varphi \text{curl } F = \oint_\alpha F$$

since curl $F = 0$ everywhere in D. The result now follows from the corollary to Theorem 9.3.

Next we consider the equation div $F = 0$. If this equation holds in a star-shaped open set, then we have observed that $F = \nabla \times G = \text{curl } G$ for some vector field G. We wish to show that if div $F = 0$ in a simply connected open set D, it is not necessarily true that $F = \nabla \times G$ on D. To do this we observe first that if φ is a spherical surface and F is any continuously differentiable vector field defined on the trace of φ, then

$$\iint_\varphi \text{curl } F = 0.$$

To see this let φ_1, φ_2 be surfaces with positively oriented borders and outward pointing normals which trace out the upper and lower hemispheres, respectively, as in Figure 21. If α_1 and α_2 are the respective borders, then they both trace the equator of the sphere but in opposite directions. If φ denotes the entire spherical surface,

$$\iint_\varphi \text{curl } F = \iint_{\varphi_1} \text{curl } F + \iint_{\varphi_2} \text{curl } F.$$

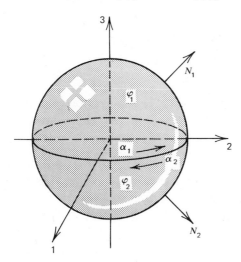

Fig. 21

However, applying Stokes' theorem we infer that

$$\iint_{\varphi_1} \text{curl } F + \iint_{\varphi_2} \text{curl } F = \oint_{\alpha_1} F + \oint_{\alpha_2} F.$$

But α_2 is equivalent to $-\alpha_1$. Hence

$$\oint_{\alpha_2} F = \oint_{-\alpha_1} F = -\oint_{\alpha_1} F.$$

Consequently, $\oint_{\alpha_1} F + \oint_{\alpha_2} F = 0$ and we are done.

Next we wish to show that div $F = 0$ throughout a simply connected open set E in $V_3(R)$ does not necessarily imply that $F = \text{curl } G$. To do this we shall exhibit a vector field F satisfying div $F = 0$ in D but $\iint_\varphi F \neq 0$ where φ is a spherical surface in D. Thus F cannot equal curl G because the integral $\iint_\varphi \text{curl } G$ always vanishes over a spherical surface φ. The appropriate vector field is $F(X) = X/|X|^3$. This is defined on the simply connected open set consisting of all of $V_3(R)$ except the origin. We leave it as an exercise to verify that div $F = 0$, yet if φ is a spherical surface about the origin,

$$\iint_\varphi F = 4\pi.$$

EXERCISES

The following sets E are the traces of surfaces φ with positively oriented borders. Assume the normal vectors $N = (\partial\varphi/\partial u) \times (\partial\varphi/\partial v)$ satisfy the given conditions. For the given vector fields F compute $\iint_\varphi \text{curl } F$ directly and by using Stokes' theorem.

1. $E = \{X : z = 1 - x^2 - y^2, \ z \geq 0\}$, $F(X) = (z^2, xy, xz)$; N has nonnegative z component.

2. $E = \{X : x^2 + y^2 + z^2 = 4, \ z \geq 0\}$, $F(X) = (y, z, x)$, N has nonnegative z component.

3. E is the surface of the cube, without the top, formed by the coordinate planes and the planes $x = 1$, $y = 1$, $z = 1$; N points outward from the cube; $F(X) = (y, z^2, x)$.

4. Construct a surface φ with positively oriented border which has the hemisphere $x^2 + y^2 + z^2 = a^2$, $z \leq 0$ as trace. Assume the normal vector N points toward the origin. Which way does the border curve run in the x, y plane? Compute $\iint_\varphi \text{curl } F$ if $F(X) = (2y, x, z)$.

5. Let φ be a simple surface with border $\partial\varphi$, and let F be a vector field defined on φ. If curl F is tangent to the surface at each point, prove $\oint_\partial F = 0$.

6. Let φ be a spherical surface of radius a. If div $F = 0$ everywhere within the sphere, prove that $\iint_\varphi F = 0$.

7. Let φ be a surface with positively oriented border having that portion of the cylinder $x^2 + y^2 = 1$ between the planes $z = 0$, $z = x + 1$ as trace. If N points out from the surface and $F(X (= (y, z, x)$, compute $\iint_\varphi \text{curl } F$.

8. Let φ be a surface with positively oriented border $\partial\varphi$. Let $N = (\partial\varphi/\partial u) \times (\partial\varphi/\partial v)$ and assume that F is a vector field on φ such that curl $F = kN/|N|$. Show that $\oint_{\partial\varphi} F = kA$ where A is the area of the surface φ.

9. Let $F(X) = X/|X|^3$ for $X = (x, y, z) \neq 0$. Show that div $F = 0$ for $X \neq 0$. If φ is a sphere centered at the origin, show that $\iint_\varphi F = 4\pi$.

10. If φ is a sphere such that the origin lies outside φ and $F(X) = X/|X|^3$, show that $\iint_\varphi F = 0$.

11. Let $F(X) = X/|X|^3$, $X = (x, y, z) \neq 0$. Show that $(\nabla \times F)(X) = 0$. Hence by Theorem 10.3, F is a gradient field on $V_3(R) - \{0\}$. Construct a function f such that $\nabla f(X) = F(X)$ if $X \neq 0$ by writing $f(X)$ as a suitable line integral along a curve α from $(0, 0, 1)$ to $\hat{X}$. [Let α be the z axis from $(0, 0, 1)$ to $(0, 0, |X|)$. From this point to X let α proceed along a circle centered at the origin.]

12. Verify formula (3), p. 395.

10.6 Gauss' Theorem

In our discussion of Green's theorem in Section 9.6 we observed that this result could be interpreted in the following way. If S is a bounded closed set in the plane with boundary curve α running in the counterclockwise direction and N is the unit outer normal, then

$$\iint_E \left(\frac{\partial F_1}{\partial x_1} + \frac{\partial F_2}{\partial x_2}\right) dx_1 \, dx_2 = \oint_\alpha (F, N) \, ds$$

for each continuously differential vector field F defined on E. We wish to extend this result from $V_2(R)$ to $V_3(R)$.

Let U be a bounded open set in $V_3(R)$ and assume that the boundary of U is the trace of a smooth surface φ. Such a surface is called a *closed surface* or a *complete surface* since the surface forms the complete boundary of the open set U. Notice that a closed surface has no border. The sphere, ellipsoid, and torus are all examples of closed surfaces. If the normal vector

$$N = \frac{\dfrac{\partial \varphi}{\partial u} \times \dfrac{\partial \varphi}{\partial v}}{\left|\dfrac{\partial \varphi}{\partial u} \times \dfrac{\partial \varphi}{\partial v}\right|}$$

points out from the closed surface φ, then the surface φ is said to be *positively oriented*. If it points inward, the surface is *negatively oriented*. If F is a continuously differentiable vector field defined on the set E consisting of U together with its boundary, then we wish to show that

$$\iiint_E \left(\frac{\partial F_1}{\partial x_1} + \frac{\partial F_2}{\partial x_2} + \frac{\partial F_3}{\partial x_3}\right) dx_1 \, dx_2 \, dx_3 = \iint_\varphi F = \iint_\varphi (F, N) \, dS.$$

Since

$$\frac{\partial F_1}{\partial x_1} + \frac{\partial F_2}{\partial x_2} + \frac{\partial F_3}{\partial x_3} = \text{div } F,$$

this result asserts that the integral of the divergence of F over the solid region has the same value as the integral of the vector field F over the boundary. This result is called the theorem of Gauss or the divergence theorem. We shall not attempt to prove this result in full generality. However, we shall give an argument valid for a rectangular tetrahedron and for a box. From this the result can be deduced for a polygonal surface. More general surfaces then can be handled by approximation techniques. However, the details of these arguments are not easy.

THEOREM 10.3 *Let φ be the positively oriented surface of a rectangular tetrahedron T. If F is a smooth vector field defined on T, then*

$$\iiint_T \left(\frac{\partial F_1}{\partial x_1} + \frac{\partial F_2}{\partial x_2} + \frac{\partial F_3}{\partial x_3}\right) dx_1 \, dx_2 \, dx_3 = \iint_\varphi F.$$

Proof. We assume that the tetrahedron T has its vertex at the origin and its perpendicular faces along the coordinate planes with the base in the first octant. See Figure 22. Let $\varphi_1, \varphi_2, \varphi_3$ be the triangular faces of the tetrahedron T in the $x_2 x_3$, $x_3 x_1$, $x_1 x_2$ planes, respectively. Let φ_4 be the triangular face

400 VECTOR FIELDS IN SPACE

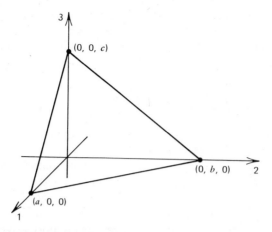

Fig. 22 Tetrahedron T.

having trace in the plane

(1) $$\frac{x_1}{a}+\frac{x_2}{b}+\frac{x_3}{c}=1.$$

By definition

$$\iint_\varphi F = \sum_{i=1}^{4} \iint_{\varphi_i} F.$$

Let E be the triangle in the (u, v) plane determined by the vectors $(1, 0)$ and $(0, 1)$. We assert

(2) $$\iint_{\varphi_1} F = -bc \iint_E F_1(0, bu, cv)\, du\, dv$$

(3) $$\iint_{\varphi_2} F = -ac \iint_E F_2(au, 0, cv)\, du\, dv$$

(4) $$\iint_{\varphi_3} F = -ab \iint_E F_3(au, bv, u)\, du\, dv$$

where F_1, F_2, F_3 are the component functions of the vector field F. To derive the first of these equations let

$$\varphi_1(u, v) = (0, bu, -cv)$$

be defined on the triangle $\tilde{E}$ in the (u, v) plane determined by the vectors $(1, 0)$ and $(0, -1)$. Then φ_1 parametrizes the face of the tetrahedron in the x_2x_3 plane. Furthermore $(\partial\varphi_1/\partial u) \times (\partial\varphi_1/\partial v) = -(bc, 0, 0)$ points outward from the tetrahedron. Hence

$$\left(F, \frac{\partial\varphi_1}{\partial u} \times \frac{\partial\varphi_1}{\partial v}\right) = -bcF_1$$

and

$$\iint_{\varphi_1} F = -bc \iint_{\tilde{E}} F_1(0, bu, -cv)\, du\, dv$$
$$= -bc \iint_E F_1(0, bu, cv)\, du\, dv.$$

The last equality follows from the change of variable formula for double integrals. Formulas (3) and (4) are derived similarly.

To derive a formula for $\iint_{\varphi_4} F$ we must parametrize the oblique face of the tetrahedron. To do this we set

$$\varphi_4(u, v) = (au, bv, c(1-u-v))$$

where (u, v) ranges over the triangle E. Then $(\partial\varphi_4/\partial u) \times (\partial\varphi_4/\partial v) = (bc, ac, ab)$ and

(5) $$\iint_{\varphi_4} F = \iint_E \left(F, \frac{\partial\varphi_4}{\partial u} \times \frac{\partial\varphi_4}{\partial v}\right) du\,dv = \iint_E (bcF_1 + acF_2 + abF_3)\,du\,dv.$$

Adding equations (2), (3), (4), and (5) we obtain

(6) $$\iint_\varphi F = \iint_E (bcF_1 + acF_2 + abF_3)\,du\,dv$$
$$- \iint_E [bcF_1(0, bu, cv) + acF_2(au, 0, cv) + F_3(au, bv, 0)]\,du\,dv.$$

Next we evaluate the volume integral $\iiint_T \mathrm{div}\,F$. To compute $\iiint_T \partial F_1/\partial x_1$ integrate first with respect to x_1 obtaining

(7) $$\iiint_T \frac{\partial F_1}{\partial x_1} = \iint_{E_1} [F_1(x_1, x_2, x_3) - F_1(0, x_2, x_3)]\,dx_2\,dx_3$$

where E_1 is the triangle in the x_2, x_3 plane determined by the vectors $(0, b, 0)$ and $(0, 0, c)$ and x_1 is obtained as a function of x_2, x_3 by solving (1) for x_1. If we set $x_2 = bu$ and $x_3 = cv$, and $x_1 = a(1-u-v)$, then we may write the right hand side of (7) as an integral over E. Using the change of variable formula for double integrals this yields.

(8) $$\iiint_T \frac{\partial F_1}{\partial x_1} = bc \iint_E (F_1(X) - F_1(0, bu, cv))\,du\,dv.$$

Similarly

(9) $$\iiint_T \frac{\partial F_2}{\partial x_2} = ac \iint_E (F_2(X) - F_2(au, 0, cv))\,du\,dv$$

and

(10) $$\iiint_T \frac{\partial F_3}{\partial x_3} = ab \iint_E (F_3(X) - F_3(au, bv, 0))\,du\,dv.$$

Adding equations (8), (9), and (10) we obtain

$$\iiint_T \mathrm{div}\,F = \sum_{i=1}^{3} \iiint_T \frac{\partial F_i}{\partial x_i} = \iint_\varphi F$$

since the sum of the integrals on the right-hand sides of (8), (9), and (10) is $\iint_\varphi F$ by equation (6). This completes the proof.

Exactly the same argument as the above proves the result for rectangular boxes. To extend the result to polyhedral solids we use the same technique as in the proof of Green's theorem. A polyhedron P may be broken up into the finite union of boxes and tetrahedra. Call these latter solids $T_1, \ldots, T_n$. Since volume integrals are additive,

$$\iiint_P \mathrm{div}\,F = \sum_{k=1}^{n} \iiint_{T_k} \mathrm{div}\,F.$$

If φ denotes the positively oriented surface of the polyhedron and $\varphi_1, \ldots, \varphi_n$ the positively oriented surfaces of the boxes and tetrahedra T_k, then

$$\iint_\varphi F = \sum_{k=1}^n \iint_{\varphi_k} F.$$

This follows since the integrals over interior plane surfaces of the solids T_k cancel each other in pairs. (The normals of a pair of such surfaces are pointing in opposite directions.) Since we have already proved that

$$\iiint_{T_k} \operatorname{div} F = \iint_{\varphi_k} F$$

we conclude that for a polyhedron P with surface φ

$$\iiint_P \operatorname{div} F = \iint_\varphi F.$$

To extend the result to more general solids we approximate these solids by polyhedra. Care must be taken performing this approximation, however. The details of this argument are beyond the scope of the treatment here. We state a form of Gauss' theorem which may be proved by these methods.

THEOREM 10.4 *Let U be a bounded open set in $V_3(R)$ such that the boundary of U is the trace of a positively oriented, piecewise smooth surface φ. If F is a smooth vector field defined on a neighborhood of the set consisting of U and its boundary φ, then*

$$\iiint_U \operatorname{div} F = \iint_\varphi F.$$

If in place of φ we use ∂U to denote the boundary of the solid region U and use ∂F to denote the divergence of the vector field, then Gauss' formula takes the symmetric form

$$\iiint_U \partial F = \iint_{\partial U} F.$$

This formula extends to solid regions in $V_n(R)$. If U is a bounded open set in $V_n(R)$ with a suitably smooth boundary surface ∂U and F is a smooth vector field defined on a neighborhood of U and ∂U, then it can be shown that

$$\int_U \partial F = \int_U F$$

where $\partial F = (\partial F_1/\partial x_1) + \cdots + (\partial F_n/\partial x_n)$ and $F_1, \ldots, F_n$ are the component functions of F. The integral on the left is an ordinary integral over a region in n-space. The integral on the right is an $(n-1)$-dimensional surface integral analogous to a 2-dimensional surface integral in 3-space. We shall usually abbreviate the volume integral $\iiint_U \operatorname{div} F\, dx_1\, dx_2\, dx_3$ by $\iiint_U \operatorname{div} F$. We may also write $\iiint_U \operatorname{div} F\, dV$ where dV is the volume element $dx_1\, dx_2\, dx_3$.

Gauss' theorem simplifies greatly the computation of surface integrals over closed surfaces.

Example 1. Let S be the unit cube $0 \leq x, y, z \leq 1$ and $F(X) = (x^2, z, y)$. If φ is the positively oriented surface of this cube, compute $\iint_\varphi F$ by using Gauss' theorem. Since $\operatorname{div} F = 2x$

$$\iint_\varphi F = \iiint_S 2x\, dx\, dy\, dz = \int_0^1 \int_0^1 \int_0^1 2x\, dx\, dy\, dz = 1.$$

Next as a corollary of the Gauss theorem we establish a formula for the divergence of F which often serves as the definition.

THEOREM 10.5 *Let F be a smooth vector field defined on an open set U and let $X \in U$. Let T_δ be a solid region containing X and contained in the ball of radius δ about X. Assume that Gauss' theorem applies to T_δ. If ∂T_δ denotes the boundary of T_δ and $v(T_\delta)$ the volume of T_δ, then*

$$(11) \qquad (\operatorname{div} F)(X) = \lim_{\delta \to 0} \frac{1}{v(T_\delta)} \iint_{\partial T_\delta} F.$$

Proof. By Theorem 10.4

$$\iint_{\partial T_\delta} F = \iiint_{T_\delta} (\operatorname{div} F).$$

We must show

$$\lim_{\delta \to 0} \frac{1}{v(T_\delta)} \iiint_{T_\delta} (\operatorname{div} F) = (\operatorname{div} F)(X).$$

This, on the other hand, depends only on the continuity of div F. Indeed if f is any function continuous at X, then for a given ϵ we may choose a corresponding δ so that $|Y - X| < \delta$ implies

$$f(X) - \epsilon < f(Y) < f(X) + \epsilon.$$

Hence if T_δ is any region inside the ball $\{Y : |Y - X| < \delta\}$, we have

$$\iiint_{T_\delta} (f(X) - \epsilon) \leq \iiint_{T_\delta} f \leq \iiint_{T_\delta} (f(X) + \epsilon).$$

But $\iiint_{T_\delta} (f(X) - \epsilon) = (f(X) - \epsilon) v(T_\delta)$ and $\iiint_{T_\delta} (f(X) + \epsilon) = (f(X) + \epsilon) v(T_\delta)$, since X is a fixed point of T_δ. Hence

$$f(X) - \epsilon \leq \frac{1}{v(T_\delta)} \iiint_{T_\delta} f \leq f(X) + \epsilon$$

and

$$\lim_{\delta \to 0} \frac{1}{v(T_\delta)} \iiint_{T_\delta} f = f(X).$$

Setting $f = \operatorname{div} F$ in the above we have (11)

We close this section with some remarks on the equation div $F = 0$. We have shown in Section 10.1 that if $(\operatorname{div} F)(X) = 0$ for all X in a star-shaped open set U, then $F = \operatorname{curl} G$ for some vector field G. However, Exercise 9 of Section 10.5 shows that $(\operatorname{div} F)(X)$ may be zero throughout a simply connected open set U and it may not be true that $F = \operatorname{curl} G$. However, if U is a connected open set, it can be shown that a necessary and sufficient condition for there to exist a vector field G satisfying $F = \operatorname{curl} G$ is that $\iint_\varphi F = 0$ for every surface φ in U which is topologically equivalent to a sphere. To say that φ is topologically equivalent to a sphere means that φ may be deformed into a sphere in a continuous one-to-one fashion.

If we couple this result with Gauss' theorem, we are able to infer the following. Suppose div $F = 0$ throughout an open set U which has the property that for each surface φ in U which is topologically equivalent to a sphere the interior of φ also lies in U. Then $F = \text{curl } G$ for some vector field G. To see this note that if V is the interior of the closed surface φ, then φ is the boundary of V and by Gauss' theorem.

$$0 = \iiint_V \text{div } F = \iint_\varphi F.$$

Hence, using the statement of the previous paragraph, we infer that $F = \text{curl } G$.

EXERCISES

In the following exercises each of the closed surfaces φ is assumed to be positively oriented.

1. Use Gauss' theorem to compute $\iint_\varphi F$ if $F(X) = (x, y, z)$ and the trace of φ is the sphere $x^2 + y^2 + z^2 = 1$.
2. Use Gauss' theorem to compute $\iint_\varphi F$ if $F(X) = (x^2y, y^2z, z^2x)$ and φ traces out the surface of the unit cube $0 \leq x, y, z \leq 1$.
3. Let φ be the tetrahedron formed by the coordinate planes and the plane $2x + y + 3z = 1$. Compute $\iint_\varphi F$ if $F(X) = (yz, zx, xy)$.
4. If U is a region to which Gauss' theorem applies show that the volume of U, $v(U)$ equals $1/3 \iint_{\partial U} (x_1 \, dx_2 \, dx_3 + x_2 \, dx_3 \, dx_1 + x_3 \, dx_1 \, dx_2)$.

Verify Gauss' theorem by computing separately $\iiint_U \text{div } F$ and $\iint_{\partial U} F$ for the following vector fields F and regions U.

5. $F(X) = (x^2, y^2, z^2)$, U is the cylinder $x^2 + y^2 \leq 4$, $0 \leq z \leq 4$.
6. $F(X) = (x, y, z)$, U is the region inside the sphere $x^2 + y^2 + z^2 = 4$ and outside the cylinder $x^2 + y^2 = 1$.
7. $F(X) = (3x, 2y, z)$, U is the region above the paraboloid $z = x^2 + y^2$ and below the plane $z = 4y$.
8. If it is known that $(\text{div } F)(X) = 0$ for all X outside the cylinder $x^2 + y^2 = 1$, can one conclude that $F = \text{curl } G$? Why? (Give a heuristic justification for your answer.)
9. If it is known that $(\text{div } F)(X) = 0$ for all X outside the sphere $x^2 + y^2 + z^2 = 1$, can one conclude that $F = \text{curl } G$ in this region?
10. Let U be an open set in $V_3(R)$ to which Gauss' theorem applies and let ∂U be its boundary. If F is a smooth vector field and f and g are smooth scalar-valued functions defined on a neighborhood of the union of U and ∂U, use Gauss' theorem to show that

    (i) $\iiint_U f(\text{div } F) = \iint_{\partial U} fF - \iiint_U (\nabla f, F)$
    (ii) $\iiint_U \nabla^2 g = \iint_{\partial U} f \nabla g - \iiint_U (\nabla f, \nabla g)$
    (iii) $\iiint_U (f \nabla^2 g - g \nabla^2 f) = \iint_{\partial U} (f \nabla g - g \nabla f)$.

    These formulas are known as Green's identities.

    In the following exercises, let U be as in Exercise 10.

11. Let f and g be harmonic in a neighborhood of the union of U and ∂U. Show that
    $$\iint_U \nabla f = 0 \text{ and } \iint_U g \nabla f = \iint_{\partial U} f \nabla g.$$

12. Vector-valued surface integrals are sometimes useful. If F is a vector field we define
    $$\iint_\varphi F \, dS = \left(\iint_\varphi F_1 \, dS, \iint_\varphi F_2 \, dS, \iint_\varphi F_3 \, dS \right)$$

where $dS = |(\partial\varphi/\partial u) \times (\partial\varphi/\partial v)|\,du\,dv$. Let T_δ be a region containing the point X to which the Gauss theorem applies. Assume the region T_δ is contained in the ball of radius δ about X. Using Gauss' theorem show that

$$(\operatorname{curl} F)(X) = \lim_{\delta \to \infty} \frac{1}{v(T_\delta)} \iint_\varphi (N \times F)\, dS$$

where

$$N = \frac{\dfrac{\partial\varphi}{\partial u} \times \dfrac{\partial\varphi}{\partial v}}{\left|\dfrac{\partial\varphi}{\partial u} \times \dfrac{\partial\varphi}{\partial v}\right|}.$$

*10.7 Applications to Fluid Mechanics

The study of the motion of fluids, that is, liquids and gases, is a portion of applied mathematics that has been under constant investigation from the 18th century to the present. The flow of water through pipes and jets or the flow of air around a reentering space vehicle are just two aspects of this vast subject. Many observable phenomena await complete mathematical treatment.

Fundamental to the study of fluid flow are the techniques of vector analysis and, in particular, the theorems of Gauss and Stokes. Using these results we may derive some fundamental equations of fluid flow. The first of these is the *equation of continuity*.

If the vector field $F(X)$ measures the velocity of a moving fluid at a point X and ρ is the density of the fluid at a point X and time t, then the equation of continuity for fluid flow states that

(1) $$\frac{\partial \rho}{\partial t} + \operatorname{div}(\rho F) = 0.$$

To derive this equation from the theorem of Gauss we proceed as follows. At each fixed point X let D be a small ball centered at X and let ∂D be its surface. The mass of the fluid in the ball D as a function of time is given by

$$\iiint_D \rho = \iiint_D \rho\, dV.$$

Hence the time rate of change of this mass is just the partial derivative

(2) $$\frac{\partial}{\partial t}\left(\iiint_D \rho\, dV\right).$$

On the other hand, assuming that no fluid is created or destroyed within D, the rate of change of mass in D is just the rate that matter enters or leaves through the surface ∂D of D. This quantity is the flux

(3) $$\iint_{\partial D} \rho F = \iint_{\partial D} (\rho F, N)\, dS.$$

If we assume that the normal N is pointing outward, then (3) is positive if fluid is leaving D. On the other hand, (2) is positive if fluid is entering the region D. Since (2) and (3) measure the rate of change of the fluid in D, we must have

$$\frac{\partial}{\partial t}\left(\iiint_D \rho\, dV\right) = -\iint_{\partial D}(\rho V, N)\, dS.$$

The left-hand side of this expression is just

$$\iiint_D \frac{\partial \rho}{\partial t} dV,$$

and if we apply Gauss' theorem to the right-hand side, we obtain

$$\iint_D (\rho F, N) \, dS = \iiint_D \operatorname{div}(\rho F) \, dV.$$

Equating the two expressions we have that for each point X and for each ball D about X

$$\iiint_D \left(\frac{\partial \rho}{\partial t} + \operatorname{div}(\rho V) \right) dV = 0.$$

From this it follows that at each point X

$$\frac{\partial \rho}{\partial t} + \operatorname{div}(\rho F) = 0.$$

and this is the equation of continuity.

We call the fluid flow *incompressible* if the density ρ is constant. Then $\partial \rho / \partial t = 0$, and the equation of continuity takes the form

$$\operatorname{div} F = 0.$$

Next let α be a simple closed path lying in the fluid. The *circulation* of the flow around α is defined to be the line integral of the velocity F around α. If

(4) $$\oint_\alpha F = 0$$

for each closed path α, then the flow is called *irrotational*. We know from the discussion in Chapter 9 that (4) will hold for each curve α if and only if

$$F = \nabla f$$

for some scalar function f. The function f is the potential of the flow. Hence an irrotational flow is equivalently called a *potential* flow. For such a flow curl $F = 0$, and we know from Stokes theorem that if the flow is occurring in a simply connected region, the condition curl $F = 0$ will guarantee that the flow is a potential flow.

If the flow is both incompressible and irrotational then

$$\operatorname{div} F = \operatorname{div} \operatorname{curl} f = \nabla^2 f = 0$$

and the velocity potential f is a solution of the Laplace equation $\nabla^2 f = 0$.

These facts form a meager introduction to the fascinating study of fluid mechanics. The literature in this subject is vast. However, for an elaboration of these ideas, and many more, the reader may consult *Fluid Mechanics*, by Landau and Lifschitz, Addison Wesley, 1959, with profit. Chapter one is an elegant presentation of the fundamental ideas of the subject.

ANSWERS TO EXERCISES

CHAPTER 1

Section 1.1 – p. 3. **1.** $(1, 7), (-2, 5), (1, -1), (10, 26), \sqrt{601}$. **2.** $(-1, 1, 0)$, $(27, 3, -15), (0, 0, -2), \sqrt{74}$. **6.** $(2, -4) = 14(1, 1) - 6(2, 3)$.
7. $(1, -1, 2) = -\frac{1}{13}(1, 0, -1) - \frac{2}{13}(2, 2, 1) + \frac{9}{13}(2, -1, 3)$. **8.** No. **9.** No.

Section 1.2 – p. 9. **1.** (a) $1/\sqrt{27}, 1/\sqrt{27}, 5/\sqrt{27}$; (b) $3/\sqrt{14}, 2/\sqrt{14}, -1/\sqrt{14}$;
(c) $-5/\sqrt{33}, 2/\sqrt{33}, 2/\sqrt{33}$. **2.** (a) $1/\sqrt{14}\,(3, 2, -1)$; (b) $1/\sqrt{21}\,(4, -1, 2)$;
(c) $1/\sqrt{5}\,(0, -1, 2)$. **3.** $1/\sqrt{5}\,(2, -1, 0); 1/3\sqrt{3}\,(1, 1, 5); 1/\sqrt{117}\,(4, 1, 10)$;
$1/\sqrt{30}\,(1, -2, -5)$. **4.** $1/\sqrt{34}\,(5, 3); 1/\sqrt{26}\,(-1, -5); 1/13\,(12, 5)$. **5.** (a) $(4, 2, 1)$;
(b) $(5, 3, -1)$; (c) $(6, 4, 4)$. **6.** (c). **7.** (a) $(-2, -3, 4)$; (b) $(-5, -3, 3)$;
(c) $(-7, 0, 2)$.

Section 1.3 – p. 14. **3.** (a) $1/\sqrt{21}\,(2, 1, 4)$; (b) $1/\sqrt{51}\,(1, 5, 5)$; (c) $(0, 0, 1)$.
4. Straight line passing through the tip of A in the direction of B. **6.** Straight line segment passing through the tips of vectors A and B. **7.** (a) Straight line passing through the tip of B in the direction of A; (b) straight line segment connecting the tips of A and B.
8. (a) $(0, 0) = OA + OB$; (b) $(2, 4) = (7/5)A + (1/5)B$. **9.** (a) $(0, 0) = OA + OB$;
(b) $(2, 4) = (2/7)A + (8/7)B$. **11.** A plane through $(1, 2, 3)$; no; $d = 0$. **12.** A plane through $(1, 0, -1)$; no; no values of d. **13.** Yes; $(-1, 0, -1) = (1/2)A - (5/2)B - C$;
$(0, 0, 0) = OA + OB + OC$. **14.** Yes; $(-1, 0, -1) = -A + 2C$; $(0, 0, 0) = OA + OB + OC$.
15. $50(-1 - 5\sqrt{2}, 5\sqrt{2}), 1/(101 + 10\sqrt{2})(-1 - 5\sqrt{2}, 5\sqrt{2})$. **17.** $F_1 = 50(-2/\sqrt{21}, 1)$, $F_2 = 50(2/\sqrt{21}, 1)$ in both cases.

Section 1.4 – p. 17. **1.** $6, 0, 2$. **2.** 1. **3.** $-1/7$. **4.** 2. **5.** 3. **6.** No; consider $A = (0, 0, 1), B = (0, 1, 0), C = (1, 0, 0)$. **7.** $(1, 1, 2)$, yes. **8.** $(2, 3, -1)$, yes.
9. $\pm 1/\sqrt{2}\,(1, 1, 0)$. **10.** $\pm 1/\sqrt{2}\,(-1, 0, 1)$. **11.** $x \neq 0, y = -5x, x = \pm 1/3\sqrt{5}$.
12. $x \neq 0, y = 2x, x = \pm 1/\sqrt{6}$. **17.** The diagonals must be equal in length. **18.** Yes.

Section 1.5 – p. 20. **1.** A and B parallel; A and D orthogonal; B and D orthogonal.
2. $0, 1/2, \sqrt{3}/2$. **3.** (a) $x = -1$; (b) For no x. **4.** (a) $x = 13$; (b) $x = -1$.
8. $\sqrt{23}/6, \sqrt{13}/6, 0$. **9.** $-\sqrt{2}/8$. **10.** $t = -|A|$. **12.** $x = -(A, B)/|B|^2$. **13.** No.
14. A diamond; a square.

Section 1.6 – p. 27. **1.** (a) $X = (1, -1) + t(2, 3); 3x_1 - 2x_2 - 5 = 0$;
(b) $X = (2, -3) + t(-3, 5); 5x_1 + 3x_2 - 1 = 0$; (c) $X = (4, 1) + t(2, 1); x_1 - 2x_2 - 2 = 0$;
(d) $X = (1, 3) + t(-3, 5); 5x_1 + 3x_2 - 14 = 0$. **2.** $X = (0, 3) + t(2, -4)$;
$X = (3/2, 1/2) + t(5/2, -7/2); X = (1/2, 3/2) + t(1/2, 1/2)$. **3.** $(1, 3) + t(1/2, 1/2)$.
4. $(-1, 1), (5, 5), (4, 6)$. **5.** $D = \sqrt{5}$. **6.** (a) $X = (1, -1, 1) + t(2, 0, -3)$;
$(x_1 - 1)/2 = (x_3 - 1)/-3$; (b) $X = (2, 1, -2) + t(1, -1, -5)$;
$(x_1 - 2)/1 = (x_2 - 1)/-1 = (x_3 + 2)/-5$; (c) $X = (1, -1, 1) + t(0, 3, 2)$;
$(x_2 + 1)/3 = (x_3 - 1)/2$. **10.** (a) $X = (5, -3, 1) + t(2, -4, 6)$;
(b) $X = (0, -1, 1/2) + t(-2, 2, -1)$. **11.** $D = \sqrt{62/21}$. **14.** $D = (8\sqrt{5})/5$.
15. (a) $(3\sqrt{30})/5$; (b) $\sqrt{3}$; (c) $(3\sqrt{66})/11$. **16.** $\sqrt{182/7}$. **17.** $\sqrt{3}$.

Section 1.7 – p. 35. **1.** (a) $\{5/2, -5, 5/3\}$; (b) $(1/\sqrt{14})(2, -1, 3)$;
(c) $X = (0, -5, 0) + s(1, -1, -1) + t(1, 2, 0)$; (d) $(1, -1, -1)$ and $(1, 2, 0)$.
2. (a) $x_1 - x_2 + x_3 = 3$; (b) $(1, 1, 0)$ and $(0, 1, 1)$; (c) $\{3, -3, 3\}$; (d) $D = \sqrt{3}$.
3. $2x_1 + 2x_2 + x_3 = 6; N = (2/3, 2/3, 1/3); A = (-3, 1, 4), B = (-1, 2, -2)$; (d) $D = 5/3$.
4. (b) and (d) are in π. **5.** $x_1 + 3x_2 + x_3 = 18$. **6.** $x_1 + 9x_2 - 2x_3 + 12 + 5\sqrt{86} = 0$;
$x_1 + 9x_2 - 2x_3 + 12 - 5\sqrt{86} = 0$. **8.** $D = \sqrt{389/62}$. **9.** $2x_1 + x_2 - x_3 = 7$.
10. (a) $D = 9/(2\sqrt{14})$; (b) $D = 8/\sqrt{11}$. **13.** $41x_1 + 3x_2 - 18x_3 = 17$.

14. $n_1 x_1 + n_2 x_2 + n_3 x_3 + n_4 x_4 = n_1 c_1 + n_2 c_2 + n_4 c_4$; $D = \sqrt{3}/3$.   **15.** $x_1 \neq 0$, $x_1 = x_2 = x_3$.
16. $x_1 \neq 0$, $x_2 = 3x_1$, $x_3 = -4x_1$, $x_4 = -3x_1$.

Section 1.8 — p. 42.   **1.** (a) $(-3, 3, 3)$;   (b) $(-3, 3, 3)$;   (c) $(0, 0, 0)$;   (d) $(-14, 14, 14)$;
(e) $(12, 0, -3)$;   (f) 15;   (g) 15;   (h) 0;   (i) 0.   **2.** (a) $(-1, 1, 1)$;   (b) $(1, 5, 3)$;
(c) $(-1, 0, 1)$.   **3.** (a) $(2, 0, 1)$;   (b) $(4, 0, -1)$;   (c) $(20, -3, 8)$.
4. $A = |(Q-P) \times (R-P)|/2$.   **5.** (a) $\sqrt{53}/2$;   (b) $\sqrt{174}/2$;   (c) $3\sqrt{11}/2$.   **7.** A is perpendicular to B.   **8.** *Hint.* The altitude is the distance from C to the plane spanned by A and B.   **9.** $t = \pm 1/28$.

Section 1.9 — p. 48.   **1.** (a) Independent;   (b) dependent;   (c) independent;
(d) dependent.   **2.** (a) $t = -7/3$;   (b) no t exists;   (c) all t.   **3.** (a) Linearly dependent, no solution;   (b) linearly dependent, no solution;   (c) $x_1 = -7/13$, $x_2 = 4/13$, $x_3 = -1/13$;   (d) $x_1 = 1/2$, $x_2 = -1$, $x_3 = 3/2$.   **4.** (a) $x_1 = 1/5$, $x_2 = -4/5$, $x_3 = 0$;
(b) $A, B, A \times B$ are linearly dependent, no solution;   (c) $x_1 = 23/43$, $x_2 = 65/43$, $x_3 = 15/43$;
(d) $x_1 = 2$, $x_2 = 0$, $x_3 = 0$.

12.
$$x_2 = \frac{\begin{vmatrix} a_1 & d_1 & c_1 \\ a_2 & d_2 & c_2 \\ a_3 & d_3 & c_3 \end{vmatrix}}{\begin{vmatrix} a_1 & b_1 & c_1 \\ a_2 & b_2 & c_2 \\ a_3 & b_3 & c_3 \end{vmatrix}} \qquad x_3 = \frac{\begin{vmatrix} a_1 & b_1 & d_1 \\ a_2 & b_2 & d_2 \\ a_3 & b_3 & d_3 \end{vmatrix}}{\begin{vmatrix} a_1 & b_1 & c_1 \\ a_2 & b_2 & c_2 \\ a_3 & b_3 & c_3 \end{vmatrix}}.$$

CHAPTER 2

Section 2.2 — p. 53.   **2.** Yes; no.   **3.** No.   **4.** Yes; yes.   **5.** When $q = 0$.
6. Yes; $\begin{vmatrix} 3 & -2 & 1 \\ 2 & 1 & -1 \\ 5 & -8 & 5 \end{vmatrix} = 0$; no.   **7.** $\begin{vmatrix} a_{11} & a_{12} & a_{13} \\ a_{21} & a_{22} & a_{23} \\ a_{31} & a_{32} & a_{33} \end{vmatrix} \neq 0$.   **13.** $S \cup T$ is not a subspace.
15. (i) Yes, (ii) no, (iii) yes, (iv) no, (v) yes.

Section 2.4 — p. 59.   (The answers to Exercises 8–12 are not unique.)   **8.** $(-1, 0, 1)$, $(0, 1, 5)$, $(0, 0, 7)$.   **9.** $(1, -1, 2)$, $(0, 2, -5)$, $(0, 0, 3)$.   **10.** $(1, 0, -1, 2)$, $(0, -1, 3, -6)$, $(0, 0, 1, 3)$.   **11.** $(1, -2, 3, 1)$, $(0, 0, 1, 1)$, $(0, 0, 0, 1)$.
12. $(1, -1, 0, 0, 2)$, $(0, -2, 0, 1, 7)$, $(0, 0, 2, -1, -11)$, $(0, 0, 0, 1, 5)$.

Section 2.5 — p. 64.   **1.** Adjoin $(0, 0, d)$, $d \neq 0$.   **2.** Adjoin $(0, 0, b, 0)$ and $(0, 0, 0, c)$, $b \neq 0 \neq c$.   **3.** Adjoin $(0, 0, d)$, $d \neq 0$.   **4.** Adjoin $(0, 0, b, 0)$ and $(0, 0, 0, c)$, $b \neq 0 \neq c$.   **5.** (The bases constructed are not unique.)   (a) $(1, -1)$, dim = 1;
(b) $(1, 2, -1)$, $(0, 5, -6)$, dim = 2;   (c) $(-1, 0, 1)$, $(0, 1, 5)$, dim = 2;   (d) $(1, 2, -1)$, $(0, 1, 2)$, $(0, 0, 1)$, dim = 3;   (e) $(1, 2, 3)$, $(0, 1, 1)$, dim = 2;   (f) $(1, -1, 1, -1)$, $(0, 2, -1, 3)$, $(0, 0, 1, -1)$, $(0, 0, 0, 1)$, dim = 4;   (g) $(1, 1, -1, -1)$, $(0, 1, -1, 0)$, dim = 2.
14. 1, 2.   **22.** (a) $(1/2)X_1 + (3/2)X_2$;   (b) $(7/5)X_1 + (1/5)X_2$;
(c) $(1/2)X_1 + (2/3)X_2 + (1/6)X_3$;   (d) $(5/3)X_1 + (11/14)X_2 - (13/42)X_3$;   (e) $X_1 + 2X_3$.

Section 2.6 — p. 72.   **1.** Yes; $(-1, 1/2, 1/2)$.   **2.** No solution.   **3.** No solution.
4. $(1, 1, -1, 0)$ and $(1/2, -1, 0, 1)$.   **5.** $(-1/2, 1/2, 1, 0, 0)$, $(-1/2, 1/2, 0, -1, 0)$, and $(0, 1, 0, 0, 1)$.   **6.** $(2, -1, 1, 0, 0, 0)$, $(0, 1, 0, 1, 0, 0)$, $(2, -1, 0, 0, -1, 0)$, and $(2, -1, 0, 0, 0, 1)$.   **7.** $(0, 2, 1, 1, 0)$ and $(0, 2, 1, 0, -1)$.
8. $(0, -1, 0, 0) + a(0, 1, 1, 0) + b(2, 1, 0, 1)$.   **9.** $(3/2, 3/2, 1, 1) + a(0, 1, 1, 0) + b(1, 0, 0, 1)$.
10. $(-1, -3, 0, -1, 0) + a(1, 4, -1, 1, 0) + b(1, 0, 1, 0, -1)$.
11. $(3, 1, -1, 0) + a(1, 1, 0, 0) + b(2, 0, -1, 1)$.
12. $(1, 0, 0, 0, 1) + a(1, 2, 0, 0, 0) + b(1, 0, -1, -1, 0) + c(0, 0, 1, 0, 1)$.
13. $(5, -2, 0, 1, 0) + a(2, -1, 1, 0, 0) + b(0, 0, 0, 1, 1)$.
14. $\begin{pmatrix} 1 & 1 & -1 & 1 & 1 \\ 0 & 2 & -2 & 2 & 1 \\ 0 & 0 & -4 & 4 & -3 \end{pmatrix}$, $(1/2, 5/4, 3/4, 0) + a(0, 0, 1, 1)$.
15. $\begin{pmatrix} 2 & -1 & 5 & 1 \\ 0 & -2 & -2 & 0 \\ 0 & 0 & 4 & 2 \end{pmatrix}$, $(-1, -1/2, 1/2)$.
16. $\begin{pmatrix} 1 & -1 & -1 & 1 & 1 & 1 \\ 0 & -2 & -2 & 2 & 2 & 2 \\ 0 & 0 & -2 & 0 & 2 & 2 \end{pmatrix}$, $(0, 0, 0, 0, 1) + a(0, 1, 0, 1, 0) + b(0, 0, 1, 0, 1)$.

CHAPTER 3

Section 3.1 – p. 79.   1. $\begin{pmatrix}-2\\1\end{pmatrix}$.   2. $\begin{pmatrix}1\\5\\1\end{pmatrix}$.   3. $\begin{pmatrix}-2\\5\\1\\0\end{pmatrix}$.   4. $\begin{pmatrix}-4\\4\end{pmatrix}$.   5. $\begin{pmatrix}-1\\3\\1\end{pmatrix}$.

6. $\begin{pmatrix}0\\3\end{pmatrix}, \begin{pmatrix}0\\6\end{pmatrix}, \begin{pmatrix}3\\3\end{pmatrix}$.   9. S does but D does not.   11. $\begin{pmatrix}3\\1\\1\end{pmatrix}$.   12. Not possible.

13. *Hint.* Express $AX = 0$ as a system of linear equations.

Section 3.2 – p. 84.   1. (a) $\begin{pmatrix}-1 & 4\\-2 & -1\end{pmatrix}$;   (b) $\begin{pmatrix}0 & -1 & -1\\-2 & 3 & 2\\-1 & 2 & 1\end{pmatrix}$;   (c) not defined;

(d) $\begin{pmatrix}6 & 3\\-3 & 3\end{pmatrix}$;   (e) not defined;   (f) not defined;   (g) $\begin{pmatrix}-2 & 1 & 0\\-4 & 7 & 5\\0 & -1 & -1\end{pmatrix}$.

2. (a) $\begin{pmatrix}1 & -2\\2 & -1\end{pmatrix}$;   (b) $\begin{pmatrix}-1 & -2\\-2 & -1\end{pmatrix}$;   (c) $\begin{pmatrix}-2 & 0\\-4 & 0\end{pmatrix}$.

3. (a) $\begin{pmatrix}-1 & 2 & 1\\3 & 1 & 3\\3 & -1 & -1\end{pmatrix}$;   (b) $\begin{pmatrix}-3 & 1 & 0\\-2 & -1 & 1\\1 & 0 & 3\end{pmatrix}$;   (c) $\begin{pmatrix}-1 & 3 & -2\\-1 & 5 & 0\\-4 & -1 & 3\end{pmatrix}$.

Section 3.3 – p. 90.   11. P_{n-2}, P_1, no, no.   12. $0, n+1$, yes, no.   13. $n+1, 0$, yes, yes.
14. Yes, yes.   15. $2, 1$.   16. $4, 0$.   17. $3, 1$.   18. $\{aI_1 + bI_2\}, \{aI_3\}$.
19. Plane through the origin perpendicular to Y, line through origin parallel to Y.
20. No, not every continuous function has a derivative.

Section 3.4 – p. 99.   1. $\begin{pmatrix}0 & 1 & 1\\1 & 1 & 0\end{pmatrix}, (1, 5)$.   2. $\begin{pmatrix}-1 & 1 & 1\\2 & -1 & 0\end{pmatrix}, (4, -3)$.

3. $\begin{pmatrix}1/2 & -1/2\\3/2 & 1/2\\-1/2 & -1/2\\0 & 1\end{pmatrix}, (3/2, 5/2, -1/2, -1)$.   4. $\begin{pmatrix}1/2 & 1/2 & -1\\3/2 & -1/2 & 0\\5/2 & 3/2 & -1\end{pmatrix}, (0, 1, 3)$.

5. $\begin{pmatrix}2 & -1 & 0\\0 & 1 & 2\end{pmatrix}, 5, 1$.   6. $\begin{pmatrix}1 & -1 & -1\\0 & 1 & 2\\1 & 2 & 0\end{pmatrix}, 2, 1, 0$.   7. $\begin{pmatrix}0 & -1\\1 & 0\end{pmatrix}, \begin{pmatrix}-1 & 0\\0 & -1\end{pmatrix}, \begin{pmatrix}0 & 1\\-1 & 0\end{pmatrix}$.

8. $\begin{pmatrix}+\sqrt{2}/2 & \sqrt{2}/2\\-\sqrt{2}/2 & \sqrt{2}/2\end{pmatrix}, \begin{pmatrix}0 & 1\\-1 & 0\end{pmatrix}$.   9. $\begin{pmatrix}0 & -1\\-1 & 0\end{pmatrix}, \begin{pmatrix}1 & 0\\0 & -1\end{pmatrix}$.   10. $(a_2, -a_1), (a_1, a_2)$.

11. $\begin{pmatrix}1 & 0 & 0\\0 & 0 & -1\\0 & 1 & 0\end{pmatrix}, \begin{pmatrix}0 & 0 & 1\\0 & 1 & 0\\-1 & 0 & 0\end{pmatrix}, \begin{pmatrix}0 & 0 & 1\\0 & 1 & 0\\1 & 0 & 0\end{pmatrix}$.   12. $\begin{pmatrix}1 & 0 & 0\\0 & 1 & 0\\0 & 0 & -1\end{pmatrix}, \begin{pmatrix}0 & 1 & 0\\1 & 0 & 0\\0 & 0 & 1\end{pmatrix}, \begin{pmatrix}1 & 0 & 0\\0 & 0 & -1\\0 & -1 & 0\end{pmatrix}$.

13. $\begin{pmatrix}0 & 0 & 2 & 0 & 0 & \cdots & 0\\0 & 0 & 0 & 6 & 0 & \cdots & 0\\ & & & & 12 & & \\ \cdot & \cdot & \cdot & \cdot & & \cdot & \\ \cdot & \cdot & \cdot & \cdot & & & n(n-1)\\ \cdot & \cdot & \cdot & \cdot & & & 0\\ 0 & 0 & 0 & 0 & 0 & & 0\end{pmatrix}$.   14. $\begin{pmatrix}1 & 0 & 2 & 0 & \cdot & \cdot & \cdot & 0\\0 & 1 & 0 & 6 & \cdot & \cdot & \cdot & 0\\0 & 0 & 1 & 0 & 12 & 0 & \cdots & 0\\0 & 0 & 0 & 1 & 0 & & & 0\\ & & & & & & & \cdot\\ \cdot & & & & & \cdot & & n(n-1)\\ \cdot & & & & & & & 0\\ 0 & & & & & & & 1\end{pmatrix}$.

15. $\begin{pmatrix}1 & 0 & 0 & 0 & \cdots & 0\\0 & 2 & 0 & 0 & \cdots & 0\\0 & 0 & 3 & 0 & & \cdot\\ \cdot & \cdot & & 0 & 4 & \cdot\\ \cdot & \cdot & & & 0 & \cdot\\ \cdot & \cdot & \cdot & \cdot & \cdot & \cdot\\0 & 0 & 0 & \cdot & \cdots & (n+1)\end{pmatrix}$.   16. $\begin{pmatrix}0 & 0 & 0 & 0 & \cdots & 0\\2 & 0 & 0 & \cdot & \cdots & 0\\0 & 1 & 0 & \cdot & \cdots & \cdot\\0 & 0 & 2/3 & \cdot & & \cdot\\ \cdot & \cdot & & 0 & \cdot & \cdot\\ \cdot & \cdot & \cdot & \cdot & \cdot & \cdot\\0 & 0 & 0 & \cdots & & 2/(n+1)\end{pmatrix}$.

17. $\begin{pmatrix} -1 & -1/2 & -1/3 & \cdots & & -1/(n+1) \\ 1 & 0 & 0 & \cdots & & 0 \\ 0 & 1/2 & 0 & \cdots & & 0 \\ 0 & 0 & 1/3 & 0 & \cdots & 0 \\ \vdots & & & 0 & & \vdots \\ 0 & 0 & & \cdots & 0 & 1/(n+1) \end{pmatrix}$.

18. $\begin{pmatrix} -2 & -8/3 & -4 & \cdots & & -2^{n+2}/(n+2) \\ 0 & 0 & 0 & & & 0 \\ 1/2 & 0 & 0 & 0 & & \cdot \\ 0 & 1/3 & 0 & & & \cdot \\ \cdot & 0 & 1/4 & & & \cdot \\ \vdots & & & & & \vdots \\ 0 & 0 & \cdots & & 0 & 1/(n+2) \end{pmatrix}$.

19. $\begin{pmatrix} 1 & 0 & 0 \\ 0 & 0 & 0 \\ 0 & 0 & 0 \end{pmatrix}$. **20.** $\begin{pmatrix} 0 & -1 & 1 & 0 \\ 1 & 0 & 0 & 1 \\ 0 & 0 & 0 & -1 \\ 0 & 0 & 1 & 0 \end{pmatrix}, \begin{pmatrix} -1 & 0 & 0 & -2 \\ 0 & -1 & 2 & 0 \\ 0 & 0 & -1 & 0 \\ 0 & 0 & 0 & -1 \end{pmatrix}$.

21. $\begin{pmatrix} 0 & 1 & 0 & 0 & 0 \\ 0 & 0 & 4 & 0 & 0 \\ 0 & 0 & 0 & 9 & 0 \\ 0 & 0 & 0 & 0 & 16 \\ 0 & 0 & 0 & 0 & 0 \end{pmatrix}, \begin{pmatrix} 0 & 0 & 0 & 0 & 0 \\ 0 & 0 & 2 & 0 & 0 \\ 0 & 0 & 0 & 6 & 0 \\ 0 & 0 & 0 & 0 & 12 \\ 0 & 0 & 0 & 0 & 0 \end{pmatrix}, \begin{pmatrix} 0 & 1 & 0 & 0 & 0 \\ 0 & 0 & 2 & 0 & 0 \\ 0 & 0 & 0 & 3 & 0 \\ 0 & 0 & 0 & 0 & 4 \\ 0 & 0 & 0 & 0 & 0 \end{pmatrix}$.

22. $\begin{pmatrix} 0 & 0 & 0 \\ 0 & 0 & -1 \\ 0 & 1 & 0 \end{pmatrix}, \begin{pmatrix} 0 & 0 & 1 \\ 0 & 0 & 0 \\ -1 & 0 & 0 \end{pmatrix}, \begin{pmatrix} 0 & -1 & 0 \\ 1 & 0 & 0 \\ 0 & 0 & 0 \end{pmatrix}$. **23.** $\begin{pmatrix} 0 & -y_3 & y_2 \\ y_3 & 0 & -y_1 \\ -y_2 & y_1 & 0 \end{pmatrix}$. **24.** $\begin{pmatrix} 1 & 0 & 0 \\ 0 & 1 & 0 \\ 0 & 0 & 0 \end{pmatrix}$.

29. $A = \begin{pmatrix} 1 & 1 & 0 \\ 1 & -1 & 0 \end{pmatrix}, B = \begin{pmatrix} 1 & 1 & 1 \\ 1 & -1 & 1 \\ 1 & 0 & 0 \end{pmatrix}, A \cdot B = \begin{pmatrix} 2 & 0 & 2 \\ 0 & 2 & 0 \end{pmatrix}$.

Section 3.5 – p. 107. **1.** $\begin{pmatrix} 1 & 0 \\ 0 & 1/2 \end{pmatrix}$. **2.** Does not exist. **3.** $\begin{pmatrix} 1/2 & 0 \\ 1/2 & -1 \end{pmatrix}$. **4.** $\begin{pmatrix} 1 & 1 \\ 0 & 1 \end{pmatrix}$.

5. $\begin{pmatrix} 1 & 0 & 0 \\ 0 & 1/2 & 0 \\ 0 & 0 & 1/3 \end{pmatrix}$. **6.** Does not exist. **7.** $\begin{pmatrix} 1/2 & -1 & -1/2 \\ 0 & 2 & 1 \\ 0 & 0 & 1 \end{pmatrix}$. **8.** $\begin{pmatrix} 1 & 0 & 0 \\ 1/2 & 1/2 & 0 \\ -2/3 & -1/3 & 1/3 \end{pmatrix}$.

9. $\begin{pmatrix} 1 & 0 & 0 \\ 0 & -1 & 0 \\ -1 & 2 & 1 \end{pmatrix}$. **10.** $\begin{pmatrix} 1 & 0 & 1 \\ 0 & 1 & 0 \\ 0 & 0 & -1 \end{pmatrix}$. **11.** Does not exist. **12.** $\begin{pmatrix} 1 & 0 & -1 & 0 \\ 0 & -1 & 0 & 1 \\ 0 & 0 & 1 & 0 \\ 0 & 0 & 0 & -1 \end{pmatrix}$.

13. $\begin{pmatrix} 1/2 & 0 & 0 & 0 \\ 0 & -1 & 0 & 0 \\ 0 & 0 & 1 & 0 \\ -1/4 & 0 & 0 & 1/2 \end{pmatrix}$. **14.** $\begin{pmatrix} 1 & 1 & 0 & 0 \\ 0 & 1 & -1 & 0 \\ 0 & 0 & -1 & -1 \\ 0 & 0 & 0 & -1 \end{pmatrix}$. **15.** (b) $\begin{pmatrix} 1 & -1 & 2 & -6 \\ 0 & 1 & -2 & 6 \\ 0 & 0 & 1 & -3 \\ 0 & 0 & 0 & 1 \end{pmatrix}$;

(c) $f(x) = 2 - 2x + x^2$, yes. **16.** $f(x) = x$, yes. **17.** $a \neq 0$.

Section 3.6 – p. 113. (1–5 are not unique.) **1.** $\begin{pmatrix} 1 & 1 \\ 0 & -2 \end{pmatrix}$. **2.** $\begin{pmatrix} 2 & -1 & 1 \\ 0 & -1 & -3 \end{pmatrix}$.

3. $\begin{pmatrix} 1 & -1 \\ 0 & -1 \\ 0 & 0 \end{pmatrix}$. **4.** $\begin{pmatrix} 2 & -1 & 1 \\ 0 & -1 & -1 \\ 0 & 0 & 2 \end{pmatrix}$. **5.** $\begin{pmatrix} 1 & 1 & -1 \\ 0 & 1 & 0 \\ 0 & 0 & -1 \\ 0 & 0 & 0 \end{pmatrix}$. **6.** $\begin{pmatrix} 1/3 & -2/3 \\ 1/3 & 1/3 \end{pmatrix}$.

7. Not invertible. **8.** $\begin{pmatrix} 1/2 & -1/2 \\ 1/2 & 1/2 \end{pmatrix}$. **9.** Not invertible. **10.** $\begin{pmatrix} -1 & -1 & 4 \\ -1 & 0 & 1 \\ 1 & 1 & -3 \end{pmatrix}$.

ANSWERS TO EXERCISES   411

11. Not invertible.   **12.** Not invertible.   **13.** $\begin{pmatrix} 0 & 1 & 0 \\ 1 & 0 & 0 \\ 0 & 0 & 1 \end{pmatrix}$.   **14.** Not invertible.

15. Not invertible.   **16.** Not invertible.   **17.** $\begin{pmatrix} 0 & 1 & 0 & 0 \\ 0 & 0 & 1 & 0 \\ 0 & 0 & 0 & 1 \\ 1 & 0 & 0 & 0 \end{pmatrix}$.

22. $\begin{pmatrix} 1 & 0 & 0 & 0 \\ 0 & 1 & -2 & 0 \\ 0 & 0 & 1 & -6 \\ 0 & 0 & 0 & 1 \end{pmatrix}$, $y = x^2 - 2x$, yes.

Section 3.7 – p. 118.   **1.** (a) $\begin{pmatrix} -1 & 1 \\ 0 & 1 \end{pmatrix}, \begin{pmatrix} -1 & 1 \\ 0 & 1 \end{pmatrix}$;   (b) $\begin{pmatrix} 1 & 0 \\ -1 & 2 \end{pmatrix}, \begin{pmatrix} 1 & 0 \\ 1/2 & 1/2 \end{pmatrix}$;

(c) $\begin{pmatrix} 2 & 1 \\ 1 & 2 \end{pmatrix}\begin{pmatrix} 2/3 & -1/3 \\ -1/3 & 2/3 \end{pmatrix}$;   (d) $\begin{pmatrix} 1 & 1 \\ -1 & 1 \end{pmatrix}\begin{pmatrix} 1/2 & -1/2 \\ 1/2 & 1/2 \end{pmatrix}$.   **2.** (a) $-5, -2; -1, 1$;

(b) $3, -7; 2, 3/2$;   (c) $4, -1; 1, 0$;   (d) $1, -5; 1/2, 3/2$.   **3.** (a) $\begin{pmatrix} 1 & 0 & 0 \\ -1 & 1 & 0 \\ 1 & 1 & 2 \end{pmatrix}, \begin{pmatrix} 1 & 0 & 0 \\ 1 & 1 & 0 \\ -1 & -1/2 & 1/2 \end{pmatrix}$;

(b) $\begin{pmatrix} 2 & 1 & 2 \\ 0 & -1 & 1 \\ 0 & 0 & -1 \end{pmatrix}\begin{pmatrix} 1/2 & 1/2 & 3/2 \\ 0 & -1 & -1 \\ 0 & 0 & -1 \end{pmatrix}$;   (c) $\begin{pmatrix} 2 & 1 & 1 \\ 1 & 0 & 1 \\ -1 & 1 & 1 \end{pmatrix}, -1/3\begin{pmatrix} -1 & 0 & 1 \\ -2 & 3 & -1 \\ 1 & -3 & -1 \end{pmatrix}$;

(d) $\begin{pmatrix} 1 & 1 & 0 \\ 0 & 1 & -1 \\ -1 & 1 & 1 \end{pmatrix}, 1/3\begin{pmatrix} 2 & -1 & -1 \\ 1 & 1 & 1 \\ 1 & -2 & 1 \end{pmatrix}$.   **4.** (a) $2, -1, 1$;   (b) $3, -2, 1$;   (c) $4, 1, -2$;

(d) $3, 2, -2$.   **5.** $\begin{pmatrix} 1 & 1 & 1 & 1 \\ 0 & 1 & 1 & 1 \\ 0 & 0 & 1 & 1 \\ 0 & 0 & 0 & 1 \end{pmatrix}, \begin{pmatrix} 1 & -1 & 0 & 0 \\ 0 & 1 & -1 & 0 \\ 0 & 0 & 1 & -1 \\ 0 & 0 & 0 & 1 \end{pmatrix}$.

6. $\begin{pmatrix} 1 & -1 & 1 & -1 \\ 0 & 1 & -2 & 3 \\ 0 & 0 & 1 & -3 \\ 0 & 0 & 0 & 1 \end{pmatrix}$.   **7.** $\begin{pmatrix} 1 & 1 \\ -1 & 1 \end{pmatrix}, \begin{pmatrix} 1/2 & -1/2 \\ 1/2 & 1/2 \end{pmatrix}$.   **8.** (a) $\begin{pmatrix} -1 & 2 \\ -3 & 3 \end{pmatrix}$;   (b) $\begin{pmatrix} 3/2 & 9/4 \\ -1 & 1/2 \end{pmatrix}$;

(c) $\begin{pmatrix} 3 & 3 \\ -2 & -1 \end{pmatrix}$;   (d) $1/2\begin{pmatrix} 1 & 3 \\ -3 & 3 \end{pmatrix}$.   **9.** (a) $\begin{pmatrix} 0 & -1 & 1 \\ -1 & 1 & 0 \\ -2 & -1 & 2 \end{pmatrix}$;   (b) $\begin{pmatrix} 0 & -1/2 & -1/2 \\ 2 & 2 & 2 \\ 0 & 0 & 1 \end{pmatrix}$;

(c) $-1/3\begin{pmatrix} -4 & 1 & 1 \\ -8 & -4 & -1 \\ 7 & 2 & -1 \end{pmatrix}$;   (d) $1/3\begin{pmatrix} 6 & -1 & -2 \\ 0 & 1 & -1 \\ 3 & 1 & 2 \end{pmatrix}$.   **10.** $\begin{pmatrix} 0 & 1 & -1 & -1 \\ 0 & 0 & 2 & -1 \\ 0 & 0 & 0 & 3 \\ 0 & 0 & 0 & 0 \end{pmatrix}$.

11. $J_1 = (a_2, -a_1), J_2 = (a_1, a_2)$.

Section 3.8 – p. 122.   **1.** $(1, 1), (-1, 1)$.   **2.** $(1, -2), (2, 1)$.   **3.** $(1, 0, 0), (0, 1, 0)$,

$(0, 0, 1), \pm 1, \begin{pmatrix} 1 & 0 & 0 \\ 0 & 1 & 0 \\ 0 & 0 & -1 \end{pmatrix}$.   **4.** $(1, 0, -1), (1, -1, 0), (1, 1, 1), \pm 1, \begin{pmatrix} 1 & 0 & 0 \\ 0 & 1 & 0 \\ 0 & 0 & -1 \end{pmatrix}$.

5. $0, 1$.   **11.** No, 0.

CHAPTER 4

Section 4.1 – p. 127.   **1.** No.   **2.** No.   **6.** $D(D_i(a)) = a, D(E_{ij}) = -1, D(F_{ij}(a)) = 1$.

Section 4.2 – p. 131.   **1.** 6.   **2.** 2.   **3.** 0.   **4.** 2.   **5.** -5.   **6.** 8.   **7.** 0.   **8.** 4.
9. -34.   **10.** $1, 2$.   **11.** $1, 0, 4$.   **12.** $0, 2$.   **13.** $0, \sqrt{2}, -\sqrt{2}$.

Section 4.3 – p. 139.   **1.** 2.   **2.** 4.   **3.** 1.   **4.** 4.   **5.** 0.   **6.** 1.   **7.** -1.
8. -1.   **9.** 2.   **10.** (a) $1, -1$;   (b) $\sqrt{2}, -\sqrt{2}$;   (c) $3, -1$;   (d) $1, -1, 2$;
(e) $-1, 1, 3$;   (f) $0, 1, 2$;   (g) $0, 2, 3$;   (h) $1, 3, -4$.   **11.** $0, 2$.   **12.** $0, 1, -2$.
13. $-2, 4$.   **14.** $1, 1, 1$.   **15.** $2, 3, 4$.   **16.** $1, 3, 7$.

Section 4.5—p. 148. 1. $1/3\begin{pmatrix} 1 & 1 \\ -1 & 2 \end{pmatrix}$. 2. $1/3\begin{pmatrix} 1 & -2 \\ 1 & 1 \end{pmatrix}$. 3. $1/3\begin{pmatrix} 1 & -1 \\ 0 & 3 \end{pmatrix}$.

4. $1/5\begin{pmatrix} 1 & 2 & -1 \\ -2 & 1 & 2 \\ 2 & -1 & 3 \end{pmatrix}$. 5. $\begin{pmatrix} 2 & 1 & -1 \\ 2 & 2 & -1 \\ -1 & -1 & 1 \end{pmatrix}$. 6. $1/3\begin{pmatrix} -2 & -1 & 2 \\ 4 & 2 & -1 \\ 1 & 2 & -1 \end{pmatrix}$.

7. $1/2\begin{pmatrix} 1 & -1 & -1 & 1 \\ 1 & 1 & -1 & 1 \\ 1 & 1 & 1 & 1 \\ -1 & -1 & -1 & 1 \end{pmatrix}$. 8. $\begin{pmatrix} 0 & 1 & 0 & 0 \\ 0 & 0 & 1 & 0 \\ 1 & 0 & 0 & 0 \\ 0 & 0 & 0 & 1 \end{pmatrix}$. 9. Not applicable. 10. $x_1 = -2$, $x_2 = 3$, $x_3 = -2$. 11. $x_1 = 2, x_2 = 2, x_3 = 3, x_4 = 0$. 12. Not applicable.
13. Not applicable. 14. $x_1 = 1/8, x_2 = -5/8, x_3 = 1/4$. 16. $-2 + x/2 + x^2/2 = p(x)$.

17. $D\begin{pmatrix} 1 & x_0 & \cdots & y_0 & \cdots & x_0^n \\ 1 & x_1 & \cdots & y_1 & \cdots & x_1^n \\ \vdots & \vdots & & \vdots & & \vdots \\ 1 & x_n & \cdots & y_n & \cdots & x_n^n \end{pmatrix}$.

$$\prod_{\substack{k,j=0 \\ k>j}}^{n} (x_k - x_j)$$

Section 4.6—p. 154. 1. $1, 4; \begin{pmatrix} 1 & 2 \\ -1 & 1 \end{pmatrix}, 1/3\begin{pmatrix} 1 & -2 \\ 1 & 1 \end{pmatrix}$. 2. $1, 0; \begin{pmatrix} 1 & 0 \\ -1 & 1 \end{pmatrix}, \begin{pmatrix} 1 & 0 \\ 1 & 1 \end{pmatrix}$.

3. $5, -2; \begin{pmatrix} 1 & 2 \\ 3 & -1 \end{pmatrix}, 1/7\begin{pmatrix} 1 & 2 \\ 3 & -1 \end{pmatrix}$. 4. $-2, -3; \begin{pmatrix} -1 & -1 \\ 2 & 3 \end{pmatrix}, \begin{pmatrix} -3 & -1 \\ 2 & 1 \end{pmatrix}$.

5. $0, 1, 2; \begin{pmatrix} 1 & 0 & 1 \\ -2 & 1 & 2 \\ -3 & 1 & 3 \end{pmatrix}, 1/2\begin{pmatrix} 1 & 1 & -1 \\ 0 & 6 & -4 \\ 1 & -1 & 1 \end{pmatrix}$. 6. $0, 1, 3; \begin{pmatrix} 1 & 1 & 1 \\ 0 & 1 & 3 \\ 0 & 1 & 9 \end{pmatrix}, 1/6\begin{pmatrix} 6 & -8 & 2 \\ 0 & 9 & -3 \\ 0 & -1 & 1 \end{pmatrix}$.

7. $0, 1, 2; \begin{pmatrix} 1 & 1 & 1 \\ 0 & 1 & 2 \\ 0 & 1 & 4 \end{pmatrix}, 1/2\begin{pmatrix} 2 & -3 & 1 \\ 0 & 4 & -2 \\ 0 & -1 & 1 \end{pmatrix}$. 8. $2i, -2i; \begin{pmatrix} i & i \\ 1 & -1 \end{pmatrix}, 1/2\begin{pmatrix} -i & 1 \\ -i & -1 \end{pmatrix}$.

9. $-1, 5; \begin{pmatrix} 1 & 1 \\ -1 & 5 \end{pmatrix}, 1/6\begin{pmatrix} 5 & -1 \\ 1 & 1 \end{pmatrix}$. 10. $0, i, -i; \begin{pmatrix} 1 & i & i \\ 0 & -1 & 1 \\ 0 & -i & -i \end{pmatrix}, 1/2\begin{pmatrix} 2 & 0 & 2 \\ 0 & -1 & i \\ 0 & 1 & i \end{pmatrix}$.

14. $y_1(t) = e^{-t}, y_2(t) = -e^{-t}$. 15. $y_1(t) = (e^{2t} + e^{4t})/2, y_2(t) = (e^{4t} - e^{2t})/2$.
16. $y_1(t) = (e^t - e^{-3t})/4, y_2(t) = (e^t + 3e^{-3t})/4$. 17. $y_1(t) = e^{2t}, y_2(t) = -3e^t + 2e^{2t}$, $y_3(t) = -3e^t + 3e^{2t}$. 18. $y_1(t) = (e^{2t} + e^{-4t})/2, y_2(t) = (e^{2t} - e^{-4t})/2, y_3(t) = (e^{6t} + e^{2t})/2$.

Section 4.7—p. 159. 1. $y(t) = e^{-t}$. 2. $y(t) = (e^{4t} + 4e^{-t})/5$. 3. $y(t) = e^{-t}\sin t$.
4. $y(t) = 5/4 + (1/12)e^{-4t} - (1/3)e^{-t}$. 5. $y(t) = ae^t + be^{-t}$. 6. $y(t) = ae^{2t} + be^{-t}$.
7. $y(t) = a + be^{-3t} + ce^t$. 8. $y(t) = ae^t + be^{-t} + ce^{2t} + de^{-2t}$.

CHAPTER 5

Section 5.1—p. 164. 1. Rotation, $\theta = \pi/4$. 2. Reflection, $X = (1, \sqrt{2} - 1)$.
3. Rotation, $\theta = -\arccos(3/5)$. 4. Reflection, $X = (-1, 2)$. 5. Rotation, $\theta = \arccos(2/\sqrt{5})$. 6. Rotation, $\theta = \pi + \arccos(5/13)$. 7. Rotation, $\theta = \arccos(1/\sqrt{5})$.
8. Reflection, $X = (3, \sqrt{13} + 2)$. 9. $X_1 = (1, \sqrt{5} - 2), X_2 = (-1, \sqrt{5} + 2)$.
10. $X_1 = (5, 25), X_2 = (-5, 1)$. 11. $X_1 = (1, \sqrt{2} + 1), X_2 = (-1, \sqrt{2} - 1)$.
12. $X_1 = (-3, 1), X_2 = (1, 3)$. 13. $\cos\theta = b/\sqrt{2-2a}$.

Section 5.2—p. 169. 1. Rotation of 240° about $J_1 = 1/\sqrt{3}\,(1, 1, 1)$, $\begin{pmatrix} 1 & 0 & 0 \\ 0 & -1/2 & \sqrt{3}/2 \\ 0 & -\sqrt{3}/2 & -1/2 \end{pmatrix}$.

2. Rotation of 120° about $J_1 = 1/\sqrt{3}\,(1, 1, 1)$, $\begin{pmatrix} 1 & 0 & 0 \\ 0 & -1/2 & -\sqrt{3}/2 \\ 0 & \sqrt{3}/2 & -1/2 \end{pmatrix}$. 3. Reflection through plane spanned by $(0, 1, 0)$ and $1/\sqrt{2}\,(1, 0, 1)$, $\begin{pmatrix} -1 & 0 & 0 \\ 0 & 1 & 0 \\ 0 & 0 & 1 \end{pmatrix}$. 4. Rotation through

180° about $J_1 = 1/\sqrt{2}\,(1, 0, 1)$, $\begin{pmatrix} 1 & 0 & 0 \\ 0 & -1 & 0 \\ 0 & 0 & -1 \end{pmatrix}$. **5.** Reflection through plane spanned by $1/\sqrt{2}\,(1, 1, 0)$ and $1/\sqrt{18}\,(1, -1, 4)$ followed by rotation of $\theta = \arccos(4/5)$ about $J_1 = 1/3\,(2, -2, -1)$, $\begin{pmatrix} -1 & 0 & 0 \\ 0 & 4/5 & -3/5 \\ 0 & 3/5 & 4/5 \end{pmatrix}$. **6.** Reflection through plane spanned by $1/\sqrt{2}\,(1, 0, 1)$ and $1/\sqrt{18}\,(1, 4, -1)$ followed by a rotation of $\theta = \pi + \arccos 4/5$ about $J_1 = 1/3\,(-2, 1, 2)$, $\begin{pmatrix} -1 & 0 & 0 \\ 0 & 4/5 & 3/5 \\ 0 & -3/5 & 4/5 \end{pmatrix}$. **7.** Reflection through plane spanned by $1/\sqrt{2}\,(-1, 1, 0)$ and $1/\sqrt{22}\,(3, 3, 2)$ followed by a rotation of $\theta = \arccos(5/6)$ about $J_1 = 1/\sqrt{11}\,(1, 1, -3)$, $\begin{pmatrix} -1 & 0 & 0 \\ 0 & 5/6 & -\sqrt{11}/6 \\ 0 & \sqrt{11}/6 & 5/6 \end{pmatrix}$.

Section 5.3 – p. 173.   **1.** $0, 2$; $1/\sqrt{2}\begin{pmatrix} 1 & -1 \\ 1 & 1 \end{pmatrix}$.   **2.** $1, -1$; $1/\sqrt{2}\begin{pmatrix} 1 & 1 \\ -1 & 1 \end{pmatrix}$.

3. $5, -5$; $1/\sqrt{10}\begin{pmatrix} 3 & -1 \\ 1 & 3 \end{pmatrix}$.   **4.** $5, 10$; $1/\sqrt{5}\begin{pmatrix} 2 & -1 \\ 1 & 2 \end{pmatrix}$.   **5.** $0, 1/2, -2$; $1/\sqrt{2}\begin{pmatrix} 0 & \sqrt{2} & 0 \\ 1 & 0 & 1 \\ 1 & 0 & -1 \end{pmatrix}$.

6. $1, 1, 10$; $\begin{pmatrix} 2/3 & 1/\sqrt{2} & 2/3 \\ -1/3 & -1/\sqrt{2} & 2/3 \\ -2/3 & 0 & 1/3 \end{pmatrix}$.   **7.** $-1, 5, 2$; $1/3\begin{pmatrix} -2 & 1 & 2 \\ 2 & 2 & 1 \\ -1 & 2 & -2 \end{pmatrix}$.

8. $0, 25, 25$; $1/5\begin{pmatrix} -3 & 4 & 0 \\ 0 & 0 & 5 \\ 4 & 3 & 0 \end{pmatrix}$.   **9.** $1, -1, 1, -1$; $1/\sqrt{2}\begin{pmatrix} 1 & 1 & 0 & 0 \\ -1 & 1 & 0 & 0 \\ 0 & 0 & 1 & 1 \\ 0 & 0 & -1 & 1 \end{pmatrix}$.

Section 5.4 – p. 179.   **1.** $\begin{pmatrix} 2 & -3/2 \\ -3/2 & 1 \end{pmatrix}$.   **2.** $\begin{pmatrix} 0 & 1/2 \\ 1/2 & -4 \end{pmatrix}$.   **3.** $\begin{pmatrix} 1 & 0 & -1/2 \\ 0 & 1 & 0 \\ -1/2 & 0 & -1 \end{pmatrix}$.

4. $\begin{pmatrix} 3 & 1/2 & 0 \\ 1/2 & 1 & -1/2 \\ 0 & -1/2 & 0 \end{pmatrix}$.   **5.** $\begin{pmatrix} 4 & 8 & 16 \\ 8 & 16 & 32 \\ 16 & 32 & 64 \end{pmatrix}$.   **6.** $\begin{pmatrix} 1 & 2 & 3 & 4 \\ 2 & 4 & 6 & 8 \\ 3 & 6 & 9 & 12 \\ 4 & 8 & 12 & 16 \end{pmatrix}$.

7. $2y_1^2 - 2y_2^2$, $\begin{pmatrix} 1/\sqrt{2} & -1/\sqrt{2} \\ 1/\sqrt{2} & +1/\sqrt{2} \end{pmatrix}$.   **8.** $5y_2^2$, $1/\sqrt{5}\begin{pmatrix} 2 & 1 \\ -1 & 2 \end{pmatrix}$.   **9.** $4y_1^2 - 2y_2^2$, $1/\sqrt{2}\begin{pmatrix} 1 & 1 \\ -1 & 1 \end{pmatrix}$.

10. $-10y_1^2 + 15y_2^2$, $1/\sqrt{5}\begin{pmatrix} 2 & -1 \\ 1 & 2 \end{pmatrix}$.   **11.** $y_1^2 + 49y_2^2$, $1/\sqrt{2}\begin{pmatrix} 1 & 1 \\ -1 & 1 \end{pmatrix}$.

12. $3y_1^2 - 2y_2^2$, $1/\sqrt{6}\begin{pmatrix} \sqrt{2} & 0 & 2 \\ \sqrt{2} & \sqrt{3} & -1 \\ -\sqrt{2} & \sqrt{3} & 1 \end{pmatrix}$.   **13.** $y_1^2 - 2y_2^2 + 4y_3^2$, $1/3\begin{pmatrix} 2 & 1 & 2 \\ 1 & 2 & -2 \\ -2 & 2 & 1 \end{pmatrix}$.

14. $y_2^2 + 4y_3^2$, $P = \begin{pmatrix} 0 & 1/\sqrt{3} & 2/\sqrt{6} \\ 1/\sqrt{2} & 1/\sqrt{3} & -1/\sqrt{6} \\ 1/\sqrt{2} & -1/\sqrt{3} & 1/\sqrt{6} \end{pmatrix}$.   **15.** Hyperbola.   **16.** Parallel lines.

17. Hyperbola.   **18.** Parallel lines.   **19.** Hyperbola.   **20.** Ellipse.
21. Parallel lines.   **22.** Hyperbola.   **23.** Parabola.

Section 5.5 – p. 187.   **1.** $z_1^2 - z_2^2 + z_3 = 0$; hyperbolic paraboloid; $(0, 0, 0) \xrightarrow{T} (0, 0, 1)$.
2. $z_1^2 + z_2^2 - z_3^2 = 5$; hyperboloid of one sheet; $(0, 0, 0) \xrightarrow{T} (-1, 0, 0)$.
3. $z_1^2 + 4z_2^2 + 2z_3^2 = -6$; empty set; $(0, 0, 0) \xrightarrow{T} (1, -1/2, 0)$.
4. $z_1^2 + 2z_2^2 + z_3 = 0$; elliptic paraboloid; $(0, 0, 0) \xrightarrow{T} (1, 0, 5)$.
5. $4z_1 + z_2^2 - 3z_3^2 = 0$; hyperbolic paraboloid; $(0, 0, 0) \xrightarrow{T} (0, -1, -9/4)$.
6. $2z_1^2 - z_2^2 - z_3^2 = 2$; hyperboloid of two sheets; $(0, 0, 0) \xrightarrow{T} (-1, 0, -1)$.
7. $z_3^2 = 5$; two parallel planes; $(0, 0, 0) \xrightarrow{T} (0, 0, 2)$.

8. $z_2^2 + 5z_1 = 0$; parabolic cylinder; $R = \begin{pmatrix} 1 & 0 & 0 \\ 0 & -3/5 & -4/5 \\ 0 & 4/5 & -3/5 \end{pmatrix}$.

9. $2z_1^2 + \sqrt{10}\,z_2 = 0$; parabolic cylinder $(0, 0, 0) \xrightarrow{T} (1, 0, 1/3)$; $R = \begin{pmatrix} 1 & 0 & 0 \\ 0 & -1/\sqrt{10} & -3/\sqrt{10} \\ 0 & +3/\sqrt{10} & -1/\sqrt{10} \end{pmatrix}$.

10. $z_1^2 + 2z_2^2 + z_3 = 0$; elliptic paraboloid $(0, 0) \xrightarrow{T} (1, 0, 2)$.   **11.** Hyperboloid of one sheet.
12. Elliptic cone.   **13.** Hyperboloid of two sheets.   **14.** Hyperbolic cylinder.
15. Hyperboloid of two sheets.   **16.** Elliptic cone.   **17.** Hyperboloid of two sheets.
18. Pair of planes.   **19.** Hyperbolic cylinder.   **20.** Two parallel planes.

CHAPTER 6

Section 6.1 – p. 193.   **1.** $(f-g)(1, 2) = (-8/3, -5)$, $(f+g)(3, 4) = (50/7, 15)$,
$(f \circ 3g)(1, 2) = (1/21, -1/3)$, $(g \circ f)(0, 1) = (0, 1)$.   **2.** $(2f+A)(1, 0) = (4, 2e-1)$,
$(f-3A)(1, 2) = (-5, e-3)$, $(A \circ f)(1, 1) = (0, e)$, $(f \circ A^2)(0, 1) = (1, 1)$.
3. $(A \circ f)(X)$ cannot be computed, $(f \circ A)(X)$ cannot be computed,
$(f \circ B)(X) = ((x_1 - x_2)(2x_2 + x_3), (2x_2 + x_3)(x_1 - x_3), (x_1 - x_2)(x_1 - x_3))$,
$(B \circ f)(X) = (x_1 x_2 - x_2 x_3, 2x_2 x_3 + x_1 x_3, x_1 x_2 - x_1 x_3)$.
4. $(f-g)(x_1, x_2) = \left(\dfrac{x_1 - x_2}{x_1 + x_2} - x_1 x_2, e^{x_1} - (x_1 + x_2)^2\right)$,
dom $(f-g) = \{(x_1, x_2) | x_1 \neq -x_2\} =$ dom $(f+g) =$ dom $(2f - g)$,
$(f+g)(x_1, x_2) = \left(\dfrac{x_1 - x_2}{x_1 + x_2} + x_1 x_2, e^{x_1} + (x_1 + x_2)^2\right)$,
$(2f - g)(x_1, x_2) = \left(\dfrac{2(x_1 - x_2)}{x_1 + x_2} - x_1 x_2, 2e^{x_1} - (x_1 + x_2)^2\right)$,
$(f \circ g)(x_1, x_2) = \left(\dfrac{x_1 x_2 - (x_1 + x_2)^2}{x_1 x_2 = (x_1 + x_2)^2}, e^{x_1 x_2}\right)$,
$(g \circ f)(x_1, x_2) = \left(e^{x_1}\left[\dfrac{x_1 - x_2}{x_1 + x_2}\right], \left[e^{x_1} + \dfrac{x_1 - x_2}{x_1 + x_2}\right]^2\right)$,
dom $(f \circ g) = \{(x_1, x_2) | x_1 x_2 + x_1^2 + x_2^2 \neq 0\}$, dom $(g \circ f) = \{(x_1, x_2) | x_1 \neq -x_2\}$.
5. $(f-g)(x_1, x_2) = (\sqrt{x_1 x_2} - x_1 x_2, x_1 - x_2 - \sqrt{x_1 - x_2})$,
$(f+g)(x_1, x_2) = (\sqrt{x_1 x_2} + x_1 x_2, x_1 - x_2 + \sqrt{x_1 - x_2})$,
$(f-3g)(x_1, x_2) = (\sqrt{x_1 x_2} - 3x_1 x_2, x_1 - x_2 - 3\sqrt{x_1 - x_2})$,
dom $(f-g) =$ dom $(f+g) =$ dom $(f-3g) = \{(x_1, x_2) | x_1 x_2 \geq 0\} \cap \{(x_1, x_2) | x_1 \geq x_2\}$,
$(f \circ 2g)(x_1, x_2) = (2\sqrt{x_1 x_2} \sqrt{x_1 - x_2}, 2x_1 x_2 - 2\sqrt{x_1 - x_2})$,
dom $(f \circ 2g) = \{(x_1, x_1) | x_1 \geq x_2\} \cap \{(x_1, x_2) | x_1 x_2 \sqrt{x_1 - x_2} \geq 0\}$.
6. $AX = (2x_1 - x_2)$, $(f \circ A)(X) = (2x_1 - x_2, (2x_1 - x_2)^2)$, $(A \circ f)(X)$ cannot be
computed for $X \in V_2(R)$. For $x \in R$, $(A \circ f)(x) = 2x - x^2$; $A(0, 1) = -1$;
$A(-1, 1) = -3$; $A(2, 1) = 3$; $(f \circ A)(0, 1) = (-1, 1)$;
$(f \circ A)(-1, 1) = (-3, 9)$, $(f \circ A)(2, 1) = (3, 9)$; $(A \circ f)(1) = 1$; $(A \circ f)(2) = 0$.
7. $AX = x_1 - x_2 + 2x_3$; $(f \circ A)(X) = (x_1 - x_2 + 2x_3, x_1 - x_2 + 2x_3 - 1, (x_1 - x_2 + 2x_3 - 2)^2)$,
$(A \circ f)(x) = x - (x-1) + 2(x-1)^2$; $A(2, 0, 1) = 4$; $(f \circ A)(2, 0, 1) = (4, 3, 4)$;
$(A \circ f)(2) = (2, 1, 0)$.   **8.** $(B \circ A \circ g)(X) = 2x_1 x_2 - \dfrac{2x_1}{x_2}$,
$(f \circ B \circ A \circ A)(X) = \left((4x_1 - 4x_2)^2, \dfrac{1}{4x_1 - 4x_2}\right)$, $(f \circ h \circ A)(X) = \left(x_2^2(2x_1 - x_2)^2, \dfrac{1}{x_2(2x_1 - x_2)}\right)$,
$(A \circ f \circ h \circ g)(X) = \left(2x_1^4 - \dfrac{1}{x_1^2}, \dfrac{1}{x_1^2}\right)$, $(B \circ A \circ g)(1, -1) = 0$, $(f \circ B \circ A \circ A)(1, -1) = (64, 1/8)$,
$(f \circ h \circ A)(1, -1) = (9, -1/3)$, $(A \circ f \circ h \circ g)(1, -1) = (1, 1)$.

Section 6.2 – p. 201.   **6.** $\{(x_1, x_2) : x_1 - 2x_2 \neq 0\}$.   **7.** $\{(x_1, x_2) : x_1^2 - x_2^2 \neq 0\}$.
8. $\{(x_1, x_2) : x_1^2 - 4x_2^2 \neq 0\}$.   **9.** $\{(x_1, x_2) : x_1 + x_2 \neq 0 \text{ and } x_1^2 + x_2^2 \neq 1\}$.
15. Not open.   **16.** Open.   **17.** Open.   **18.** Not open.   **19.** Not open.   **20.** Open.
21. Not open.   **22.** Open.

Section 6.3 – p. 206.   **1.** $f_{x_1} = 4(2x_1 + x_2)$; $f_{x_2} = 2(2x_1 + x_2)$.   **2.** $f_{x_1} = \dfrac{-2x_2}{(x_1 - x_2)^2}$;
$f_{x_2} = \dfrac{2x_1}{(x_1 - x_2)^2}$.   **3.** $f_{x_1} = \dfrac{\log x_2 [1 - x_1]}{e^{x_1}}$; $f_{x_2} = \dfrac{x_1}{e^{x_1} x_2}$.   **4.** $f_{x_1} = \dfrac{\cos(x_1/x_2)}{x_2}$;
$f_{x_2} = \dfrac{-x_1 \cos(x_1/x_2)}{x_2^2}$.   **5.** $f_{x_1} = \dfrac{-1}{(x_1 + x_2)^2 + 1}$; $f_{x_2} = \dfrac{-1}{(x_1 + x_2)^2 + 1}$.

6. $f_x = \dfrac{-3y}{(x-y)^2}, f_y = \dfrac{3x}{(x-y)^2}, f_{xx} = \dfrac{6y}{(x-y)^3}, f_{yy} = \dfrac{6x}{(x-y)^3}, f_{xy} = \dfrac{-3(x+y)}{(x-y)^3}$.

7. $f_x = \cos y, f_y = -x \sin y, f_{xx} = 0, f_{yy} = -x \cos y, f_{xy} = -\sin y$.

8. $f_x = \dfrac{e^{xy}}{x-y} + y \log(x-y) e^{xy}$, $f_y = \dfrac{-e^{xy}}{x-y} + x \log(x-y) e^{xy}$,

$f_{xx} = e^{xy} \left[\dfrac{-1 + 2y(x-y)}{(x-y)^2} + y^2 \log(x-y) \right]$, $f_{yy} = e^{xy} \left[\dfrac{-1 - 2x(x-y)}{(x-y)^2} + x^2 \log(x-y) \right]$,

$f_{xy} = e^{xy} \left[1 + \dfrac{1}{(x-y)^2} + (1+xy) \log(x-y) \right]$.   9. $f_x = a \cos(ax + by)$,

$f_y = b(\cos(ax+by)), f_{xx} = a^2 \sin(ax+by), f_{yy} = -b^2 \sin(ax+by), f_{xy} = -ab \sin(ax+by)$.

10. $f_x = \dfrac{1}{\sqrt{y^2 - x^2}}, f_y = \dfrac{-x}{y\sqrt{y^2 - x^2}}, f_{xx} = \dfrac{x}{(\sqrt{y^2-x^2})^3}, f_{yy} = x\left[\dfrac{1}{(\sqrt{y^2-x^2})^3} + \dfrac{1}{y^2\sqrt{y^2-x^2}}\right]$,

$f_{xy} = \dfrac{-y}{(\sqrt{y^2-x^2})^3}$.   11. $-1/4$.   12. $\pm 1/6$.

13. $f_{xy} = -x^2 y (x^2 + y^2)^{-3/2} + y(x^2+y^2)^{-1/2} = f_{yx}$.   14. $f_{xy} = \dfrac{y-x}{(x+y)^3} = f_{yx}$.

15. $f_{xy} = \cos y - \sin x = f_{yx}$.   24. $\dfrac{\partial f}{\partial x_1} = \left(\dfrac{-2x_2}{(x_1-x_2)^2}, e^{x_1-x_2} \right)$, $\dfrac{\partial f}{\partial x_2} = \dfrac{2x_1}{(x_1-x_2)^2}, e^{x_1-x_2}$,

$\left(\dfrac{\partial f}{\partial x_1}\right)(2,1) = (-2, e); \left(\dfrac{\partial f}{\partial x_2}\right)(2,1) = (4, -e)$.   25. $\dfrac{\partial f}{\partial x_1} = \left(\dfrac{1}{x_2} \cos \dfrac{x_1}{x_2}, -x_2 \sin x_1 x_2, 0 \right)$,

$\dfrac{\partial f}{\partial x_2} = \left(-\dfrac{x_1}{x_2^2} \cos \dfrac{x_1}{x_2}, -x_1 \sin x_1 x_2, -\dfrac{x_3}{x_2^2} \sec^2 \dfrac{x_3}{x_2} \right), \dfrac{\partial f}{\partial x_3} = \left(0, 0, \dfrac{1}{x_2} \sec^2 \dfrac{x_3}{x_2} \right), \dfrac{\partial f}{\partial x_1}(\pi, 1, \pi) = (-1, 0, 0)$,

$\dfrac{\partial f}{\partial x_2}(\pi, 1, \pi) = (\pi, 0, -\pi), \dfrac{\partial f}{\partial x_3}(\pi, 1, \pi) = (0, 0, 1)$.

Section 6.4 — p. 211.   1. $f'(X) = (2(x_1 - x_2), -2(x_1 - x_2))$.   2. $f'(X) = \begin{pmatrix} 1 & 1 \\ 1 & -1 \end{pmatrix}$,

$f'(X) = \begin{pmatrix} x_3 & 0 & x_1 \\ -1 & 1 & 0 \end{pmatrix}$.   4. $f'(X) = \begin{pmatrix} x_2 & x_1 & \\ -1/x_1^2 & 0 & \\ -2x_2/(x_1+x_2)^2 & 2x_1/(x_1+x_2)^2 & \end{pmatrix}$.

5. $f'(t) = \begin{pmatrix} 1 \\ \cos t \\ -\sin t \end{pmatrix}$.   6. $f'(X) = \dfrac{1}{|X|}(x_1, x_2, \ldots, x_n) = \dfrac{X}{|X|}$.

7. $\begin{pmatrix} \dfrac{\partial^2 g}{(\partial x_1)^2} & \dfrac{\partial^2 g}{\partial x_2 \partial x_1} & \cdots & \dfrac{\partial^2 g}{\partial x_n \partial x_1} \\ \dfrac{\partial^2 g}{\partial x_1 \partial x_2} & \dfrac{\partial^2 g}{\partial x_2 \partial x_2} & \cdots & \dfrac{\partial^2 g}{\partial x_n \partial x_2} \\ \vdots & & & \vdots \\ \dfrac{\partial^2 g}{\partial x_1 \partial x_n} & & \cdots & \dfrac{\partial^2 g}{\partial x_n \partial x_n} \end{pmatrix}$.   8. 0.   9. $2\sqrt{2}$.   10. $2\sqrt{2}$.   11. $9/\sqrt{10}$.

12. $\dfrac{\sqrt{\pi^2 - 4\pi + 8}}{\pi}$.   13. .48 sq. in.   14. $10^{-4}/14$ cu. in.   15. $.09\pi$ cu. in.

16. $10^4/16$ ft.   17. $1.25(2+\sqrt{3})$.   18. .5.   22. To show $\partial f/\partial x_j, j=1,2$, exist at $(0,0)$, use the definition of partial derivative. To show f is not differentiable at $(0,0)$, use Theorem 6.6 and known facts.

Section 6.5 — p. 218.   1. $L(t) = (3, 12) + (t-2)(4, 12)$.

2. $L(t) = (2, 0) + (t - \pi/2)(0, -4)$.   3. $L(t) = (1, 0, 1) + (t-1)(2, 1, 3)$.

4. $L(t) = (\pi/4, \sqrt{2}/2, \sqrt{2}/2) + (t - \pi/4)(1, \sqrt{2}/2, -\sqrt{2}/2)$.

5. $L(t) = (1, -1, 1, -1) + (t+1)(0, 1, -2, 3)$.   6. $4x_1 + 4x_2 - x_3 + 1 = 0$.

7. $-15x_1 - 3x_2 + x_3 = -18$.   8. $x_1 - x_2 - 2x_3 = 0$.   9. $x_1 + x_2 + 2x_3 = 2$.

10. $y_3 = 0$.   11. $y_1 + y_2 + 2y_3 = 4$.   12. $y_1 + y_3 = 0$.   13. $-\sqrt{2} y_2 + y_3 = 0$.

14. $(12, 9, 4)$.   16. $(X - C, \partial f/\partial x_1 \times \partial f/\partial x_2) = 0$, where $C = (c_1, c_2, c_3)$, c_i scalars.

17. $Y = \begin{pmatrix} 1 \\ 1 \\ 1 \\ 3 \end{pmatrix} + \begin{pmatrix} 1 & 1 & 0 \\ 0 & 1 & 1 \\ 1 & 0 & 1 \\ 1 & 1 & 1 \end{pmatrix} \begin{pmatrix} x_1 - 1 \\ x_2 - 1 \\ x_3 - 1 \end{pmatrix}$.   18. $(1, 1, 1, -2); y_1 + y_2 + y_3 - 2y_4 = -3$.

20. $(\partial g/\partial x_1, \partial g/\partial x_2, \ldots, \partial g/\partial x_n, -1)$.

Section 6.6 – p. 225.   **1.** -4.   **2.** 0.   **3.** $-1/4$.   **4.** $\cos(x_1 - x_2)$.
5. $-2(x_2 + 2x_1)/(x_1 + x_2)^2$.   **6.** e^{-x_1/x_2}.   **7.** $3x_1 - x_2 + 3x_3$.   **8.** $8(x_1 + x_2 - x_3)$.
9. $\dfrac{\cos(x_1 + x_2)\cos(x_2 - x_3) + \sin(x_1 + x_2)\sin(x_2 - x_3)}{\cos^2(x_2 - x_3)} = \dfrac{\cos(x_1 + x_3)}{\cos^2(x_2 - x_3)}$.
10. $2x - 2y + z$; $y^2 + z - 2xy + x$.   **11.** $4x + 2y - 2z$; $2x^2 + 2y^2 + 2z^2$; 20.
12. $(-1, 3)$.   **13.** $(3\sqrt{2}/2, -3)$.   **14.** $(1, -2, -1)$.   **15.** $24/\sqrt{13}$.
16. $(3 - \sqrt{2})/\sqrt{6}$.   **17.** $(20\sqrt{3} - 48)/\sqrt{22}$.   **18.** $Y = \pm 1/5 \,(4, -3)$.
19. $\pm 1/\sqrt{145}\,(9, -8)$.   **20.** $Y = -1/\sqrt{10}\,(8, 1)$.   **21.** $X_0 = (3/2, 1)$.
22. $X_0 = (1/\sqrt{3}, 1)$.   **23.** $X_0 = (0, -1/2, 1)$.   **24.** $X_0 = (3/2, 0, 1)$.

Section 6.7 – p. 228.   **1.** $\nabla f = (2x_1, -2x_2)$.   **2.** $\nabla f = (-2x_2/(x_1 - x_2)^2, 2x_1/(x_1 - x_2)^2)$.
3. $\nabla f = (x_1 \sec^2(x_1 + x_2) + \tan(x_1 + x_2), x_1 \sec^2(x_1 + x_2))$.
4. $\nabla f = (-(x_2 + x_3)/x_1^2 x_2 x_3, -(x_1 + x_3)/x_1 x_2^2 x_3, -(x_1 + x_2)/x_1 x_2 x_3^2)$.
5. $\nabla f = \left(\dfrac{x_3[\cos(x_1 - x_2) - 2x_1 \sin(x_1 - x_2)]}{e^{x_1^2 + x_2^2}}, \right.$
$\left. \dfrac{-x_3[\cos(x_1 - x_2) + 2x_2 \sin(x_1 - x_2)]}{e^{x_1^2 + x_2^2}}, \dfrac{\sin(x_1 - x_2)}{e^{x_1^2 + x_2^2}} \right)$.
6. $\nabla f = \left(\dfrac{2x_1}{\tan(1 + x_3^2)(x_1^2 + x_2^2)}, \dfrac{2x_2}{\tan(1 + x_3^2)(x_1^2 + x_2^2)}, -2x_3 \log(x_1^2 + x_2^2) \csc^2(1 + x_3^2) \right)$.
7. $(-1/(x^2 + y^2)^{1/2})(x, y)$.   **8.** $(1/\sqrt{146})(1, 8, -9)$.   **9.** $(-1/3\sqrt{2})(4, 1, 1)$.

Section 6.8 – p. 234.   **1.** $h'(t) = 2(t^3 - t^2 + 1)(3t^2 - 2t)$, $h'(2) = 80$.
2. $u'(t) = \cos(t\sqrt{1+t} - t^2)[(t/2\sqrt{1+t}) + \sqrt{1+t} - 2t]$.
3. $\partial h/\partial x_1 = 4(1 - x_2)(x_1 + x_2 - x_1 x_2)[(x_1 + x_2 - x_1 x_2)^2 - 3]$,
$\partial h/\partial x_2 = 4(1 - x_1)(x_1 + x_2 - x_1 x_2)[(x_1 + x_2 - x_1 x_2)^2 - 3]$, at $(2, 1)$, $\partial h/\partial x_1 = 0$ and $\partial h/\partial x_2 = 8$.
5. 1.   **6.** $h'(2, 1) = \begin{pmatrix} 5 & -4 \\ 1 & 4 \end{pmatrix}$, $P'(1, -2) = \begin{pmatrix} 0 & 0 \\ -1/4 & 1/8 \end{pmatrix}$.
7. $(f \circ g)'(x, y) = \begin{pmatrix} 3\cos(3x - y) & -\cos(3x - y) \\ -\sin(x + 3y) & -3\sin(x + 3y) \end{pmatrix}$,
$(g \circ f)'(x, y) = \begin{pmatrix} 2\cos(x + y) - \sin(x - y) & 2\cos(x + y) + \sin(x - y) \\ \cos(x + y) + 2\sin(x - y) & \cos(x + y) - 2\sin(x - y) \end{pmatrix}$.
8. $\begin{pmatrix} -3 & 0 \\ 4 & 2 \end{pmatrix}$.

Section 6.9 – p. 244.   **1.** $\dfrac{1}{3}\left(\dfrac{\partial g}{\partial x_1} + \dfrac{\partial g}{\partial x_2}\right)$, $\dfrac{1}{3}\left(-\dfrac{\partial g}{\partial x_1} + 2\dfrac{\partial g}{\partial x_2}\right)$.   **2.** $2\dfrac{\partial f}{\partial y_1} - \dfrac{\partial f}{\partial y_2}$, $\dfrac{\partial f}{\partial y_1} + \dfrac{\partial f}{\partial y_2}$.
3. $\dfrac{1}{9}\left(\dfrac{\partial^2 g}{(\partial x_1)^2} + \dfrac{\partial^2 g}{\partial x_2 \partial x_1} + \dfrac{\partial^2 g}{\partial x_1 \partial x_2} + \dfrac{\partial^2 g}{(\partial x_2)^2}\right)$.   **4.** $\dfrac{1}{9}\left(\dfrac{\partial^2 g}{(\partial x_1)^2} - 2\dfrac{\partial^2 g}{\partial x_2 \partial x_1} - 2\dfrac{\partial^2 g}{\partial x_1 \partial x_2} + 4\dfrac{\partial^2 g}{(\partial x_2)^2}\right)$,
$\dfrac{1}{9}\left(\dfrac{-\partial^2 g}{(\partial x_1)^2} - \dfrac{\partial^2 g}{\partial x_2 \partial x_1} + 2\dfrac{\partial^2 g}{\partial x_1 \partial x_2} + 2\dfrac{\partial^2 g}{(\partial x_2)^2}\right)$.   **5.** $4\dfrac{\partial^2 f}{(\partial y_1)^2} - 2\dfrac{\partial^2 f}{\partial y_2 \partial y_1} - 2\dfrac{\partial^2 f}{\partial y_1 \partial y_2} + \dfrac{\partial^2 f}{(\partial y_2)^2}$,

$\dfrac{\partial^2 f}{(\partial y_1)^2} + 2\dfrac{\partial^2 f}{\partial y_2 \partial y_1} + 2\dfrac{\partial^2 f}{\partial y_1 \partial y_2} + \dfrac{\partial^2 f}{(\partial y_2)^2}$.   **6.** $\cos^2\theta \dfrac{\partial^2 u}{(\partial x_1)^2} + 2\cos\theta \sin\theta \dfrac{\partial^2 u}{\partial x_1 \partial x_2} + \sin^2\theta \dfrac{\partial^2 u}{(\partial x_2)^2}$,

$-\sin\theta \dfrac{\partial u}{\partial x_1} + \cos\theta \dfrac{\partial u}{\partial x_2} - r\sin\theta\cos\theta \dfrac{\partial^2 u}{(\partial x_1)^2} + r(\cos^2\theta - \sin^2\theta)\dfrac{\partial^2 u}{\partial x_1 \partial x_2} + r\cos\theta\sin\theta \dfrac{\partial^2 u}{(\partial x_2)^2}$.
12. $\dfrac{\partial^2 u}{\partial r^2} = \cos^2\theta \sin^2\varphi \dfrac{\partial^2 u}{\partial x_1^2} + 2\sin\theta \cos\theta \sin^2\varphi \dfrac{\partial^2 u}{\partial x_1 \partial x_2}$

$+ 2\cos\theta \cos\varphi \sin\varphi \dfrac{\partial^2 u}{\partial x_3 \partial x_1} + \sin^2\theta \sin^2\varphi \dfrac{\partial^2 u}{\partial x_2^2} + 2\sin\theta \sin\varphi \cos\varphi \dfrac{\partial^2 u}{\partial x_3 \partial x_2} + \cos^2\varphi \dfrac{\partial^2 u}{\partial x_3^2}$,

$\dfrac{\partial^2 u}{\partial \theta^2} = -\dfrac{\partial u}{\partial x_1} r\cos\theta \sin\varphi - \dfrac{\partial u}{\partial x_2} r\sin\theta \sin\varphi + r^2 \sin^2\theta \sin^2\varphi \dfrac{\partial^2 u}{\partial x_1^2} - 2r\sin\theta\cos\theta \sin^2\varphi \dfrac{\partial^2 u}{\partial x_2 \partial x_1}$

$+ r^2 \cos^2\theta \sin^2\varphi \dfrac{\partial^2 u}{\partial x_2^2}$, $\dfrac{\partial^2 u}{\partial \varphi^2} = -\dfrac{\partial u}{\partial x_1} r\cos\theta \sin\varphi - \dfrac{\partial u}{\partial x_2} r\sin\theta \sin\varphi - \dfrac{\partial u}{\partial x_3} r\cos\varphi$

$+ r^2 \cos^2\theta \cos^2\varphi \dfrac{\partial^2 u}{\partial x_1^2} + r^2 \sin^2\theta \cos^2\varphi \dfrac{\partial^2 u}{\partial x_2^2} + r^2 \sin^2\varphi \dfrac{\partial^2 u}{\partial x_3^2}$

$+ 2r^2 \sin\theta \cos\theta \cos^2\varphi \dfrac{\partial^2 u}{\partial x_1 \partial x_2} - 2r^2 \cos\theta \sin\varphi \cos\varphi \dfrac{\partial^2 u}{\partial x_3 \partial x_1} - 2r^2 \sin\theta \sin\varphi \cos\varphi \dfrac{\partial^2 u}{\partial x_3 \partial x_2}$.

14. $\begin{pmatrix} \cos\theta \cos\gamma & -r\sin\theta \cos\gamma & -r\cos\theta \sin\gamma \\ \sin\theta \cos\gamma & r\cos\theta \cos\gamma & -r\sin\theta \sin\gamma \\ \sin\gamma & 0 & r\cos\gamma \end{pmatrix}$.

15. $\begin{pmatrix} \frac{e^\theta + e^{-\theta}}{2} & \frac{r(e^\theta - e^{-\theta})}{2} \\ \frac{e^\theta - e^{-\theta}}{2} & \frac{r(e^\theta + e^{-\theta})}{2} \end{pmatrix}$.

19. $\frac{1}{\sqrt{2}}\begin{pmatrix} 1 & -1 \\ 1 & 1 \end{pmatrix}$. **20.** $\frac{1}{\sqrt{2}}\begin{pmatrix} 1 & -1 \\ 1 & 1 \end{pmatrix}$. **21.** $\frac{1}{\sqrt{5}}\begin{pmatrix} 1 & -2 \\ 2 & 1 \end{pmatrix}$.

Section 6.10 – p. 253.   **1.** $D \neq 0$, all X; $\begin{pmatrix} 1/3 & 1/3 \\ -1/3 & 2/3 \end{pmatrix}$.   **2.** All (x_1, x_2) such that $x_1^2 \neq x_2^2$;

$\begin{pmatrix} \frac{x_2}{x_2^2 - x_1^2} & \frac{-x_1}{2x_2^2 - 2x_1^2} \\ \frac{-x_1}{x_2^2 - x_1^2} & \frac{x_2}{2x_2^2 - 2x_1^2} \end{pmatrix}$.   **3.** All (x_1, x_2) such that $x_2 \neq 0$ and $3x_1^2 x_2 - x_1^3 - 2x_1 x_2^2 \neq 0$;

$\frac{1}{6x_1^2 x_2 - 2x_1^3 - 4x_1 x_2^2} \begin{pmatrix} -x_1^2 & 2x_2^2(x_1 - x_2) \\ -2x_1 x_2 & 2x_2^2(x_1 - x_2) \end{pmatrix}$.

4. All (x_1, x_2) such that $\sin(x_1 + x_2) \cos(x_1 - x_2) \neq 0$;

$\begin{pmatrix} \frac{-1}{2\sin(x_1+x_2)} & \frac{1}{2\cos(x_1-x_2)} \\ \frac{-1}{2\sin(x_1+x_2)} & \frac{-1}{2\cos(x_1-x_2)} \end{pmatrix}$.

5. All x; $\frac{1}{5}\begin{pmatrix} 1 & 1 & -1 \\ -2 & 4 & 1 \\ -1 & 2 & -2 \end{pmatrix}$.

6. All (x_1, x_2, x_3) such that $x_1 x_2 x_3 \neq 0$; $\begin{pmatrix} \frac{1}{2x_2} & \frac{-x_1}{2x_2 x_3} & \frac{1}{2x_3} \\ \frac{1}{2x_1} & \frac{1}{2x_3} & \frac{-x_2}{2x_1 x_3} \\ \frac{-x_3}{2x_1 x_2} & \frac{1}{2x_2} & \frac{1}{2x_1} \end{pmatrix}$.

8. All (x_1, x_2) such that $x_2 \neq 0$; $-2x_1/3x_2$.   **9.** For all (x_1, x_2); $-4x_1/(3x_2^2 + 1)$.
10. For all (x_1, x_2) such that $-e^{x_1-x_2} + 2x_2 \neq 0$; $e^{x_1-x_2}/(e^{x_1-x_2} - 2x_2)$.
11. For all (x_1, x_2) such that $\sin(x_1 + x_2) + \frac{2x_2}{x_1^2 + x_2^2} \neq 0$; $-1 - \frac{2x_1 - 2x_2}{(x_1^2 + x_2^2)\sin(x_1+x_2) + 2x_2}$.
12. (8) All X such that $x_1 \neq 0$, $-3x_2/2x_1$;   (9) All X such that $x_1 \neq 0$, $-(3x_2^2 + 1)/4x_1$;
(10) All X, $(e^{x_1-x_2} - 2x_2)/e^{x_1-x_2}$; (11) All X such that $\sin(x_1 + x_2) + 2x_1/(x_1^2 + x_2^2) \neq 0$,
$-1 - \frac{2x_2 - 2x_1}{(x_1^2+x_2^2)\sin(x_1+x_2)+2x_1}$.   **13.** $(-3, 3)$.   **14.** $(1, -2)$.   **15.** $(0, 1)$.
16. $\nabla f = 0$, so Theorem 6.16 does not apply.

Section 6.11 – p. 259.   **1.** $(3, 1, -1)$.   **2.** $(0, 0, 1)$.   **3.** $\nabla f = 0$, so Theorem 6.18 does not apply.   **4.** $(0, 0)$.   **7.** $L(t) = (1, 1, 1) + t(1, -1, -1)$.   **8.** $x_1 - x_3 = 0$.
10. If $(Y) = \nabla f_1 \times \nabla f_2$, $T = \pm Y/|Y|$; $L(t) = A + t(\nabla f_1 \times \nabla f_2)$.
12. $L(t) = A + t(1, e/(1+e), -1/(1+e))$.   **13.** No.   **14.** $a^2 - r^2 = 1$.
15. $a = \pm\sqrt{3}$.   **16.** $y_0 z_0 x + x_0 z_0 y + x_0 y_0 z = 3x_0 y_0 z_0$.   **18.** $\partial v/\partial x = 4$, $\partial u/\partial y = 1$.

CHAPTER 7

Section 7.1 – p. 265.   **1.** $3 + 2(x - 1) + 2(y - 1) + 2(x - 1)(y - 1)$.
2. $3 + 4(x - 2) + 4(y - 1) + (x - 2)^2 + 4(x - 2)(y - 1) + (x - 2)^2(y - 1)$.
3. $-1 - (x - 1) + 3(y + 1) - 3(y + 1)^2 + 3(x - 1)(y + 1)$
$+ (y + 1)^3 - 3(x - 1)(y + 1)^2 + (x - 1)(y + 1)^3$.   **4.** $1 + x - y + \frac{1}{2}(x^2 - 2xy + y^2)$.
5. x.   **6.** $1 + x + y + x^2 + 2xy + y^2$.   **7.** $x - y - x^2/2 + y^2/2$.
9. $(x - 2y) - (x - 2y)^3/3! + (x - 2y)^5/5! + \cdots + (-1)^n((x - 2y)^{2n+1}/(2n+1)!)$.
10. $1 + (2x + y) + (2x + y)^2/2! + \cdots + (2x + y)^n/n!$.
11. $(x + y) - (x + y)^3/3 + (x + y)^5/5 - \cdots + (-1)^n((x + y)^{2n+1}/2n + 1)$.
12. $1 - (3x + 2y) + (3x + 2y)^2 - \cdots + (-1)^n(3x + 2y)^n$.   **13.** y^2.
14. $-6(x - 1)^3 + 18(x - 1)^2(y + 1)$.   **15.** $-6x^3 - 18x^2 y - 18x^2 y - 18xy^2 - 6y^3$.

418 ANSWERS TO EXERCISES

Section 7.3 – p. 276. **1.** Parabolic for all (x, y). **2.** Parabolic for all (x, y). **3.** Elliptic if $y < -1/24$, concave up; hyperbolic if $xy > -1/24$; parabolic if $xy = -1/24$. **4.** Elliptic if $x > 0$, concave up; hyperbolic if $x < 0$; parabolic if $x = 0$. **5.** Assume $x \neq 0$; hyperbolic if $x \neq 0$. **6.** Assume $y \neq 0$; parabolic for all (x, y). **7.** Assume $x \neq 0$; elliptic if $x, y > 0$ or $x, y < 0$; concave up if $x > 0$, down if $x < 0$; hyperbolic if $x > 0, y < 0$ or $x < 0, y > 0$; parabolic if $y = 0$. **8.** Elliptic if $a, b > 0$ or $a, b < 0$; concave up if $a > 0$, down if $a < 0$; hyperbolic if $a > 0, b < 0$ or $a < 0, b > 0$; parabolic if $a = 0$ or $b = 0$. **9.** Concave up if $z < 0$, down if $z > 0$. **10.** Hyperbolic. **11.** Elliptic.

Section 7.4 – p. 282. **1.** No local extreme at $(0, 0)$. **2.** No local extreme at $(0, 0)$, maximum at $(-4, -4)$. **3.** No local extreme at $(0, 0)$, maximum at $(-\sqrt[3]{4}, \sqrt[3]{2})$. **4.** Maximum at $(1, 0)$. **5.** Minimum at $(\sqrt[3]{4}, -\sqrt[3]{2})$. **6.** No local extreme at $(0, 0)$. **7.** No local extreme at $(0, \sqrt{6})$ and $(0, -\sqrt{6})$, minimum at $(2\sqrt{6}/5, \sqrt{6}/5)$, maximum at $(-2\sqrt{6}/5, -\sqrt{6}/5)$. **8.** Minimum at $(2 \cdot 4^{1/3}, 4^{1/3})$. **9.** Saddlepoint at $(0, 0, 0)$. **10.** Minimum at $(0, 0, 0)$. **11.** Minimum at $(0, 0)$, maximum at $(0, 2)$. **12.** Minimum at $(-1, -2)$, maximum at $(-1, 2)$. **13.** Minimum at $(0, 0)$, maximum at $(\pm 1, \pm 1)$. **14.** Minimum if $x + y = 1$, or $x = 0$, or $y = 0$, maximum at $(1/3, 1/3)$. **15.** $4096/27$. **16.** 12. **17.** $256\sqrt{3}/9$. **18.** $96\sqrt[3]{2}$. **19.** $8\sqrt{3}\,abc/9$.

Section 7.5 – p. 289. **1.** $2\sqrt{5}/5$. **2.** $\sqrt{2}$. **3.** Minimum $= 1$, maximum $= 3$. **4.** $(\sqrt{2}, -\sqrt{2}), (-\sqrt{2}, \sqrt{2})$. **5.** $2\sqrt{3}(-1, 1, -1)$. **6.** $(0, 0, \pm 2)$. **8.** $P = 4(a^2 + b^2)/\sqrt{a^2 + b^2}$; rectangle is $a^2/\sqrt{a^2 + b^2}$ by $b^2/\sqrt{z^2 + b^2}$. **9.** $V = a^3/12^3$. **10.** $r = \sqrt{2}a/2, h = \sqrt{2}a$. **11.** $D = 3\sqrt{2}/8$. **12.** $D = \sqrt{39}/9$. **14.** $V = 250{,}000/27$ cu. in.

CHAPTER 8

Section 8.2 – p. 297. **5.** (1) $F(x_1) = 0 \quad -1 \leq x_1 < 0$ (2) $F(x_1) = 0 \quad 0 \leq x_1 < 2$
$\qquad\qquad\qquad\qquad\quad = 3 \quad\; 0 \leq x_1 < 1 \qquad\qquad\qquad\;\; = 5 \quad\; 2 \leq x_1 < 4$

(3) $F(x_1) = 0 \quad 0 \leq x_1 < 1$ (4) $F(x_1) = 0 \quad 0 \leq x_1 \leq 1/2$
$\qquad\;\; = -1 \quad 1 \leq x_1 < \sqrt{2} \qquad\qquad\quad = 2 \quad 1/2 \leq x_1 < 1$
$\qquad\;\; = -2 \quad \sqrt{2} \leq x_1 < \sqrt{3} \qquad\qquad\;\; = 4 \quad 1 \leq x_1 < 3/2$
$\qquad\;\; = -3 \quad \sqrt{3} \leq x_1 < 2 \qquad\qquad\quad\;\; = 6 \quad 3/2 \leq x_1 < 2$
$\qquad\qquad\qquad\qquad\qquad\qquad\qquad\qquad\; = 8 \quad 2 \leq x_1 < 5/2$
$\qquad\qquad\qquad\qquad\qquad\qquad\qquad\quad\;\; = 10 \quad 5/2 \leq x_1 < 3.$

6. (1) $G(x_2) = 0 \quad 0 \leq x_2 < 1$ (2) $G(x_2) = 2 \quad 0 \leq x_2 < 1/2$
$\qquad\qquad\;\; = 1 \quad 1 \leq x_2 < 2 \qquad\qquad\qquad\;\; = 4 \quad 1/2 \leq x_2 < 1$
$\qquad\qquad\;\; = 2 \quad 2 \leq x_2 < 3 \qquad\qquad\qquad\;\; = 6 \quad 1 \leq x_2 < 3/2$
$\qquad\qquad\qquad\qquad\qquad\qquad\qquad\qquad\;\; = 8 \quad 3/2 \leq x_2 < 2$

(3) $G(x_2) = 3(\sqrt{2} + \sqrt{3} - 5) \quad\;\; -1 \leq x_2 < -2/3$ (4) $G(x_2) = 15/2 \quad 0 \leq x_2 \leq 2$.
$\qquad\;\; = 2(\sqrt{2} + \sqrt{3} - 5) \quad\;\; -2/3 \leq x_2 < -1/3$
$\qquad\;\; = (\sqrt{2} + \sqrt{3} - 5) \quad\;\;\;\; -1/3 \leq x_2 < 0$
$\qquad\;\; = 0 \qquad\qquad\qquad\qquad\quad 0 \leq x_2 < 1/3$
$\qquad\;\; = -\sqrt{2} - \sqrt{3} + 5 \qquad\;\; 1/3 \leq x_2 < 2/3$
$\qquad\;\; = 2(-\sqrt{2} - \sqrt{3} + 5) \quad\; 2/3 \leq x_2 < 1$

7. (1) $\int_R f = 3$, (2) $\int_R f = 10$, (3) $\int_R f = (\sqrt{2} + \sqrt{3} - 5)$, (4) $\int_R f = 15$. **8.** No.

Section 8.3 – p. 304. **1.** 44. **2.** 0. **3.** $(e - 1)^2/2$. **4.** $8/3$. **5.** 8. **6.** $e^4 - 1$. **7.** $32/5$. **8.** 54. **9.** $-15/2$. **10.** $6546/24$.

11. $\int_0^2 \left(\int_{-\sqrt{4-y^2}}^{\sqrt{4-y^2}} \sqrt{x^2 + y^2}\, dx \right) dy,\; \int_{-2}^{2} \left(\int_0^{\sqrt{4-x^2}} \sqrt{x^2 + y^2}\, dy \right) dx.$

12. $\int_{-\sqrt{2}}^{\sqrt{2}} \left(\int_y^{\sqrt{4-y^2}} dx \right) dy + \int_{-2}^{-\sqrt{2}} \left(\int_{-\sqrt{4-y^2}}^{\sqrt{4-y^2}} dx \right) dy,$

$\int_{-\sqrt{2}}^{\sqrt{2}} \left(\int_{-\sqrt{4-x^2}}^{x} dy \right) dx + \int_{\sqrt{2}}^{2} \left(\int_{-\sqrt{4-x^2}}^{\sqrt{4-x^2}} dy \right) dx.$

13. $\int_{-9}^{-1} \left(\int_{-2}^{(y+5)/2} x\, dx \right) dy + \int_{e^{-2}}^{e^2} \left(\int_{\log y}^{2} x\, dx \right) dy + \int_{-1}^{e^2} \left(\int_{-2}^{2} x\, dx \right) dy,\; \int_{-2}^{2} \left(\int_{2x-5}^{e^x} x\, dy \right) dx.$

14. $\int_0^{2/3} \left(\int_{1-y}^{y+1} x\,dx \right) dy + \int_{2/3}^{2} \left(\int_{1-y}^{(4-y)/2} x \right) dx\,dy + \int_2^4 \left(\int_{(y-4)/2}^{(4-y)/2} x\,dx \right) dy,$

$\int_{-1}^0 \left(\int_{1-x}^{4+2x} x^2\,dy \right) dx + \int_0^1 \left(\int_{1-x}^{4-2x} x^2\,dy \right) dx + \int_1^{5/3} \left(\int_{x-1}^{4-2x} x^2\,dy \right) dx.$ **15.** $2(b^3-a^3)/3.$

Section 8.4 — p. 312. **1.** $9/2.$ **2.** $(7+8\sqrt{2})/6.$ **3.** $16/3.$ **4.** $13/3.$
5. $24.$ **6.** $1/12.$ **7.** $16.$ **8.** $28\pi.$ **9.** $16a^3/3.$ **11.** $c/6;$ $(2/5, 2/5).$
12. $c/3;$ $(1/4, 3/8).$ **13.** $c/60;$ $(2/5, 2/5).$ **14.** $4c/3;$ $(-1/4, -1/4).$ **15.** $(1/5, 1/5).$
17. $(11/10, 23/10);$ $(2/3, 13/6).$ **18.** $c/20, c/20.$ **19.** $c/24.$

Section 8.5 — p. 318. **1.** $32/15.$ **2.** $(1/2, 3/4, 3/2).$ **3.** $2.$ **4.** $(a/4, b/4, c/4).$
5. $\int_{-\sqrt{7}}^{\sqrt{7}} dx \int_{-\sqrt{7-x^2}}^{\sqrt{7-x^2}} dy \int_{(x^2+y^2)/7}^{\sqrt{8-(x^2+y^2)}} dz.$
6. $\int_{-\sqrt{6/2}}^{\sqrt{6/2}} dx \int_{-\sqrt{6-4x^2}}^{\sqrt{6-4x^2}} dy \int_{-\sqrt{(6-4x^2-y^2)/3}}^{\sqrt{(6-4x^2-y^2)/3}} dz - \int_{-\sqrt{3/2}}^{\sqrt{3/2}} dx \int_{-\sqrt{3-4x^2}}^{\sqrt{3-4x^2}} dy \int_{(4x^2+y^2)/3}^{\sqrt{(6-4x^2-y^2)/3}} dz.$
7. $b^2a^3/6.$ **8.** $abc(b^2+c^2)/3$ with respect to side having length $a.$
9. $M = c\int_{-1}^{1} dx \int_{1-x}^{2-x/2} dy \int_{-\sqrt{1-y^2}}^{\sqrt{1-y^2}} \sqrt{x^2+y^2+z^2}\,dz,\ \bar{x} = c/M \int_{-1}^{1} dx \int_{1-x}^{2-x/2}$
$\times dy \int_{-\sqrt{1-y^2}}^{\sqrt{1-y^2}} \sqrt{x^2+y^2+z^2}\,dz.$ **10.** $c\int_{-\sqrt{2}}^{\sqrt{2}} dx \int_{-\sqrt{2-x^2}}^{\sqrt{2-x^2}} dy \int_{2x^2+y^2}^{4-y^2} \sqrt{x^2+y^2}(x^2+z^2)\,dz.$

Section 8.6 — p. 328. **1.** $2.$ **2.** $(0, 7/2).$ **3.** $(0, 0, 3a/8).$ **4.** $(2a^3\pi/3)$ arc tan $a.$
5. $20\sqrt{3}\pi a^3/3.$ **6.** $\pi a^4/2, \pi a^4/4.$ **7.** $3(e^3-1)/2.$ **8.** $\bar{x}=4/3, \bar{y}=10/3.$
9. (c) $0;$ (d) no. **10.** $4.$ **11.** log $5.$ **12.** $1/48.$ **13.** $4/3\pi(abc).$
14. $\bar{z} = 3(r_1^4 - r_2^4)/8(r_1^3 - r_2^3).$ **15.** $8\pi a^5/15.$ **16.** **17.** $\pi a/4.$ **18.** $2.$

CHAPTER 9

Section 9.1 — p. 334. **1.** (a) $\alpha(t) = (2\cos t, 2\sin t), 0 \le t \le 2\pi;$
(b) $\alpha(t) = (2\cos 2\pi t, \sin 2\pi t), 0 \le t \le 1;$ (c) $\alpha(t) = (e^t, t), -\infty < t < +\infty;$
(d) $\alpha(t) = (\cos^3 t, \sin^3 t), 0 \le t \le 2\pi.$
(e) $\alpha(t) = \left(\frac{1}{\sqrt{6}}\cos t + \frac{1}{\sqrt{10}}\sin t, -\frac{1}{\sqrt{6}}\cos t + \frac{1}{\sqrt{10}}\sin t \right), 0 \le t \le 2\pi.$
2. (a) $\alpha(t) = (-2 + 3t, 5 - 5t, t), -\infty < t < +\infty;$ (b) $\alpha_1(t) = (4/3 - (4/3)t, 4t, 0),$
$0 \le t \le 1, \alpha_2(t) = (0, 4 - 4t, 2t), 0 \le t \le 1, \alpha_3(t) = (4t/3, 0, 2 - 2t), 0 \le t \le 1;$
(c) $\alpha(t) = (\sqrt{3}\cos t, \sqrt{3}\sin t, 1), 0 \le t \le 2\pi;$ (d) $\alpha(t) = (2\sqrt{2}\cos 2\pi t, \sqrt{2}\sin 2\pi t, 1),$
$0 \le t \le 1;$ (e) $\alpha(t) = ((1/3)\cos t - (\sqrt{3}/3)\sin t + 1/3, (1/3)\cos t + (\sqrt{3}/3)\sin t + 1/3,$
$-(2/3)\cos t + 1/3), 0 \le t \le 2\pi.$ **3.** $(5^{3/2} - 1)/12.$ **4.** $0.$ **5.** $(145^{3/2} - 1)/54.$
6. $0.$ **7.** $\sqrt{5}/2.$ **8.** $\sqrt{5}/24.$ **9.** $128/3.$ **10.** $16/3 + 4\pi.$ **11.** $-4\sqrt{2}/3.$
12. $(12\pi - 8)/3.$ **13.** $0.$ **14.** $\sqrt{6}/96.$ **15.** $0.$ **16.** $8\pi\sqrt{3}.$ **17.** $2\pi\sqrt{6}/9.$

Section 9.2 — p. 336. **1.** $(10\sqrt{10})/81, (7/10, 1/10).$ **2.** $\sqrt{13}, (3/2, 1).$
3. $(4\sqrt{3} + 8)/3.$ **4.** $2\sqrt{2}\pi^2, (0, 1/\pi, 4\pi/3), 2\sqrt{2}\pi^2.$ **5.** $a\pi(0, 2a/\pi).$
6. $2\pi\sqrt{2/3}.$ **7.** $2, 4, (0, (2+\pi)/8).$

Section 9.3 — p. 343. **1.** $114/5.$ **2.** $-8/3.$ **3.** $17/2.$ **4.** $-17/12.$ **5.** $-8\pi.$
6. $0.$ **7.** $-3/2.$ **8.** $0.$ **9.** $1.$ **10.** $1/2.$ **11.** $-38.$ **12.** $43/6.$ **13.** $-2\pi a^2.$
14. $-2\pi.$ **15.** $0.$ **16.** $1.$ **17.** $0.$ **18.** $0.$

Section 9.4 — p. 347. **1.** $\alpha_1(t) = (a\cos t, a\sin t), 0 \le t \le 2\pi,$
$\alpha_2(t) = (a\cos t, -a\sin t), 0 \le t \le 2\pi, \alpha_3(t) = (a\cos 2t, a\sin 2t), 0 \le t \le 2\pi.$
2. $\alpha_1(t) = \left(\frac{1}{2}\cos t, \frac{\sqrt{2}}{2}\sin t \right), 0 \le t \le 2\pi, \alpha_2(t) = \left(\frac{1}{2}\cos t, \frac{\sqrt{2}}{2}\sin t \right), 0 \le t \le 6\pi.$

3. $\alpha_1(t) = \begin{cases} (1/2 - (1/2)t, 0), & \text{if } 0 \le t \le 1 \\ (0, t-1), & \text{if } 1 \le t \le 2 \\ ((1/2)t - 1, 3 - t), & \text{if } 2 \le t \le 3 \end{cases}$

$\alpha_2(t) = \begin{cases} ((1/2)t, 0), & \text{if } 0 \le t \le 1 \\ ((2-t)/2, t-1), & \text{if } 1 \le t \le 2 \\ (0, 3-t), & \text{if } 2 \le t \le 3 \\ ((t-3)/2, 0), & \text{if } 3 \le t \le 4 \\ ((5-t)/2, t-4), & \text{if } 4 \le t \le 5 \\ (0, 6-t), & \text{if } 5 \le t \le 6 \end{cases}$

$\alpha_3(t) = \begin{cases} (0, t), & \text{if } 0 \le t \le 1 \\ ((t-1)/2, 2-t), & \text{if } 1 \le t \le 2 \\ ((3-t)/2, 0), & \text{if } 2 \le t \le 3 \\ (0, t-3), & \text{if } 3 \le t \le 4 \\ ((t-4)/2, 5-t), & \text{if } 4 \le t \le 5 \\ ((6-t)/2, 0), & \text{if } 5 \le t \le 6 \\ (0, t-6), & \text{if } 6 \le t \le 7 \\ ((t-7)/2, 8-t), & \text{if } 7 \le t \le 8 \\ ((9-t)/2, 0), & \text{if } 8 \le t \le 9. \end{cases}$

420 ANSWERS TO EXERCISES

4. (a) $\alpha_1(t) = ((1/2)\cos t, \sin t), 0 \leq t \leq \pi/2$
$\alpha_2(t) = \alpha_1((\pi/2)t), \quad 0 \leq t \leq 1$
$\beta(t) = \alpha_2(1-t)$
$\alpha_1 \sim \alpha_2, \alpha_2$ not equivalent to β;

(b) $\alpha_1(t) = (t, t^2), \quad 0 \leq t \leq 4$
$\alpha_2(t) = \alpha_1(2t), \quad 0 \leq t \leq 2$
$\beta(t) = \begin{cases} (t, t^2), & \text{if } 0 \leq t \leq 4 \\ (8-t, (8-t)^2), & \text{if } 4 \leq t \leq 8 \end{cases}$
$\alpha_1 \sim \alpha_2, \alpha_1$ not equivalent to β;

(c) $\alpha_1(t) = \begin{cases} (t, 0), & \text{if } 0 \leq t \leq 1 \\ (2-t, t-1), & \text{if } 1 \leq t \leq 2 \\ (0, 3-t), & \text{if } 2 \leq t \leq 3 \end{cases}$
$\alpha_2(t) = \alpha_1(2t), \quad 0 \leq t \leq 3/2$
$\beta(t) = \alpha_1(3-t), 0 \leq t \leq 3$
$\alpha_1 \sim \alpha_2, \alpha_1$ not equivalent to β;

(d) $\alpha_1(t) = \begin{cases} (1-t, t), & \text{if } 0 \leq t \leq 1 \\ (1-t, 2-t), & \text{if } 1 \leq t \leq 2 \\ (t-3, 2-t), & \text{if } 2 \leq t \leq 3 \\ (t-3, t-4), & \text{if } 3 \leq t \leq 4 \end{cases}$
$\alpha_2(t) = \alpha_1(2t), \quad 0 \leq t \leq 2$
$\beta(t) = \alpha_1(4-t), 0 \leq t \leq 4$
$\alpha_1 \sim \alpha_2, \alpha_1$ not equivalent to β.

5. $\alpha(t) = \begin{cases} (1-t), (1/2)/t, 0), & 0 \leq t \leq 1 \\ (0, (2-t)/2, t-1), & 1 \leq t \leq 2 \\ (t-2, 0, 3-t), & 2 \leq t \leq 3 \end{cases}$
$(-\alpha)(t) = \alpha(3-t), 0 \leq t \leq 3$

Section 9.5 – p. 356.   **1.** $x^2/2 - xy$.   **2.** Not a gradient field.   **3.** Not a gradient field.   **4.** x^2yz.   **5.** Not a gradient field.   **6.** $x^2y - y^2z$.   **7.** $x^2yz + xy^2 + z^2$ **8.** (a) No; (b) 1/2.   **10.** $f = |X|^{n+2}/(n+2)$.   **11.** (a) If $r_1 < a < r_2$ and $\alpha(t) = (a\cos t, a\sin t)$, $\int_\alpha F = 2\pi$.   **14.** (a) $F(X) = -X/|X|^3$;   (b) 0.

Section 9.6 – p. 363.   **1.** $-2/3$.   **2.** -36.   **3.** 1/2.   **4.** π.   **5.** 2.   **6.** 2.   **7.** $-2\pi a^2$.   **8.** $2\pi(b^2 - a^2)$.

CHAPTER 10

Section 10.1 – p. 372.   **1.** $F(X) = (1/2)(z - 3y, 3x - 2z, 2y - x)$.
2. $F(X) = (1/3)(3yz - z^2, -3xz, xz)$.
3. $(1/3)(z^2 + xz - y^2 - yz, xy + 2xz - yz, y^2 - xy - xz - x^2)$.
5. $(z^2/3 - xyz/2 + 2xy^2z/5, x^2z/4 - 3x^2yz/5, x^2y^2/5 - xz/3 + x^2y/4) + \nabla f$ where f is some differentiable scalar function.

Section 10.2 – p. 378.   **1.** φ is not 1:1.   **2.** $(u, 0); (0, v); (2\pi, v)$.
3. $z = x^2/a^2 + y^2/b^2; 0 \leq u \leq 2\pi; 0 \leq v < \infty; (u, 0), (0, v), (2\pi, v)$.
4. $y^2/a^2 + z^2/b^2 = 1; 0 \leq u \leq 2\pi; -\infty < v < \infty; (0, v); (2\pi, v)$.
5. $N = (a_1, a_2, a_3) \times (b_1, b_2, b_3) \neq 0$.   **6.** $f(v)(\cos u, \sin u, f'(v)); f(v) = 0$.
7. $(R + r\cos v)(r\cos v \cos u, r\cos v \sin u, r\sin v)$.   **8.** $(1/\sqrt{2}\cos u \sin v, 1/2 \sin u \sin v, \cos v); 0 \leq u \leq 2\pi; -\pi/2 \leq v \leq \pi/2$.   **9.** $(1/\sqrt{2} v \cos u, v \sin u, v)$; $0 \leq u \leq 2\pi, -\infty < v < \infty$.   **10.** $(v\cos u, v\sin u, v^2); 0 \leq u \leq 2\pi; 0 \leq v < \infty$.
11. $((2 + \cos v)\cos u, (2 + \cos v)\sin u, \sin v); 0 \leq u, v \leq 2\pi$.
12. $(u + v, -u + v, -2)$; everywhere.   **13.** $(6(uv^2 - vu^2), 3(u^2 - v^2), 2(v - u)); u \neq v$.
14. $\begin{pmatrix} a_1 \ldots a_4 \\ b_1 \ldots b_4 \end{pmatrix}$ has rank = 2; two.

Section 10.3 – p. 383.   **1.** $4\sqrt{6}$.   **2.** $\pi a^2 \sqrt{3}/4$.   **3.** $\pi r^2/\sin\theta$.
4. $[17\sqrt{17} - 1]\pi/6$.   **5.** $\pi\sqrt{2}$.   **6.** $64\sqrt{2}$.   **8.** $4\pi^2 rR$.
9. $(a(t - \sin t)\cos u, a(t - \sin t)\sin u, a(1 - \cos t))$
$0 \leq u \leq 2\pi, 0 \leq t \leq 2\pi$

$$A = 2\pi a^2 \int_0^{2\pi} \sqrt{2 - 2\cos t}\, dt.$$

10. $(v\cos u, v\sin u, f(v)); 0 \leq u \leq 2\pi; a \leq v \leq b$.   **11.** $2\pi ad$.

ANSWERS TO EXERCISES

Section 10.4 – p. 392.   **1.** $(a/2, a/2, a/2)$.   **2.** $(0, 0, 2h/3)$.   **3.** $1/12\sqrt{14}$.   **4.** 3.   **5.** $\sqrt{3}/6$.   **6.** $\pi\sqrt{2}/4$.   **7.** $\bar{x} = \bar{y} = \bar{z} = (2+\sqrt{3})/(9+3\sqrt{3})$; $1/2 + \sqrt{3}$.   **8.** $1/2$.

Section 10.5 – p. 398.   **1.** 0.   **2.** -4π.   **3.** 1.   **4.** πa^2.   **7.** $-\pi$.   **11.** $f(X) = 1 - 1/|X|$.

Section 10.6 – p. 404.   **1.** 4π.   **2.** $3/2$.   **3.** 0.   **5.** 64π.   **6.** $20\sqrt{3}\pi$.   **7.** 48π.   **8.** No.   **9.** No.

INDEX

Absolute maximum, 280
Absolute minimum, 280
Acceleration, 13
   geometric representation, 13
   vector, 13
Adjoint of a matrix, 145
Affine subspace, 68
Algebra, 84
Algebraic isomorphism, 102
Alternating function, 126
Angle, direction, 5
   between two vectors, 16
Approximation, by polynomials, 262
   by total differential, 211
Area, of a plane region, 305
   surface, 380
Arithmetic mean, 283
Augmented matrix, 67
Axiom, of choice, 58
Axioms for vector space, 52

Back substitution, 106
Ball, open unit, 194
Bang, Thøger, 384
Basis, 61
   orthogonal, 65
   orthonormal, 162
Bilinear form, 174
Binomial coefficients, 264
Border of simple surface, 374
Bott, Raoul, 10
Boundary curves, 384

Canonical basis, 61
Canonical form, for an orthogonal matrix, 166
Cartesian product of two sets, 293
Center of mass, of a region, 310
   of a curve, 336
Centroid, of a curve, 336
   of a region, 312
   of a surface, 387
Chain rule for derivatives, 229, 232, 236
Change of basis matrix, 115
Change of parameter, in line integrals, 345
   in surface integrals, 388
Change of variable in multiple integrals, 319
Characteristic function, 305, 307
Characteristic polynomial, 139
Circulation, 406
Closed curve, 350

Closed set, 281
Coefficient matrix, 66
Cofactor, 130
   matrix, 145
Column rank, 75
Column vector, 66
Complex numbers, 3
Component, of a vector, 1
Composite function, 199
   continuity of, 199
   differentiability of, 232
Composition of transformations, 88
Concave down, 273
   up, 273
Connected open set, 349
Conservative field, 353
Constrained extrenal problems, 283
Continuity for vector function, 196
Continuity theorems, 198, 199
Continuous in each variable, 200
Continuously differentiable function, 211
Convex set, 226
Coordinate axes, 4
Coordinate function, 190
Coordinate planes, 4
Coordinate transformations, 236
Coordinate vector, 9
Coordinates, cylindrical, 239
   geographical, 243
   polar, 237
   spherical, 239
Cramer's rule, 47, 147
Cross product, 37
Counterclockwise direction, 358
Curl of a vector field, 267
Curve, 330
Cycloid, 383
Cylindrical coordinates, 239

Definite matrix, 272
Del, 368
Dependent set of vectors, 44, 56
Derivative, directional, 221
   partial, 202
   with respect to a vector, 219
Descartes' rule of signs, 186
Determinant, expansion formulas, 130
   function, 126
   of a matrix, 127
Diagonal matrix, 120
Diagonalization, of quadratic form, 176

of symmetric matrix, 170
Differential equations, linear, 151-159
Differential, 208
  second, 263, 270
Differentiable function, 207, 208
Differentiation, matrix for, 86
  under integral sign, 357
Dimension of vector space, 58, 61
Direction, angles, 5, 7
  cosines, 5, 7
  of a vector, 8
Directional derivative, 221
Distance, 25
  from point to a plane, 34
Divergence of vector field, 367
Domain, of a function, 189
  of transformation, 88
Dot product, 15
Double integral, 296, 301
  applications, 305

Echelon form for vectors, 58
Echelon matrix, 70, 108
Eigenvalues, 120, 138, 149
  complex, 150
  of symmetric matrix, 170
Eigenvector, 120
Elementary matrix, 110
Elementary row operation, 109
Ellipsoid, 181
Elliptic cone, 184
Elliptic cylinder, 183
Elliptic paraboloid, 185
Elliptic surface, 274
Equation, for a plane, 34
  of continuity, 405
Equivalent curves, 345
  surfaces, 388
Error, in Taylor's formula, 266
Euler's theorem for homogeneous functions, 235
Even permutation matrix, 143
Existence of determinant, 128
Expansion of a determinant, 130
Extension of a basis, 61
Extreme value theorem, 281

Field, gradient, 348, 361
  scalar, 53
  vector, 337
Finitely generated vector space, 56
Fluid mechanics, 405
Flux, 388
Forces as vectors, 13
Frobenius, 10
Full linear group, 165
Function, composition, 192
  continuously differentiable, 211
  differentiable, 207
  domain, 189
  harmonic, 366

range, 189
vector valued, 189
Fundamental solutions, 157
Fundamental theorem of calculus for line integrals, 351
Fundamental vector product, 375

Gauss' theorem, 399
Generators, of a vector space, 55
Geographical coordinates, 243
Geometric mean, 283
Geometric vector addition, 10
Geometric vectors, 4
Gradient, 227
Gradient field, 348, 361
Graph of quadratic equation, 176
Green's identities, 365, 404
  theorem, 359, 363
Group, 165

Hamilton, W. R., 10
Harmonic function, 366
Helical wire, 336
Homogeneous functions, 235
Homogeneous linear equations, 29, 68
Hyperbolic cylinder, 183
Hyperbolic surface, 275
Hyperboloid, of one sheet, 181
  of two sheets, 182
Hyperplane, 34, 64
Hypersurface, 219

Identity, matrix, 83
  transformation, 102
Implicit function theorem, 251, 252
  vector version, 254
Incompressible flow, 406
Independence, linear, 44, 56
  of path, 351
Inequality, between geometric-arithmetic mean, 283
  Schwatrz, 16, 18
  Triangle, 19
Inhomogeneous linear equations, 68
Initial point of a vector, 5
Inner product, 15
Integrable function, 292, 298
Integral, 298
  Lebesgue, 307
  line, 332
  surface, 387
Integrand, 315
Integration, matrix for, 96
Intercepts of plane, 32
Interior of a set, 195
Interior point, 194
Inverse function, 247
Inverse function theorem, 248
Inverse matrix, 84
Invertible linear transformation, 102
Invertible matrix, 84, 104

Irrotational vector field, 371
Irrotational flow, 406
Isomorphism, algebraic, 102
   linear, 90
Iterated integral, 296, 316

Jacobian, determinant, 247
   matrix, 209
Jordan, Camille, 358
Jordan curve theorem, 358

Kinetic energy, 353

Lagrange expansion, 38
Lagrange identity, 39
Lagrange interpolation theorem, 149
Lagrange multipliers, 289
Laplace equation, 245
Latitude, 243
Lebesgue, Henri, 307
Lebesgue integral, 307
Left-hand coordinate system, 41
Length of a vector, 3
Level set, 249
Limit of a function, 195
Limit point of a set, 281
Limit theorems, 197
Line, 21
Line integrals, applications of, 335, 341
   fundamental theorems for, 351
   independence of path, 351
   notations for, 339
   of scalar functions, 330, 332
   of vector field, 338
Line through A parallel to B, 21
Linear combination, 2
Linear differential equations, 151
Linear equations, 28, 65
Linear independence, 44, 56
Linear transformation, 79, 85
Local Extreme, 277
Longitude, 243
Lower triangular matrix, 105

Main diagonal, 105
Mapping, 189
Matrix, 76
   adjoint, 145
   change of basis, 115
   cofactor, 145
   determinant of, 127
   diagonal, 149
   invertible, 84
   multiplication, 81
   of differentiation, 95
   of integration, 96
   of a linear transformation, 92, 94
   of linear transformation, examples, 93, 95
   of projection, 100
   of quadratic form, 175
   of reflection, 97
   of rotation, 96
   orthogonal, 163
   sum, 76
   transpose, 140
Maximum, absolute, 280
   existence of, 281
   local, 277
Mean value theorem, 224
Milnor, John W., 10
Minor, 130
Möbius band, 395
Moment, first, 313
   of inertia, 313, 317
Multiplication, matrix, 81
Multiple integrals, 315
Multipliers, Lagrange, 289

Negative definite matrix, 271
Negative definite quadratic form, 271
Negative rotation, 96
Negatively oriented curve, 358
Neighborhood, 248
Newtonian potential, 357
n-linear, 126
Nonorientable surface, 396
Norm, of a vector, 3, 21
   non-Euclidean, 21
Normal vector, 24, 32
   to surface, 375
   to tangent plane, 216
Null space, 89

Odd permutation matrix, 143
One-to-one, 89
Open set, 195
Open unit ball, 194
Ordinate set, 307
Orientation, of border of surface, 384
Oriented curve, 358
Orthogonal basis, 65
Orthogonal matrix, 163
Orthogonal surfaces, 260
Orthogonal transformation, 160, 165
Orthogonal vectors, 17
Orthonormal basis, 162
Orthonormal set of vectors, 162

Pappas, theorem of, 383
Parabolic cylinder, 186
Parabolic surface, 275
Paraboloid, 185
Parallel vectors, 10, 17, 19
Parameter, 330
Parametric equations of plane, 34
Parametric equation of line, 22
Partial derivatives, 202
Partition, 293
Path independence, 351
Permutation matrix, 114, 143
Perpendicular vectors, 17

Piecewise regular curve, 331
Piecewise smooth surface, 390
Plane through C spanned by A and B, 28
Plank problem, 384
Polar coordinates, 237
Polygonally connected, 227
Polynomial, characteristic, 139
   Taylor, 263
Positive definite matrix, 173, 271
Positive definite quadratic form, 271
Positive rotation, 97
Positively oriented curve, 358
Potential energy, 354
Potential function, 348, 354
Product, of matrices, 81
   of transformations, 88
   rule for determinants, 137
Projection, 87
Pythagorean theorem, 18

Quadratic form, 175
   diagonalization of, 176
   negative definite, 271
   positive definite, 271
Quadric surface, 180
Quaternions, 10

Range, of a function, 189
   of a linear transformation, 88
Rank of a matrix, 67, 73, 74, 90
Refinement, of a partition, 294
Reflection, 86, 161
Regular curve, 330
Reparametrization, of a curve, 346
   of a surface, 377
Replacement principle, 59
Replacement theorem, 60
Right-hand coordinate system, 41
Rotation, 86, 160
Row equivalence, 109
Row rank, 75
Row vector, 66

Saddle shaped, 275
Scalar multiplication, 2
Scalar product, 15
Schwartz inequality, 16, 18
Second differential, 270
Second partial derivatives, 205
Separated sets, 292
Set of constancy, 249
Sets, Cartesian product, 293
Similar matrices, 120, 149
Simple closed curve, 350
Simple smooth surface, 374
Simply connected, 396
Simply connected open set, 361
Smooth curve, 330
Solenoidal vector field, 371
Solutions to linear equations, 29, 66
Span, 12, 45, 55

Spherical coordinates, 240
Spherical surface, 376
Square matrix, 83
Star shaped, 354
Stationary point, 278
Step function, 291, 293
Step polygonal curve, 353
Stokes' theorem, 393
Subgroup, 165
Subspace, 53, 88
   proper, 53
   zero, 53
Sum, matrix, 76
Surface, 374
   area, 379
   of revolution, 377
Surface integral of vector field, 387
   application, 387
   notations for, 387
Symmetric matrix, 141, 170
Systems of linear differential equations, 151

Tangent, hyperplane, 219
   line, 213
   plane, 216
   space, 214, 218
Tarski, Alfred, 384
Taylor polynomial, 263
Taylor's theorem, for function of one variable, 262
   for function of two variables, 267
Terminal point of a vector, 5
Torus, 379
Total differential, 211
Trace of curve, 330
Transformation, 78
   linear, 79, 85
Transpose of a matrix, 140
Transposition, 145
Triangle inequality, 19
Trivial solution, 29
Two dimensional surface, 215

Uniqueness of determinant function, 132
Unit sphere, 194
Unit tangent, 213
Unit vector, 8
Upper triangular matrix, 105

Variable, 190
Veblen, Oswald, 358
Vector, 1
   field, 337
   product, 37, 87
   space, definition, 52
   sum, 1
   valued integral, 370
Velocity field, 338
Velocity of a curve, 330
Velocity vector, 213
   geometric representation of, 13

Volume, 307

Work, 341

Zero transformation, 88